Recent Developments in Crop Science

Recent Developments in Crop Science

Editor: Corey Aiken

CALLISTO REFERENCE

www.callistoreference.com

Callisto Reference,
118-35 Queens Blvd., Suite 400,
Forest Hills, NY 11375, USA

Visit us on the World Wide Web at:
www.callistoreference.com

ISBN: 978-1-64116-065-0 (Hardback)

Trademark Notice: Registered trademark of products or corporate names are used only for explanation and identification without intent to infringe.

Cataloging-in-Publication Data

Recent developments in crop science / edited by Corey Aiken.
    p. cm.
Includes bibliographical references and index.
ISBN 978-1-64116-065-0
1. Crop science. 2. Crops–Development. 3. Agronomy. I. Aiken, Corey.
SB91 .R43 2019
631--dc21

Table of Contents

Permissions

List of Contributors

Index

Preface

This book was inspired by the evolution of our times; to answer the curiosity of inquisitive minds. Many developments have occurred across the globe in the recent past which has transformed the progress in the field.

Crop science is the study of crops and the aspects related to them such as crop breeding, genetics, ecology, etc. It also delves into the development of modern agronomic methods and techniques to increase crop production or to build innovative models for crop management. Some prominent areas of study under crop science include horticulture, pest management, soil science, etc. This book presents concepts and theories fundamental to crop science in a comprehensive manner. It elucidates the principles and innovative models around prospective developments with respect to crop science. It includes some of the vital pieces of work being conducted across the world, on various topics related to this discipline. As this field is emerging at a rapid pace, the contents of this book will help the readers understand the modern applications of the subject.

This book was developed from a mere concept to drafts to chapters and finally compiled together as a complete text to benefit the readers across all nations. To ensure the quality of the content we instilled two significant steps in our procedure. The first was to appoint an editorial team that would verify the data and statistics provided in the book and also select the most appropriate and valuable contributions from the plentiful contributions we received from authors worldwide. The next step was to appoint an expert of the topic as the Editor-in-Chief, who would head the project and finally make the necessary amendments and modifications to make the text reader-friendly. I was then commissioned to examine all the material to present the topics in the most comprehensible and productive format.

I would like to take this opportunity to thank all the contributing authors who were supportive enough to contribute their time and knowledge to this project. I also wish to convey my regards to my family who have been extremely supportive during the entire project.

Editor

Advances in Understanding Cold Sensing and the Cold-Responsive Network in Rice

Qi Zhang[1#], Nan Jiang[1#], Guo-Liang Wang[1,2], Yahui Hong[1*], Zhilong Wang[1*]

[1]Hunan Provincial Key Laboratory of Crop Germplasm Innovation and Utilization, College of Biological Science and Technology, College of Agronomy, Hunan Agricultural University, Changsha, Hunan 410128, China
[2]Department of Plant Pathology, Ohio State University, Columbus, Ohio 43210, USA
[#]These authors contributed equally to this work

Abstract

Cold stress reduces the growth and production of many crops including rice (*Oryza sativa L.*), which is a staple food crop and model monocot plant. During the past two decades, significant progresses have been made in understanding the cold-responsive network in rice. Various genes and gene products functioning in cold sensing and transcriptional regulation and post-transcriptional processing of the response to cold have been identified and these include *OsDREBs, OsNACs, OsMAPs, OsCDPKs,* miR-171, and miR-444a. The important roles of calcium, Reactive Oxygen Species (ROS) and Abscisic Acid (ABA) in cold sensing and signaling during both vegetative and reproductive stages have also been revealed. The new findings summarized in this review will facilitate the breeding of cold tolerance in rice and will also be useful for studying and understanding cold sensing and signaling in other crops.

Keywords: Cold stress; Sensing; Responsive network; Rice

Introduction

Temperature severely affects plant growth and basically determines the geographical distribution of plants. The changes in mean annual temperatures predicted by climate change models will affect both soil organic matter turnover and cropping patterns in agriculture [1,2]. Rice is one of the most important stable crops globally and also a monocot model plant for molecular research. Like other plant species originating in tropical or sub-tropical areas, rice is sensitive to low temperatures with prolonged exposure resulting in chilling injury [3]. In tropical areas, especially in Indonesia and Malaysia, cold stress rarely occurs and is not the major problem for rice breeding in those areas. Nevertheless, rice is cultivated well beyond tropical environments to temperate and high altitude areas. In these areas, temperatures fluctuate greatly across seasons and rice is often exposed to temperatures below 20°C during the most sensitive stages of development (i.e., germination, seedling, and reproduction) [4]. Low temperature impairs seed germination, reduces seedling vigor, weakens photosynthetic ability by inducing leaf discoloration, reduces plant height, produces degenerated spikes, delays days to heading, reduces spikelet fertility, causes irregular grain maturity, and poor grain quality [5]. About 30.7 million ha of rice in China are grown over a wide area from 53°27'N to 18°90'N, and almost the entire area can be harmed by cold injury caused by low temperatures. Annual losses are 3-5 million tonnes. Low temperature during the reproductive stage of rice in the Republic of Korea caused 17, 78, and 20% damage to the total rice area in 1971, 1980, and 1993, respectively, with maximum yield loss of milled rice of 3.9 t/ha in 1980. Yield loss of 1-2 t/ha due to low temperature during the reproductive stage in 1995-96 was also reported in Australia and an annual yield loss of 3-5 million tons was recorded in China [5,6]. The mechanisms of cold stress in rice have been extensively investigated in the past two decades, and cold-responsive strategies that differ from those in other plant species such as *Arabidopsis* and tomato have been detected in rice. The identification of QTLs for cold stress shows that different loci are involved in cold tolerance at different growth stages in rice [7,8]. And these findings are useful to facilitate the selection and development of improved cold-tolerant genotypes with high percent seed set for their cultivation in temperate environments and high-altitude areas. According to common cold-responsive network in various plants,

previous reviews provide useful information, but the review which specifically focuses on the studies in crops, especially in rice, is scarce. In this review, we summarize recent advances on cold sensing and transcriptional regulation in rice. In addition, we discuss intriguing findings on the role of post-transcriptional regulation in chilling and/ or freezing tolerance in rice.

Cold sensing

Plant cells can sense cold stress based on changes in membrane rigidity, changes in the physical state of membrane proteins, and changes in the concentrations of metabolites. At low temperatures, an increase in electrolyte leakage (EL) commonly occurs in rice cultivars because of increased membrane rigidity [9-11]. Ca^{2+} influx into the cystol is an important and early event caused by cold stress, an influx which may be mediated by membrane rigidification-activated mechano-sensitive or ligand-activated Ca^{2+} channels [12]. Subsequently, the Ca^{2+} signatures are interpreted and amplified by calcium sensors, such as Calmodulin (CaM) and Calmodulin-like Proteins (CMLs), Calcineurin B-like Proteins (CBLs), and Calcium-Regulated Protein Kinases (CDPKs) (Figure 1). The CDPK gene *OsCDPK7* is induced by chilling in rice, and overexpression of this gene enhances cold tolerance in transgenic rice

*Corresponding authors: Zhilong Wang, Hunan Provincial Key Laboratory of Crop Germplasm Innovation and Utilization, College of Biological Science and Technology, College of Agronomy, Hunan Agricultural University, Changsha, Hunan 410128, China, E-mail: zhilongwang@126.com

Yahui Hong, Hunan Provincial Key Laboratory of Crop Germplasm Innovation and Utilization, College of Biological Science and Technology, College of Agronomy, Hunan Agricultural University, Changsha, Hunan 410128, China, E-mail: yahuihong@vip.sina.com

[13]. Other calcium-regulated protein kinase genes such as *OsCDPK13*, *OsCBF1*, and *OsNAC5* are also cold-induced and involved in the calcium signaling pathway [14-16].

As a cold-responsive signaling in cold-stressed plants, abscisic acid (ABA) accumulates and then initiates the cold-sensing period (Figure 1) [17]. An ABRE (ABA-responsive elements)-binding bZIP transcription factor OsABF2 plays a role in the cold response, and significantly increases at the transcriptional level [18]. In a rice mutant deficient in the ABA-responsive pathway *ABI5-Like1* (*abl1*), over one-third of the genes are down-regulated in response to cold and other abiotic stresses, indicating that ABA may play a positive and complex role in the sensing of cold stress in rice [19].

Reactive Oxygen Species (ROS) accumulate in cells when plants are challenged with biotic or abiotic stresses. ROS, which include superoxide (O_2^-), hydrogen peroxide (H_2O_2), and the hydroxyl radical, play two roles under cold stress: they induce ROS scavengers and initiate protective mechanisms [20]. When plants are exposed to biotic and abiotic stimuli, an increase in ROS level can damage membranes, induce alterations of cellular ion conductance, and trigger mitogen-activated protein kinase (MAPK) cascades (Figure 1) [19]. Transgenic rice overexpressing *OsAPXa* (ascorbate peroxidase) exhibits an elevated cold tolerance that is negatively correlative with the levels of H_2O_2 and lipid peroxidation, which have been scavenged by increased ascorbate peroxidase activity during cold treatment [21]. Moreover, ROS regulate the *OsMKK6* (MAPK kinase)-OsMPK3 (MAPK) pathway through redox control of OsTRX23 [22]. Interestingly, an analysis of the transcriptional regulatory network in japonica rice subjected to cold treatment revealed that oxidative-mediated clusters (several H_2O_2-induced gene such as *bZIP*, *ERF*, and *MYB* genes) are activated earlier than ABA-mediated clusters (factors acting on ABRE-like enriched clusters), and that the former may play a more important role in early cold sensing [8,23-25].

CDPKs pathway

Calcium-Dependent Protein Kinases (CDPKs) belong to a family of Ser/Thr protein kinases that were first discovered in plants. By playing a central role in calcium-dependent pathways and by controlling the calcium content in the cell, CDPKs have been thought to coordinate the sensing of and responses to abiotic stress [26,27]. Expression by all 29 CDPK (CPK) genes in rice has been evaluated under various abiotic stresses. Nine CDPK genes (*OsCPK7, OsCPK13, OsCPK15, OsCPK17, OsCPK20, OsCPK21, OsCPK23, OsCPK24, and OsCPK29*) are induced by chilling temperatures in rice seedlings [28]. Moreover, *OsCDPK7* and *OsCDPK13* overexpression transgenic lines exhibit higher recovery rates after cold stress than the vector control. An *in situ* hybridization localized the expression of *RAB16A*, one of the proposed target genes regulated by the *OsCDPK7* signaling pathway. CDPKs have also been identified as downstream components in Gibberellin (GA) and Brassino Steroids (BRs) signaling pathways [29-31]. With transgenic analysis in rice, the expression of *OsCDPK13* is up-regulated in response to both GA3 and cold treatment, suggesting that *OsCDPK13* may mediate the GA-induced signaling and the Ca^{2+} influx [32]. A reduction in GA content of cold-treated *Arabidopsis*, modulated through the stimulation

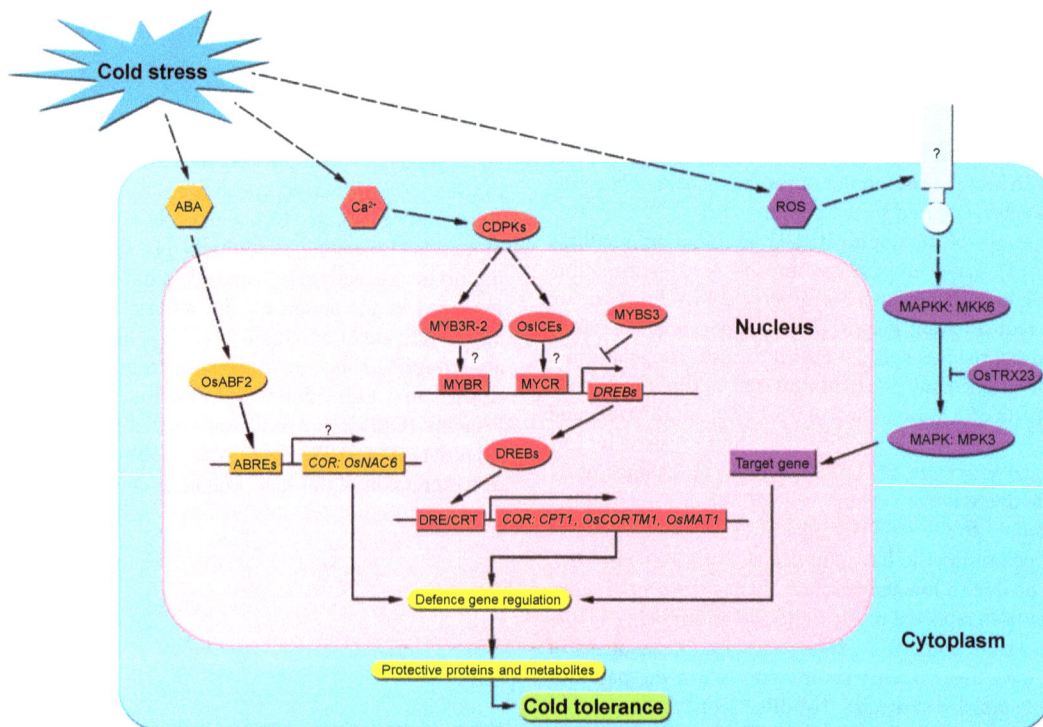

Figure 1: In response to low temperature stress, rice upregulates the expression of protective proteins, which is initiated by the ABA signaling pathway, the DREBs pathway, and the MAPK cascade. Solid arrows indicate direct activation; broken arrows indicate indirect activation; lines ending with a bar indicate negative regulation. ABA: abscisic acid; ABRE: ABA-responsive elements; ABF: ABRE-binding factor; CDPK: calcium-dependent protein kinase; *COR*: cold-responsive gene; *CORTM1*: one of the cold-regulated (*COR*) genes in rice; *CPT*: one gene that is activated by *OsMYB3R-2*; CRT: C-repeat elements; DRE: dehydration-responsive elements; DREB: dehydration-responsive element-binding protein; MAPK (MPK): mitogen-activated protein kinase; MAPKK: MAP kinase kinase; ICE: inducer of CBF expression; MAT1: malonyltransferase 1; MYB: myeloblastosis; MKK6: MAPK kinase 6; MYB3R-2: a R1R2R3 MYB transcription factor; MYBR: MYB recognition site; MYCR: MYC recognition site; NAC: NAM-ATAF-CUC; ROS: reactive oxygen species; OsTRX23: *Oryza sativa* thioredoxin.

of expression of GA-inactivating GA 2-oxidase (*GA2ox*) genes is associated with the accumulation of a green fluorescent protein-tagged DELLA protein [33]. Because CBF1 acts downstream in the CDPK pathway, a question arises about whether there is a feedback regulation between CBFs, GA, and CDPKs that explains their changes in response to cold stress. Accordingly, further research in the function of the rice *CDPK* genes would be useful for increasing our understanding of signal transduction pathways in cold stress responses.

OsMAPKs

The MAPK pathway is a well-characterized signal transduction cascade system in animals, yeast, and plants. The MAPK cascade consists of three subsequently interactive phosphorylation kinases: MAP kinase kinase kinase (MAPKKK), MAP kinase kinase (MAPKK), and MAPK [34,35]. MAPK, the last component of the phosphorylation cascade, is activated by MAPKK, which is in turn activated by MAPKKK. The MAPK cascade regulates cell division, development, and differentiation in response to stress stimuli in animals and yeast [36,37]. In recent years, a series of genes encoding MAPKs, MAPKKs, and MAPKKKs have been identified from different plant species. An increasing body of evidence has shown that MAPKs play important roles in signal transduction in response to drought, ROS, pathogen attack, wounding, and low temperature in plants [38-42].

Of the 20 MAPK genes in *Arabidopsis*, *AtMPK3* and *AtMPK6*, have been implicated in the tolerance to multiple abiotic and biotic stresses [39,43]. Nine MAPK genes have been identified from rice, and each *MAPK* encodes a distinct protein kinase that plays a role in mediating abiotic tolerance [44]. *BWMK1*, which was the first MAPK gene found in rice and which was detected in the indica-type cultivar IR36, is induced by infection by the blast pathogen (*Magnaporthe oryzae*) and by mechanical wounding [45]. Recently, a time-based transcriptional regulation mechanism controlled by *OsTRX23* has been reported for *OsMPK3/OsMAP1* and *OsMPK6/OsSIPK* which are strongly induced by low temperature treatment (12°C) at the transcriptional level [22]. In the first 24 h of cold treatment, transcripts of *OsMPK3* and *OsMPK6* rapidly increase, and then, with an increasing transcription of *OsTRX23*, they sequentially decrease, suggesting that the MAPK pathway could be a rapid stress-responsive element associated with the transient accumulation of H_2O_2 during chilling stress. These results also indicate that some unknown late-responsive element mediated by the inhibition of *OsMPK3* and *OsMPK6* by *OsTRX23* leads to cold tolerance in rice. Recently, a yeast two-hybrid screen designed to identify partner MAPKs for *OsMKK6* revealed specific interactions of OsMKK6 with OsMPK3 and OsMPK6. A constitutively active form of OsMKK6 and OsMKK6DD, showed elevated phosphorylation activity against OsMPK3 and OsMPK6 *in vitro*. *OsMPK3*, but not *OsMPK6*, was constitutively activated in transgenic plants overexpressing *OsMKK6DD*, indicating that *OsMPK3* is an *in vivo* target of *OsMKK6*. Chilling tolerance was enhanced in transgenic plants overexpressing *OsMKK6DD* [46]. Taken together, these data suggest that the MAPK signaling cascade is one of the low-temperature signaling pathways in rice and positively regulates cold stress tolerance.

The transcriptional-regulatory network associated with cold response

The transcriptional regulatory network involved in low temperature response prolongs the expression of cold acclimation genes in rice. Of the 2,604 genes that are upregulated in response to chilling in japonica rice, about 6% (148) have been estimated to be Transcription Factors (TFs)

based on the classification of the Database of Rice Transcription Factors. These putative TFs include members of AP2/ERF, bZIP, MYB, WRKY, bHLH, and NAC families, and are characterized by waves of induction at different time periods. For example, the 'early rapid response' group (phase-1) is induced during the initial 6 h of cold treatment, and the 'early slow response' group (phase-2) is induced between 6 and 24 h. A few TFs exhibited 'late response' profiles (phase-3) with no significant induction until after 24 h. Most TFs that are upregulated by chilling are activated at phase-1 and phase-2, indicating that early responses are critical for cold stress tolerance and that late responses may be involved in unknown mechanisms that protect rice growth after 24 h of exposure to low temperatures [10]. In the last two decades, researchers have made significant progress in clarifying the function of key TFs responding to cold stress, as discussed in the next section.

DREB-CRT/DRE pathway

DREBs (dehydration-responsive element-binding proteins, also known as CBFs for C-repeat binding factors) have been the most intensively studied of the TFs involved in plant responses to cold. In rice, seven members in the DREB family have been studied in depth: OsDREB1A/CBF3, OsDREB1B/CBF1, OsDREB1C/CBF2, OsDREB1D, OsDREB1F, OsDREB2A, and OsDREB2B [47]. Overexpression of DREBs in transgenic rice significantly increases cold tolerance; relative to wild-type plants, the transgenic plants exhibit increased survival and growth, and changes in ROS scavenging, membrane transport, hormone metabolism, proline concentration, and accumulation of sugars such as raffinose, sucrose, glucose, and fructose [14,48].

DREBs belong to the APETALA2/Ethylene response factor (AP2/EREBP) family of TFs. These TFs can bind to GCC-box and C-repeat/dehydration-responsive elements (CRT/DRE) to promote the Cold responsive (*COR*) gene. To analyze the DNA-binding specificity of OsDREBs, researchers have selected *OsDREB1A* as a typical *DREB1*-type gene and *OsDREB2A* as a typical *DREB2*-type gene. The ability of the OsDREB1A and OsDREB2A fusion proteins to bind the wild-type or mutated DRE sequences in the *rd29A* promoter has been identified by the gel mobility shift assay. Furthermore, the transactivating DRE-dependent transcription capability of *OsDREB1A* and *OsDREB2A* in rice cells has been determined by transactivation experiments. Protoplasts are co-transfected with a β-glucuronidase (GUS) reporter gene fused to the dimeric 75-bp fragments containing the DRE motif and the effector plasmid. The GUS activity is distinctly upregulated in the presence of *OsDREB1A* and *OsDREB2A* relative to the vector control, indicating that the DREB1A and DREB2A proteins act as transcription activators in rice protoplasts [49].

In *Arabidopsis*, ICE1 (Inducer of CBF expression1), a MYC-type basic helix-loop-helix transcription factor, is a common TF that promotes the expression of *DREB1A*. ICE1 binds to MYC recognition (MYCR) elements in the *DREB1A* promoter region and is important for the expression of *DREB1A* during cold acclimation [12]. So far, there is no evidence that any ICE can function as a transcriptional factor for DREBs in rice. The Basic Helix-loop-Helix (bHLH) protein gene *OrbHLH001*, encoding an ICE1-like protein containing multiple homopeptide repeats, has been characterized and isolated from Dongxiang Wild Rice (*Oryza rufipogon*). Overexpression of *OrbHLH001* enhances the tolerance to freezing stress in transgenic *Arabidopsis*. Examination of the expression of cold-responsive genes in transgenic *Arabidopsis* shows that the effect of *OrbHLH001* in cold response differs from that of *ICE1* and is independent of the CBF/DREB1 cold-response pathway [50]. Recently, two *ICE* homologs in rice, OsICE1 and OsICE2, have been identified by BLAST searches with the nucleotide sequence of *Arabidopsis ICE1* in the Rice Annotation Project Data Base [51]. Unlike the increased expression of *Arabidopsis ICE1* under cold stress, the expression of *OsICE1* and *OsICE2* remains constant under cold stress as indicated by semi-quantitative RT-PCR; immunoblot with the anti-ICE specific antibody showed, however, that the levels of OsICE1 and OsICE2 proteins are upregulated by both cold and salt stresses. Although the expression of OsDREB1B increases under cold treatment, no evidence of a direct interaction between OsDREB1B and OsICE1/OsICE2 has been provided, There is little information available concerning the role of the MYC-type transcription factor in the DREB pathway but many genes have been found that upregulate the expression of *DREB* in rice [10,52,53]. OsMYB3R-2, a nuclear-localized R1R2R3 MYB TF, plays a role in cold response in rice [54]. *Arabidopsis* transgenic plants overexpressing *OsMYB3R-2* show increased tolerance to cold, drought, and salt stresses, and the expression of some cold-related genes such as *DREB2A* and *CBF1/2/3* is increased in the *OsMYB3R-2*-overexpression plants. These results suggest that OsMYB3R-2 acts as a master switch in cold tolerance. A more recent study has shown that overexpression of *OsMYB3R-2* in transgenic rice leads to higher transcript levels of several G2/M phase-specific genes and *OsCPT1*, which are putatively targeted by the *DREB* genes [53]. The role of R2R3-type MYB TF (OsMYB2), which is localized in the nucleus and which has transactivation activity, in tolerance to cold stress has been functionally characterized by generating overexpression and RNAi transgenic rice plants of *OsMYB2*. In the *OsMYB2*-overexpression plants, the expression of *OsDREB2A* is upregulated compared with the wild type under low temperature treatment, suggesting that OsMYB2 is an upstream regulon for *OsDREB2A* [55].

A negative regulator of the CBF regulon, MYBS3, operates via a distinct pathway to help rice plants tolerate cold stress. MYBS3 is a single DNA-binding repeat MYB TF. By using genotypes that overexpress or underexpress *MYBS3* to identify genes in the MYBS3-mediated cold signaling pathway, Su et al. [56] determined that *MYBS3* responds slowly to cold stress and enables rice seedling to tolerate 4°C for at least 1 week. Surprisingly, MYBS3 represses the well-known *DREB1/CBF* fast-acting and short-term signaling pathway at the transcriptional level in rice [56]. By repressing the expression of the *DREB1* regulon, the slow acting MYBS3 may prevent unnecessary expression of *DREB1*.

NAC TFs

NAC proteins represent one of the largest families of TFs. There are at least 105 putative NAC TFs in *Arabidopsis*, 140 in rice, 205 in soybean, and 152 in tobacco (*Nicotiana tabacum*) [10]. NAC TFs are key regulators of stress sensation and developmental programs, and contain an N-terminal NAC domain. Although the NAC TF family is widely distributed in plants, it has not been found in other eukaryotes [57,58].

Analysis of the NAC domain has been conducted by database searches with a comprehensive analysis of NAC family genes both in rice and *Arabidopsis* [59]. Five subdomains, each typically ~50 amino acids long, have been identified in the DNA-binding domain of a typical NAC protein. A comprehensive *in silico* analysis of the NAC TF family in rice recently identified 36 putative motifs within the NAC family. Based on the pattern of these motifs, the NAC family of rice can be classified into 15 types (types A-O). Most rice NAC proteins (97 of 140) contain a complete NAC DNA-binding domain with five major subdomains, which are classified into types A-E [60,61]. The sequences in the 1.5 kb promoter region of *OsNAC6* include ABREs,

MYBRS, MYCRS, W-boxes, GCC boxes, and as-1 motifs [24]. ABREs, MYBRS, and MYCRS play important roles in ABA signaling and abiotic stress response. The W-boxes and GCC boxes are recognition sites for WRKY and ERF transcription factors, respectively, and are involved in the regulation of various plant-specific physiological processes such as pathogen defense and senescence. The as-1 motifs are known as oxidative stress-responsive elements. These sequences may function as abiotic and/or biotic stress-responsive cis-acting elements in the *OsNAC6* promoter, and they may sense cold stress from the upstream transcription factors induced by ABA, ROS, and other metabolite changes.

Regarding the *OsNAC5*-dependent tolerance to abiotic stress in rice, research has demonstrated that the overexpression line is more tolerant of cold stress and that the *OsNAC5* RNAi line is less tolerant of cold stress than the wild type [16,62]. Moreover, knockdown and overexpression of *OsNAC5* enhances and reduces, respectively, the accumulation of malondialdehyde and H_2O_2, suggesting that knockdown of *OsNAC5* renders RNAi plants more sensitive to oxidative damage. In addition, the sensitivity of seed germination to ABA is increased in overexpression plants and decreased in RNAi plants. These results indicate that *OsNAC5* is important for cold tolerance in rice and that the effect of *OsNAC5* is associated with an ABA-related transcriptional regulation network.

However, the target genes of NAC TFs in rice remain largely unknown. The rice *NAC* gene *SNAC1* (stress-responsive NAC 1) is predominantly induced in guard cells by drought and encodes NAC TF with transactivation activity [63]. *SNAC1*-overexpression plants show a greater sensitivity to ABA and an increased rate of stomatal closure that reduces water loss. In the field, the drought tolerance of transgenic plants is significantly greater than that of wild-type plants at anthesis, and the increased drought tolerance is not associated with other phenotypic changes or with yield reduction. The distinctive binding character of SNAC1 suggests that a special NAC binding site exists in rice.

Other TFs

In addition to the well-characterized rice TF genes in the cold responsive pathway that were described in the previous sections, many other genes also display changes in transcriptional expression upon low temperature treatment and are involved in the cold-sensing cascade.

Proteins with the A20/AN1 zinc-finger domain are present in all eukaryotes and are well characterized as common elements in the stress response of animals and plants [64]. The stress-associated proteins (SAPs) contain the AN1 domain, which has a dimetal (zinc)-bound alpha/beta fold and often combines with A20 zinc finger domains (SAP8) or C2H2 domains (SAP16) including ascidian posterior end mark 6 (PEM-6) protein and human AWP1 protein (associated with PRK1) [65,66]. The human AWP1 protein is expressed during early embryogenesis, and mutations in *SMUBP-2* (human immunoglobulin mu binding protein 2) cause muscular atrophy with respiratory distress type 1 [67]. *OsiSAP8* is an early cold-responsive gene in rice, and the OsiSAP8 protein fused to GFP is specifically localized in the cytoplasm, indicating that, unlike many zinc-fingers containing proteins, OsiSAP8 is a cytoplasmic protein. Furthermore, the A20 and AN1 type zinc-finger domains of OsiSAP8 interact with each other in yeast two-hybrid assays, and *OsiSAP8* overexpression transgenic rice plants grow better than wild-type plants under cold stress [68]. According to a microarray analysis of the rice A20/AN1-type zinc finger genes, four genes (*ZFP177, ZFP181, ZFP176, ZFP173*), two genes (*ZFP181* and *ZFP176*), and one gene (*ZFP157*) are significantly induced by cold, drought, and H_2O_2 treatments, respectively [69]. The promoter region of *ZFP177*, which is a specific cold-induced A20/AN1-type zinc finger gene, contains no stress-associated cis-acting elements other than HSE, suggesting that the heat response of *ZFP177* might be regulated by heat shock factors but that the cold response is not directly mediated by CBF/DREB transcription factors.

The TFIIIA-type zinc finger protein was first detected in *Xenopus* oocytes [70] and contains at least one TFIIIA-type zinc finger motif with the consensus of $CX_{2-4}CX_3FX_5LX_2HX_{3-5}H$. A number of TFIIIA-type zinc finger proteins are related to stress response [71]; for example, *Arabidopsis* STZ/ZAT10 and ZAT7 are involved in salt tolerance [72-74], soybean SCOF-1 is involved in cold tolerance [52], and *Arabidopsis* ZAT12 is involved in cold and oxidative stress [75]. The first TFIIIA-type zinc finger protein identified in rice, ZFP245, is involved in cold and drought tolerance [9]. Overexpression of *ZFP245* in rice leads to increased cold tolerance and increased sensitivity to exogenous ABA treatment, suggesting that *ZFP245* may play a role in the ABA signal transduction pathway during stress responses. Another TFIIIA-type zinc finger protein, ZFP182, helps protect cell membrane integrity under cold stress (as indicated by reduced electrolyte leakage) and is associated with the DREB pathway [76].

Recent research has also demonstrated the involvement of many other genes in the cold signal pathway, including *bHLH1, OsP5CS2, LIP19, OSPGYR, ZFP245, OVP1, bZIP52/RISBZ5, TEF1* and *SlCZFP1*. The CBF-independent gene *OrbHLH001* from Dongxiang wild rice can enhance tolerance to low temperature, suggesting that wild rice may have a distinct stress responsive pathway that enables a perennial life history [9,23,50,77-82].

Post-transcriptional Regulation

mRNA processing

Pre-mRNA processing is an important mechanism of post-transcriptional regulation of gene expression in eukaryotes. For conversion into mature mRNA, pre-mRNA requires various modifications such as the addition of a 5' methyl cap and poly (A) tail and intron splicing. Precursor mRNA with more than one intron can undergo alternative splicing to produce functionally different proteins from a single gene. In plants, about 20% of genes require alternative splicing [83]. Alternative splicing affects many important processes and characteristics in plants including photosynthesis, flowering, grain quality in cereals, and stress responses. For example, the *Arabidopsis COR15A* gene, which encodes a chloroplast stromal protein with cryoprotective activity, plays an important role in conferring freezing tolerance to chloroplasts; the *Arabidopsis stabilized1* (*sta1*) mutant is defective in the splicing of the cold-induced *COR15A* pre-mRNA and is hypersensitive to chilling, ABA, and salt stress [84,85]. *STA1* encodes a nuclear pre-mRNA splicing factor and is upregulated by cold stress. STA1 catalyzes the splicing of *COR15A*, which is necessary for cold tolerance. Furthermore, pre-mRNAs of serine/arginine-rich (SR) proteins, which are involved in the regulation or execution of mRNA splicing, also undergo alternative splicing under cold and heat stresses in *Arabidopsis*.

In rice, recent studies on the mRNA processing of *OsDREB2B*, a member of the DREB-CRT/DRE pathway, have revealed an efficient mRNA processing strategy for cold acclimation [86,87]. To recognize the two types of *OsDREB2B* transcripts, the authors of these papers designed two primers that are specific for each transcript. In the non-stress plants, *OsDREB2B1* is approximately five times more abundant than *OsDREB2B2*. However, in the cold stress plants, *OsDREB2B2* accumulates to a level equal to or greater than that of *OsDREB2B1*. To determine whether *OsDREB2B2* is modified from *OsDREB2B1* or is spliced alternatively from the *OsDREB2B* gene, the authors introduced *OsDREB2B1* or *OsDREB2B2* fused to the synthetic green fluorescent protein (sGFP) genes into two kinds of transgenic rice plants. RT-PCR analysis did not detect *sGFP-OsDREB2B2* in the Ubi:sGFP:OsDREB2B1 plants but did detect the accumulated endogenous *OsDREB2B2* transcript. This result indicated that *OsDREB2B2* is produced directly from the *OsDREB2B2* gene and is not modified from *OsDREB2B1*. In addition, a transcriptional analysis showed that transcription is greater for *OsDREB2B2* than for *OsDREB2A* under stress conditions. These data suggest that mRNA processing is an important regulatory strategy for *OsDREB2B* expression as part of the abiotic stress response in rice.

Small RNAs

Small RNAs (sRNAs) are sequence-specific regulatory elements that mediate endogenous gene silencing in eukaryotes. Plant sRNAs have been divided into four classes based on their origins and structures: microRNAs (miRNAs) and three types of small interfering RNAs (siRNAs), including trans-acting siRNAs (ta-siRNAs), natural cis-antisense transcripts-derived siRNAs (nat-siRNAs), and repeat-associated siRNAs (ra-siRNAs) [88-90]. One major difference between miRNAs and siRNAs is that miRNAs result from the processing of a single-stranded hairpin precursor while siRNAs are generated from long double-stranded RNAs (dsRNAs) [91]. Plant miRNAs, a class of short non-coding RNAs (~22 nucleotides), are processed from primary miRNA transcripts through two sequential cleavages by Dicer-like1 (DCL1), and mature miRNAs are loaded into argonaute proteins to guide cleavage of target mRNAs or translational repression [92]. The biogenesis of ta-siRNAs is initiated by miRNA-mediated cleavage of non-coding transcripts. The cleaved RNAs are copied into dsRNAs by RNA-dependent RNA polymerase 6 (RDR6) and are processed by DCL4 into phased siRNAs from the end defined by miRNA-mediated cleavage. The production of ra-siRNAs requires the activity of DCL3, RDR2, and polymerase (Pol) IV, a plant-specific DNA-dependent RNA polymerase [93,94].

The rice genome contains approximately 250,000 transposable elements (TEs), constituting approximately 35% of the genome sequence [95,96]. TEs can lead to the folding of RNA sequences into hairpin structures (reminiscent of the pre-miRNAs), and many miRNAs derived from TEs have been found in rice [97]. Global expression profiling of rice miRNAs showed that 18 miRNAs respond rapidly to cold stress. Interestingly, most of these miRNAs are down-regulated, indicating that the expression of target genes controlled by these miRNAs is induced in an adaptive response to cold stress [96].

miRNAs also affect plant hormones, which regulate many important aspects of growth and development as well as responses to environmental stresses. miR-167 helps regulate the auxin signal by cleaving two auxin-response factors and might play a role in cold tolerance by affecting auxin-signaling pathways [96]. miR-171 is a large and conserved miRNA family. A microarray-based analysis has revealed that 6 h of cold stress upregulated miR-171a in *Arabidopsis* but down-regulated miR-171a in rice [98]. These results demonstrate that miRNAs belonging to the same family can display opposite patterns in response to cold stress, suggesting that they may perform different functions. The results also suggest that differences in their expression in rice vs. *Arabidopsis* may represent species-specific differences in the response to cold stress. Generally, miRNA accumulation negatively correlates with the level of target transcripts. Two MADS-box genes, *MADS 57 and MADS 27,* are targets of miR-444a [96,99]. How miR-444a differentially regulates the two *MADS* genes under cold stress warrants further investigation.

Conclusion

Intensive research has greatly increased our understanding of cold sensing, transcriptional networks, post-transcriptional regulation, and other responses of rice to cold stress. Rice plants use cold-induced calcium influx and changes in levels of ABA and ROS as signals to activate a response to cold stress. The DREB-CRT/DRE pathway has an important effect on the rice response to cold, and all DREB1 and DREB2 family members in the pathway have been identified and characterized by transgenic analysis in rice. The MAP kinase pathway and ABF/AREB pathway, whose effects have been well characterized in responses to other biotic and abiotic stress such as pathogen attack, heat, hyperosmotic stress, and oxidative stress, are involved in the cascade of cold stress responses in rice. Recent microarray analysis has revealed that small RNAs are important for cold responses. The recent research on cold acclimation also indicates that these cold-relevant genes can be used for molecular-assisted selection and transgenic research [100].

However, development of breeding techniques for rice shows that breeding of cold tolerant rice is still a challenge. Although many QTLs in cold tolerant rice varieties are identified, those QTLs are mostly stage specific and only effect germination stage, vegetative stage, or reproductive stage. Concerning current breeding techniques, varieties which tolerate to cold stress in all stages are hard to breed for the reason that the convergence of those QTLs into one rice variety is time and labor consuming. It is important, therefore, to develop alternative strategies for the breeding of crops and to discover more cold tolerant varieties and put them into breeding practices, including those wild rice varieties which live in high altitude areas and have an inborn tolerance under cold stress [101]. As the cold-responsive network is complex, characterizations of key factors and components among different cold signal transduction pathways are also useful for the breeding of cold tolerance in rice.

Although the number of cold-responsive genes found continues to increase, the biological complexity underlying cold acclimation has not decreased. An increased understanding of the molecular basis of cold tolerance in rice should facilitate the development of varieties of rice and of other crops that are tolerant to cold stress.

Acknowledgments

This work was supported, in part, by grants from the National Transgenic Project (2012ZX08009001), the National 973 Project (2012CB723000), and the National Natural Science Foundation of China (31071674). Z. W. and G. L. W. were also supported by Program for Innovative Research Team in University (IRT1239), the Aid Program for Science and Technology Innovative Research Team in Higher Educational Institutions of Hunan Province, the Hunan Provincial Key Laboratory of Crop Germplasm Innovation and Utilization (11KFXM01) and Hunan Agricultural University (11YJ13).

References

1. Hedhly A, Hormaza JI, Herrero M (2009) Global warming and sexual plant reproduction. Trends in Plant Science 14: 30-36.

2. Thomsen IK, Lægdsmand M, Olesen JE (2010) Crop growth and nitrogen turnover under increased temperatures and low autumn and winter light intensity. Agric Ecosyst Environ 139: 187-194.

3. Lee TM, Lur HS, Chu C (1995) Abscisic acid and putrescine accumulation in chilling-tolerant rice cultivars. Crop Sci 35: 502-508.

4. Kim SI, Tai TH (2011) Evaluation of seedling cold tolerance in rice cultivars: a comparison of visual ratings and quantitative indicators of physiological changes. Euphytica 178: 437-447.

5. Suh J, Jeung JU, Lee JI, Choi YH, Yea JD, et al. (2010) Identification and analysis of QTLs controlling cold tolerance at the reproductive stage and validation of effective QTLs in cold-tolerant genotypes of rice (Oryza sativa L.). Theor Appl Genet 120: 985-995.

6. Xu LM, Zhou L, Zeng YW, Wang FM, Zhang HL, et al. (2008) Identification and mapping of quantitative trait loci for cold tolerance at the booting stage in a japonica rice near-isogenic line. Plant Science 174: 340-347.

7. Koseki M, Kitazawa N, Yonebayashi S, Maehara Y, Wang ZX, et al. (2010) Identification and fine mapping of a major quantitative trait locus originating from wild rice, controlling cold tolerance at the seedling stage. Mol Genet Genomics 284: 45-54.

8. Saito K, Hayano-Saito Y, Kuroki M, Sato Y (2010) Map-based cloning of the rice cold tolerance gene Ctb1. Plant Science 179: 97-102.

9. Huang J, Sun SJ, Xu DQ, Yang X, Bao YM, et al. (2009) Increased tolerance of rice to cold, drought and oxidative stresses mediated by the overexpression of a gene that encodes the zinc finger protein ZFP245. Biochem Biophys Res Commun 389: 556-561.

10. Yun KY, Park MR, Mohanty B, Herath V, Xu F, et al. (2010) Transcriptional regulatory network triggered by oxidative signals configures the early response mechanisms of japonica rice to chilling stress. BMC Plant Biol 10: 16.

11. Tian Y, Zhang H, Pan X, Chen X, Zhang Z, et al. (2011) Overexpression of ethylene response factor $TERF_2$ confers cold tolerance in rice seedlings. Transgenic Res 20: 857-866.

12. Chinnusamy V, Zhu J, Zhu JK (2006) Gene regulation during cold acclimation in plants. Physiol Plant 126: 52-61.

13. Saijo Y, Hata S, Kyozuka J, Shimamoto K, Izui K (2001) Over-expression of a single Ca^{2+}-dependent protein kinase confers both cold and salt/drought tolerance on rice plants. Plant J 23: 319-327.

14. Lee SC, Huh KW, An K, An G, Kim SR (2004) Ectopic expression of a cold-inducible transcription factor, CBF1/DREB1b, in transgenic rice (Oryza sativa L.). Mol Cells 18: 107-114.

15. Morsy MR, Almutairi AM, Gibbons J, Yun SJ, De Los Reyes BG (2005) The OsLti6 genes encoding low-molecular-weight membrane proteins are differentially expressed in rice cultivars with contrasting sensitivity to low temperature. Gene 344: 171-180.

16. Song SY, Chen Y, Chen J, Dai XY, Zhang WH (2011) Physiological mechanisms underlying OsNAC5-dependent tolerance of rice plants to abiotic stress. Planta 234: 331-345.

17. Oliver SN, Dennis ES, Dolferus R (2007) ABA regulates apoplastic sugar transport and is a potential signal for cold-induced pollen sterility in rice. Plant Cell Physiol 48: 1319-1330.

18. Hossain MA, Cho JI, Han M, Ahn CH, Jeon JS, et al. (2010) The ABRE-binding bZIP transcription factor OsABF2 is a positive regulator of abiotic stress and ABA signaling in rice. J Plant Physiol 167: 1512-1520.

19. Yang X, Yang YN, Xue LJ, Zou MJ, Liu JY, et al. (2011) Rice ABI5-Like1 regulates abscisic acid and auxin responses by affecting the expression of ABRE-containing genes. Plant Physiol 156: 1397-1409.

20. Triantaphylidès C, Havaux M (2009) Singlet oxygen in plants: production, detoxification and signaling. Trends in Plant Science 14: 219-228.

21. Sato Y, Masuta Y, Saito K, Murayama S, Ozawa K (2011) Enhanced chilling tolerance at the booting stage in rice by transgenic overexpression of the ascorbate peroxidase gene, OsAPXa. Plant Cell Rep 30: 399-406.

22. Xie G, Kato H, Sasaki K, Imai R (2009) A cold-induced thioredoxin h of rice, OsTrx23, negatively regulates kinase activities of OsMPK3 and OsMPK6 in vitro. FEBS Lett 583: 2734-2738.

23. Shimizu H, Sato K, Berberich T, Miyazaki A, Ozaki R, et al. (2005) LIP19, a basic region leucine zipper protein, is a Fos-like molecular switch in the cold signaling of rice plants. Plant Cell Physiol 46: 1623-1634.

24. Nakashima K, Tran LS, Van Nguyen D, Fujita M, Maruyama K, et al. (2007) Functional analysis of a NAC-type transcription factor OsNAC6 involved in abiotic and biotic stress-responsive gene expression in rice. Plant J 51: 617-630.

25. Liu C, Wu Y, Wang X (2012) bZIP transcription factor OsbZIP52/RISBZ5: a potential negative regulator of cold and drought stress response in rice. Planta 235: 1157-1169

26. Sheen J (1996) Ca^{2+}-dependent protein kinases and stress signal transduction in plants. Science 274: 1900-1902.

27. Allwood EG, Smertenko AP, Hussey PJ (2001) Phosphorylation of plant actin-depolymerising factor by calmodulin-like domain protein kinase. FEBS Lett 499: 97-100.

28. Wan B, Lin Y, Mou T (2007) Expression of rice Ca^{2+}-dependent protein kinases (CDPKs) genes under different environmental stresses. FEBS Lett 581: 1179-1189.

29. Abo-El-Saad M, Wu R (1995) A rice membrane calcium-dependent protein kinase is induced by gibberellin. Plant Physiol 108: 787-793.

30. Sharma A, Matsuoka M, Tanaka H, Komatsu S (2001) Antisense inhibition of a BRI1 receptor reveals additional protein kinase signaling components downstream to the perception of brassinosteroids in rice. FEBS Lett 507: 346-350.

31. Sharma A, Komatsu S (2002) Involvement of a Ca^{2+}-Dependent Protein Kinase Component Downstream to the Gibberellin-Binding Phosphoprotein, RuBisCO Activase, in Rice. Biochem Biophys Res Commun 290: 690-695.

32. Abbasi F, Onodera H, Toki S, Tanaka H, Komatsu S (2004) OsCDPK13, a calcium-dependent protein kinase gene from rice,is induced by cold and gibberellin in rice leaf sheath. Plant Mol Biol Report 55: 541-552.

33. Achard P, Gong F, Cheminant S, Alioua M, Hedden P, et al. (2008) The cold-inducible CBF1 factor-dependent signaling pathway modulates the accumulation of the growth-repressing DELLA proteins via its effect on gibberellin metabolism. Plant Cell 20: 2117-2129.

34. Robinson MJ, Cobb MH (1997) Mitogen-activated protein kinase pathways. Curr Opin Cell Biol 9: 180-186.

35. Kyriakis JM, Avruch J (2001) Mammalian mitogen-activated protein kinase signal transduction pathways activated by stress and inflammation. Physiol Rev 81: 807-869.

36. Schaeffer HJ, Weber MJ (1999) Mitogen-activated protein kinases: specific messages from ubiquitous messengers. Mol Cell Biol 19: 2435-2444.

37. Keyse SM (2000) Protein phosphatases and the regulation of mitogen-activated protein kinase signalling. Curr Opin Cell Biol 12: 186-192.

38. Jonak C, Kiegerl S, Ligterink W, Barker PJ, Huskisson NS, et al. (1996) Stress signaling in plants: a mitogen-activated protein kinase pathway is activated by cold and drought. Proc Natl Acad Sci U S A 93: 11274.

39. Kovtun Y, Chiu WL, Tena G, Sheen J (2000) Functional analysis of oxidative stress-activated mitogen-activated protein kinase cascade in plants. Proc Natl Acad Sci U S A 97: 2940.

40. Ichimura K, Shinozaki K, Tena G, Sheen J, Henry Y, et al. (2002) Mitogen-activated protein kinase cascades in plants: a new nomenclature. Trends Plant Sci 7: 301-308.

41. Gupta R, Luan S (2003) Redox control of protein tyrosine phosphatases and mitogen-activated protein kinases in plants. Plant Physiol 132: 1149-1152.

42. Xiong L, Yang Y (2003) Disease resistance and abiotic stress tolerance in rice are inversely modulated by an abscisic acid–inducible mitogen-activated protein kinase. Plant Cell 15: 745-759.

43. Droillard MJ, Boudsocq M, Barbier-Brygoo H, Laurière C (2002) Different protein kinase families are activated by osmotic stresses in Arabidopsis thaliana cell suspensions:: Involvement of the MAP kinases AtMPK3 and AtMPK6. FEBS Lett 527: 43-50.

44. Reyna NS, Yang Y (2006) Molecular analysis of the rice MAP kinase gene family in relation to Magnaporthe grisea infection. Mol Plant-Microbe Interact 19: 530-540.

45. He C, Fong SHT, Yang D, Wang GL (1999) BWMK1, a novel MAP kinase induced by fungal infection and mechanical wounding in rice. Mol Plant-Microbe Interact 12: 1064-1073.

46. Xie G, Kato H, Imai R (2012) Biochemical identification of the OsMKK6-OsMPK3 signalling pathway for chilling stress tolerance in rice. Biochem J 443: 95-102.

47. Agarwal PK, Agarwal P, Reddy M, Sopory SK (2006) Role of DREB transcription factors in abiotic and biotic stress tolerance in plants. Plant Cell Rep 25: 1263-1274.

48. Wang Q, Guan Y, Wu Y, Chen H, Chen F, et al. (2008) Overexpression of a rice OsDREB1F gene increases salt, drought, and low temperature tolerance in both Arabidopsis and rice. Plant Mol Biol 67: 589-602.

49. Dubouzet JG, Sakuma Y, Ito Y, Kasuga M, Dubouzet EG, et al. (2003) OsDREB genes in rice, Oryza sativa L., encode transcription activators that function in drought-, high-salt-and cold-responsive gene expression. Plant J 33: 751-763.

50. Li F, Guo S, Zhao Y, Chen D, Chong K, et al. (2010) Overexpression of a homopeptide repeat-containing bHLH protein gene (OrbHLH001) from Dongxiang Wild Rice confers freezing and salt tolerance in transgenic Arabidopsis. Plant Cell Rep 29: 977-986.

51. Nakamura J, Yuasa T, Huong TT, Harano K, Tanaka S, et al. (2011) Rice homologs of inducer of CBF expression (OsICE) are involved in cold acclimation. Plant Biotechnol 28: 303-309.

52. Kim JC, Lee SH, Cheong YH, Yoo CM, Lee SI, et al. (2001) A novel cold-inducible zinc finger protein from soybean, SCOF-1, enhances cold tolerance in transgenic plants. Plant J 25: 247-259.

53. Ma Q, Dai X, Xu Y, Guo J, Liu Y, et al. (2009) Enhanced tolerance to chilling stress in OsMYB3R-2 transgenic rice is mediated by alteration in cell cycle and ectopic expression of stress genes. Plant Physiol 150: 244-256.

54. Dai X, Xu Y, Ma Q, Xu W, Wang T, et al. (2007) Overexpression of an R1R2R3 MYB gene, OsMYB3R-2, increases tolerance to freezing, drought, and salt stress in transgenic Arabidopsis. Plant Physiol 143: 1739-1751.

55. Yang A, Dai X, Zhang WH (2012) A R2R3-type MYB gene, OsMYB2, is involved in salt, cold, and dehydration tolerance in rice. J Exp Bot 63: 2541-2556.

56. Su CF, Wang YC, Hsieh TH, Lu CA, Tseng TH, et al. (2010) A novel MYBS3-dependent pathway confers cold tolerance in rice. Plant Physiol 153: 145-158.

57. Kim HS, Park BO, Yoo JH, Jung MS, Lee SM, et al. (2007) Identification of a calmodulin-binding NAC protein as a transcriptional repressor in Arabidopsis. J Biol Chem 282: 36292-36302.

58. Lin R, Zhao W, Meng X, Wang M, Peng Y (2007) Rice gene OsNAC19 encodes a novel NAC-domain transcription factor and responds to infection by Magnaporthe grisea. Plant Sci 172: 120-130.

59. Ooka H, Satoh K, Doi K, Nagata T, Otomo Y, et al. (2003) Comprehensive analysis of NAC family genes in Oryza sativa and Arabidopsis thaliana. DNA Res 10: 239-247.

60. Jeong JS, Kim YS, Baek KH, Jung H, Ha SH, et al. (2010) Root-specific expression of OsNAC10 improves drought tolerance and grain yield in rice under field drought conditions. Plant Physiol 153: 185-197.

61. Nuruzzaman M, Manimekalai R, Sharoni AM, Satoh K, Kondoh H, et al. (2010) Genome-wide analysis of NAC transcription factor family in rice. Gene 465: 30-44.

62. Takasaki H, Maruyama K, Kidokoro S, Ito Y, Fujita Y, et al. (2010) The abiotic stress-responsive NAC-type transcription factor OsNAC5 regulates stress-inducible genes and stress tolerance in rice. Mol Genet Genomics 284: 173-183.

63. Hu H, Dai M, Yao J, Xiao B, Li X, et al. (2006) Overexpressing a NAM, ATAF, and CUC (NAC) transcription factor enhances drought resistance and salt tolerance in rice. Proc Natl Acad Sci U S A 103: 12987-12992.

64. Vij S, Tyagi AK (2008) A20/AN1 zinc-finger domain-containing proteins in plants and animals represent common elements in stress response. Funct Integr Genomics 8: 301-307.

65. Satou Y, Satoh N (1997) posterior end mark 2 (pem-2), pem-4, pem-5, and pem-6: Maternal Genes with Localized mRNA in the Ascidian Embryo. Dev Biol 192: 467-481.

66. Duan W, Sun B, Li TW, Tan BJ, Lee MK, et al. (2000) Cloning and characterization of AWP1, a novel protein that associates with serine/threonine kinase PRK1 in vivo. Gene 256: 113-121.

67. Liepinsh E, Leonchiks A, Sharipo A, Guignard L, Otting G (2003) Solution Structure of the R3H Domain from Human Sμbp-2. J Mol Biol 326: 217-223.

68. Kanneganti V, Gupta AK (2008) Overexpression of OsiSAP8, a member of stress associated protein (SAP) gene family of rice confers tolerance to salt,

drought and cold stress in transgenic tobacco and rice. Plant Mol Biol 66: 445-462.

69. Huang J, Wang MM, Jiang Y, Bao YM, Huang X, et al. (2008) Expression analysis of rice A20/AN1-type zinc finger genes and characterization of *ZFP177* that contributes to temperature stress tolerance. Gene 420: 135-144.

70. Miller J, McLachlan A, Klug A (1985) Repetitive zinc-binding domains in the protein transcription factor IIIA from Xenopus oocytes. EMBO J 4: 1609.

71. Ciftci-Yilmaz S, Mittler R (2008) The zinc finger network of plants. Cell Mol Life Sci 65: 1150-1160.

72. Lippuner V, Cyert MS, Gasser CS (1996) Two classes of plant cDNA clones differentially complement yeast calcineurin mutants and increase salt tolerance of wild-type yeast. J Biol Chem 271: 12859-12866.

73. Sakamoto H, Maruyama K, Sakuma Y, Meshi T, Iwabuchi M, et al. (2004) Arabidopsis Cys2/His2-type zinc-finger proteins function as transcription repressors under drought, cold, and high-salinity stress conditions. Plant Physiology 136: 2734-2746.

74. Ciftci-Yilmaz S, Morsy MR, Song L, Coutu A, Krizek BA, et al. (2007) The EAR-motif of the Cys2/His2-type zinc finger protein Zat7 plays a key role in the defense response of *Arabidopsis* to salinity stress. J Biol Chem 282: 9260-9268.

75. Davletova S, Schlauch K, Coutu J, Mittler R (2005) The zinc-finger protein Zat12 plays a central role in reactive oxygen and abiotic stress signaling in *Arabidopsis*. Plant Physiol 139: 847-856.

76. Huang J, Sun S, Xu D, Lan H, Sun H, et al. (2012) A TFIIIA-type zinc finger protein confers multiple abiotic stress tolerances in transgenic rice (*Oryza sativa* L.). Plant Mol Biol 80: 337-350.

77. Wang YJ, Zhang ZG, He XJ, Zhou HL, Wen YX, et al. (2003) A rice transcription factor *OsbHLH1* is involved in cold stress response. Theor Appl Genet 107: 1402-1409.

78. Hur J, Jung K-H, Lee CH, An G (2004) Stress-inducible *OsP5CS2* gene is essential for salt and cold tolerance in rice. Plant Sci 167: 417-426.

79. Li H, Yang J, Wang Y, Chen Z, Tu S, et al. (2009) Expression of a novel *OSPGYRP* (rice proline-, glycine- and tyrosine-rich protein) gene, which is involved in vesicle trafficking, enhanced cold tolerance in E. coli. Biotechnol Lett 31: 905-910.

80. Liu C, Wu Y, Wang X (2011) bZIP transcription factor *OsbZIP52/RISBZ5*: a potential negative regulator of cold and drought stress response in rice. Planta 235: 1157-1169.

81. Zhang X, Guo X, Lei C, Cheng Z, Lin Q, et al. (2011) Overexpression of *SICZFP1*, a novel TFIIIA-type zinc finger protein from tomato, confers enhanced cold tolerance in transgenic *Arabidopsis* and rice. Plant Mol Biol Report 29: 185-196.

82. Paul P, Awasthi A, Rai AK, Gupta SK, Prasad R, et al. (2012) Reduced tillering in Basmati rice T-DNA insertional mutant *OsTEF₁* associates with differential expression of stress related genes and transcription factors. Funct Integr Genomics 12: 291-304.

83. Chinnusamy V, Zhu JK, Sunkar R (2010) Gene regulation during cold stress acclimation in plants. Methods Mol Biol 639: 39-55.

84. Lee B, Kapoor A, Zhu J, Zhu JK (2006) STABILIZED1, a stress-upregulated nuclear protein, is required for pre-mRNA splicing, mRNA turnover, and stress tolerance in *Arabidopsis*. Plant Cell 18: 1736-1749.

85. Chinnusamy V, Zhu J, Zhu JK (2007) Cold stress regulation of gene expression in plants. Trends Plant Sci 12: 444-451.

86. Matsukura S, Mizoi J, Yoshida T, Todaka D, Ito Y, et al. (2010) Comprehensive analysis of rice *DREB2*-type genes that encode transcription factors involved in the expression of abiotic stress-responsive genes. Mol Genet Genomics 283: 185-196.

87. Maruyama K, Todaka D, Mizoi J, Yoshida T, Kidokoro S, et al. (2012) Identification of cis-acting promoter elements in cold- and dehydration-induced transcriptional pathways in *Arabidopsis*, rice, and soybean. DNA Res 19: 37-49.

88. Hamilton AJ, Baulcombe DC (1999) A species of small antisense RNA in posttranscriptional gene silencing in plants. Science 286: 950-952.

89. Ghildiyal M, Zamore PD (2009) Small silencing RNAs: an expanding universe. Nat Rev Genet 10: 94-108.

90. Ameres SL, Horwich MD, Hung JH, Xu J, Ghildiyal M, et al. (2010) Target RNA-Directed Trimming and Tailing of Small Silencing RNAs. Science 328: 1534-1539.

91. Lee YS, Nakahara K, Pham JW, Kim K, He Z, et al. (2004) Distinct roles for *Drosophila* Dicer-1 and Dicer-2 in the siRNA/miRNA silencing pathways. Cell 117: 69-81.

92. Jones-Rhoades MW, Bartel DP (2004) Computational identification of plant microRNAs and their targets, including a stress-induced miRNA. Mol Cell 14: 787-799.

93. Akbergenov R, Si-Ammour A, Blevins T, Amin I, Kutter C, et al. (2006) Molecular characterization of geminivirus-derived small RNAs in different plant species. Nucleic Acids Res 34: 462-471.

94. Chen X (2009) Small RNAs and their roles in plant development. Annu Rev Cell Dev Biol 25: 21-44.

95. Khurana P, Gaikwad K (2005) The map-based sequence of the rice genome. Nature 436: 793-800.

96. Lv DK, Bai X, Li Y, Ding XD, Ge Y, et al. (2010) Profiling of cold-stress-responsive miRNAs in rice by microarrays. Gene 459: 39-47.

97. Piriyapongsa J, Jordan IK (2008) Dual coding of siRNAs and miRNAs by plant transposable elements. RNA 14: 814-821.

98. Liu HH, Tian X, Li YJ, Wu CA, Zheng CC (2008) Microarray-based analysis of stress-regulated microRNAs in *Arabidopsis thaliana*. RNA 14: 836-843.

99. Ratcliffe OJ, Kumimoto RW, Wong BJ, Riechmann JL (2003) Analysis of the Arabidopsis *MADS AFFECTING FLOWERING* gene family: *MAF₂* prevents vernalization by short periods of cold. Plant Cell 15: 1159-1169.

100. Phan Tran LS, Nishiyama R, Yamaguchi-Shinozaki K, Shinozaki K (2010) Potential utilization of NAC transcription factors to enhance abiotic stress tolerance in plants by biotechnological approach. GM Crops 1: 32-39.

101. Sanghera GS, Wani SH, Hussain W, Singh N (2011) Engineering Cold Stress Tolerance in Crop Plants. Curr Genomics 12: 30-43.

2

Effect of Four Mycorrhizal Products on Squash Plant Growth and its Effect on Physiological Plant Elements

Al-Hmoud G* and Al-Momany A

University of Jordan, Department of Plant Protection, Amman, Jordan

Abstract

Vesicular arbuscular mycorrhizal fungi (VAM) are a symbiotic fungi belonging to phylum Glomeromycota, which interact with the root system of higher plants by producing external and internal hyphae, vesiculars and arbuscules. This study aimed to determine the efficiency of VAM fungi in increasing squash plant growth, plant root surface and its effect on physiological plant elements. Four mycorrhizal products; Bacto_Prof, Endomyk_Basic, Endomyk_Conc and Endomyk_Prof were imported from Terrabioscience Company, Germany, contained Glomus intraradices. Each product was used in three doses; half, recommended and double dose according to the application rates and instructions indicated by the manufacturer. Endomyk_Basic was the most effective product in improving plant growth. Height of squash plants in the recommended dose treatment was enhanced significantly by all mycorrhizal products; Bacto_Prof, Endomyk_Basic, Endomyk_Conc and Endomyk_Prof by 17, 18, 19 and 13% in comparison to control plants, respectively. Double dose was the most effective treatment which increased squash plant growth and root weight. Concentration of nitrogen (N) and phosphorus (P) were markedly higher in root system than in shoot system. Mycorrhizal products increased N in roots by 6, 21, 17 and 11%, while P was enhanced by 21, 4, 9 and 8% for Bacto_Prof, Endomyk_Basic, Endomyk_Conc and Endomyk_Prof, respectively. Concentrations of fat in shoot was higher than root system, while in crude protein and fiber, the plant root was significantly higher than shoot system by 6 and 52%, respectively. For the carbohydrate concentration, shoot system was higher than root system by 97% increase. From the results of this study, it was concluded that all mycorrhizal products were effective on physiological plant content and plant growth more than non-treated plants.

Keywords: Symbiotic; Glomeromycota; Vesicular arbuscular mycorrhizal fungi

Introduction

Mycorrhiza is a symbiotic association between a fungus and the root system of vascular plants belonging to phylum Glomeromycota [1,2]. Symbiosis termed by De Bary as the mutual beneficial association. Mycorrhizal fungi are the most widespread fungal symbionts that colonize the root system of over 90% of plant species to the mutual benefit of both the plant host and fungus [3,4] either exteracellularly as in ectomycorrhizal fungi or intercellularly as in endomycorrhizal fungi [arbuscular mycorrhizal (AM) fungi]. There are different types of fungi that form these symbiotic associations, but for agriculture, the arbuscular mycorrhizal fungi are highly important [2], which colonizes the root system of most cultivated crops and horticultural plants; usually it invades the different layers of the outer root cortex [5].

Vesicular arbuscular mycorrhiza(VAM) colonizes plant roots and extend into the surrounding bulk soil to the root depletion zone around the root system [6]. VAM fungi has no sexual stage so, the only means for gene transfer among individuals is through the vegetative fusion of mycelia [7]. Different crops exhibited different VAM species and different stages of fungal invasion ranging from hyphae, arbuscules and vesicles or combinations of all structures [8].

VAM fungi are associated with improved growth of many plant species due to increased nutrients uptake, production of growth promoting substances, tolerance to drought and salinity and transplant shock and synergistic interaction with other beneficial microorganisms such as nitrogen fixers and phosphorus solubilizers [9]. The major role of VAM fungi is to supply plant roots with phosphorus, because phosphorus is an extremely immobile element in soils [10], due to the fungal extraradical mycelium ability; it grows beside the phosphate depletion zone that quickly develops around the root [3] and can be extended up to 9 cm in the soil [11]. Hyphae of VAM fungi explore

a larger volume of soil and phosphate solubilization from unavailable sources present in the soil [12]. Plant phosphate is often the main controlling factor in the plant-fungal relationship [13] and this will be associated with increased plant growth and yield [14].

Therefore, this work was conducted to determine the effect of half of the dose, double of the dose in addition to the recommended application dose of different commercial mycorrhizal products on squash growth and the effect of mycorrhizal products on physiological plant elements.

Materials and Methods

Source of mycorrhizal products

Four different mycorrhizal products were exported from Terrabioscience UG. Bernbug-Germany; Bacto_Prof, Endomyk_Conc, Endomyk_Prof and Endomyk_Basic. Each product was used in different doses according to the application rates and instructions indicated by the manufacturer; Bacto_Prof (1 g/L soil), Endomyk_Conc (1 g/L soil), Endomyk_Prof (2 g/L soil) and Endomyk_Basic (8 g/L soil). All the mycorrhizal products were in powder form except Endomyk_Basic, the mycorrhizae was produced in granule form. We used the recommended

*Corresponding author: Al-Hmoud G, Department of Plant Protection, University of Jordan, Amman, Jordan, E-mail: eng_ghina2009@ yahoo.com

dose in addition to half dose and double dose for each product to test its effect on plant growth.

Glasshouse work

Seedlings of squash plant (*Cucurbita maxima*) var. Yasmina F1 were prepared in trays 2-4 weeks before inoculation with mycorrhizal products. The experiments were carried out under normal environmental conditions 25-32°C under protected cages in the glasshouse, and each pot was irrigated with 100-150 ml tap water day after day.

Mixture of soil: sand (1:2 v/v) ratio was used in this research; soil was sterilized in the oven for 6 hours at 70°C, and then mixed with pure sand in the same day. Seedlings were inoculated with commercial mycorrhizal products and planted in new and clean plastic pots of one liter size and 12 cm depth (Table 1). One fourth liter soil mixture was put at the bottom of the pot then added the mycorrhizal product to the rest amount, to avoid product leaching with irrigation water. The rest of soil mixture with the mycorrhizal products was mixed carefully; then added to the pot around the seedling.

For each crop, five treatments; Bacto_Prof, Endomyk_Conc, Endomyk_Prof, Endomyk_Basic and non inoculated (control) were conducted in three doses in separate experiment. Each treatment was comprised of five replicate pots, grown under glasshouse conditions until harvest time (7-8 weeks). At the harvest day; plants were cut, shoot length was measured, fresh shoot and root weights were recorded. One gram of fine feeder roots from each treatment was taken to examine its mycorrhization according to Philips and Hayman method [15]. Fresh shoots were dried for 24 hrs at 70°C to record its weight. All data were statistically analyzed by using the SAS program, comparison between means was done according to LSD at 5% level.

Laboratory experiments

Evaluation of VAM roots colonization: Roots were washed carefully to remove soil particles, then heated in 10% KOH at 90°C for half an hour to destroy cell cytoplasm, so the plant cell will be cleared, then washed for 2-3 minutes in running tap water gently, then stained with 0.05% trypan blue in lactophenol according to Philips and Hayman method [15]. Ten root segments with 1 cm length were mounted on each slide and examined microscopically. The incidence and intensity of root colonization with mycorrhiza were calculated using a scale of (0-10) where zero means no colonization, 5 means 50% mycorrhizal root colonization and 10 means 100% mycorrhizal root colonization [16], the readings were taken from the average percentage of thirty roots for each treatment.

Analysis of plant tissues

Squash plants inoculated with the recommended dose of mycorrhizal products for 7-8 weeks, were used in analysis of plant tissues, in order to distinguish the differences between the mycorrhizal treated plants and non-treated plants in nutrient absorption; nitrogen, phosphorus, proteins and carbohydrates.

Determination of moisture: Samples were dried immediatͰly after harvesting in electrical oven on 70°C for 48 hours then on 103 ± 2°C for 24 hours. To determine the moisture, the samples were in homogenous form, by using the mill with 1 mm sieve under the same conditions. Weight the sample before and after drying to calculate the moisture % and the dry matter % [17].

Calculation: $\% \ Moisture = \dfrac{(A - B) \ \times \ 100}{A}$

Treatment	Dose(g/Lsoil)	Height(cm)	FSW(g)	DSW(g)	FRW(g)
Bacto_Prof	0.5	29.9 d	10.5 cd	2.2 b	4.6 cde
	1	31.7 cd	11 bcd	2.5 b	4.8 bcd
	2	33.7 abc	13.6 a	3 a	5.3 abcd
Endomyk_Basic	4	30.9 d	12 b	2.4 b	4.9 bcd
	8	32 cd	13.7 a	2.4 b	5.4 abcd
	16	35.3 a	14.2 a	3.4 a	6 a
Endomyk_Conc	0.5	29.9 d	10.9 bcd	2.4 b	4.9 bcd
	1	32.2 bcd	11.6 bc	2.6 b	5.5 abc
	2	34.7 ab	13.5 a	3 a	5.6 ab
Endomyk_Prof	1	30 d	10.2 d	2. b	4.5 de
	2	30.6 d	10.7 bcd	2.4 b	4.7 cde
	4	33.8 abc	13.5 a	3 a	5.5 abc
Control	0	24.1 e	10.6 d	2.2 b	3.4 e

Values are average of five plants, values within each column followed by the same letter are not significantly different (P<0.05) according to LSD.

Table 1: Effect of three doses of four mycorrhizal products on squash growth compared with control.

$\% \ Dry \ Matter = \dfrac{(A - B) \ \times \ 100}{B}$

Where: A is the original sample weight, while B is the sample weight after drying.

Determination of crude fat by ether extracts: Weight the moisture free sample, clean dry receiving flask was used to place the thimble with the sample, 30-40 ml petroleum solution (organic solution) was added to the receiving flask and placed in sample container then placed under the condenser of the soxhlet apparatus. The extraction lasted for 16 hours, after that the sample was transferred to the oven for drying at 105°C for 30 minutes; then cooled in desicator at room temperature and weighed to calculate the crude fat % [17].

Calculation: $\% \ crude \ fat = \dfrac{(M_2 - M_1)}{M_0} \times 100$

Where: M0: original weight of sample

M1: weight of receiving flask before extraction (empty).

M2: weight of receiving flask and fat after extraction.

Determination of nitrogen and protein by Kjeldahl method: 0.2 gm dry sample; was prepared by adding 3.50-4.00 gm of the digestion mixture. Put them in a flask, and then add 25 ml sulfuric acid H_2SO_4, then put it in the digestor. Digest the mixture until the solution becomes colorless, wait for 30 minutes until cool down, then transfer the solution to volumetric flask (100 ml) and complete to 100 ml with distilled water (Table 2).

Place a flask containing 30 ml of 4% boric acid (pH 3.50-4.00) with 3 drops indicator 5:1 (screened methyl red indicator solution) on shelf at the outlet of the condenser of the Kjeldahl distillator. Transfer 5 ml of the filtrated solution to the mixing chamber, and then add 10 ml NaOH 50%, then rinse several times with small amount of distilled water. Free ammonia NH_3 from the sample will react with NH_4OH and trapped by boric acid to make ammonia biurate with dark blue color between 7-10 minutes. After that; the flask was taken a side for titration, the used volume of HCl must change the blue color to light orange color and this volume of HCl titrated applied to the calculation formula to calculate the N% and protein% [17].

Calculation:

$Calculation: \ \%N = \dfrac{(vol. HCl \ sample - HCl \ blank) \times No. of \ HCl \ \times 14.007}{1000 \times wt. of \ sample} \times \dfrac{100}{5} \times 100$

Treatment	Bacto_Prof			Endomyk_Basic			Endomyk_Conc			Endomyk_Prof		
Dose (g/L)	0.5	1	2	4	8	16	0.5	1	2	1	2	4
Squash	35%	45%	52%	36%	46%	56%	28%	50%	54%	30%	40%	52%

Table 2: Mycorrhizal root colonization of the three doses of four mycorrhizal products affected on squash plant growth.

% protein = % Nitrogen × 6.25

Determination of ash: This method was used to burn off all organic materials and keeping the inorganic materials called ash in order to determine the phosphorus and acid insoluble ash. The crucible used must be clean, so heat it for 1 hour in muffle furnace at 600°C. Cool and weight as quickly as possible. After that weight the sample and place it in muffle furnace on 600°C, leave it for 6-8 hours. After heating, transfer it to the desicator and cool at room temperature, then weight and use the following equation to calculate the ash content;

$$\% \ Ash \ = \frac{wt. \ of \ ash(g)}{w.t \ of \ sample(g)} \times 100$$

% Organic matter = 100 - % (water + ash)

Determination of phosphorus (Colorimetric method): In the crucible with the ash; add 5 ml HCl with 5 drops of nitric acid HNO_3 and dissolve it on hot plate, then filtrate into 200 ml volumetric flask, and dilute it with distilled water. Take 1 ml of the filtered solution to another volumetric flask, dilute it with distilled water, and then add 4 ml ammonium molybdate dissolved in H_2SO_4. Dissolve 0.25 gm of stannous chloride ($SnCl_2$) in 10 ml of concentrated HCl, then add 1 ml for each sample from this solution, then bring up to 100 ml with distilled water. Shake the solution then let it stand for 10 minutes. In this time the color of the solution will turn blue.

Calculate the phosphorus of each sample by measuring the absorbance at 650 nm with spectrophotometer. Distilled water was used as blank, to test the spectrophotometer on zero reading (AOAC, 1995).

$$\% \ Phosphorus \ = \frac{Reading \ from \ curve}{1000000} \times dilution \ factor \times \frac{100}{(w.t. \ of \ sample)}$$

Determination of fibers: Fibers are defined as the organic fraction remaining after digestion with standard solutions of sulfuric acid H_2SO_4 (1.25%) and Sodium hydroxide NaOH (1.25%) under controlled conditions. Weight the filter bag then put in the bag 0.50 gm of dried sample.

Ankom apparatus (01/02) was used in this method. Put the sealed bag inside the Ankom fiber analyzer vessel, add H_2SO_4 1.25% until it covers the bag, close tightly and turn agitate and heat on for 45 minutes. After that, add NaOH 1.25% until it covers the bag, close tightly for 45 minutes. Then take off the bag, rinse with distilled water, dry the bag in oven at 105°C for 2 hours, then recalculate the bag weight [17].

$$Calculation: \ \% \ Fiber \ = \frac{(loss \ weight \ (A - B) \ gm)}{(mass \ of \ sample \ gm)} \times 100$$

Where: A is weight after extraction process; B is the bag weight empty.

Results

Effect of different commercial mycorrhizal products on squash growth

Effect of the four mycorrhizal products on squash growth planted with three doses; half dose, recommended dose and double dose was summarized in Table 3. In half dose treatments; there were significant differences in plant height, FSW and FRW. Plant height was improved

significantly by all mycorrhizal products over the control plants; Bacto_Prof, Endomyk_Conc and Endomyk_Prof by 11% and Endomyk_Basic by 14%. Plant FSW was increased significantly only in Endomyk_Basic product by 22% more than control treatment. Squash FRW was enhanced significantly in Endomyk_Basic and Endomyk_Conc by 25 and 24% above control plants, respectively.

In the recommended dose; there were significant differences in plant height, FSW and FRW. Height of squash plants was enhanced significantly by all mycorrhizal products; Bacto_Prof, Endomyk_Basic, Endomyk_Conc and Endomyk_Prof by 17, 18, 19 and 13% in comparison to control plants, respectively. Plant FSW was increased significantly in Endomyk_Basic and Endomyk_Conc products by 38 and 17%, respectively more than control plants, while plant FRW were increased by Bacto_Prof, Endomyk_Basic and Endomyk_Conc products significantly more than control plants; by 43, 58 and 61%, respectively.

In double dose treatments; there were significant differences in plant height, FSW, DSW and FRW. Plant height was increased by all mycorrhizal products more than control plants; Bacto_Prof, Endomyk_Basic, Endomyk_Conc and Endomyk_Prof by 40, 46, 44 and 40%, respectively. Plant FSW were increased by mycorrhizal products; Bacto_Prof, Endomyk_Basic, Endomyk_Conc and Endomyk_Prof by 29, 34, 27 and 27%, respectively more than non-treated plants, but there were no differences between the mycorrhizal products in general. The increase in DSW was 37, 57, 43 and 37%, respectively by Bacto_Prof, Endomyk_Basic, Endomyk_Conc and Endomyk_Prof compared to control plants. However, plant FRW was highly significant in mycorrhizal products compared to non-mycorrhizal plants; by 56, 76, 65 and 61%, respectively for Bacto_Prof, Endomyk_Basic, Endomyk_Conc and Endomyk_Prof products.

Evaluations of VAM root colonization

Mycorrhizal root colonization of the three doses of different products affected squash crop was presented in Table 2. All mycorrhizal products were efficient in root colonization in different intensity. The examination of control plant roots for a possible contamination with mycorrhizal fungi was negative. Endomyk_Basic was more efficient

Treatment	Ncontent (mg/100 gmDM)	Pcontent (mg/100 gmDM)
Bacto_Prof (Shoot)	1350.4 h	145.6 f
Endomyk_Basic (Shoot)	1777.6 b	139.2 g
Endomyk_Conc (Shoot)	1620.5 d	139.4 g
Endomyk_Prof (Soot)	1616.7 f	139.1 g
Control (Shoot)	1254.2 j	130.7 h
Bacto_Prof (Root)	1578.9 g	194.6 a
Endomyk_Basic (Root)	1806.5 a	167.6 d
Endomyk_Conc (Root)	1753.1 c	176 b
Endomyk_Prof (Root)	1654 e	175 c
Control (Root)	1494 i	162 e

Values within each column followed by the same letter are not significantly different (P<0.05) according to LSD.

Table 3: Concentration of nitrogen and phosphorus in shoot and root systems of squash plants inoculated with four mycorrhizal products.

than others by 56%. However; the three doses were close to each other in percentage of mycorrhization, and that means this VAM fungi was effective on this crop even in low doses.

Effect of four mycorrhizal products on physiological plant elements

Treated plants with mycorrhizal products were highly significant than control in nitrogen and phosphorus content as shown in Table 3. Concentration of nitrogen and phosphorus were markedly higher in root system than in shoot system according to Table 4.

Concentration of nitrogen in shoot system was increased significantly by mycorrhizal products; by 8% for Bacto_Prof, 42% for Endomyk_Basic, 29% for both Endomyk_Conc and Endomyk_Prof more than non-mycorrhizal plants. While in root system, mycorrhizal products increased the concentration of nitrogen by 6, 21, 17 and 11%, respectively for Bacto_Prof, Endomyk_Basic, Endomyk_Conc and Endomyk_Prof.

Concentration of phosphorus in shoot system was raised by mycorrhizal products by 11, 6% for Bacto_Prof and Endomyk_Prof, respectively and 7% for Endomyk_Basic and Endomyk_Conc. In root system of squash plant, the concentration of phosphorus was enhanced by 21, 4, 9 and 8%, respectively for Bacto_Prof, Endomyk_Basic, Endomyk_Conc and Endomyk_Prof products.

There were highly significant differences in ash%, fat%, crude protein%, crude fiber% and carbohydrate% between mycorrhizal products and control plants in all treatments (Table 5). Concentration of ash was higher in control plants than in mycorrhizal products in both shoot and root systems. Concentration of fat was markedly higher in Bacto_Prof followed by Endomyk_Basic and Endomyk_Conc, than Endomyk_Prof in both shoot and root systems. Concentration of crude protein was higher in shoot and root systems by mycorrhizal products; in Endomyk_Basic by 56, 24% followed by Endomyk_Conc 28, 17% then Endomyk_Prof by 28, 6% and Bacto_Prof by 6, 4% in shoot and root systems, respectively. Concentration of crude fiber was different, where Endomyk_Prof product recorded the highest in shoot, while Bacto_Prof product has recorded the highest in root system. However; the concentration of carbohydrates in shoot and root systems was higher in Endomyk_Basic than all other products by 35, 128% followed by Endomyk_Conc 43, 80% then Endomyk_Prof 13, 45% and Bacto_Prof by 17, and 22% in shoot and root systems, respectively.

Comparison between plant shoot and root system within each treatment was summarized in Table 6. Concentration of ash in root system was higher than shoot system by 11%, while the concentrations of fat in shoot system was highly increased more than root system. In crude protein and fiber, the plant root system was significantly higher than plant shoot system by 6 and 52%, respectively. For the carbohydrate concentration, shoot system was higher than root system by 97% increase.

Discussion

Four mycorrhizal products produced by Terrabioscience Company, Germany, they were effective on different crop plants, in several levels under experimental conditions. The most effective product was Endomyk_Basic in improving plant growth and mycorrhizal root

colonization. All contain *Glomus intraradices* with some exceptions of containing other microorganisms such as Bacillus, Algae or other ingredients. Increasing the dose in general enhanced plant growth and increased the root growth by increasing the uptake of water and nutrients.

All mycorrhizal products affected positively on squash crop by producing huge amount of external and internal hyphae, vesicles and arbuscules which will increase the root area surface, thus enhance the growth of the whole plant. Similar results have been reported in Al-Karaki and Al Raddad Al Momany; Al Raddad Al-Momany and Al-Saket [18,19]. The huge network of mycorrhizal hyphae, which spread into the surrounding soil, influence soil fertility and plant nutrition by changing the physico-chemical characteristics of soils stabilizing agents in the formation and maintenance of soil structure [20].

Plant height, shoot fresh weight, shoot dry weight and root fresh weight were significantly different in mycorrhizal than non-mycorrhizal plants in all products. *Glomus intraradices* increased plant yield, height and shoot fresh weight. Mycorrhizal effect is highly dependent on the

Treatment	N %	P %
Shoot	1523.9 b	138.8 b
Root	1657.3 a	174.9 a

Values within each column followed by the same letter are not significantly different (P<0.05) according to LSD.

Table 4: Concentration of nitrogen and phosphorus in shoot and root systems of squash plants.

Treatment	Ash %	Fat %	Crude Protein%	Crude Fiber%	CHO %
Bacto_Prof (Soot)	33.7 g	3.8 a	8.4 g	22.6 g	31.6 c
Endomyk_Basic (Shoot)	31.2 h	3.6 b	12.4 a	16.5 h	38.7 a
Endomyk_Conc (Shoot)	30.7 i	3.6 b	10.2 d	16.9 h	36.4 b
Endomyk_Prof (Shoot)	29.3 j	3.4 bc	10.2 d	27.2 e	30.6 c
Control (Shoot)	34.5 f	3.4 c	7.9 h	26.5 f	27 d
Bacto_Prof (Root)	37.6 b	1.8 de	9.8 f	37.5 a	13.1 h
Endomyk_Basic (Root)	31.7 e	1.9 d	11.8 b	30.2 c	24.6 e
Endomyk_Conc (Root)	32.7 d	1.7 e	11.1 c	35.1 b	19.4 f
Endomyk_Prof (Root)	36. c	1.7 e	10 d	35 b	15.6 g
Control (Root)	39.3 a	1.2 f	9.5 e	28.7d	10.8 i

Values within each column followed by the same letter are not significantly different (P<0.05) according to LSD.

Table 5: Concentration of ash, fat, crude protein, crude fiber and carbohydrates in shoot and root systems of squash plant inoculated with four mycorrhizal products.

Treatment	Ash %	Fat %	Crude Protein%	Crude Fiber%	CHO %
Shoot	31.9 b	3.6 a	9.8 b	21.9 b	32.9 a
Root	35.5 a	1.7 b	10.4 a	33.3 a	16.7 b

Values within each column followed by the same letter are not significantly different (P<0.05) according to LSD.

Table 6: Comparison of the concentration of ash, fat, crude protein, crude fiber and carbohydrates in shoot and root systems of squash plants.

type of root system. Plants with few fine roots depend on the network of hyphae which acts as a bridge for nutrients uptake from the soil to the plant cells through its connections with root system. Wheat with extensive fine root system would not respond to VAM fungi except in phosphorus deficient soils. Other crops with less hairy roots such as onion and citrus will be highly respond to VAM fungi even in soils with moderate phosphorus levels [21]. Al Raddad Al Momany reported that length of the growing season and type of root system makes the rhizosphere more favourable to spore propagation and good mycorrhizal colonization [8].

All the mycorrhizal products were efficient in affecting physiological plant contents more than in control treatment, except in ash; it was the highest in control plants due to less nutrients present in the dry shoot and root system. Bacto_Prof treated plants contained more fat than other products; however Endomyk_Basic induced more protein and carbohydrates. Concentration of fat and carbohydrates were more in the shoot due to chlorophyll synthesis. While ash, protein and fibers presented in the root were more than in plant shoot. Endomycorrhizal plants contained less carbohydrate than non-mycorrhizal, but roots contained higher content of protein than shoots. The fungus absorbs P from soil and takes carbohydrate from plant cells as a source of energy [22]. Arbuscules are believed to function in bidirectional transfer of nutrients; essentially transfer carbohydrates from plant cell to fungus and minerals especially phosphorus from fungus to host cells [22]. Al Raddad Al Momany [19] recorded that the addition of P fertilizer decreased number of spores on fruit trees, where soil samples were taken from the University farm in Jordan Valley.

Concentration of nitrogen in shoot and root systems of squash were the best in Endomyk_Basic product treatment, while phosphorus showed the best absorption for shoot and root system by Bacto_Prof product. The VAM fungi can increase P uptake in plants that is documented in many researches [9,12,23-25]. P and N uptake were higher in mycorrhizal plants than in control treatment. Plant P is the main controlling factor in the plant-fungal relationship, which plays a significant role in increasing the total uptake of nutrients which leads to the increase in growth and yield [2]. VAM inoculation stimulated the plant growth and it was attributed to enhanced photosynthesis which associated with increased P uptake in leaves stems and flower heads in wheat [18]. Low phosphorus soil showed significant increase in mycorrhizal maize considering P content and total dry weight [26].

Some researchers have indicated that AMF inoculation tends to decrease pH in the rhizosphere, and leads to produce more carbon dioxide (CO_2) [12]. AMF has been assumed to be a major mechanism through increasing carbon inputs to soil and protecting organic carbon from decomposition by aggregation [27]. Many reports have shown that VAM fungi are able to avoid soil erosion by increasing the stability of soil aggregates through the combined action of extraradical hyphae and their exudates. Glomalin is a fungal component, insoluble and hydrophobic proteinaceous substance, which has been reported to improve the stability of soil by avoiding disaggregation by water, so VAM could be used as an indicator of soil and bio-fertilizer in agroecosystems [20,28]. Soil type is a very important factor in the introduction and reproduction of VAM spores, such as clay loam sandy soils which will facilitate the rapid buildup of *Glomus* populations in rainfed areas [8,29].

Conclusions

All mycorrhizal products were effective and significantly different from non-treated plants. VAM fungi increased plant height, fresh shoot weight and fresh root weight in squash crop. Double dose was the most effective treatment, which increased the plant growth and root weight. Endomyk_Basic was the best product in enhancing squash growth. Nitrogen, phosphorus, proteins and carbohydrates were absorbed more by mycorrhizal plants, which enhanced plant growth.

References

1. Kirk PM, Cannon PF, David JC, Stalpers J (2001) Ainsworth and Bisby's Dictionary of the Fungi. 9th edn. Wallingford, UK.

2. Gosling P, Hodge A, Goodlass G, Bending GD (2006) Arbuscular mycorrhizal fungi and organic farming. Agriculture Ecosystems and Environment 113: 17-35.

3. Smith SE, Read DJ (1997) Mycorrhizal Symbiosis. 2nd edn. Academic Press, London, UK.

4. Garmendia I, Goicoechea N, Aguirreolea J (2004) Effectiveness of three Glomus species in protecting pepper (Capsicum annuum L.) against verticillium wilt. Biological Control 31: 296-305.

5. Tisdale SL, Nelson WL, Baton JD (1995) Soil Fertility and Fertilizers. Macmillan Publishing Company, USA.

6. Bethlenfalvay GJ, Barea JM (1994) Mycorrhizae in sustainable agriculture. I. Effects on seed yield and soil aggregation. American Journal of Alternative Agriculture 9: 157-161.

7. Purin S, Morton JB (2011) In situ analysis of anastomosis in representative genera of arbuscular mycorrhizal fungi. Mycorrhiza 21: 505-514.

8. Al-Raddad, Al-Momany A (1993) Distribution of different Glomus species in rainfed areas in Jordan. Dirasat 20: 165-182.

9. Al-Raddad, Al-Momany A (1990) Response of bean, broadbean and chickpea plants to inoculation with Glomus species. Scientia Horticulturae 46: 195-200.

10. Wetterauer DG, Killorn RJ (1996) Fallow- and flooded-soil syndromes: effects on crop production. Journal of Production Agriculture 9: 39-41.

11. Sylvia DM (1998) Activity of external hyphae of vesicular-arbuscular mycorrhizal fungi. Soil Biology and Biochemistry 20: 39-43.

12. Goussous SJ, Mohammad MJ (2009) Comparative Effect of two arbuscular mycorrhizae and N and P fertilizers on growth and nutrient uptake of onions. International Journal of Agriculture and Biology 11: 463-467.

13. Graham JH (2000) Assessing cost of arbuscular mycorrhizal symbiosis in agroecosystems. In: Podola GK, Douds DD (eds.), Current Advances in Mycorrhizal Research. APS Press, St Paul, NM, pp: 127-140.

14. Koide R (1991) Nutrient supply, nutrient demand and plant-response to mycorrhizal infection. New Phytologist Journal 117: 365-386.

15. Phillips J, Hayman D (1970) Improved procedures for clearing roots and staining parasitic and vesicular-arbuscular mycorrhizal fungi for assessment of infection. Translocations of the British Mycological Society 55: 158-161.

16. Bierman B, Linderman R (1981) Quantifying vesicular-arbuscular mycorrhizae: proposed method towards standardization. New Phytologist Journal 87: 63-67.

17. AOAC (1995) Official methods of analysis. 13th edn. Association of official agriculture chemists. Washington DC, USA.

18. Al-Karaki GN, Al-Momany A (1997) Effect of arbuscular mycorrhizal fungi and drought stress on growth and nutrient uptake of two wheat genotype differing in drought resistance. Mycorrhiza 7: 83-88.

19. Al-Raddad, Al-Momany A (1989) Occurrence of vesicular arbuscular mycorrhizal fungi on crop plants under irrigation. Research Journal of Aleppo University 13: 31-44.

20. Bedini S, Pellegrino E, Avio L, Pellegrini S, Bazzoffi P, et al. (2009) Changes in soil aggregation and glomalin-related soil protein content as affected by the arbuscular mycorrhizal fungal species Glomus mosseae and Glomus intraradices. Soil Biology and Biochemistry 41: 1491-1496.

21. Al-Raddad, Al-Momany A (1987) Effect of three vesicular arbuscular mycorrhizal isolates on growth of tomato, eggplant and pepper in a field soil. Dirasat 14: 161-168.

22. Al-Ameiri NS (1987) Interaction between vesicular arbuscular mycorrhizal fungi and Fusarium root rot of tomato. Master's Dissertation, University of Jordan, Amman, Jordan.

23. Al-Karaki GN, Al-Momany A (1996) Effects of water stress and inoculation with VA mycorrhizal fungi on growth and nutrient uptake in wheat. Mu'tah Journal for Research and Studies 11: 213-232.

24. Karajeh M, Al-Raddad Al-Momany A (1999) Effect of VA Mycorrhizal fungus (Glomus mosseae Gerd and Trappe) on Verticillium dahliae Kleb. Dirasat, 26: 338-341.

25. Bouwmeester HJ, Roux C, Lopez-Raez JA, Becard G (2007) Rhizosphere communication of plants, parasitic plants and AM fungi. Plant Science 12: 224-230.

26. Katbeh MR (1993) The role of endomycorrhizal in improvement of barley productivity in arid-zone. Master's Dissertation, University of Jordan, Amman, Jordan.

27. Cheng L, Booker FL, Tu C, Burkey KO, Zhou L, et al. (2012) Arbuscular Mycorrhizal Fungi Increase Organic Carbon Decomposition under Elevated CO_2. Science 337: 1084-1087.

28. Liu Y, Mao L, He X, Cheng G, Ma X, (2012) Rapid change of AM fungal community in a rain-fed wheat field with short-term plastic film mulching practice. Mycorrhiza 22: 31-39.

29. Al-Momany AM, Al-Saket I (1989) Effect of endomycorrhizal fungi on maximizing the efficiency of olive cakes as fertilizer for young olives. Research Journal of Aleppo University 13: 31-47.

Distribution of Wheat Stem Rust *(Puccinia Graminis F. Sp. Tritici)* in West and Southwest Shewa Zones and Identification of its Phsiological Races

Alemayehu Hailu[1]*, Getaneh Woldeab[2], Woubit Dawit[3] and Endale Hailu[3,4]

[1,2,4]*Ethiopian Institute of Agricultural Research, Plant Protection Research Center P.O.Box 37, Ambo, Ethiopia*

[3]*Ambo University, P.O.Box 19, Ambo, Ethiopia*

Abstract

Stem rust (black rust) caused by Puccinia graminis f.sp.tritici is one of the most important air borne diseases of wheat (*Triticum aestivum*) in the central high lands of Ethiopia, including west and southwest Shewa zones. The pathogen is capable to produce new physiological races that attack resistant varieties and develop epidemic under favorable environmental conditions which results in a serious yield loss. However, information on the status of stem rust distribution and races in west and southwest Shewa zones is lacking. Therefore, the present studies were based on stem rust survey to compute the prevalence and intensity of disease; race analysis via inoculation of stem rust isolates and multiplication of single-pustule of the pathogen and race designation by inoculating on wheat differential lines. Eighty six wheat fields were assessed in 12 districts of west and south west Shewa zones with altitude ranges between1925-2915 m.a.s.l. Seventy five (87.2%) wheat fields infected with stem rust had the overall mean of 33% incidence and 10.8% severity. The mean prevalence of stem rust was 96.3% in southwest and 83.1% in west Shewa zones, whereas, the mean incidence was 34.7% and 31.2% in west and southwest Shewa zones, respectively. Similarly, mean severity was 14.5% in west and 7.1% in southwest Shewa zones. Forty five stem rust samples collected during the survey were analyzed on the twenty standard stem rust differentials and resulted in identification of 5 races (TTTTH, TTKSK, TKTTF, HKPPF & HKNTF). Of these, 88.4% of the isolates were TKTTF (Digalu race) followed by 4.7% of the isolates by TTKSK (Ug99). Among the five races, the most virulent, which made 18 *Sr* genes non-effective was TTTTH. TKTTF and TTKSK races were virulent on 85% of *Sr* genes. Differential host carrying *Sr*24 was an effective gene which confers resistance to all of the races identified in the area. On the other hand, the wheat differential hosts carrying the resistance genes *Sr* McN, *Sr*10, *Sr*9a, *Sr*30, *Sr*9g, *Sr*8a, *Sr*6, *Sr*7b and *Sr*21 were ineffective to 100% of the isolates tested. Hence, the *Sr* resistance gene *Sr*24 can be used as sources of resistance in wheat breeding program.

Keywords: Wheat stem rust; Race; *Puccinia graminis f.sp.tritici*; *Sr* genes; Disease prevalence; Disease severity; Disease incidence

Introduction

Ethiopia is the largest wheat producer in sub-Saharan Africa [1].West and southwest Shewa zones are among the major wheat producing areas in Oromia region [2]. Wheat is the staple food for 4.5 billion people in the world [3]. Its popularity comes from the versatility of its use in the production of a wide range of food products, such as "Injera", breads, cakes, Pastas, cookies, etc .,[4].

Although the productivity of wheat has increased in the last few years in Ethiopia, it is still very low as compared to other wheat producing countries. The national average productivity is estimated to 2.4 tons/ha [2], which is by far below the world's average of 3.3 tons/ha [5]. The low productivity is attributed to a number of factors including: Biotic (Diseases, insect pests, and weeds), abiotic (moisture, soil fertility, etc.,) [6]. Among biotic factors, rusts are the most important diseases of wheat, cause up to 60% loss of wheat yield for leaf or stripe (yellow) rust and 100% loss for stem rust [7].Wheat and rusts have co-evolved for thousand years and resulted in the accumulation of wide spectrum of the pathogens in Ethiopia [8].

However, Stem rust or black rust (caused by *Puccinia graminis* f. sp. *tritici*) is a serious wheat disease causing a decrease of wheat production in many areas of the world [9]. Yield loss due to stem rust in Ethiopia was estimated to reach up to 100% on susceptible wheat varieties at times of disease epidemics [10].According to Leppik [11] and Singh et al. [12] the highland of Ethiopia is considered as a hot spot for the development of stem rust races diversity.

In a study conducted in Germany, Admassu et al. [13] reported

22 stem rust races from 152 collections made in Ethiopia in 2006. Similarly, due to lack of infrastructure, race analyses of stem rust samples collected in Ethiopia 56 was done in St. Paul and Winnipeg. Surveys made from 1996 to 2005 in Bale indicated that stem rust was the most damaging to the crop with severity levels of 40% in 'Genna' and 90% in 'Bona' [14]. Similarly, Wheat stem rust disease was recorded with 44.1% prevalence, 19.2% incidence and 11.3% of severity in west and southwest Shewa zones in 2008 cropping season [15]. Moreover, due to sudden changes in stem rust race patterns, commercial varieties tend to become vulnerable. Hence, detailed information on the wheat stem rust status and physiological race variability have been essential in the west and southwest Shewa zones of Oromia region.

Materials and Methods

Description of the study area

The wheat stem rust survey was carried out in West and Southwest

*Corresponding author: Alemayehu Hailu, Ethiopian Institute of Agricultural Research, Plant Protection Research Center P.O.Box 37, Ambo, Ethiopia E-mail: alemayehuhailu65@yahoo.com.

Shewa zones of Oromia Regional State in Ethiopia. West Shewa zone is located at 8°57′N latitude and 38°07′ E longitude and within elevation ranges between 1380-3300 m.a.s.l. Annual mean maximum and minimum rain fall is 1900 mm and 600 mm, respectively. The mean minimum and maximum air temperature of the area is 11.7°C and 25.4°C, in that order. Southwest Shewa zone is located at 8°16-9° 56′ N latitude and 37° 05′-38° 46′ E longitude and altitude ranging from 1600-3576 m.a.s.l. It receives annual rainfall ranging from 900 -1900 mm. The mean minimum and maximum air temperature of the area is 10°C and 35°C, respectively. Stem rust race analysis was done in Ambo Plant Protection Research Center (APPRC). It is located at 08° 96′ 885″ N latitude and 37° 85′ 923″ E longitude and at an altitude of 2147 m.a.s.l. The annual average temperature and rain fall is 27.54°C and 1077.68 mm, respectively.

Wheat stem rust field survey

A total of twelve districts that included seven from West Shewa zone (Ambo, Dendi, Chelia, Tokaye Kutaye, Dire Inchine, Dawo, Ejere) and five from Southwest Shewa zone (Woliso, Suden Sodo, Bechio, Amaya and Wonchi) were surveyed. The districts were selected based on wheat area coverage and followed systematic sampling every 5-10 km intervals. The survey was conducted following main and feeder roads on pre-planned routes in areas where wheat is predominantly grown. Stem rust assessment was made once at the vital growth stage of the crop per field, along the two diagonals (in an ''X'' pattern) of the field at five points using 0.5 m × 0.5 m (0.25 m²) quadrant. In each field, wheat plants within the quadrant were counted and recorded as diseased/infected and healthy/non-infected and intensity of stem rust was calculated. The incidence of stem rust was calculated by using the number of infected plants and expressed as a percentage of the total number of plants assessed and recorded the average incidence.

Plant disease incidence (%)=<u>Number of diseased plants</u> × 100

Total Number of plants in quadrant

The disease severity under field condition was recorded as percentage of leaf/stem area covered by rust disease followed modified Cobb's scale as developed by Peterson et al.[16] According to this scale, at 100% disease severity, the actual leaf/stem area covered by rust pustules is 37%. Disease severity was assessed by selecting 10 plants from a single quadrant and five quadrants were used for the estimation of disease severity from a single wheat field.

Disease severity (%) = <u>Area of plant tissue affected</u> x100

Total area

The prevalence of rust disease was measured by using the number of fields affected divided by total number of fields and expressed in percentage. It is calculated as:

Disease prevalence (%) = <u>No. of infected fields</u> × 100

Total number of fields assessed

In addition, data on geographical information (latitude, longitude and elevation) of each field was recorded using GPS (e Trex Legend GPS system, Garmin). Crop growth stage was assessed based on the decimalized key developed by Zadoks et al. [17].

Collection of stem rust samples

Stems and/or leaf parts of wheat plants infected with stem rust were cut in to small pieces of 5-10 cm using scissors and put in paper bags after the leaf sheath was separated from the stem in order to keep stem and/or leaf sheath dry. The samples collected in the paper bags were tagged with the name of the Zone, district, variety and date of collection. The samples within the paper bags were air dried and kept in refrigerator at 4°C for race analysis purpose in the greenhouse until the survey in all districts between zones completed. A total of 45 stem rust samples (27 and 18 from West and Southwest Shewa zones of Oromia regions, respectively) were collected.

Isolation and multiplication of single-pustules

The inoculum was multiplied and maintained on standard rust susceptible variety" McNair " which does not carry stem rust resistant genes [18]. Five seedlings of this variety for each samples were raised in suitable 8 cm diameter clay pots that was filled with a mixture of steam sterilized soil, sand and manure in the ratio of 2:1:1, respectively. Seven-day old seedlings or when the primary leaves were fully expanded and the second leaves beginning to grow, the leaves were rubbed gently with clean (disinfected with 97% alcohol) moistened (with distilled water) fingers.

Green house inoculations were carried out using the methods and procedures developed by Stakman et al. [19]. Uredio spores of the stem rust were collected from the diseased wheat parts by using motorized spore collector in a capsule container and diluted by using lightweight mineral oil (SolTrol 130) chemicals and then [20] to make rust uredial spore more uniform. These were sprayed on to the seedlings of Mc Nair from a distance with clean motorized stem rust inoculator. For incubation, inoculated plants were moistened with fine droplets of distilled water by using atomizer after twenty minutes of inoculation and placed in dew chamber for 18 hr dark period at 18-22°C followed by exposure to light at least for 4 hr to provide favorable condition for stem rust infection. Seedlings were allowed to dry/remove their dew/moisture for about 3-4 hr. Following this, the seedlings were transferred from dew chamber to glass compartments in the green house where conditions were regulated at 12 hr photoperiod, at temperature range of 18-25°C and RH of 60-70%.

After seven to ten days of inoculation (when the flecks/symptoms was clearly visible) leaves containing single fleck that produce single pustule was selected from the base of the leaves and the remaining seedlings within the pots were eliminated using hand scissors. Only 2-3 leaves which contain single pustule were left and each of them was covered with cellophane bag (145 × 235 mm) and tied up at the base with a rubber band to avoid cross contamination [21].

After two weeks of inoculation (when the monopustule was well developed) each monopustule was sucked using electric power operated machine (vacuum pump) and collected in capsule container separately. A suspension, prepared by mixing urediospores of the monopustule in lightweight mineral oil, was inoculated on seven-day-old seedlings of the susceptible variety 'McNair' for multiplication purpose on the separate pots. Soon after inoculation, the seedlings were placed in a humid chamber in dark condition and transferred to a green house following the earlier mentioned procedure.

After inoculation of 15 days, the spores of each monopustule/isolate were collected in separate test tubes and stored at 4°C until they were inoculated on the standard differential lines. This procedure was repeated till sufficient amount of spores are produced in order to inoculate the stem rust differential lines. By following this procedure a total of 43 monopustules/isolates were developed from 45 wheat stem rust samples. A schematic overview of the general protocol used for race analysis in the greenhouse has been given in appendix (Figure 1).

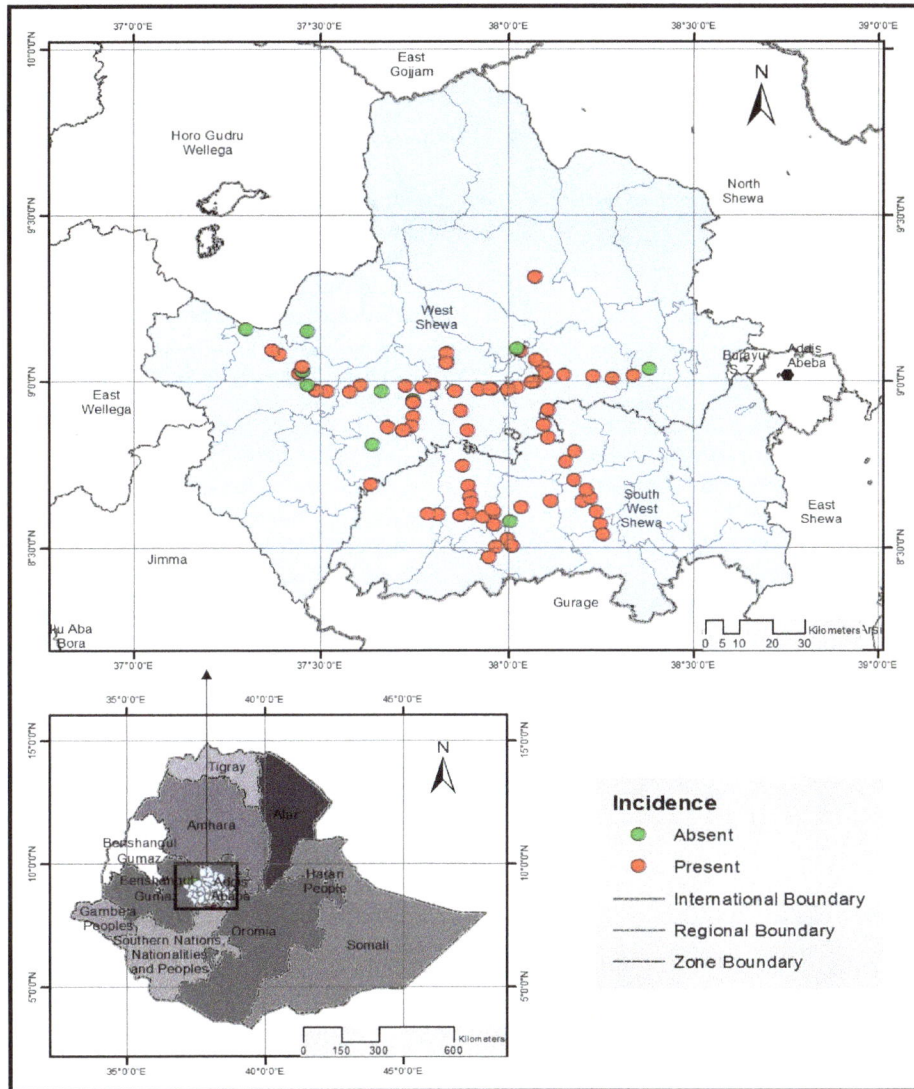

Figure 1: Map showing wheat stem rust survey areas in west and southwest Shewa zones of Oromia region in 2014.

Inoculation of wheat stem rust isolates on the differential lines

Five seeds each of the 20 stem rust differential lines including the susceptible variety (Table 9) were grown in 3 cm diameter pots separately in the growth chamber. The Susceptible variety was used to determine the viability of spores inoculated on the differential hosts and as a check. The single pustule spores/ isolate/ mixed with lightweight mineral oil (approximately 4 mg of spores per 1 ml) was sprayed/inoculated on to seven-day-old seedlings. Similar methods of inoculation, incubation and green house condition were applied as mentioned in section 2.4. Natural day light was supplemented with additional 4 hr/day that emitted by cool white fluorescent tubes arranged directly above plants in the green house.

Stem rust infection types were scored 14 days after inoculation using the 0-4 scale (Table 1) of Stakman et al. [19]. Infection types were grouped in to two, where, Low (resistance) = incompatibility (infection phenotype 0, 0; (fleck), 1, 2, and 2⁺) and High (Susceptible) = compatibility (infection phenotype, 3⁻, 3⁺ & 4).

Designation of races

Race designation was done by grouping the 20 differential lines in to five subsets in the following order (Table 2).

Each isolates was assigned a five letter race code based on its reaction on the differential lines [21]. For example, low infection types on the four lines in a set is assigned with the letter 'B' while high infection types on the four lines is assigned with letter 'T'. Hence, if an isolate produces low infection type (resistant reaction) on the 20 differential lines, the race will be designated with a five letter race code 'BBBBB'. Similarly, an isolate which produces a high infection type (susceptible reaction) on the 20 wheat differential lines will have a race code 'TTTTT'. If an isolate produces a low infection type on *Sr11, Sr24,* and *Sr31,* but a high infection type on the remaining 17 differential lines, the race will be designated as TKTTF (Table 2). The experiment was repeated once, and only differential lines that produced similar infection types in the two experiments were considered for the data analysis. When there was infection type 0 (immune reaction), the test was done again to exclude the possibility of disease escape.

Class	IT	Description of symptoms
Immune	0	No sign of infection on the naked eye
Very Resistant	0	No uredia, but distinct flakes of varying size, usually a chlorotic yellow but occasionally necrotic
Resistant	1	Small uredia surrounded by yellow chlorotic and necrotic area.
Moderately Resistant	2	Small to medium sized uredia, typically in a dark green island surrounded by a chlorotic area
Mesothentic/ Heterogeneous	x	A range of infection type from resistant to susceptible scattered randomly on a single leaf caused by a single isolate not mixture
Moderately Susceptible	3	Medium sized Uredia. Usually surrounded by a light green chlorotic
Susceptible	4	Large uredia with a limited amount of chlorosis: may be diamond shaped
Modified characters		
Lower uredia	=	Uredia much smaller than typical and at the lower limit of the infection type
Small Uredinia	−	Uredia smaller than normal
Large Uredinia	+	Uredia larger than normal
Largest Uredinia	++	Uredia much larger than typical and at the upper limit for the infection type

IT=infection type.

Table 1: Description of infection types used in classifying the reactions of stem rust on leaves of wheat seedlings.

Infection phenotype of pathogen and wheat *Pgt* gene					
	Set1	5	21	9e	7b
	Set2	11	6	8a	9g
Pgt-code	Set3	36	9b	30	17
	Set4	9a	9d	10	Tmp
	Set5	24	31	38	McN
B		Low	Low	Low	Low
C		Low	Low	Low	High
		Low	Low	High	Low
D		Low	Low	High	High
F		Low	High	Low	Low
G		Low	High	Low	High
		Low	High	High	Low
H		Low	High	High	High
J		High	Low	Low	Low
K		High	Low	Low	High
L		High	Low	High	Low
		High	Low	High	High
M		High	High	Low	Low
N		High	High	Low	High
P		High	High	High	Low
Q					
R		High	High	High	High
S					
T					

Low/Resistant infection type (0 to 2+), High/ Susceptible infection type (3- to 4).

Table 2: Code for the 20 differential lines for *P. graminis* f.sp. *tritici* in ordered sets of five.

Data analysis

Survey data (prevalence, incidence and severity) were analyzed by using the descriptive statistical analysis (means) over districts, varieties, altitude range and crop growth stages. Similarly, race analysis was analyzed using the descriptive statistics.

Result and Discussion

Survey of wheat stem rust in west and southwest shewa zones of oromia region

Survey of wheat stem rust was carried out in west and southwest Shewa zones in October, 2014. A total of 86 wheat fields were surveyed mainly for assessment of wheat stem rust intensity. During the surveys, the crop was at flowering to hard dough growth stages (Table 3). From 86 fields inspected, 21 (24.4%), 42 (48.8%), 7 (8.1%), 4 (4.7%) and 12 (14%) of wheat fields were at flowering, milk, soft dough, dough and hard dough stages, respectively. In the same order, stem rust was observed in 17 (81%), 38 (90.5%),7(100%), 4 (100%) and 9 (75%) of 21, 42, 7, 4 and 12 wheat fields inspected in the mentioned growth stages. Thirteen wheat varieties were grown by farmers such as Digelu, Kakaba, Danda'a, Kubsa, ET-13A2, Shorima, Kulutu, Kilinto, Roma awn less, Hidasie, Bedu Gela, Gisoo, and Chofero (Table 3). Out of 86 inspected wheat fields, 48 (55.8%), 14 (16.3%), 7 (8.1%) and 5 (5.8%) fields were sown by Digalu, Kakaba, Danda'a and Kubsa, respectively. ET-13A2, Shorima and Kulutu were sown in two fields (2.3%) each. Similarly, six varieties (Roma awn less, Hidasie, Bedu Gela, Gisoo, and Chofero) were planted with 1 (1.2%) of assessed fields for each. Thirteen wheat varieties have been grown in west Shewa zone whereas only 3 varieties (Digelu, Kakaba, Kubsa) were sown in southwest Shewa zone. Disease survey was carried out at altitude ranges of 1925-2915 m.a.s.l in west and 1935-2859 m.a.s.l in southwest Shewa zones.

Intensity of stem rust across locations: Of the 86 wheat fields assessed in the two zones, 87.2% were infected by the stem rust disease (Figure 1). The mean field prevalence of stem rust was 96.3% in southwest and 83.1% in west Shewa zones (Table 4). Whereas, the mean incidence was 34.7% and 31.2% in west and southwest Shewa zones, respectively. Similarly, mean severity was 14.5% in west and 7.1% in southwest Shewa zones. The assessed wheat fields showed susceptible (S), moderately susceptible (MS) and resistance (R) types of responses to stem rust infection. Hundred percent stem rust prevalence was recorded from 7 districts i.e., 4 districts from southwest and 3 from west Shewa zones. The least field prevalence was observed in west Shewa zone from Ejere (50%) and Chelia (54.5%) districts, respectively (Table 4).The mean incidence of stem rust in the areas varied between 1.4% in Chelia to 78% in Bechio districts (Table 4).

The overall mean incidence of wheat stem rust in both zones was 33%. The highest stem rust incidences (100%) were recorded in Ambo (Senkale locality), Bechio (Soyoma Guenji), Dawo (Uluma Busa, Girmi), Dendi (Cherto Kogn, jemjem lagabatu, Arera Kurae, Degawuchi), Ejere (Temoye, Kalana Imbortu) , and Tokaye Kutaye (Birbisana Duguma), while the lowest (zero) were recorded in Woliso (Obi,), Ejere (Chere, Tosegne gefere) , Tokaye Kutaye (Kele Boredu, Birbsana duguma), Chelia (Wegdi Kortu, Chobi tulu ,Tulu goseru,Mida kegn) and Dire Inchine (Woledo Hign) Districts.

The mean severity of stem rust ranged from 0.9% in Chelia to 35% in Dawo district. The overall mean severity of the disease was 10.8%. A maximum disease severity of 80% was recorded in Dendi (jemjem lagabatu locality) followed by Tokay Kutaye (Birbisana Duguma & Koleba) and Dawo (Girmi) districts with 60% of each. In general, most of the assessed wheat fields lied between 1 to 20% for stem rust severity (Figure 2).

Intensity of stem rust in different altitude ranges: The survey was conducted in the altitude ranges between 1925-2915 m.a.s.l. Based on CSA [22] altitude agro-ecology classification, out of 86 wheat fields observed, 5 (5.8%), 69 (80.2%) and 12 (14%) fields were found at low-altitude (1500-2000), mid-altitude (2001-2500) and high-altitude (2501-3560 m.a.sl.), respectively. Of the 5 wheat fields inspected in the altitude ranges between 1500-2000 m.a.s.l, stem rust was observed in 5 (100%) wheat fields which had 30.8% mean incidence and 9.6% mean severity. Of the 69 wheat fields surveyed in the elevation that ranges between 2001-2500 m.a.s.l, black rust was recorded in 62 (89.9%) fields, with mean incidence and severity of 36.3% and 13.5%, following the same order mentioned. Similarly, 66.7% of stem rust disease prevalence was recorded at high-altitude which had 3.4% mean incidence and 1.5% mean severity. The survey result indicated that, mean incidence and severity increased from low-altitude to mid-altitude and decreased at high altitude (Table 5). Maximum stem rust disease severity (80%) was recorded at mid-altitude followed by 40% at low-altitude. Similarly, maximum stem rust incidence (100%) was recorded at mid and low-altitude. The highest level of stem rust infection has been cited in literatures in the altitude ranges of 1600 and 2500 m.a.s.l. Ayele et al [23]. Abebe et al. [24] showed that stem rust occurred in the altitude ranges of 1494-1800 m.a.s.l in southern Tigray. Dagnatchew [25] also mentioned as stem rust of wheat disease was very important at altitude below 2300 m.a.s.l. In Kenya, stem rust had been recorded and known to occur mainly in the low altitude areas of 1800 m.a.s.l [26] Even though stem rust has been seemed more important at mid and low-altitude, it also occurred at higher elevation as shown below in the data. This indicates, wheat stem rust has been extensive in the wide altitude ranges through times; and this might be due to climate change, widely cultivation of susceptible varieties and appearance of new races. Hence, wheat stem rust survey could be carried out in the wide altitude ranges in order to know disease distribution and race variability before going to out of control.

Intensity of stem rust in different wheat varieties grown in the surveyed areas: Of the 86 assessed wheat fields, only six (7%) fields were covered by five different local varieties and the remaining 80 (93%) were covered by eight different released varieties. The local varieties such as Bedu Gela and Chofero have shown resistance response to stem rust infection during survey in west Shewa zone. The absence of stem rust in local varieties may probably be due to their relative resistance and/or may be cultivated at a relatively higher altitude (≥ 2810 m.a.s.l) (Table 6), where stem rust disease is not a threat to

Zone	District	No. of fields observed	Varieties	Altitude range (m.a.s.l.)	Growth stage
SWS	Wonchi	7	Digelu, Kakaba, Kubsa	2079-2859	FS-MS
	Amaya	2	Digelu,	2009, 2038	MS
	Woliso	11	Kubsa, Digelu, Kakaba	1935-2353	FS-HDS
	Bechio	5	Digelu, Kakaba	2172-2223	MS-HDS
	Suden Sodo	2	Digelu	2268,2360	MS-HDS
WS	Ambo	12	Digelu, Danda'a, Kakaba, Kubsa, Kilito,ET-13A2	1925- 2904	FS-HDS
	Dawo	3	Digelu,	2173-2399	FS-MS
	Dendi	12	Roma, Digelu, Kakaba, Kubsa	2172-2773	FS-HDS
	Ejere	4	Kakaba, Digelu, Bedu Gela	2149-2915	FS-DS
	Tokaye Kutaye	13	Danda'a, Kakaba, ET-13A2, Digelu, Gisoo,	1949-2399	MS-HDS
	Chelia	11	Digelu, Kakaba, Hidasie, Shorima, Kulutu, Chofero,	2261-2891	FS-MS
	Dire Inchine	4	Digelu, Shorima, Kakaba,	2373-2462	FS-MS
Total		86		1925-2915	FS-HDS

FS: Flowering stage; MS: Milk stage; DS: Dough stage; HDS: Hard dough stage;
SWS: South west Shewa; WS: West Shewa

Table 3: Number of fields, varieties, altitude ranges and growth stages of wheat by zones and districts, 2014.

Zone	District	No. of fields inspected	Prevalence (%)	Incidence (%)		Severity (%)		Host response
				Range	Mean	Range	mean	
SWS	Wenchi	7	100	1-20	6.3	1-5	2.3	MS
	Amaya	2	100	2,10	6	2,5	3.5	MS
	Woliso	11	90.9	0-30	5.8	0-10	2.3	MS-R
	Bechio	5	100	50-100	**78**	10-30	20	MS
	Suden Sodo	2	100	60,60	60	5,10	7.5	MS
Subtotal/ mean		27	96.3	0-100	31.2	0-30	7.1	MS-R
WS	Ambo	12	100	1-100	33.3	1-40	12.7	MS-R-S
	Dendi	12	100	1-100	54.3	1-80	20.2	MS-S
	Ejere	4	50	0-100	50	0-50	20	MS-R-S
	Dawo	3	100	30-100	76.7	5-60	35	MS-S
	Tokaye kutaye	13	84.6	0-100	22.2	0-60	10.5	MS-R-S
	Chelia	11	54.5	0-10	**1.4**	0-5	0.9	MS-R
	Dire Inchine	4	75	0-15	5.3	0-5	2	MS-R
Subtotal/ mean		59	83.1%	0-100	34.7	0-80	14.5	MS-R-S
Grand total/Mean		86	87.2	0-100	33	0-80	10.8	MS-R-S

MS: Moderately Susceptible; R: Resistance; S: Susceptible; SWS: South west Shewa; WS: West Shewa.

Table 4: Intensity of wheat stem rust in 12 districts of west and southwest Shewa zones, Oromia region in 2014.

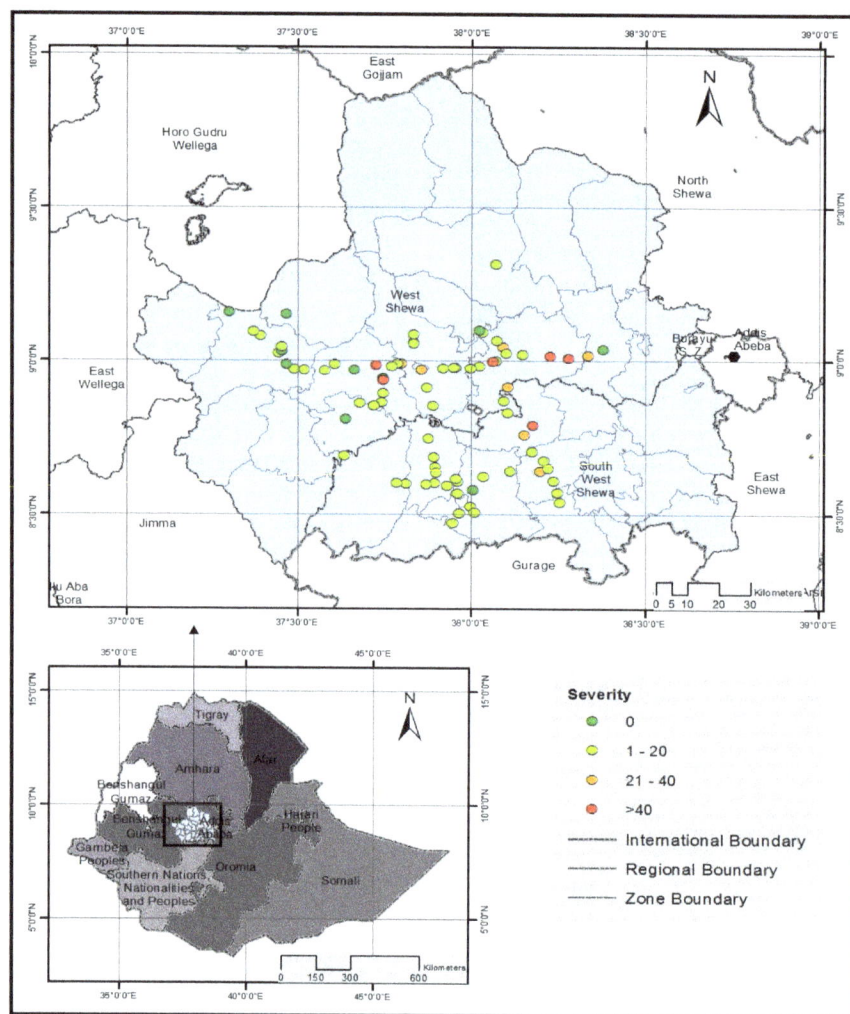

Figure 2: Map showing stem rust severity in west and southwest Shewa zones of Oromia region in 2014.

Altituderange (m.a.s.l)	Class Name (Traditional)	No. of fields inspected	Prevalence		Incidence (%)		Severity (%)	
			No	%	Range	Mean	Range	Mean
1500-2000	Low- altitude	5	5	100	1-100	30.8	1-40	9.6
2001-2500	Mid-altitude	69	62	89.9	0-100	36.3	0-80	13.5
2501-3560	High-altitude	12	8	66.7	0-20	3.4	0-5	1.5

Table 5: Intensity of stem rust based on different altitude ranges.

wheat crop [25] The most widely grown wheat variety was Digalu and it covered 55.8% of surveyed wheat fields in west and south west Shewa zones, Oromia region with 1 to 80% ranges of stem rust severity (Table 6). It showed susceptible to moderately susceptible reactions with 39.2% mean incidence and 14.5 mean severity. The second commonly grown variety Kakaba was also infected with stem rust at different intensity levels and its coverage was 16.3% surveyed wheat fields in both Zones. This variety showed moderately susceptible to resistance stem rust reaction with mean incidence and severity of 12.5% and 3.4%, respectively. Variety Danda'a was the third widely grown (8.1% wheat fields) in west Shewa zone only and it also showed similar field response as Kakaba for stem rust disease with 12.7% mean incidence and 2.1% mean severity. Hidasie variety was released in 2012 by KARC/EIAR and was not widely cultivated in the assessed areas except one field in

Chelia district. This variety has shown resistance response for stem rust disease in west Shewa zone during surveying time. Most improved wheat varieties have shown moderately susceptible type of reaction to wheat stem rust disease in surveyed areas in 2014 main crop growing season.

Of the 48 inspected Digalu variety fields, 100% stem rust incidence was recorded in 12 (25%) fields. Similarly, the highest disease severity of 80% was recorded on Digalu variety followed by Kilinto with 40%. From eight improved wheat varieties in the assessed areas, stem rust disease was observed on 7 (87.5%); and of the five local varieties, stem rust appeared on 3(60%) varieties. Likewise, Out of 75 (87.2) infected wheat fields, 72 (83.7%) stem rust disease prevalence was recorded on the improved wheat varieties whereas 3 (3.5%) recorded on the local varieties.

In general, the survey result indicated that the intensity of stem rust varied across locations, elevation, varieties, growth stage. In addition, out of 75 (87.2%) stem rust disease infected wheat fields, only 2 (2.7%) wheat fields were sprayed with fungicide (Tilt 250EC). This low percentage use of fungicide by farmers was due to lack of awareness, unaffordable price of the fungicide and low technical support from agricultural experts according to farmers.

Physiological races and virulence diversity of stem rust on wheat in west and southwest zones

Race analysis is done based on the reaction of differential lines which contain 20 monogenic resistance genes. These genes are race specific and they show different response for various race groups. Race analysis provided essential information in determining the range of pathogenic variation in a specific region, screening for resistance in varieties, confirming that host responses are due to race changes, understanding the mechanism of variation as well as in determining the direction of research and breeding programs before the pathogen became a threat to wheat crop production in a specific region (District).

In this study, of the total 45 stem rust samples, 43 from farmer's fields and 2 from experimental plot of Ambo Plant Protection Research Center (APPRC) were collected. Of these, 2 samples from west Shewa

zone did not yield viable spores at the time of inoculation on the susceptible check McNair701 in the green house. Forty-three viable isolates were identified and further multiplied on differential line for final race analysis.

Virulence and physiological race composition of wheat stem rust: Of the 43 isolates tested, 5 races were identified from west and southwest Shewa zones. The result showed that, most of the isolates collected from different wheat fields belonged to the same race group, except Ambo and Woliso districts .Three races namely TKTTF, TTKSK and TTTTH were identified from west Shewa zone. Similarly, 4 races (TKTTF, HKPPF,TTKSK and HKNTF) were identified from southwest Shewa zone. TKTTF is common race and detected from all districts of the two zones (Table 7). It was identified from 38 isolates while 4 (TTKSK, TTTTH, HKPPF, and HKNTF) races were identified only from 5 isolates from those particular districts (Ambo and Woliso). Four races were identified from Woliso followed by Ambo district (3 races). Among the identified races, 4 races such as TKTTF, TTTTH, HKPPF and HKNTF were identified for the first time in the sampling zones. The most important race TTKSK (Ug99) was isolated from two fields grown with ET-13A2 in Ambo Plant protection research center, on station experimental plots and Kakaba in Woliso district. Out of 43 viable stem rust collected wheat fields, 88.4% fields were infected

Zone	Variety	Elevation Range (m.a.s.l.)	No. Of fields inspected	Prevalence (%)	Incidence (%)		Severity (%)		Variety Responses
					Range	Mean	Range	Mean	
W & SW Shewa	Digelu	1925-2904	48	100	1-100	39.2	1-80	14.5	MS-S
W & SW Shewa	Kakaba	2059-2575	14	71.4	0-50	12.5	0-20	3.4	R-MS
W. Shewa	Danda'a	1949-2460	7	85.7	0-50	12.7	0-5	2.1	R-MS
W & SW Shewa	Kubsa	1935-2291	5	100	1-10	3	1-5	2	MS
W. Shewa	ET-13A2	2147, 2373	2	50	0,60	30	0, 30	15	R-S
W. Shewa	Shorima	2434, 2495	2	50	0, 1	0.5	0, 1	0.5	R-MS
W. Shewa	Kulutu	2821, 2891	2	50	0, 1	0.5	0, 1	0.5	R-MS
W. Shewa	Roma	2588	1	100	20	20	5	5	MS
W. Shewa	Bedu gela	2915	1	0	0	0	0	0	R
W. Shewa	Hidasie	2493	1	0	0	0	0	0	R
W. Shewa	Chofero	2810	1	0	0	0	0	0	R
W. Shewa	Gisoo	2202	1	100	1	1	1	1	MS
W. Shewa	Kilinto	2154	1	100	90	90	40	40	S

Table 6: Stem rust intensity in different wheat varieties grown in west and southwest Shewa zones of Oromia region in 2014.

Zone	District	Race	No.Of isolates	Altitude (masl)	Variety
WS	Ambo	TKTTF	7	1925-2460	Danda'a, Kakaba, Digalu, Kilinto
		TTKSK	1	2147	ET-13A2
		TTTTH	1	2154	Kilinto
	Dawo	TKTTF	1	2199	Digalu
	Dendi	TKTTF	7	2172-2588	Digalu, Kakaba, Roma awn less, Kubsa
	Ejere	TKTTF	1	2160	Digalu
	Tokaye kutaye	TKTTF	4	1949-2332	Digalu, Danda'a
	Chelia	TKTTF	1	2261	Digalu
	Dire Inchine	TKTTF	2	2417, 2434	Digalu, Shorima
SWS	Wonchi	TKTTF	5	2079-2575	Digalu, Kakaba, Kubsa
	Amaya	TKTTF	2	2009, 2038	Digalu
	Woliso	TKTTF	4	2005-2326	Digalu
		HKPPF	1	2054	Digalu
		TTKSK	1	2059	Kakaba
		HKNTF	1	2073	Digalu
	Bechio	TKTTF	3	2172-2223	Digalu, Kakaba
	Suden Sodo	TKTTF	1	2360	Digalu

WS: West Shewa; **SWS**: South west Shewa
Table 7: Races of wheat stem rust across district, altitude and wheat variety in west and southwest Shewa zones.

by TKTTF race and the remaining 11.6% fields infected by other races such as TTKSK, TTTTH, HKPPF, and HKNTF. Twenty three and fifteen sampled wheat fields were infected with TKTTF in west and southwest Shewa zones, in the mentioned order. Out of 27 samples taken from Digalu variety, 25 (92.6%) fields were infected with TKTTF. Similarly, 5 (83.3%), 3 (75%), 2 (100), 1 (100%), 1 (100) and 1 (100%) of Kakaba, Danda'a, Kubsa, Kilinto, Roma and Shorima sampled wheat fields were infected with TKTTF, respectively. On the other hand, other three new races such as TTTTH, HKPPF, and HKNTF were detected only at single location of each (Table 7). HKNTF and HKPPF races were identified from Digalu; and TTTTH race identified from Kilinto (Durum wheat type).

In general, the new, TKTTF race was distributed in the altitude range of 1925-2588 m.a.s.l in 12 districts of west and south west Shewa zones, Oromia region. This showed that, TKTTF is the most virulent race on wheat varieties and it is rapidly spreading to a wide altitude ranges. This might be due to favorable environmental conditions as well as cultivation of susceptible wheat varieties in those districts. In contrast, other 3 new races were found in the altitude range of 2054-2154 m.a.s.l. only from two districts (Table 7). The race TTKSK (Ug99) was detected from elevations of 2059 and 2147 m.a.s.l.

Out of 5 races, the most frequently and predominantly occurred race was TKTTF with a frequency of 88.4% (Table 8). The second frequently race was TTKSK with a frequency of 4.7%. This might be widely growth of resistant variety like Digalu for this race in those districts and/or it might be dominated by virulent race like TKTTF. However, it was reported by Admassu et al.[13] reported that TTKSK race was dominant throughout the country including west and south west Shewa zones at a frequency of 26.6%. The least frequently occurring races were TTTTH, HKPPF and HKNTF with a frequency of 2.3% each.

The observed/recorded virulence spectrum varied between 13-18 Sr genes (Table 8). The most wide virulence spectrum was recorded on the race of TTTTH that exhibited virulence on 18 Sr genes. The second broad virulence spectrum was recorded on the TKTTF and TTKSK races that showed virulence on 17 Sr genes. The most devastating stem rust race TTKSK (commonly known as Ug99) virulence on gene Sr31 was first detected in Uganda in 1999 [27] and had been spread to most of the wheat growing areas of Kenya in 2002 and Ethiopia in 2003 [28]. In 2005, Ethiopian reports confirmed its presence in six dispersed locations [29] and was spread to most of wheat growing areas in the country and becoming the main threat for wheat production [30]. Similarly, TTKSK has been reported by Teklay in southern Tigray zone with a virulent spectrum on the 17 resistance gene of differential lines [24] The least virulence spectrum was recorded on the HKPPF and HKNTF races that they caused 13 stem rust resistance genes ineffective each (Table 8).

TTKSK (Ug99) was avirulent to Sr36, Sr24, and SrTmp (Table 8). In the same way, the new race TKTTF (Digalu race) was avirulent to Sr11, Sr31, and Sr24. Virulence on the resistance gene **SrTmp** is

considered the main factor behind the complete susceptibility of the variety "Digalu" to this new race. This race, before the present study, had not been detected in the 2 zones. The assumption therefore is that this is either a foreign incursion (most likely by wind) or a mutation in-country. At present, very little is known about the regional and global distribution of Pgt race TKTTF and members of this genetic lineage. The race was reported in Turkey previously [31] TTTTH, HKNTF and HKPPF races were avirulent to Sr24, Sr38; Sr5, Sr9e, Sr11, Sr9b, Sr17, Sr24, Sr31; and Sr5, Sr9e, Sr11, Sr9b, Sr9d, Sr24, Sr31 genes, respectively (Table 8).

Generally, the identified races had wider range of virulence in the study areas (Table 8). High virulence diversity of stem rust races were reported earlier in Ethiopian [8,30,32] .Co-evolution of Pgt along with wheat being the reason for high virulence diversity in Ethiopian Pgt populations [33]. This might be due to variation over location and time, as the races found in a specific season and region depend on the type of wheat varieties grown [29] and to some extent on the predominant environmental conditions, especially temperature [18]Virulence diversities within Pgt were also reported from countries such as South Africa, Mexico, USA and Canada [34].

The race spectrum in Ethiopia was clearly different from other parts of the world. For example surveys in Canada [21,34-36] USA, Russia and South Africa detected fewer races such as 15, 5, 6 and 7, respectively. Whereas, more races were identified from Ethiopia, i.e. 60, 41, 17, 44, 22 and 20 [24, 30, 31, 37-39] at different times and locations in the country.However, the present study is dissimilar to the previous works that have been done in Ethiopia. It is evident that only 5 races have been identified from two zones and TKTTF was the most dominant across the locations and it covered 88.4% of the race frequency occurrence in those 12 districts.

Most of Ethiopian races varied from one another by single gene/step changes Belayneh et al.[30] Abebe et al.[24] also reported that, 40% of the races that were identified from Southern Tigray in 2010 cropping season varied by single gene changes. Such single step changes in virulence were reported to be the main process of evolutionary change in P. graminis f. sp. tritici populations [40].However, the present study showed that all identified races were not varied by single step changes (Table 8). There might be other factors for race variation in the studied area like parasexualism, migration, selection pressure and gene combination.

Virulence frequency of *P. graminis* f. sp. *tritici* isolates to Sr resistance genes: The results showed that the majority of the stem rust resistance genes were found ineffective against most of the isolates tested in this study. 85% of the Sr genes were ineffective to 88.4% of the isolates. The wheat differential line that carry the resistance genes SrMcN, Sr10, Sr9a, Sr30, Sr9g, Sr8a, Sr6, Sr7b and Sr21 were ineffective to 100% of the isolates tested (Table 9). In the same way, three differential lines that carry resistance genes Sr17, Sr9d and Sr38 were ineffective to 97.7% of the tested isolates each. However, two differential lines carrying resistance genes Sr31 and Sr11 were ineffective with the least

Race	Virulence (ineffective Sr genes)/Avirulence (effective Sr genes) spectrum	No.Of isolates	Frequency (%)
TKTTF	5, 21, 9e, 7b, 6, 8a, 9g, 36, 9b, 30, 17, 9a, 9d, 10, Tmp, 38, McN/11, 24, 31	38	88.4
TTKSK	5, 21, 9e, 7b, 11, 6, 8a, 9g, 9b, 30, 17, 9a, 9d, 10, 31, 38, McN/36,Tmp, 24	2	4.7
TTTTH	5,21,9e, 7b,11, 6, 8a, 9g, 36, 9b, 30, 17, 9a, 9d, 10, Tmp, 31, McN/24, 38	1	2.3
HKPPF	21, 7b, 6, 8a, 9g, 36, 30, 17, 9a, 10, Tmp, 38, McN/5, 9e, 11, 9b, 9d, 24 , 31	1	2.3
HKNTF	21, 7b,6, 8a, 9g, 36, 30, 9a, 9d, 10, Tmp,38, McN/5, 9e, 11, 9b, 17, 24, 31	1	2.3
Total		43	100

Table 8: Virulence/Avirulence spectrum and frequency of races of *P. graminis* f. sp. *tritici* collected from west and south west Shewa zones of Oromia region in 2014.

virulence frequency of 7% each to the tested isolated. Belayneh et al. [30] reported similar finding that McNair 701 (SrMcN) was susceptible to all of the races identified. According to these authors five stem rust resistance gene in the differential lines; Sr9a, Sr9g, Sr10, Sr7b and Sr9d were infective for more than 96% of isolates that were collected from Shewa, Arsi, Bale, and northwest regions of Ethiopia, during 2006-2007 cropping season. Similarly, Abebe et al. (2010) also reported that, McNair 701 (SrMcN) was susceptible to 95% of the races identified and about 55% of the Sr genes were ineffective to more than 60% of the isolates. This report indicated that, Six differential lines carrying resistance genes Sr9d, Sr21, Sr6, Sr10, Sr9g and Sr9b were ineffective with virulence frequency of 65.6, 78.1, 75, 81.2, 87.5 and 93.8% to the isolates tested, respectively. Roelfs et al. [9] also reported that Virulence for Sr6, Sr9a, and Sr9d are common worldwide. However, the present study showed that, the virulence frequency of stem rust identified races are a little bit higher on the most tested differential line genes than earlier studies. This could be due to emerge of new virulent stem rust races and extensive cultivation of susceptible varieties in west and southwest Shewa zones of Oromia region as well as in the country.

In contrast, the stem rust resistance gene Sr24 was found effective to all 43 stem rust isolates collected from west and south west Shewa zones of Oromia region (Table 9). This was previously confirmed by the reports of Roelfs et al.[9] Abebe et al.[24] and CIMMYT [41] as Sr24 gene is amongst the effective genes in different countries. Even though, in Kenya 2006, a lineage of Ug99 called TTKST added virulence on stem rust gene Sr24 has further increased the vulnerability of wheat to the rust worldwide [42]. Based on the present study, Sr11 and Sr31 resistance genes were found to be effective against most of stem rust races detected in both zones. Differential lines that carry Sr31 and Sr11 were resistant to 93% of the isolates tested. Sr31 and Sr11 genes were resistant to 3 common (TKTTF, HKPPF, and HKNTF) races (Table 8). Whereas, differential lines that carry Sr36, Sr9e, Sr9b, SrTmp and Sr5 showed resistance to 4.7% of the isolates tested. It was found to be effective to 59.4% of the isolates collected from Southern zone of Tigray [24] even though there was a historical damage of Sr36 by the race emerged in Ethiopia in the variety Enkoy in 1993/94, CIMMYT [39] and Belayneh et al.[30] also reported in their finding that Sr36 and SrTmp were effective for 81.6 and 76.3% of the isolates tested, respectively, for samples collected during 2006-2007 cropping season in Arsi, Shewa, Bale and northeast regions of Ethiopia. But, these authors reports were not similar to the present study due to more effectively resistance of Sr36 and SrTmp genes to their isolates. Therefore, the effective genes such as Sr11, Sr31 and Sr24 can be used as a source of resistance genes, in wheat breeding programs in west and southwest Shewa zones of Oromo region as well as in Ethiopia. Besides, genes that confer seedling and/or adult plant resistance to Ug99 include Sr2, Sr13, Sr14, Sr22, Sr28, Sr29, Sr32, Sr33, Sr35, Sr37, Sr39, Sr40 and Sr44 [43] are used as a source of genetic material in breeding program.

Conclusion

The study confirmed the presence of high virulence spectrum among the five identified wheat stem rust races. This indicated that, West and southwest Shewa zones are hot spot areas for appearance of virulent genetic diversity of stem rust races. Therefore, regular assessment and physiological stem rust race identification will be mandatory for virulence and/or avirulence information in west and southwest Shewa zones. Sr24 gene was the only effective gene that showed resistant for all identified races. Hence, the Sr resistance gene Sr24 can be used as sources of resistance in wheat breeding program.

Differential line	Sr gene	Frequency (%)
ISe5-Ra	5	95.3
CnS-T-mono-deriv	21	100
Vernsteine	9e	95.3
ISr7b-Ra	7b	100
ISr11-Ra	11	7
ISr6-Ra	6	100
ISr8a-Ra	8a	100
CnsSr9g	9g	100
W2691SrTt-1	36	95.3
W2691Sr9b	9b	95.3
BtSr30Wst	30	100
Combination V	17	97.7
ISr9a-Ra	9a	100
ISr9d-Ra	9d	97.7
W2691Sr10	10	100
CnS SrTmp	Tmp	95.3
LeSr24Ag	24	0
Sr31 (Benno)/6*LMPG	31	7
VPM1	38	97.7
McNair701	McN	100

Table 9: Virulence frequency of P. graminis f. sp. tritici isolates on the 20 Sr genes.

Acknowledgement

I am deeply grateful and indebted to EIAR for allowing me to pursue postgraduate study at Ambo University. In this regard, I would like to express my deep and heartfelt gratitude to Dr. Asenak Fikre, for his positive support to start my M.Sc. on time. I would like to thank East African Agricultural Productivity Project (EAAPP) for partial financial support to conduct this M.Sc. thesis work. In this regard, I also owe my deepest gratitude to Dr. Alemayehu Asefa and Mr. Endale Hailu for their role to attach me to the project.

References

1. FAOSTAT (2014). FAO Statistical database.

2. CSA (2014). Agricultural Sample Survey. Report on Area and production of Major crops.

3. Braun HJ, Atlin G, Payne T (2010). Multi-location testing as a tool to identify plant response to global climate change. Climate Change and Crop Production, edn. MP Reynolds 7: 115–38.

4. Pena RJ (2002). Wheat for bread and other foods. FAO Corporation document Repository.

5. FAO (2007).Crop prospects and food situations: Global cereal production brief: 4.

6. Zegeye T, Taye G, Tanner D, Verkuiji H, Agidie A, et al. (2001). Adoption of improved bread wheat varieties and inorganic fertilizer by small-scale farmers in Yelmana Densa and Farta districts of Northwestern Ethiopia. EARO and CIMMYT.

7. Park RF, Bariana HS, Wellings CS (2007).Stem rust of wheat in Australia. Preface Australian Journal of Agricultural Research 58: 469.

8. Mengistu H, Getaneh W, Yeshi A, Rebka D, Ayele B (1991).Wheat pathology research in Ethiopia. Wheat research 173-218.

9. Roelfs AP (1978). Estimated losses caused by rust in small grain cereals in the united states 1918-76. Miscellaneous publication 1363.

10. Bechere E, Kebede H, Belay G (2000). Durum wheat in Ethiopia: An old crop in an ancient land. Institute of Biodiversity Conservation and Research (IBCR) :68.

11. Leppik EE (1970). Gene centers of plants as sources of disease resistance. Ann Rev Phytopathol 8: 323-344.

12. Singh RP, Hodson DP, Jin Y, Huerta-Espino J, Kinyua MG, et al. (2006). Current status, likely migration and strategies to mitigate the threat to wheat production from race Ug99 (TTKS) of stem rust pathogen. CAB Reviewes 1: 054.

13. Admassu B, Lind V, Friedt W, Ordon F (2009) Virulence analysis of Puccinia

graminis f. sp. tritici populations in Ethiopia with special consideration of Ug99. Plant Pathol 58: 362-369

14. SARC (2004). Progress report for year 2004. Department of cereal pathology, Sinana, Ethiopia.

15. APPRC (2010). Progress report for the year 2010. Department of cereal pathology, Plant Protection Research Center, Ambo, Ethiopia.

16. Peterson R.F, Campbell AR, Hannah AE (1948). A diagrammatic scale for estimating rust intensity on leaves and stem of cereals. Canadian Journal Research 26: 490-500

17. Zadoks JC, Chang TT, Kanzak CF (1974) a decimal code for the growth stage of cereals. Weed Research 14: 415-421.

18. Roelfs AP, Singh, RP, Saari EE (1992). Rust Diseases of Wheat: Concept and Methods of Disease Management. CIMMYT: 81.

19. Stakman EC, Steward DM, Loegering WQ (1962). Identification of physiologic races of Puccinia graminis var. tritici. Agric Res Serv E-617: 1-53.

20. Jin Y, Singh RP, Ward RW, Wanyera R, Kinyua M, et al. (2007). Characterization of seedling infection types and adult plant infection responses of monogenic Sr gene lines to race TTKS of Puccinia graminis f.sp. tritici. Plant Dis 91: 1096–1099

21. Fetch TG, Dunsmore KM (2004). Physiological specialization of P. graminis on wheat, barley, and oat in Canada in 2001. Canadian Journal of Plant Pathology 26: 148-55.

22. Central Statistical Authority (CSA). (2008). Agricultural Sample Survey 1998/99. Report on Area and Production of Major Crops Volume 1. Statistical Bulletin 200. CSA, Addis Ababa, Ethiopia 111.

23. Ayele B, Eshetu B, Betelehem B, Bekele H, Melaku D, et al. (2008). Review of two decades of research on diseases of small cereal crops. In: Abrham Tadesse (eds). Increasing crop production through improved plant protection volume I. Proceedings of 14th annual conference of plant protection society of Ethiopia (PPSE) 19-22 December. 2006 Addis Ababa, Ethiopia 375-416.

24. Abebe T, Woldeab G, Dawit W (2010) Distribution and Physiologic Races of Wheat Stem Rust in Tigray, Ethiopia. J Plant Pathol Microb 3:142.

25. Dagnatchew Y (1967). Plant disease of economic importance in Ethiopia. Haileslassie I University, College of Agriculture, Environmental station bulletin . Addis Ababa, Ethiopia: 30.

26. Wanyera R, Macharia JK, Kilonzo SM, Kamundia JW (2009). Foliar fungicides to Control wheat stem rust, race TTKS (Ug99), in Kenya. Plant Disease 93: 929-932.

27. Pretorius ZA, Singh RP, Wagoire WW, Payne TS (2000). Detection of virulence to wheat stem rust gene Sr31 in Puccinia graminis f. sp. tritici in Uganda. Phytopathology 84: 203.2.

28. Wanyera R, Kinyua MG, Jin Y, Singh RP (2006). The spread of stem rust caused by Puccinia graminis f. sp. tritici, with virulence on Sr31 in wheat in Eastern Africa. Plant Dis 90: 113.

29. Singh RP (1991). Pathogenicity variation of Puccinia recondita f. sp. tritici and P.graminis f. sp. tritici in wheat growing areas of Mexico during 1988-1989. Plant Disease 75: 790-794.

30. Belayneh A, Lind V, Friedt W, Ordon F (2009). Virulence analysis of Puccinia graminis f. sp. Tritici populations in Ethiopia with special consideration of Ug99. Plant Pathol 58: 362-369.

31. Mert Z, Karakaya A, Dusunceli F, Akan K, Cetin L (2012). Determination of Puccinia graminis f.sp. tritici races of wheat in Turkey. Turk J Agric For 36: 107-120.

32. Belayneh A, Emebet F (2005). Physiological races and virulence diversity of P. graminis f. sp. tritici on wheat in Ethiopia. Phytopathol. Mediteer 44: 313-318.

33. Van Ginkel M, Getinet G, Tesfaye T (1989). Stripe, stem and leaf rust races in major wheat producing areas in Ethiopia. IAR Newslett. Agric Res 3: 6-8.

34. Jin Y (2005). Races of Puccinia graminis identified in the United States in 2003. Plant Disease 75: 1125-1127.

35. Lekomtseva SN, Volkova VT, Zaitseva LG, Skolotneva ES (2007). Races of Puccinia graminis f. sp. tritici in the Russian Federation in 2007. Moscow, Russian Federation. Annual Wheat Newsletter 55: 178-179.

36. Pretorius ZA, Bender CM, Visser B, Terefe T (2010). First report of a Puccinia graminis f. sp. tritici race virulent to the Sr24 and Sr31 wheat stem rust resistance genes in South Africa. Plant Dis 94: 784-785.

37. SPL (1988). Annual report for the period of 1985-1988. Ambo, Ethiopia. 5-16.

38. Ayele B, Alemtaye A, Bedada G, Payne T (2001). Double sources of resistance to Puccinia striformis and P. graminis f. sp. tritici. In: CIMMYT bread wheat lines. Proceeding of 9th Annual Conference 22-23 June, CSSE and EIAR. Addis Ababa, Ethiopia. Sebil 9: 11-19

39. Serbessa N (2003). Wheat Stem Rust (P. graminis f. sp. tritici) Intensity and Pathogenic Variability in Arsi and Bale zones of Ethiopia. M.Sc. Thesis. Alemaya University Ethiopia 92.

40. Green GJ (1975). Virulence changes in Puccinia graminis f. sp. tritici in Canada. Canadian Journal of Botany 53: 1377-1386.

41. CIMMYT (2005). Sounding the alarm on global stem rust: an assessment of race Ug99 in Kenya and Ethiopia and the potential for impact in neighboring countries and beyond. Mexico city, Mexico.

42. Jin Y, Szabo LJ, Pretorius ZA, Singh RP, Ward R, et al. (2008). Detection of virulence to resistance gene Sr 24 within race TTKS of Puccinia graminis f. sp. tritici. Plant Dis 92: 923-926.

43. Singh RP, Hodson DP, Huerta-Espino J, Jin Y, Njau P, et al. (2008). Will stem rust destroy the world's wheat crop? Adv. Agron 98: 271–309.

4

Effect of Drying Off Period and Harvest Age on Quality and Yield of Ratoon Cane (*Saccharium officinarium L.*)

Hadush Hagos[1]*, Walelign Worku[2] and Abuhay Takele[1]

[1]*Ethiopian Sugar Corporation, Research and Training Division, Sugarcane Production Research Directorate, Agronomy and Protection Research Team, Wonji Research Center, P.O.Box 15, Wonji, Ethiopia*
[2]*Hawassa University, College of Agriculture, Hawassa, Ethiopia*

Abstract

Field experiment was conducted at Metahara Sugar Estate during the 2011/2012 cropping period to determine the effect of drying off period and harvest age on quality and yield of ratoon sugarcane (Saccharium officinarium L.). Major variety B52-298 was used under the four levels of drying off periods (25, 45, 65 and 85 days) and harvest ages (12, 13, 14 and 15 months) in a completely randomized block design with 4*4*3 factorial treatment arrangements. All data's were collected at the end of each levels of drying off periods and harvest ages. Analysis of variance (ANOVA) showed that drying off period significantly influenced maturity testing parameters (sheath moisture content and handrefractometer brix), quality parameters (brix, pol, purity and CCS) and yield parameters (sugar yield) ($P < 0.001$). Brix, pol, juice purity, estimated recoverable sucrose and sugar yield were significantly increased when the level of drying off period was increased with a peak at 65 days. Brix, pol, juice purity, estimated recoverable sucrose and sugar yield were significantly increased with increasing level of drying off period up to 65 days. In contrast, soil moisture content, sheath moisture content, plant height and stalk diameter were reduced with increasing drying off period. Sugar yield increased by 20.97 % with extending drying off period from 25 days to 65 days with no further increase at 85 days drying off period. Effect of harvest age also significantly influenced maturity testing parameters (hand refractometer brix), quality parameters (brix, pol, purity and ERS) and yield parameters (plant height, cane yield and sugar yield) ($P < 0.001$). Increase in harvest age, significantly increased brix, pol, juice purity, estimated recoverable sucrose, plant height, cane yield and sugar yield. Optimum yield was recorded on 65 days drying of period and 15 months harvest age with economically acceptable marginal rates of return 2544% and 135%, respectively. Therefore, adjusting the drying off period to 65 days and the harvest age to 15 months for variety B52-298 under the tested soil condition for ratoon cane is recommended to increase sugar yield at Metahara Sugar Estate.

Keywords: Drying off period, harvest age, maturity, ratoon crop (sugarcane cutting)

Introduction

In ratoon sugarcane, irrigation is often withheld before harvest to reduce soil compaction from harvesting machinery and to enhance quality parameters (brix%, pol%, purity and estimated recoverable) to be deposited preferentially in sugarcane stalks [1,2]. The days required for pre-harvest drying-off to improve sucrose accumulation in sugarcane could range from 30 to 150 days depending on low to high water holding capacity of the soil [3]. The complete suspension of irrigation for the final two months before harvest gave the best results of soluble solids and sucrose content in South African sugarcane industries [2]. Australian researchers found highest sucrose content and sugar yield at 56 days drying off period retarding stalk height and shortening vegetative growth of sugarcane as compared to the shorter drying off period (35 days) [3]. In 61% of the drying-off treatments in South Africa, there was a significant increase in the soluble solids together with dehydration throughout the stalk. However, sucrose yields only increase if water stress reduces stalk biomass by less than 4% or unchanged [4]. However, severe stress could develop in crops grown in low moisture available soils when completely suspending irrigation for a long period of time before harvest [5].

Changes in sheath moisture content can well reflect the effects of water stress on plants [6]. Sheath moisture content is measured to determine the level of water stress as influenced by drying off period. As a result, after 35 days of drying off period the morphology was changed because sugar cane sheaths showed signs of wilting and dramatically decreased its moisture content [5]. Sheath moisture content was high in unstressed treatments but after 19 days after irrigation was withheld it was reduced from 80 to 75% [7]. The standard sheath moisture of sugarcane to have peak sucrose concentration was 68-74% [8,9].

On the other hand, harvest age was one of the factors that determine maturity of sugarcane. Sheath moisture content was reduced as harvest age increased because at later ages of growth there was a possibility that the crop may have lost some potential of root activities [10]. At a later age old root system of cane stalk gradually ceases to function and decay with time resulting to decreased sheath moisture content of the plant [9]. High sheath weight per stalk in young canes may be used as an indicator of the water status of the crop [7]. The percentage of sheath moisture content of sugarcane in old cane could be decreased because at senescence green leaves are dried and decrease in number [1].

Several maturity testing schemes have been proposed during the growing and harvest season. The common practice is to test the standing cane in the field for brix with a hand refractometer, which is a measure of the amount of sucrose in the cane [11]. Sugarcane matures when top/bottom brix ratio approached unity. It can also be done by using bottom minus top brix and bottom minus middle brix together and recommended to harvest when the two indices approach zeros [12]. The highest hand refractometer brix reading of top to bottom portion of ripened stalk is required to have 19 and 20% (a ratio of 0.95) brix %, respectively [13]. With stalk maturation, more and more internodes reach the same condition and a progressive increase in total soluble solids to include sucrose [14].

The quality parameters (brix%, pol%, purity and estimated

*****Corresponding author:** Hadush Hagos, Ethiopian Sugar Corporation, Research and Training Division, Sugarcane Research Directorate, Agronomy and Pretction Research Team, Wonji Research Center, P.O.Box 15, Wonji, Ethiopia
E-mail: hadgos@gmail.com

recoverable sucrose) in juice were improved with increasing harvest age which indicates the function time on the storage of sugar due to the effect of diluting the structural cell wall components [15,16]. Estimated recoverable sucrose and sugar yield were increased in a first ratoon crop, for harvest age ranging between 8 to 14 months [17]. The growth and yield parameters such as stalk height, cane yield and sugar yield were increased with increasing age of sugarcane suggesting that more growth and sucrose accumulation was achieved in the longer harvest ages [18]. From the research conducted on different harvest ages of ratoon cane in South Africa (10 months, 12 months, 14 months and 16 months) longer height, high cane and sugar yield was obtained from the two longer harvest ages (14 and 16 months) [19,20].

The current sugar production of the Ethiopian Sugar Industry covers only 60% of the annual demand for domestic Consumption while the deficient is imported from abroad. In order to make the country self-sufficient in sugar and export the surplus sugar and produce ethanol and other by-products, the Federal government of Ethiopia is working to establish sugarcane plantation on 325,000 ha in addition to the vast expansion project of the previously established farms with erection of high crashing capacity 10 new sugar mills [21].

However, improper pre-harvest practices; drying off period (wet or excessive drying) of fields before harvest and harvesting many fields without considering crop age are common constraints in sugarcane production in Ethiopian Sugar Estates [22]. The importance of determining yield potentials for sugarcane has been noted by many scientists with goals to aim for barriers to be broken. Law of the minimum suggests that there is always some factor limiting yield. Therefore, yield potential need to be defined in terms of the limiting factor [23].

Efforts have been made in the past to address the effects of drying off period [23,24] and harvest age [19,25] on first cuttings of sugarcane in various countries. However, there are no studies dealing with yield and quality response of ratoon crops as influenced by different drying off period and harvest age on Ethiopian Sugar Estates. So, this experiment is initiated with the following specific objectives:

(1) To determine quality and yield responses that will be attained under various drying off period of ratoon cane

(2) To examine the influence of harvest age on ratoon cane quality and yield parameters

Material and Method

The study was conducted on clay soils of Metahara Sugar Factory. The clay soils cover more than 90% of the estate and they are grouped into four distinct textural groups as heavy clay, clay, clay loam and loam soil groups [26]. The experiment has two factors namely, drying off period (25, 45, 65 and 85 days before harvest) and harvesting age (12, 13, 14, and 15 months) with a factorial combination giving 16 treatments. Optimum drying off period and harvest age of ratoon crops was not yet studied at Ethiopia Sugar Industries. To address this problem, a standard sugar cane variety B52-298, was selected because of its high yielding potential and high area coverage which is 23% of the total area covered by various commercial varieties in the Sugar Estate.

The experiment was carried out on a first ratoon cane using randomized complete block design (RCBD) with three replications in the cool season. Each plot had five rows with 10 m length and 1.45 m width for each row (10 m x 1.45 m x 5 rows) having an area of 72.5 m² for a single plot. The distance between plots was 2.9 m while it was 4.35 m between replications. The harvested plot consisted of three rows with

10 m length and 1.45 m width each (10 m x 1.45 m x 3 rows) with an area of 43.5 m².

Cultural practices such as weeding, fertilizer application, molding, pesticide application and irrigation frequency of the experimental field were based on the current practice practices of Metahara Sugar Estate except manipulating harvest age and drying off period before harvest. Irrigation was applied with hydroflume application system delivering water to the furrows at an average inflow rate of 5 l/s with 89 minutes cut off time for the 100 m length furrow [26].

Economic analysis was done using partial budget analysis procedures [27]. The average sugar yield was adjusted down ward by 20% to reflect the yield difference between the experimental yield and commercial yield which was deteriorated by pre-harvest burning, delaying at field to reach the milling center, yield loss with impurities and yield loss by low factory efficiency.

Result and Discussion

Effect of drying off period on soil moisture content

Drying off period affected soil moisture content negatively (Figure 1). Although soil sample was taken from two depths 0-30 cm and 0-60 cm, their average was taken for discussion because both depths showed similar decline in soil moisture content with increasing days of drying off period. Shorter drying off periods (25 days and 45 days) had a relatively higher mean soil moisture content of 38.77% and 31.47% respectively, at the end of drying (Figure 1). An initial soil moisture content of 45.58% was recorded two days after irrigation at field capacity as a reference to compare with the moisture loss by drying off periods. Thus, a sharp decline of soil moisture content was observed when drying off period was increased to 65 and 85 days which had 24.94% and 19.58%, respectively. The soil moisture content of 24.94% obtained from 65 days drying off period could be considered as adequate for maximum sucrose (pol) accumulation as indicated in Table 2. Similarly, previous studies reported that the average soil moisture content of 24.0 to 27.5% was adequate for high sucrose accumulation [24]. The average reduction of soil moisture content over 65 days and 85 days drying off periods was 26.18% and 55.45%, respectively as compared to the control drying off period (45 days). Another studies also confirmed that there is high soil moisture content in the short drying off periods [5]. However, soil moisture was reduced with increasing days of drying off period. Thus, longer drying off period was needed as the water holding capacity of soils increase to improve sucrose accumulation in sugarcane stalks (Figure 1).

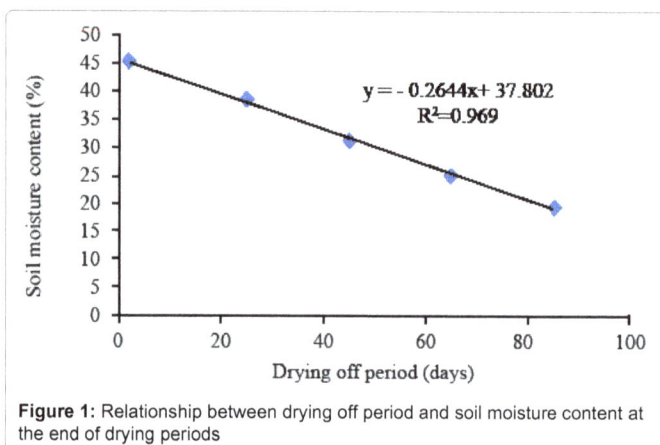

$$y = -0.2644x + 37.802$$
$$R^2 = 0.969$$

Figure 1: Relationship between drying off period and soil moisture content at the end of drying periods

Effect of drying off period and harvest age on sheath moisture content

Sheath moisture content of ratoon cane was significantly influenced by drying off period (p < 0.001) and harvest age (p < 0.01) while the interaction effect was not significant. Highest mean sheath moisture content was recorded on the two short drying off periods (25 and 45 days) while lower sheath moisture content was recorded from the two longer drying off periods (65 and 85 days) (Table 1). Previous studies confirmed that, as drying off period increased, sheath moisture content was reduced [3]. Drying off period creates depression of leaf water potential by the dehydrative effect of water stress due to excessive evapo-transpiration [28].

The highest percent sheath moisture content was obtained at the early age (12 months) of the crop and the lowest sheath moisture content was recorded under the longest harvest age (15 months) of the crop (Table 1). However, there was no significant difference between the two upper age levels. An increase of one to three months harvest age from the lowest (12 months) decreased sheath moisture content by 2.3%, 2.9% and 5.6%, respectively. In general, significant decrease in percent sheath moisture of ratoon sugarcane was noted with increased age in this study. In line to this, decreasing sheath moisture content with increasing cane age was obtained from previous studies [10]. This could be due to loss of some potential activities of root resulting in lower sheath moisture content of the crop. Old root system of cane stalk gradually ceases to function and decay with time enhancing water stress in the photosynthesis area of the plant [9].

Effect of drying off period and harvest age on hand refractometer brix% test of sugarcane cane parts at maturity

Drying off period (p < 0.001) and harvest age (p < 0.01) significantly influenced hand refractometer brix test of sugarcane parts but not by their interaction. The highest hand refractometer brix was obtained on 65 days drying off period in all the bottom, middle and top parts of the stalk (Table 1). In this drying off period, all the stalk parts accumulated peak soluble solids indicating the sign of maturity to harvest. The lowest hand refractometer brix in all stalk parts was obtained on the shortest drying off period (25 days). Comparatively, longer drying off periods (65 and 85 days) showed greater refractometer brix content than the

shorter drying off periods (25 and 45 days). For the middle part of the plant, there was no significant difference between the two upper drying off levels. In general, sequential accumulation of soluble solids was observed from the base to the top of the stalk on different parts of the stalk. High brix content was obtained from the moderately dried than fully irrigated treatments in all parts of the stalk with sequential accumulation of soluble solid from bottom to the top part of the stalk [29]. The low brix content of the top part is consistent with the fact that this part is more active in non-soluble solid storing metabolism (e.g. processes involving respiration and growth). Drying off period enhances brix content of the cane segments with high brix content at the bottom segment fallowed by the middle and top segments [1].

Harvest age affected hand refractometer brix and increased levels of harvest age enhanced hand refractometer brix of the stalk parts. The highest hand refractometer brix percent was obtained on the longest harvest age (15 months) in all stalk parts (Table 1). Similarly, the lowest mean value was recorded on the shortest harvest age (12 months). However, there was no significant difference between the first two (12 and 13 months) harvest ages in all parts of the stalk. Similar to drying off period, harvest age also showed sequential storage of soluble solids. The bottom part stored highest amount of brix followed by the middle and top part of the stalk. In agreement with this result, studies reported that brix percent of all the bottom, middle and top parts of the stalk increase with cane age [11]. In milling operations, the preferred varieties are those with high and nearly equal percentage of hand refractometer brix on its all stalk parts at maturity (bottom, middle and top parts [14].

Effect of drying off period and harvest age on quality parameters of ratoon cane

Percentage of soluble solids, pol, purity and estimated recoverable sucrose were significantly (p < 0.001) affected by drying off period and harvest age. The highest percent soluble solid, pol, purity and estimated recoverable sucrose were obtained at 65 days drying off period and the lowest yield of all the quality parameters were recorded at 25 days drying off period (Table 2). The increase of the percentage of quality parameters until 65 days drying off period was due to the decrease of the proportion of reducing sugars resulting in an increase of sucrose content. Because biomass was stored preferentially as sucrose rather than being drawn into the production of reducing sugars or non-sucrose materials (fiber).Withholding irrigation water beyond 65 days resulted in a decline of the value of all the quality parameters. This might be due to the fact that the stored assimilates were remobilized to supply the damaged part of the plant during severe stress. Sugarcane researchers indicated that moderate water stress in cane tissue was recognized as a means of cane ripening because it decreased vegetative growth and hasten the quality parameters (brix and pol) during ripening period [3,30]. The greatest contribution to change in sucrose dry weight concentration with drying off is from an increase in the dry weight concentration of soluble solids and sucrose content [2].

Increased levels of harvest age also enhanced all quality parameters. The highest percent soluble solid, pol, purity and estimated recoverable sucrose were obtained at the longest harvest age (15 months) (Table 2). This might be due to the dilution effect of sugarcane enzymes changing the reducing sugars and non-sucrose materials (fiber) to sucrose or it could be due to positive impact of harvest age on the yield components (plant height and cane yield) which allow accumulation of additional soluble solid or sucrose by delaying harvest age. Percent of soluble solids, percent pol, purity and percent of estimated recoverable sucrose significantly increased as age of sugarcane increased [19]. However,

Drying off period (D)	SHMC (%)	BHRB (%)	MHRB (%)	THRB (%)
25	75.54ᵃ	13.09ᵈ	12.75ᶜ	10.86ᶜ
45	74.07ᵃ	14.39ᶜ	13.69ᵇ	11.58ᶜ
65	66.70ᵇ	19.38ᵃ	18.92ᵃ	18.71ᵃ
85	63.67ᶜ	18.50ᵇ	18.20ᵃ	16.99ᵇ
Harvest Age (MAP)				
12	72.12ᵃ	13.74ᶜ	13.45ᶜ	11.32ᶜ
13	70.54ᵃᵇ	14.36ᶜ	14.07ᶜ	11.93ᶜ
14	68.52ᵇᶜ	17.60ᵇ	17.20ᵇ	16.14ᵇ
15	66.80ᶜ	19.66ᵃ	19.35ᵃ	18.75ᵃ
LSD (5%)	2.85	0.85	0.86	0.77
CV (%)	4.92	6.25	6.45	6.32

Means followed by the same letter within a column are not significantly different. D, days; MAP, months after planting; SHMC, sheath moisture content; BHRB, bottom hand refractometer brix; MHRB, middle hand refractometer brix; THRB, top hand refractometer brix; ns, non-significant

Table 1: Mean comparison of maturity testing parameters as influenced by drying off period and harvest age at Metahara Sugar Estate

in over aged ratoon cane, the sucrose content is reduced due to heavy lodging and remobilization to supply the unproductive bull shoots (newly growing shoots) [11].

Effect of drying off period and harvest age on cane and sugar yield of ratoon cane

Cane yield was significantly influenced by drying off period (Table 3). The reason could be reduction of plant height and stem diameter under extended drying off periods might have been offset by greater accumulation of soluble solids. The research result in South Africa reported that, from a total of 53 treatments applied to drying off period 24 treatments (i.e. about one-half) showed an increase in sucrose concentration with no significant fall in cane yield [2]. Sugar yield was significantly ($p < 0.001$) influenced by drying off period. The highest value of sugar yield was obtained at 65 days drying off period. The increase of sugar yield observed under moderate drying off period could be due to the positive impact of drying off period on all quality parameters (Table 2). Further increase of the drying off period to 85 days reduced sugar yield. This could be attributed to the negative effect of the severe stress on growth and quality parameters. From the research result of previous studies, high sugar yield was recorded on the moderate drying off period (56 days) than the shorter drying off treatment (35 days) [3]. However, severe (91 days) drying off period reduced sugar yields when the benefit from the higher sucrose content is negatively affected by more severe stress [4].

Cane yield and sugar yield were highly significantly ($p < 0.001$) influenced by harvest age. The highest cane yield and sugar yield were recorded on 15 months harvest age, followed by 14 months harvest age. This might be due to the increasing effect of longer harvest ages on yield components (plant height, cane yield) and quality parameters (brix, pol, purity and estimated recoverable sugar). The lowest cane yield and sugar yield was obtained from earliest harvest age (12 month) (Table 3). Significant increase in cane yield was recorded with an increase in harvest age from 8 to 16 months [19]. The major drop in sugar yield with an age restriction of below 14 months might be due to many hectares of crop being forced to be harvested when expected yields are extremely low as well as older crops being disallowed [15].

Economic Analysis

The partial budget analysis for drying off period showed that 25 and 45 days drying off period were dominated (Table 4). Marginal rate of return for 65 days drying off period was 2544%. Decreasing drying off period below 65 days will lead to increase in additional costs without compensating benefit. The marginal rate of return obtained at 65 days drying off period was above the 100% of the CIMMYT's minimum rate of return required for adoption of agronomic practices. The 2544% MRR recorded at 65 days drying off period indicated that for every one dollar invested in ratoon crop it could give a net return of 25.4 USD Dollars. Comparative results were reported in Australia indicating that 56 days pre-harvest drying off period was profitable for ratoon canes [3].

The profitability of sugar yield within various harvest ages considers a time value. So, the partial budget analysis for harvest age was computed in terms of t/ha/month (Table 4). The important parameters of maximizing sugar yield and net revenue in relation to harvest date and crop age is expressed by t/ha/month as an index of time value of sugarcane crop [8]. Accordingly, the partial budget analysis for harvest age gives MRR of 107% at 14 and 135% at 15 months which were both above the CIMMIT's minimum requirements of 100% MRR (Table 4). Even thought, both 14 and 15 months harvest ages give MRR above

the CIMMIT's minimum requirements, fifteen months harvest age is more profitable and advisable to first ratoon cane because it gives opportunity to additional profit from investing additional cost.

Conclusions

Improper drying off period and improper harvest age are recurrent problems of pre-harvest cultural practices, which severely affect quality and yield of ratoon cane. The economic analysis indicated that 65 days drying off period and 15 months harvest age gave the highest net benefit of 10,910.8 $ /ha and 350.2 $ /ha /month with acceptable MRR of 2544 and 135%, respectively. From this economic analysis, it is possible to conclude that, 65 days drying off period and 15 months harvest age can be recommended for ratoon cane production at Metahara to obtain maximum sucrose and sugar yield with optimum maturity. Moreover, further research could be undertaken for other popular varieties used by the estate under different soil types with extending harvest age beyond 15 months.

Acknowledgements

We would like to thank Ethiopian Sugar Corporation for financing the research and Hawassa University College of Agriculture for their unreserved material and other facilities support.

Drying off period (D)	Brix %	Pol %	Purity (%)	ERS (%)
25	20.52[c]	16.99[d]	87.64[c]	11.12[d]
45	20.58[c]	18.21[c]	88.48[bc]	12.57[c]
65	22.04[a]	20.16[a]	91.44[a]	14.27[a]
85	21.30[b]	19.05[b]	89.39[b]	13.25[b]
LSD (5%)	0.59	0.54	1.41	0.45
Harvest age (MAP)				
12	20.18[c]	17.86[d]	88.45[b]	12.03[c]
13	20.77[c]	18.16[c]	88.61[b]	12.42[c]
14	21.42[b]	18.88[b]	89.28[ab]	12.99[b]
15	22.08[a]	19.77[a]	90.61[a]	13.77[a]
LSD (5%)	0.59	0.54	1.41	0.45
CV (%)	3.36	3.48	1.89	4.19

Means followed by the same letter within a column are not significantly different. D, days ; Brix %, Percentage of refractometer brix; Pol%, percentage of sacharometer pol; ERS, estimated recoverable sucrose; MAP, months after planting; ns, non-significant

Table 2: Quality parameters of first ratoon cane as influenced by drying off period and harvest age at Metahara Sugar Estate

Drying off period (D)	PH (m)	SD (cm)	CY (t/ha)	SY (t/ha)
25	2.99[a]	2.99[a]	139.33	15.56[c]
45	2.97[a]	2.98[a]	138.70	17.48[b]
65	2.66[b]	2.77[b]	137.78	19.69[a]
85	2.65[b]	2.75[b]	136.83	18.20[b]
LSD (5%)	0.08	0.19	Ns	1.14
Harvest Age (MAP)				
12	2.58[d]	2.80	129.12[c]	15.53[c]
13	2.73[c]	2.84	132.54[c]	16.50[c]
14	2.90[b]	2.92	141.11[b]	18.31[b]
15	3.09[a]	2.93	149.89[a]	20.59[a]
LSD (5%)	0.08	NS	7.72	1.14
CV (%)	3.52	7.54	6.70	7.72

Means followed by the same letter within a column are not significantly different. D, days ; PH, plant height; SD, stalk diameter; CY, cane yield; SY, sugar yield; MAP, months after planting; ns, non-significant

Table 3: Yield parameters of ratoon cane as influenced by drying off period and harvest age at Metahara Sugar Estate

Treatment	Adjusted sugar yield (t/ha)	Gross field benefit ($ USD /ha)	Total variable cost ($ USD)	Net benefit ($ USD /ha)	Change in net benefit ($ USD /ha)	MRR (%)
DRP (days)						
85	14.56	10,115.4	0	10,115.4		
65	15.75	10,942.1	31.3	10,910.8	795.4	2541.2
45	13.39	9,302.5	62.5	9,240.0^d		
25	12.45	8,649.5	93.8	8,555.7^d		
HA(month)	t/ha/month	$ USD /ha/month	$ USD /ha/month	$ USD /ha/month	$ USD /ha/month	MRR (%)
12	0.41	287.7	0	287.7		
13	0.44	305.7	31.3	274.4		
14	0.49	339.2	31.3	307.9	33.5	107.1
15	0.55	381.5	31.3	350.2	42.3	135.2

DRP, drying off period; HA, harvest age; D, dominated; MRR, Marginal rate of return
Table 4: Partial budget analysis of the ratoon cane as influenced by drying off period and harvest age

References

1. Inman-Bamber NG (2004) Sugarcane water stress criteria for irrigation and drying off in Australia. Field Crops Research 89: 107–122.

2. Robertson MJ, Donaldson RA (1998) Changes in the components of cane and sucrose yield in response to drying-off of sugarcane before harvest. Field Crops Research 55: 201-208.

3. Inman-Bamber NG, Robertson MJ, Muchow RC, Wood AW (1999) Efficient use of water resources in sugar production: A Physiological basis for crop response to water supply. Sugar Research and Development Corporation, Australia. 1-33.

4. Donaldson RA, Bezuidenhout CN (2000) Determining the maximum drying off periods for sugarcane grown in different regions of the South African industry. Proceeding South African Sugar Technology Association 74: 162-166.

5. Olivier FC, Danaldson RA, Singels A (2006) Drying of sugarcane soils with low water holding capacity. South African Sugar Technology Association 80: 1-184.

6. Havaux M, Canaani O, Malkin S (1986) Photosynthetic responses of leaves to water stress, expressed by photoacoustics and related methods. Probing the photoacoustic method as an indicator for water stress in vivo. Plant Physiology 82: 827-833.

7. Inman-Bamber, NG (1986) Effect of water stress on growth, leaf resistance and canopy temperature in field grown sugarcane. Proceedings of the South African Sugar Technologists' Association, South Africa. 8: 15-29.

8. Bakker H (1999) Sugarcane cultivation and management. kluwer academic/plenum publisher, New York: 5-10.

9. Smith DM, Inman-Bamber NG Thorburn PJ (2005) Growth and function of the sugarcane root system. Field Crops Research 92:169-184.

10. Mequanint Y (2010) Effect of time of harvest on yield and sugar quality of sugarcane (Saccharum officinarum L.) varieties at Metahara Sugar Estate. MSc. Thesis. Haromaya University of Agriculture, Ethiopia: 1-65.

11. Qudsieh HY, Yosuf S, Osman A, Rahman RA (2001) Physico-chemical changes in sugarcane and the extracted juice at different portions of the stem during development and maturation. Faculty of Food Science, Malaysia. Journal of Food Chemistry 75: 131-137.

12. Miller JD, James NI (1977) Maturity testing of sugarcane. Proceeding American Society of Sugarcane Technologists 7: 101-111.

13. Sankaranarayanan P, Natarajan BV and Marimuthammal S (1986). Sugarcane varieties under cultivation in india, their morphological descriptions and agricultural characteristics, New Delhi.

14. Wagih ME, Ala A, Musa Y (2004) Evaluation of sugarcane varieties for maturity earliness and selection for efficient sugar accumulation. Sugarcane Agriculture. Sugar Technology North Australia 6: 297-304.

15. Muchow RC, Higgins AJ, Rudd AV, Ford AW (1998) Optimizing harvest date in sugar production: A Case study for mossman mill region in Australia. Sensitivity to crop age and crop class distribution. Field Crops Research 57: 243-251.

16. Donaldson RA, Redshaw KAR, Rhodes R, Antwerpen VR (2008) Season effects on productivity of some commercial South African sugarcane cultivars and trash production. Proceeding South African Sugar Technology Association 81: 528-538.

17. Higgins AJ, Muchow RC, Rudd AV, Ford AW (1998) Optimizing harvest date in sugar production: A Case study for the Mossman mill region in Australia, development of operations research model and solution. Field Crops Research 57:153–162.

18. Ramburan S, Sewpersad C, Mcelligott D (2009) Effects of variety, harvest age and eldana on coastal sugarcane production in South Africa. Proceeding South African Sugar Technology Association 82: 580-588.

19. Rostron H (1972) Effects of age and time of harvest on productivity of irrigated sugarcane. South African Sugar Association Experiment Station, South Africa: 142-150.

20. Lonsdale JE, Gosnell JM (1975) Effects of age and harvest season on yield and quality of sugarcane. Proceeding South African Sugar Association: 177-181.

21. Tolera B, Diro M, Belew D (2014) Response of sugarcane (Saccharum officinarum L.) varieties to BAP and IAA on in vitro shoot multiplication. Adv Crop Sci Tech 2: 126.

22. Eshete T, Tafesse A, Dametie A, Abejehu G, Negi T (1995) Remarks on the application of sugarcane plantation management standards. a review of the plantation manual. Metahara Sugar Estate, Ethiopia: 2-13.

23. Inman-Bamber NG (1995) Climate and water as constraints to production in the South African sugar industry. South African Sugar Association Experiment Station, South Africa 12: 18-34.

24. Negi T, Getaneh A, Ayele N (2010) Effect of length of pre-harvest drying-off period on cane quality in Metahara Sugarcane Factory. Biennial Conference Report of Ethiopian Sugar Development Agency Research Directorate. Wonji, Ethiopia: 1-16.

25. Teferi Y (2005) Effect of planting date and age of harvest of sugarcane cultivars on cane yield and yield components in Ethiopian Sugar Estates. Research Report, Ethiopia: 1-82.

26. Tate B (2009) Re-evaluation of the plantation soils at Metahara Sugar Factory, Ethiopia: 1-50.

27. CIMMYT (International maize and wheat improvement center) (1988) An economic training manual: from agronomic data to farmer's recommendations. CIMMYT, Mexico: 1-79.

28. Gilani S, Wahid A, Ashraf M, Arshad M (2008) Changes in growth and leaf water status of sugarcane (Saccharum officinarum L.) during heat stress and recovery. Department of botany, university of agriculture, Pakistan. International Journal of Agricultural Biology 10: 191-195.

29. Siswoyoa TA, Oktavianawatia ID, Murdiyantob U, Sugihartoa B (2007) Changes of sucrose content and invertase activity during sugarcane stem storage. University of Jember, Indonesia. Indonesian Journal of Agricultural Science 8: 75-81.

30. Singels A, Kennedy AJ, Bezuidenhout CN (2000) Effect of water stress on sugarcane biomass accumulation and partitioning. Proceeding South African Sugar Technology Association 74: 169-172.

5

Effect of First Irrigation Period on Sugarcane (*Saccharium officinarium L.*) Establishment in the Drought Areas of Tendaho, Ethiopia

Hadush Hagos[1*], Leul Mengistu[2], Yusuf Kedir[3] and Kidane Tesfamicheal[1]

[1]Ethiopian Sugar Corporation, Research and Training, Agronomy and Crop Protection Research Team, Wonji Research Center, P.O.Box 15, Wonji, Ethiopia
[2]Ethiopian Sugar Corporation, Research and Training, Sugarcane Production Research Directorate, Wonji Research Center, P.O.Box 15, Wonji, Ethiopia
[3]Ethiopian Sugar Corporation, Research and Training, Soil and Irrigation Research Team, Wonji Research Center, P.O.Box 15, Wonji, Ethiopia

Abstract

Field experiment was conducted to determine the effect of first irrigation period on sugarcane (*Saccharium officinarium* L.) establishment in the drought areas of Tendaho. Four levels of first irrigation application periods; 5 days pre-planting irrigation (DPI), 1 day after planting (DAP), 4 days after planting (DAP) and 8 days after planting(DAP) and four major varieties; N-14, NCO-334, CO-680 and B52-298 with high area coverage were used in a completely randomized block design with 4*4*3 factorial treatment arrangements. Sprout and Tillering data's were collected at 30-45 and 45-90 days after planting, respectively. Analysis of variance (ANOVA) revealed that there was a significant interaction effect between the sugarcane varieties and time of first irrigation (TFI) (Varieties*TFI=p<0.0001) on sprouting rate, number of tillers per hectare, root establishment rate and number of dead setts buds in both sugarcane varieties. The maximum sprouting rate, number of tailoring root establishment were recorded with the application of 5 days pre-planting irrigation and 1 day after planting irrigation in all varieties while the lowest sprouting rate, number of tillers, root establishment was recorded on the 8 days after planting first irrigation application in all varieties. Therefore, applying first irrigation, in early times (5 pre-planting irrigation and 1 days after planting) was highly recommended with better sprouting rate, number of tillers and root establishment to all varieties expecting high cane yield and sugar yield in the Tropical areas of Tendaho.

Keywords: First irrigation; Establishment; Sprouting rate; Number of tillers

Introduction

Ethiopia is endowed with favorable climatic and soil conditions for sugarcane production. To exploit such an immense resource, besides expanding the existing ones, the country is on the verge of establishing new sugar factories with large tract of sugarcane plantations. Tendaho is one of the new factories under establishment in the tropical areas of the country with 50,000 hectares suitable area for sugarcane production under full irrigation system [1]. In some countries like Australia, Sudan and South Africa about 60% of sugar produced is grown in irrigated areas, a practice which always results in production cost increase [2].

The ultimate goal of commercial sugarcane production system both in Ethiopia and else where in the world, is to increase cane and sugar production per unit area. In order to attain it, implementation of standard cultural practices is imperative. Experiences of some countries indicated that usage of high yielding varieties; provision and timely application of agricultural inputs, availability of sufficient soil moisture/irrigation water, and good soil condition have great shares [3]. Limited water resources restrict increasing the amount of sugarcane establishment and growth in many regions throughout the world because sugarcane requires substantial amounts of water during early stages [4]. Therefore it is important to apply irrigation water efficiently as possible during sugarcane establishment to produce maximum yields.

Plants are exposed to adverse environmental conditions, and drought is the major abiotic factor that can damage its growth and development. Drought also limits the areas suitable to agriculture. It is known that, as for any crop, during establishment water is essentially required to obtain maximum yield and drought events in this stage can significantly decrease productivity [5]. Sugarcane is among the crops which produce a higher amount of biomass per unit of cultivated area

and water requirement varies throughout the developing stages, thus for higher sprout, tillering and development of culms, there is a higher water requirement than during the maturation stage [5].

Sugarcane crop is produced from stalk cuttings denominated sets that contain one or more axillary buds. Annual growth of sugarcane can be divided into the following development phases: germination and tillering; stalk growth; and maturation [6]. The optimum temperature for germination of sugarcane is 27-33°C while good tiller production occurs when the temperature is about 30°C. A day temperature of below 18°C lengthens the tailoring period thus resulting in uneven maturity of the canes [7]. Warm, moist soil ensures rapid germination, germination results in an increased respiration and hence good soil aeration are important; therefore open structured porous soils facilitate better germination [8].

There are many reasons for lower productivity of sugarcane but the most pertinent is lack of knowledge on sugarcane management practices [9]. In sugarcane, germination denotes activation and subsequent sprouting of the vegetative bud, the germination of bud is influenced by the external as well as internal factors, the external

*Corresponding author: Hadush Hagos, Ethiopian Sugar Corporation, Research and Training Division, Agronomy and Crop Protection Research Team, Wonji Research Center, P.O. Box 15, Wonji, Ethiopia
E-mail: hadgos@gmail.com

factors are the soil moisture, soil temperature and aeration. The internal factors are the bud health, sett moisture, sett reducing sugar content and sett nutrient status [10]. The germination (shoot emergence from soil) is a critical event in the plant life to assure a good harvest and it is initially dependent on the set nutrients and water, developing its own root system after about three weeks under proper conditions [11]. The crop establishment phase and formative phase (sprouting, tillering and grand growth stages, have been identified as the critical water demand period [12]. This is mainly because 70-80% of cane yield is produced during this phase [13]. Water shortage results a negative impact on establishment of the crop, especially if the drought duration exceeds the capacity of drought tolerance of the plant species [14].

In the drought areas of Tendaho serious consideration should be given to improve the existed germination failure. As explained earlier, temperature of the area during planting time reaches up to 37°C that is good enough to increase the soil temperature. Planting seed cane into such soil (which coincides with the onset of the hottest season of area) even for few days without supplementing water will result in complete dehydration (Figure 1). In support of the fact, the time of first irrigation after planting was delayed at least a minimum of six days. Moreover, delay of first irrigation after planting and absence of pre-plant irrigation were taken as the main constraints of the drought area of Tendaho Sugar Project. This causes many gap areas in the plantation fields and the re-work in land preparation and planting activities. To alleviate the above problems, this study was initiated with the following objective: To study the effect of first irrigation period on sugarcane establishment in the drought areas of Ethiopia

Material and Methods

Site description

The experiment was conducted at Tendaho Sugar Factory Project in Afar Regional Estate in the Rift Valley of Ethiopia at an altitude and longitude ranging between 110 30' to 110 50' N and 400 45' to 410 03' E, respectively, with elevation ranging from 365 m to 340 m. The area has a mean maximum and minimum temperature of 37.20 and 21.88°C, respectively, with long-term average annual rainfall and relative humidity of 220 mm and 60.4%, respectively. The area has mean sunshine hours of 8.9 hr per day and the mean annual ETo of 7.9 mm/day [1]. The soil type of the experimental field is clay soils.

Treatment and design

Four levels of first irrigation periods; 5 days pre-planting irrigation (DPI), 1 day after planting (DAP), 4 days after planting (DAP) and 8 days after planting(DAP) and four major varieties; N-14, NCO-334, CO-680 and B52-298 which cover 90% of the area were used in a

Figure 1: Dehydrated setts and sett without root system in delaying first irrigation to 8 days after planting

completely randomized block design with 4*4*3 factorial treatment arrangements. Each plot had six rows with 8s m length and 1.45 m width having plot area of 69.6 m² for a single plot. The distance between plots was 2.9 m while it was 4.35 m between replications. Data was collected from four entire rows with 6 m length and 1.45 m width having plot area of 46.4 m².

Data collection and analysis

Sprouting is counted weekly starting 7 days after planting until 45 days. It is important to check that tillers are not confused with primary shoots during counting. This is because; there is an overlap in occurrence of tillering prior to the termination of full sprouting. Tillering is the underground branching of the cane. Tiller count will be made starting from 45 days after planting fortnightly (every 15 days) until 3 months crop age to estimate the number of tiller per ha. Number of buds that establish root and number of dead buds was collected at 30 and 45 days after planting.

Finally, sprouting rate, percent of sets that establish root and percent of dead sets calculated using the following formula [15].

$$\% \text{ Sprout} = \frac{\text{Number of sprouting seedlings}}{\text{Total number of planted buds}} \times 100$$

$$\% \text{ Buds that establish root} = \frac{\text{Number of buds that establish root}}{\text{Total number of planted buds}} \times 100$$

$$\% \text{ Dead buds} = \frac{\text{Number dead buds}}{\text{Total number of planted buds}} \times 100$$

The effect of time of first irrigation on establishing sugarcane varieties was analyzed using the appropriate analytical software (SAS 9.2). Mean separation was conducted using Least Significant Difference (LSD) at 5% probability level whenever significant differences were detected in the F-test. All cultural practices were executed based on the current practices of Tendaho Sugar Project except first irrigation.

Result and Discussions

To avoid the confusion in terms of soil fertility status of the area, soil analysis was done by taking composite samples from the study site. The soil analysis result indicated that although the soil of the farm has high pH and low ECe (indicating sodic soils) with low fertility status, it has similar nature with other fields of the area which have good crop stand (Table 1). The soil did not show any peculiar characteristic that hinder germination of sugarcane. For germination, nutrient availability of the seed cane within itself is the limiting factor than that of the soil [10] Sugarcane does not require any specific type of soil as it can be successfully raised on diverse soil types ranging from sandy soils to clay loams & heavy clays [7].

Analysis of variance (ANOVA) revealed that there was a significant interaction effect between the sugarcane varieties and first irrigation period (FIP) (Varieties*FIP=p<0.0001) on sprouting rate, number of tillers, root establishment rate and number of dead sett buds in both sugarcane varieties. N-14 gave highest sprouting rate (90.88 ± 9.73[a]), Number of tillers/ha (756.33 ± 120.10[a]), and root establishment rate (93.73 ± 6.79[a]) under 5 days pre-planting irrigation applications. The maximum sprouting rate, number of tillers, root establishment rate was also recorded with the application of 5 days pre-planting irrigation in all varieties while the lowest sprouting rate, number of tillers and root establishment rate was recorded on the 8 days after planting first irrigation application in all varieties (Table 2). On the other hand, the highest percentage of dead sett buds were recorded on the 8 days after planting first irrigation application in all sugarcane varieties tested in the area (Table 2). The result of the study indicated that, 91-93% of the sugarcane sett buds were not sprouting during the delayed of first irrigation for 8 days after planting and only 7-10% were sprouting in extremely scattered manner in all sugarcane varieties (Table 2). Of the total un-sprouting sett buds, almost all of the sett buds did not develop root system and were highly dehydrated (shrinked). This showed that, out of the various factors that influence sprouting of sugarcane sett buds under field conditions, water content of the soil is very important for sugarcane establishment. Therefore, maintaining optimum moisture during the crop establishment period may be useful for obtaining optimum cane yield in drought areas.

Drought is one of the most important environmental stress factors limiting sugarcane establishment as well as its production worldwide [13]. Water-deficit stress at early stage alters a variety of growth and physiological processes in sugarcane, which cause decreased yields [16,17]. Water deficit during establishment can trigger a negative impact upon growth and development of the crop, compromising

plant productivity especially if the drought duration exceeds the capacity of drought tolerance of the plant species [14,18]. Moderate water deficit causes significant morphological and physiological changes in sugarcane establishment [19], while severe deficit may lead to plant death [20]. Increasing levels of water stress that occur as a soil dries out, affect processes in the sugarcane crop at different stages [21]. The germination-ability decreased as the soil moisture was reduced, although a dependence of the response to cultivars was ranged from 10 to 59% [22]. Superior germination of cane irrigated at planting and loss of germination with delayed irrigation has been reported in Hawaii [23]. Well watered setts germinated at about twice the rate of un-watered setts [24]. The need for moisture to trigger the shift of the bud from dormancy to activity could explain why setts irrigated at planting to have the highest germination percentages [23]. Delaying first irrigation after planting for more than three days will result in poor germination and unsatisfactory crop stands [25] (Tables 1 and 2).

Conclusion

Soil moisture management was one of the major and critical factors

Depth	PH	EC (1:5)	TN	Av. P	Av. K
(Cm)	(1:2.5)	(ds/m)	%	ppm	ppm
0-30	9.03	0.57	0.036	2.26	248
30-60	8.97	0.63	0.035	2.22	228
0-30	9.05	0.71	0.034	1.72	212
30-60	8.84	0.93	0.025	1.28	192
0-30	8.95	0.92	0.027	0.90	238
30-60	8.86	1.12	0.025	0.62	223
0-30	8.16	1.46	0.052	2.26	238
30-60	8.68	0.97	0.025	2.10	172

Table 1: Soil chemical properties of the field at Tendaho: EC, electric conductivity; TN, total nitrogen; AV.P, Average phosphorus; AV.P, average potassium

Varieties	FIP	Sprouting Rate	Number of tillers/ha	% Setts that establish root	% Dead setts
B52-298	5 DPI	82.46 ± 16.05 [ab]	600.00 ± 78.58 [bcd]	85.33 ± 13.47 [ab]	13.63 ± 16.05 [de]
	1 DAP	74.31 ± 16.64 [ab]	517.67 ± 97.14 [cde]	78.93 ± 15.15 [ab]	21.79 ± 16.64 [de]
	4 DAP	10.33 ± 5.48 [cd]	88.00 ± 48.87 [g]	15.80 ± 5.46 [d]	85.77 ± 5.48 [ab]
	8 DAP	3.04 ± 1.59 [d]	71.33± 31.50 [g]	8.53 ± 1.58 [d]	93.06 ± 1.59 [a]
CO-680	5 DPI	83.25 ± 12.10 [ab]	463.00 ± 140.42 [e]	88.73 ± 12.08 [ab]	12.85 ± 12.10 [de]
	1 DAP	66.06 ± 5.08 [b]	397.33± 16.50 [e]	71.53 ± 5.04 [b]	30.03 ± 5.08 [d]
	4 DAP	16.32 ± 13.58 [cd]	166.33 ± 86.56 [fg]	21.77 ± 13.57 [cd]	79.77 ± 13.59 [abc]
	8 DAP	2.25 ± 0.54 [d]	47.00 ± 7.94 [g]	7.73 ± 0.51 [d]	93.84 ± 0.54 [a]
NCO-334	5 DPI	83.42 ± 12.34 [ab]	625.00 ± 31.10 [bc]	86.30 ± 7.81 [ab]	13.19 ± 11.44 [de]
	1 DAP	79.51 ± 4.05 [ab]	489.33 ± 20.98 [de]	84.97 ± 4.07 [ab]	16.59 ± 4.05 [de]
	4 DAP	28.47 ± 2.42 [c]	235.00 ± 89.65 [f]	33.93 ± 2.41 [c]	67.62 ± 2.42 [cb]
	8 DAP	5.04 ± 1.50 [d]	122.00 ± 17.44 [fg]	10.47 ± 1.50 [d]	91.06 ± 1.51 [a]
N-14	5 DPI	90.88 ± 9.73 [a]	756.33± 120.10 [a]	93.73 ± 6.79 [a]	6.43 ± 8.18 [e]
	1 DAP	83.68 ± 13.46 [ab]	706.00 ± 92.63 [ab]	86.57 ± 8.96 [ab]	13.45 ± 11.66 [de]
	4 DAP	29.08 ± 25.48 [c]	234.67 ± 29.31 [f]	34.57 ± 25.47 [c]	67.02 ± 25.48 [c]
	8 DAP	4.34 ± 2.10 [d]	91.00 ± 36.86 [g]	9.80 ± 2.08 [d]	91.75 ± 2.11 [a]
LSD		19.09	121.10	17.26	18.69
CV%		24.68	20.71	20.22	22.47

Table 2: FIP, first irrigation period; DPI, days pre-planting irrigation; DAP, days after planting *Values for sprouting rate, Number of tillers and percentage of root establishment and percentage of dead seed setts given as mean ± SD. *Numbers with in the same column with different letter(s) are significantly different from each other at p ≤ 0.05 according to LSD.

playing great role during sugarcane establishment in the drought areas. Delay in first irrigation period significantly reduced the germination rate, tillering capacity and root establishment of all varieties forcing to re-work of land preparation and planting activities. Therefore, applying first irrigation, in early times (5 pre-planting irrigation or 1 days after planting) was highly recommended with better germination rate, tillering capacity and root establishment of all varieties expecting high cane yield and sugar yield.

Acknowledgements

We would like to thank Ethiopian Sugar Corporation for financing the research and Tendaho Sugar Factory Project for their unreserved material and other facilities support.

References

1. WWDSE (2005) Draft final report of soil survey and evaluation studies: Tendaho Dam and Sugar Project, feasibility and detail design report.

2. Inman-Bamber NG, Smith DM (2005) Water relations in sugarcane and response to water deficits. FIELD CROP RES, 92: 185-202.

3. Blackwell W, Clowes M (1998) Zimbabwe Sugar Cane Production Manual, Experiment Station, Zimbabwe. pp 1-80.

4. Wiedenfeld B, Enciso J (2008) Sugarcane Responses to Irrigation and Nitrogen in Semiarid South Texas Agronomy Journal 100: 665-671.

5. Sonia MZ, Fabiana AR, Jose PG, Livia MP, Mirian VL (2012) Sugarcane Responses at Water Deficit Conditions, Water Stress, Brazil. 256-275.

6. Ellis RD, Lankford BA (1990) The Tolerance of sugarcane to water stress during its main development phases. Agricultural Water Management, Amsterdam. 17: 117-128.

7. Inman-Bamber NG (1995) Climate and water as constraints to production in the South African Sugar Industry. South African Sugar Association Experiment Station, South Africa. 12:18-34.

8. Seiichi M, Uddin SMM, Nose A, Kawamitsu Y (1990) Effect of agronomical practices on sugarcane yield. Department of Agronomy, College of Agriculture. Ryukyus University 37: 52-98.

9. Hayamichi Y (1988) Studies on the germination of sugarcane seed pieces. Part 1: On the characteristics of the germination of buds and seed pieces from different nodes of sugarcane stalks. J Agric Sci, Cambridge 33: 139-148.

10. Tarimo AJ, Takamura YT (1998) Sugarcane production, processing and marketing in Tanzania. Center for African Area Studies, Kyoto University. African Study Monographs 19: 1-11.

11. Divino RM, Victor JMC (1997) Effect of soil moisture content and the irrigation frequency on the sugarcane germination. pp 1-10.

12. Ramesh P (2000) Effect of different levels of drought during the formative phase on growth parameters and its relationship with dry matter accumulation in sugarcane. Sugarcane Breeding Institute, Coimbatore, India. J. Agron. Crop Sci., 185: 83-89.

13. Duli Z, Barry G, Jack CC (2010) Sugarcane response to water-deficit stress during early growth on organic and sand soils. AJABS 5: 403-414.

14. Smit MA, Singels S (2006) Response of sugarcane canopy development to water stresses. Field Crops Research, 98: 91-97.

15. Chinheya CC, Mutambara-Mabveni ARS, Chinwada P (2009) Assessment of damage due to Eldana saccharina Walker (Lepidoptera: Pyralidae) in sugarcane. Proceedings of the South African Sugar Technologists Association 82:446-456.

16. Zhang MQ, Li GJ, Chen RK (2001) Photosynthesis characteristics in eleven cultivars of sugarcane and their responses to water stress during the elongation stage. Proc. ISSCT., 24: 642-643.

17. Silva MDA, Jifon JL, Silva JAG, Sharma V (2007) Use of physiological parameters as fast tools to screen for drought tolerance in sugarcane. Braz. J. Plant Physiol., 19: 193-201.

18. Inman-Bamber NG (2004) Sugarcane water stress criteria for irrigation and drying off. Field Crops Res 89:107–122.

19. Creelman RA, Mason HS, Bensen RJ, Boyer JS, Mullet JE (1990) Water Deficit and Abscisic Acid Cause Differential Inhibition of Shoot versus Root Growth in Soybean Seedlings : Analysis of Growth, Sugar Accumulation, and Gene Expression. Plant Physiol 92: 205-214.

20. Cheng Y, Weng J, Joshi CP, Nguyen HT (1993) Dehydration stress-induced changes in translatable RNAs in sorghum. Jpn J Crop Sci. 33: 1397-1400.

21. Singels A, Kennedy AJ, Bezuidenhout CN (2000) Effect of water stress on sugarcane biomass accumulation and partitioning. SASTA 74: 169-172.

22. Yang SJ, Chen J (1980) Germination response of sugarcane cultivars to soil moisture and temperature. Manila ISSCT, 1:30-37.

23. Humbert RP (1968) The Growing of sugarcane. Amsterdam, Elsevier Publishing Company. pp 1-20.

24. Choudhry Jk (1960) Effect of irrigation with Ammonium Sulphate on the growth, yield and quality of sugar cane (Co 453). Indian Agriculturist 4: 33-43.

25. Abayomi YA, Etejere EO, Fadayomi O (1990) Effect of stalk section, coverage depth and date of first irrigation on seed cane germination of two commercial sugarcane cultivars in Nigeria. Turrialba. 40: 1.

6

Characterization of Soil Management Groups of Metahara Sugar Estate in Terms of their Physical and Hydraulic Properties

Zeleke Teshome[1]* and Kibebew Kibret[2]

[1]Sugar Corporation, Research and Training, P. O. Box 15, Wonji, Ethiopia
[2]Haramaya University, P.O.Box 138, Dire Dawa, Ethiopia

Abstract

A study was conducted on soil management groups of Metahara Sugar estate in order to characterize them in terms of their physical and hydraulic properties, and develop pedotransfer functions for estimating water contents at field capacity (FC) and permanent wilting point (PWP). Soils of Metahara were classified in to six textural soil management groups (soil classes) on the basis of soil moisture content at pF2 and texture to determine irrigation intervals. These are class 1, 2, 3, 4, 5, and 6 with pF2 moisture contents of <35, 35-45, 45-55, 55-65, 65-75, and >75%, respectively. pF2 is the water content at -10 kPa matric potentials. Ninety eight disturbed and undisturbed samples were taken from surface and subsurface layers. The soil analyses result indicated that mean values of the estate soils varied from class to class and with depth in which bulk density varied from 1.01 to 1.43 g/cm^3, particle density from 2.23 to 2.76 g/cm^3, total porosity from 40.91 to 61.42%, sand content from 10 to 40%, silt content from 13 to 36%, clay content from 33 to 77%, and organic matter content from 1.18 to 2.69%. The available water holding capacity varied from 99.71 to 212.01 mm/m. The mean saturated hydraulic conductivity varied from 0.96 to 5.95 µm/s while the basic infiltration rate varied from 0.43 to 3.68 cm/hr. The soil water retention characteristic curves (SWRCC) indicate the presence of three distinct groups of soils in the Estate instead of six groups. Water retention at any of the matric potential points considered increased from group 1 (classes 1 and 2) to group 3 (classes 5 and 6). Furthermore, the equation developed using clay content and bulk density as predictor variables was found to be the best equation for predicting gravimetric water content at field capacity and permanent wilting point with reasonable accuracy. Based on the results, the existing irrigation scheduling should be revised for the respective three soil groups.

Keywords: Matric potential; Soil-water retention characteristic curve; Pedotransfer functions

Introduction

Irrigation is the major practice for successful cultivation of sugarcane in all Ethiopian Sugar estates. Rational management of irrigation water is an important aspect not only for successful cane production, but also to ensure a sustainable high sugar yield [1]. For such a reason, the available water should be so planned that the water requirement of the crop is met and at the same time the system does not produce deleterious effects like water logging and salinity [1,2]. Thus, knowledge of soil physical and hydraulic properties is indispensable for solving such soil and water management problems [3].

In Metahara Sugar estate, attempts have been made to improve the irrigation system to minimize water-logging and water deficit in the farm as one of the measures to increase cane productivity. However, old soil classification system has been used by the estate, established based on soil moisture content at pF2 and texture, to group the soils into six classes for the determination of irrigation intervals. Currently, working theoretical irrigation intervals, which were revised about two decades ago, are wider than the ones established at the start of plantation development in late 1960's. In addition, in the production of sugarcane, experiencing different cultural practices and tillage operations for many years, occurrence of such changes is expected [4]. Cultivation practices can alter soil structure and porosity, as well as hydraulic conductivity and moisture retention curves significantly [5].

The characterization of the estate soils in terms of their physical and hydraulic properties is, therefore, expected to provide basic information on the drainage characteristics, water retention and transmission characteristics, and available water holding capacity that are required for various water management activities. Nonetheless, the direct measurement of hydraulic properties is expensive and time consuming, and hence, indirect methods (Pedotransfer functions) are increasingly used to predict hydraulic properties from easily measurable soil properties namely soil texture, organic matter and bulk density. Therefore, this study was initiated with the following specific objectives:

- To characterize soil management groups of Metahara Sugar estate in terms of their physical and hydraulic properties,
- To develop pedotransfer functions for selected hydraulic properties.

Materials and Methods

Metahara Sugar estate is located in Oromia region at about 200 km southeast of the capital city, Addis Ababa. It is situated at 80 53' N and 39⁰ 52' E with an altitude of 950 meters above sea level (m.a.s.l). The area has a semi arid climatic condition [6]. Six textural soil management groups or "soil classes" have been identified on the basis of soil moisture content at pF2 and texture to determine irrigation intervals. These are class 1, 2, 3, 4, 5, and 6 with pF2 moisture contents of <35, 35-45, 45-55, 55-65, 65- 75, and >75%, respectively [1].

*Corresponding author: Zeleke Teshome, Sugar Corporation, Research and Training, P. O. Box 15, Wonji, Ethiopia
E-mail: zeleketeshome@gmail.com

A total of 98 undisturbed core samples, 98 disturbed samples, and 13 composite samples were collected for determinations of soil-water retention characteristic curve and bulk density, particle size distribution and organic matter, and particle density, respectively. The samples were taken from surface and subsurface horizons in duplicate at each sampling site. The surface layer was considered up to the end of the top soils (0-35 cm) and the underlying layer (35 - 80 cm in most cases) as subsurface layer. Soil-water retention characteristic curve (SWRCC) data were obtained using sand box and pressure plate apparatus. Saturated hydraulic conductivity (Ksat) was measured in situ in the surface layer using Guelph permeameter (Model 2800 KI) while basic infiltration rate using double ring infiltrometer. Available water capacity (AWC) between field capacity and permanent wilting point (PWP) were calculated using Equation 1.

$$AWC = \theta_{FC} - \theta_{PWP} \qquad (1)$$

where: $\theta_{FC=}$ water content at FC (v/v) and $\theta_{PWP=}$ water content at PWP(v/v)

For a given soil depth (D, in mm), the available water was calculated as:

$$AW = AWC \times D \qquad (2)$$

The soil physical properties that were measured include particle size distribution, bulk density, and particle density. Particle size distribution was determined using the Bouyoucos hydrometer method [7], bulk density (ρ_b) was measured by the core method, and particle density (ρ_s) was measured by the pycnometer method. Porosity was also estimated from bulk density and particle density data. Organic carbon was determined by the Walkley-Black method (19). Point pedotransfer functions were developed to estimate gravimetric water contents at field capacity and permanent wilting point from salient soil properties. Regression equations, using the predictor variables and their combinations, were used to develop the best equation for predicting water contents at the two potentials with reasonable accuracy for Metahara Sugar estate soils. The best equation was then tested and validated using an independent data set. Fifty two paired measurements of independent and dependent variables were used for developing the equations for both FC and PWP. Multiple regression techniques were used to work out the coefficients in the equations and evaluate the relative importance of the soil properties on water content at FC and PWP. To see the relations among soil properties, correlation analysis was made using SPSS.

Results and Discussion

Soil physical properties and organic matter

The mean bulk density values, for the different soil management classes, varied from 1.01 (class 4) to 1.43 g/cm³ (class 1) and 1.06 (class 6) to 1.43 g/cm³ (class 1) for surface and subsurface layers, respectively. In general, though not consistent, bulk density increased with depth for most of the management classes. In addition to the anticipated variations in bulk density values among the different management classes, there were some differences among the chosen sampling sites within the same class. For example, the bulk density values were slightly higher than previous study. Bulk densities were 1.1 g/cm³ for classes 1 to 4 and 1.0 g/cm³ for classes 5 and 6 [8]. This variation could be attributed to the differences in management practices that have been in operation over the years and spatial variability of the sampling sites. Soil compaction, caused by wheel traffic, is generally perceived as a problem in sugarcane cultivation [9].

The particle density of soils in the different management classes varied from 2.54 (class 4) to 2.76 g/cm³ (class 2) for the surface layers and 2.23 (Class 6) to 2.64 g/cm³ (class 3) for the subsurface layers, respectively, as presented in Table 1. The particle density values did not show any consistent trend with class. Nonetheless, with the exception of class 4, it decreased with depth. Different literature sources [10,11] indicated that particle density depends on mineralogical composition, crystal structure of the mineral particles and organic matter content. However, the results of particle density did not show any consistent trend with organic matter.

In contrast to bulk density, except for class 5, the total porosity for the surface layers increased from class 1 (46.62%) to class 6 (61.42%) whereas no specific trend with class was found for the subsurface layers. Generally, the total porosity decreased with an increase in depth. The decreasing in total porosity is apparently due to increasing bulk density with depth. Except for extreme layer of class 1 (40.91%) which had a slightly lower porosity than it should normally deserve for clay loam and clay texture, the total porosity values of the different classes were in the range of values that do not affect soil properties and, hence, root growth. The range of porosity to affect soil properties and root growth depends on texture. For instance, sands with a total pore space less than about 40% are liable to restrict root growth due to excessive strength whilst, in clay soils, limiting total porosities are higher, and less than 50% can be taken as the corresponding approximate value [12] (Table 1).

The particle size distribution showed marked differences between lower and higher classes. For example, the mean sand content varied from 14% (class 6) - 40% (class 1) and 10% (class 6) to 39% (class 1) for surface and subsurface layer, respectively. The mean silt content of the surface layer varied from 15% (class 6) to 26% (class 2 soils). In the subsurface layer, it varied from 13% to 36%. The lowest (33%) and highest (77%) mean clay contents were found in the subsurface layers of classes 2 and 6 soils, respectively. The mean clay content of the surface layers increased consistently from class 1 through 6. The texture, four out of six soil management classes, both at the surface and subsurface, was clay. For classes 1 and 2, it varied from silt loam, sandy clay loam, silt clay loam, and clay loam to clay.

The average organic matter content of the surface soils of the management groups (classes) ranged from 1.82 in class 1 to 2.69% in class 4 soils (Table 1). It varied from 1.18 (class 2) to 1.47% (class 3) in the subsurface layer. Except for classes 5 and 6, the average organic matter content for the surface layers increased with class number. Soil

Class	Depth (cm)	Particle size distribution (%)			ρ_b (g/cm³)	ρ_s (g/cm³)	Porosity (%)	OM (%)
		Sand	Silt	Clay				
1	0-35	40	23	37	1.43	2.66	46.62	1.82
	35-80	39	22	39	1.28	2.46	47.97	1.42
2	0-35	36	26	38	1.15	2.76	58.33	2.08
	35-80	31	36	33	1.21	2.57	52.92	1.18
3	0-35	25	19	56	1.08	2.70	60.00	2.39
	35-80	26	15	59	1.32	2.64	50.00	1.47
4	0-35	17	20	63	1.01	2.54	60.24	2.69
	35-80	19	18	63	1.09	2.60	58.08	1.19
5	0-35	17	16	67	1.04	2.60	60.00	2.67
	35-80	16	14	70	1.15	2.33	50.64	1.26
6	0-35	14	15	71	1.03	2.67	61.42	2.56
	35-80	10	13	77	1.06	2.23	52.47	1.26

Table 1: Range of selected physical and chemical properties of surface and subsurface layers of the soil management classes of Metahara Sugar estate.

organic matter tends to increase as the clay content increases. The organic matter content showed a decreasing pattern with depth [13]. The result revealed that all the soil management classes had very low organic matter content.

The results of the correlation analysis (Table 2) indicated bulk density had significant negative correlation with water content at field capacity (r=-0.439**) and permanent wilting point (r=-0.351*). This negative correlation implies that water retention at these two points increases as bulk density decreases and vice versa. As it can be seen from the correlation matrix, the effect of bulk density on water content at permanent wilting point was less strong. This is because at this point, it is the surface property (specific surface) which is more important than pore size distribution. Similarly, sand and silt contents showed significantly negative correlation with water content at FC and PWP whereas clay and silt plus clay (Si+Clay) contents revealed significantly positive correlation with water content at FC and PWP. Organic matter content, on the other hand, showed significantly positive correlation with water content at FC only. Soil organic matter enhances soil water retention because of its hydrophilic nature and its positive influence on soil structure [14].

Hydraulic properties

The soil water retention characteristic curves for the six management classes, mean of the mean surface and subsurface volumetric water contents, were plotted against the specific matric potential values as indicated in the Figure 1. The differences were more distinct in the wet range of the curve, reflecting the differences in structural conditions of the classes, than in the dry range. These groups of water retention curves indicate that those classes that had about the same water retention characteristic curve near the wet range and dry range also have the same drainage requirement and irrigation water management scenarios. Hence, the six soil management groups can be reduced to three groups.

As the soils of the estate are dominantly clay in texture, the water release characteristics of the soils showed slight changes in water content for successive applications of matric suctions. In a clayey soil, the pore-size distribution is more uniform, and more of the water is adsorbed, so that increasing the suctions causes a more gradual decrease in water content [15,16]. More generally, the slow and gradual release behavior of clay could be an asset for the estate because it reduces irrigation

	OM	ρ_b	Sand	Silt	Clay	Si+Clay	FC	PWP	AWC
OM	1								
ρ_b	-0.408**	1							
Sand	-0.208*	0.647**	1						
Silt	-0.186	-0.031	0.193	1					
Clay	0.279**	-0.437**	-0.813**	-0.714**	1				
Si+Clay	0.233*	-0.637**	-0.983**	-0.183	0.818**	1			
FC	0.211*	-0.439**	-0.763**	-0.447**	0.800**	0.756**	1		
PWP	0.201	-0.351*	-0.621**	-0.594**	0.774**	0.620**	0.770*	1	
AWC	0.101	-0.275*	-0.633**	-0.178	0.550**	0.636**	0.800*	0.234	1

* = significant at p < 0.05; ** = significant at p< 0.01; and FC, PWP, and AWC are volumetric water content

Table 2: The Pearson's correlation matrix among measured properties of the six soil management classes.

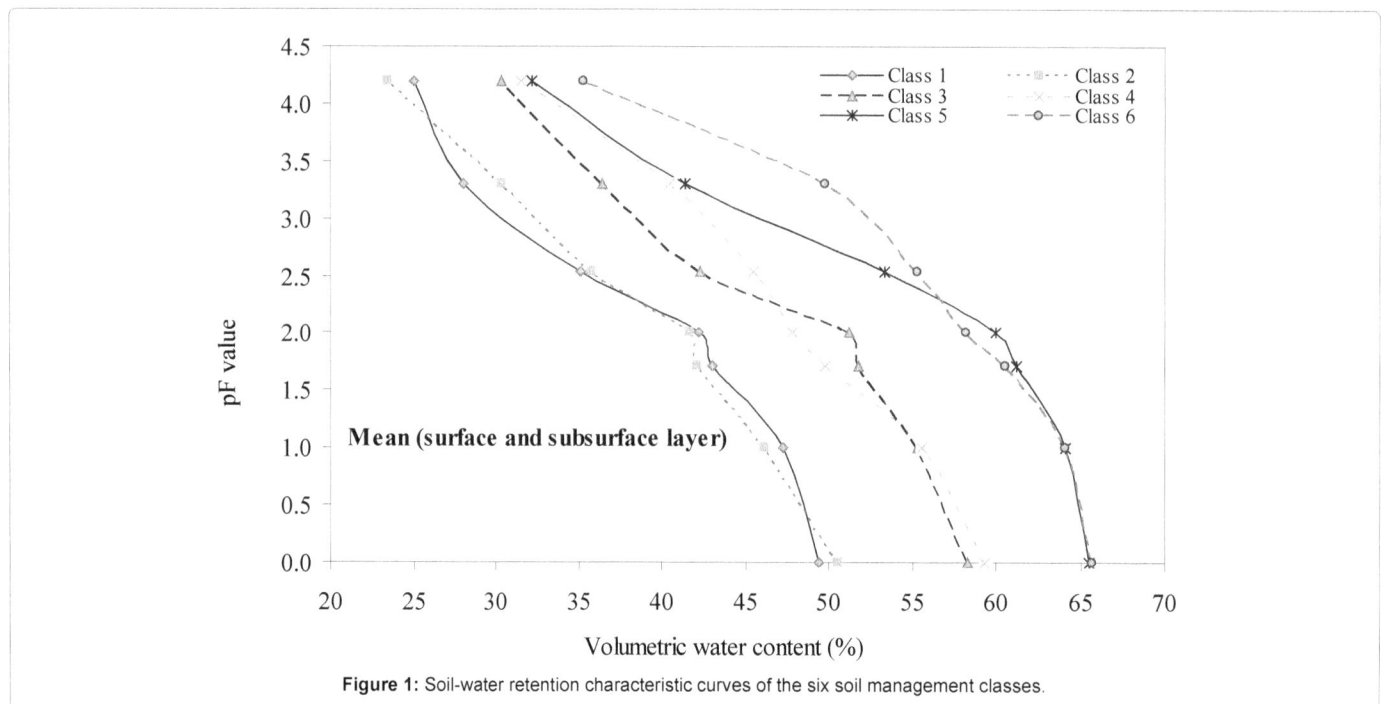

Figure 1: Soil-water retention characteristic curves of the six soil management classes.

frequencies by increasing the irrigation intervals that has direct cost implications. Similarly, abundance of clayey soils in the plantation is an advantage to ensure higher productivity provided that due effort is made to properly manage the accompanying limitations like water logging and difficulty in tillage operations [1].

The available water holding capacity, between FC and PWP, of the surface layers increased consistently with class except for class 3 soils ranging from 109.79 mm/m for class 1 soils to 216.88 mm/m for class 6 soils (Table 3). This implies that the potentially plant available water increased as the texture of the soils become finer. On the other hand, the AW of the subsurface layers varied inconsistently with class ranging from 91.86 mm/m for class 1 to 211.70 mm/m for class 5 soils. Surprisingly enough, for the subsurface layers it was not the class with the highest clay content that had the highest total available water. This could indicate the existence of threshold value for clay content beyond which available water does not increase with further increase in clay content.

Depth of the soil is a key parameter affecting its available water. To show the differences in available water for the six classes, the available water capacity per the entire sampling depth for each class were considered. This was done by calculating the AWC for the upper 80 cm of each class and then converting the values into mm/m. This AW increased inconsistently with class ranging from 99.71 mm/m for class 1 to 212.01 mm/m for class 5 soils. According to established ratings AW for irrigation suitability [16], class 1 was grouped in to low, classes 2, 3 and 4 to medium, and classes 5 and 6 to high.

The saturated hydraulic conductivity varied from 0.96 μm/s for class 6 soils to 5.95 μm/s for class 1 soils. Also, except for class 3 soils, saturated hydraulic conductivity decreased as class number increases. The general decrease in saturated hydraulic conductivity with class could be the result of decrease in the proportion of macro pores in the higher classes. As a result, when saturated the coarse textured soils have high conductivity than the finer textured soils. In few areas of the Sugar estate, soils where class 5 and 6 are found, shallow water table accompanied with low Ksat, are already giving some indications of drainage needs. On the basis of saturated hydraulic conductivity ratings established [17], the six classes can be categorized in to slow (class 6), moderately slow (classes 2, 4 and 5) and moderate (classes 1 and 3).

The basic infiltration rate, alike the saturated hydraulic conductivity values, showed a decreasing trend with class except for class 3 soils (Table 4). Class 1 had the highest basic infiltration rate (3.68 cm/hr) followed by class 3 soils (2.73 cm/hr) whereas class 6 soils had the lowest basic infiltration rate (0.43 cm/hr). On the basis of basic infiltration, the six soil classes can be grouped in to three rating classes of slow (class 6), moderately slow (classes 2, 4 and 5) and moderate (classes 1 and 3). Some of the classes showed greater infiltration rate regardless of their texture. Report indicated that infiltration rate varies greatly with soil structure and stability, even beyond the normal ranges [18].

Point pedotransfer functions

Of the different equations developed, which are shown in Table 5, the use of clay content and bulk density as predictor variables produced statistically significant (P<0.01) coefficients. However, the use of all or more than two variables produced better R^2 values indicating that the use of more number of soil properties can explain large proportion of the variations in soil water content at these two matric potential points. From the R^2 values in Table 5, about 86% of the variations in soil water content at FC were explained by the variations in soil texture, bulk density and organic matter content. On the other hand, about 85% and

81% of the variations in water content at FC and PWP, respectively, were explained by the variations in clay content and bulk density alone. From this, the improvement in precision by using all the variables was only 5% for FC. These equations show the strong impact of bulk density on water retention at FC and PWP.

From the different regression equations developed for predicting water content at FC and PWP, the following pedotransfer functions were established:

$$\theta_{FC} = 58.849 + 0.408(\%clay) - 36.198 \times \rho_b \quad R^2=0.849 \quad (3)$$

$$\theta_{PWP} = 38.431 + 0.244(\%clay) - 22.433 \times \rho_b, \quad R^2=0.811 \quad (4)$$

Soil Class	Depth (cm)	Percent volumetric water retained at matric potential (kPa)							AW (mm/m)
		0	-1	-5	-10	-33	-200	-1500	
1	0-35	47.25	45.77	42.21	41.45	37.17	28.49	26.19	109.79
	35-80	51.63	48.78	43.94	43.02	33.10	27.56	23.91	91.86
2	0-35	49.90	44.25	39.84	39.26	34.65	30.50	23.19	114.56
	35-80	51.27	48.12	44.48	44.04	36.89	30.27	23.66	132.34
3	0-35	62.32	58.30	52.66	51.86	41.85	34.87	30.46	113.90
	35-80	54.32	52.40	50.87	50.59	42.90	38.05	30.38	125.18
4	0-35	59.44	55.94	48.86	46.64	44.72	39.41	31.14	135.87
	35-80	59.25	55.31	50.73	48.98	46.40	41.59	32.06	143.39
5	0-35	67.24	65.46	61.76	60.18	53.85	42.97	32.61	212.40
	35-80	63.88	62.73	60.59	59.74	52.88	39.81	31.73	211.70
6	0-35	67.31	66.19	61.25	58.15	55.45	49.93	33.76	216.88
	35-80	64.05	62.02	59.83	58.30	55.24	49.72	36.83	184.10

Table 3: Average percent volumetric water contents at different matric potential points and available water (mm/m) of the six soil management classes.

pF2 Soil class	Infiltration rate (cm/hr)	Rating class	Ksat (μm/s)	Rating class
1	3.68	Moderate	5.95	Moderate
2	1.87	Moderately slow	5.14	Moderately slow
3	2.73	Moderate	5.73	Moderate
4	1.96	Moderately slow	3.94	Moderately slow
5	1.38	Moderately slow	1.47	Moderately slow
6	0.43	Slow	0.96	Slow

Table 4: Basic infiltration rate and saturated hydraulic conductivity of the six soil management classes.

Predicted variables	Intercept	Predictor variables						R²
		Sand	Silt	Clay	Si+clay	OM	ρ_b	
FC	74.447	-0.347	-0.217	0.079	0.092	0.782	-30.096	0.86071
	76.398	-0.367	-0.297	-	0.151	0.797	-30.08	0.86069
	74.377	-0.346	-0.125	0.172	-	0.766	-30.106	0.86068
	62.301	-0.226	-	0.294	-	0.746	-30.134	0.86056
	64.451	-0.203	-	0.307	-	-	-31.97	0.85854
	58.849	-	-	0.408	-	-	-36.198	0.84937
	86.412	-0.596	-	-	-	-	-27.943	0.80476
	60.043	-0.849	-	-	-	-	-	0.69466
PWP	75.000	-0.310	-0.294	-	-0.113	0.247	-23.637	0.81902
	63.771	-0.199	-0.293	-	-	0.244	-23.683	0.81867
	75.956	-0.305	-0.297	-	-0.111	-	24.241	0.81845
	68.339	-	-	0.296	-0.332	-	-24.155	0.81805
	45.615	-	-	0.296	-0.102	-	-24.434	0.81652
	38.431	-	-	0.244	-	-	-22.432	0.81087
	6.9489	-	-	0.343	-	-	-	0.57281

Table 5: Pedotransfer functions developed for predicting water content at field capacity and permanent wilting point from basic soil properties for the six soil management classes.

Where: $\theta_{FC}=$ water content at field capacity (% by weight), $\theta_{PWP 4}=$ water content at permanent wilting point (% by weight), and $\rho_{b=}$ bulk density (g/cm³)

Conclusion

The soil analyses result indicated that the mean values of the estate soils varied from class to class and with depth in which bulk density varied from 1.01 to 1.43 g/cm³, particle density from 2.23 to 2.76 g/cm³, total porosity from 40.91 to 61.42%, sand content from 10 to 40%, silt content from 13 to 36%, clay content from 33 to 77%, and organic matter content from 1.18 to 2.69%.

The pertinent physical and hydraulic properties were determined using standard laboratory and/or field procedures. The similarity of soil water retention characteristic curves between some classes of the six soil management classes indicated that the six management groups can be reduced to three groups (classes). It is, therefore, the existing irrigation scheduling should be revisited taking the present findings of the soil-water relations into account. The developed equation through pedotransfer functions can be used to predict gravimetric water contents at field capacity and permanent wilting point to obtain available water capacity with reasonable accuracy for the estate soils and other similar areas with soil properties in the range where the model equation developed. Periodic revision of the soil classes is important as there will be cumulative effects due to cultural practices, tillage operations and addition of amendments that cause changes in chemical, physical, and hydraulic properties of the estate soils. Additional information including other chemical and physical properties of the estate soils are required to fully understand the potentials and limitations of the soils and use according to their suitability for a given land use.

Acknowledgments

The authors would like to thank the staff of Soil and Water Management Department of the former Ethiopian Sugar Development Agency the then Sugar Corporation, particularly Ato Abiy Fantaye, for their assistance, professional comments and suggestions. Great appreciation also goes to W/ro Sintayehu Temesgen from Haramaya University for her unreserved support during the physical soil laboratory analysis. Our special gratitude is also extended to the Ethiopian Sugar Development Agency (ESDA) for sponsoring the study.

References

1. Agrima Project Engineering and Consultancy Services (APECS) (1987) A Report on the Agricultural Research Services of the Ethiopian Sugar Corporation (ESC). Matha Private, Bombay. 1-414.

2. Swinford JM, Boevey TMC (1984) The effects of soil compaction due to infield transport on ratoon cane yields and soil physical characteristics. Proceedings of South African Sugar Technologists Association, 198-203.

3. CornelisWM, Ronsyn J, Meirvenne MV, Hartmann R (2000) Evaluation of pedotransfer fuctions for predicting the soil moisture characteristic curve. Soil Sceince Society of American Journal.

4. Ambachew D (2005) Revision of sugarcane cropping cycle of Metahara Sugar Factory. Project and Productivity Improvement office. Metahara, Ethiopia.

5. Ndiaye B, Molinat J, Hallaire V, Gascuel C, Hamon Y (2007) Effects of agricultural practices on hydraulic properties and water movement in soils of brittany (France). Soil and Tillage Research. 93: 251-263.

6. Michael M, Seleshi B (2007) Irrigation practices in Ethiopia: Characteristics of selected irrigation schemes. Colombo, Srilanka. Integrated water management institute.

7. Boyoucous GJ (1962) Hydrometer method improved for making particle-size analysis of soils. Agronomy Journal, 54: 463-465.

8. Habib D (2000) Review of irrigation and water management research in Ethiopian Sugar estates. In: Ambachew D and Girma A (eds.), 2000. Review of sugarcane research in Ethiopia: Soils, Irrigation and Mechanization (1964-1998). Ethiopian Sugar Industry Support Center Share Company, Wonji, Ethiopia. 219.

9. Walkley A, Black CA (1934) An examination of the Degtjareff method for determining soil organic matter and a proposed modification of the chromic acid titration method. Soil Sci. 37: 29-38.

10. Brady NC (1990) The Nature and Properties of Soils. 10th ed., Macmillan Publishing Company, USA.

11. Foth HD (1984) Fundamentals of soil science. 7th Ed. John Wiley and Sons, Inc. Canada.

12. Harrod MF (1975) Field experience on light soils, pp: 22-51. In: Soil physical conditions and crop production. MAFF Tech Bulletin 29, HMSO, London.

13. Bot A, Benites J (2005) The importance of soil organic matter: Key to drought-resistant soil and sustained food production. FAO, Rome.

14. Huntington TG (2008) Available water capacity and soil organic matter. Taylor and Francis group, USA.

15. Hillel D (1982) Introduction to soil physics. Academic Press, San Diago, California

16. Landon JR (1984) Booker tropical soil manual: A hand book for soil survey and agricultural land evaluation in the tropics and subtropics. Booker Tate limited, England.

17. Ghildyal BP, Tripathi R (1987) Soil physics. Wiley Eastern Limited, Pantiager, UP, India. 654.

18. Cuenca RH (1989) Irrigation System Design: An engineering approach. Prentice hall, Inc. U.S.A.

7

Effect of Herbicide Application on Weed Management in Green Gram [*Vigna radiata* (L.) Wilczek]

Diwash Tamang*, Rajib Nath and Kajal Sengupta

Department of Agronomy, Bidhan Chandra Krishi Viswavidyalaya, Mohanpur, Nadia-741252, West Bengal, India

Abstract

A field experiment was carried out at Bidhan Chandra Krishi Viswavidyalaya (Nadia, West Bengal) during 2012 and 2013 (during March-May) in upland situation to judge the efficacy of the herbicides against weed flora in green gram crop field and also to find out the effect of herbicides on growth, yield and benefit cost ratio of green gram [*Vigna radiata* (L.) Wilczek] crop. The soil of experimental site was sandy loam in texture having neutral in soil reaction. The experiment was conducted with 14 treatments and laid out in Randomized Block Design with 3 replications. The green gram variety used was IPM-2-3.

It was observed that hand weeding resulted in significantly lower weed density and dry weight and gave better seed yield of green gram. Most of the herbicides were found effective in controlling weeds and maximizing seed yield of green gram. These treatments were at par with hand weeding twice at 20 and 40 DAS. Total weed free treatment showed the best performance in respect of yield and yield attributes of green gram crop and weeds management. The herbicidal treatments Fenoxaprop-p-ethyl@50 g *a.i.* ha^{-1} and @ 100 g a.i. ha^{-1} were found less effective for controlling weeds. Maximum benefit: cost ratio was obtained from Vellore 32(Pendimethalin 30 EC+Imazethapyr 2 EC)@1.00 kg *a.i.* ha^{-1}. Hand weeding treatments, though significantly reduced weed biomass and improved the grain yield, gave less benefit: cost ratio owing to higher cost of farm labour.

Keywords: Herbicides; Hand weeding; Green gram; Weed; Yield

Introduction

The pulses constitute an important group of crops and have been the main stay in Indian Agriculture, as they improve physical condition of soil and provide nutritious food and fodder. India has a distinction of being world's largest producers of pulses. However, India needs to make immediate strides in pulse production programme taking into account the extreme relevance of pulses in our diet. Increasing yield of pulse crops should be the top priority to fill up the existing gap in the requirement and availability of pulses. This will not only ensure food security but will also provide nutritional security, particularly to the large vegetarian population of our country. Among the grain legumes, green gram ranks third after chickpea and pigeon pea among the pulses in respect of production, and it can be grown throughout the year. In India, there is substantial scope of summer green gram after harvesting of winter crops due to its short duration in nature and deep rooted, it can be grown with limited irrigation. However weed infestation is one of the major constraints in green gram cultivation. The loss of yield due to weeds is quite high, ranges from 40-68%. In view of severe infestation of annual and perennial weeds in summer green gram, the potential yield is generally not realized. The available pre and post-emergence herbicide, pendimethalin, oxyfluorfen, fenaxaprop-p-ethyl and quizalofop-ethyl are able to check the emergence and growth of annual grasses and broadleaved weeds. Keeping the above in view and the known possible reasons, the present study was taken up with the following objectives: i) To determine the effect of herbicide on weed population in green gram field; ii) To determine the effect of herbicide application on growth, yield and benefit cost ratio of green gram crop.

Material and Methods

The field experiment, carried out at University farm during 2012 and 2013 in the New alluvial soil (*Inseptisol*) of Nadia (22° 93′ N, 88° 53′ E, 9.75 m above mean sea level), West Bengal, to study the effect of herbicide application on weed management in green gram [*Vigna radiata* (L.) Wilczek] was consisting of 14 treatments with three replications each; conducted in RBD (each plot size was 5 m × 2 m). The data obtained were analyzed statistically by the analysis of variance method [1]. The experimental soil were sandy loam under upland situation with good drainage facility, having soil pH 7.10, organic carbon 0.71 %, total nitrogen 0.08 %, available phosphorus 27.34 kg ha^{-1}, available potassium 132.15 kg ha^{-1} and were estimated by combined glass electrode pH meter method, Walkey and Black's rapid titration method, Modified macro Kjeldahl method, Olsen's method and Flame photometer method respectively [2]. The green gram variety IPM-2-3 was used for the experiment with recommended dose of fertilizer 20-40 kg^{-1} NPK, the entire dose of fertilizer was applied as basal, and then they were thoroughly mixed with the soil). The source of NPK was in the form of urea, single super phosphate and muriate of potash respectively. Population of different categories of weeds at 30 and 40 DAS was recorded by placing the quadrate of 50 cm × 50 cm at pre-determined locations in the sampling area. The values were converted to number/m^2. Grasses, broadleaved and sedges were counted separately and their sum was used to obtain total weed population. The weed samples for dry weight were collected from the area used for weed count on the same dates (30 and 40 DAS). The different weed management treatments being T_1: No weeding (weedy check), T_2: Hand weeding at 20 and 40 DAS, T_3: Total Weed free (Regular weeding as and when required), T_4: Quizalofop-p-ethyl@37.50 g *a.i.* ha^{-1}, T_5: Quizalofop-p-ethyl@75.00 g *a.i.* ha^{-1}, T_6: Pendamethalin@1 kg

*Corresponding author: Diwash Tamang, Department of Agronomy, Bidhan Chandra Krishi Viswavidyalaya, Mohanpur, Nadia, 741252, West Bengal, India E-mail: diwashh.tamang@gmail.com

a.i. ha[-1], T_7: sPendamethalin @2 kg *a.i.* ha[-1], T_8: Imazethapyr@25 g *a.i.* ha[-1], T_9: Imazethapyr@40 g *a.i.* ha[-1], T_{10}: Imazethapyr @55 g *a.i.* ha[-1], T_{11}: Fenoxaprop-p-ethyl@50 g *a.i.* ha[-1], T_{12}: Fenoxaprop-p-ethyl@100 g *a.i.* ha[-1], T_{13}: Vellore 32 (Pendamethalin 30 EC+Imazethapyr 2 EC)@0.75 kg *a.i.* ha[-1], T_{14}: Vellore 32 (Pendamethalin 30 EC+Imazethapyr 2 EC)@1.00 kg *a.i.* ha[-1]. Generally, the time of application of herbicides were Quizalofop-p-ethyl at 16 DAS, Pendamethalin at 2 DAS, Imazethapyr at 16 DAS, Fenoxaprop-p-ethyl at 16 DAS and Vellore 32 at 2 DAS.

Results and Discussion

Weed flora

Eleven weed species were observed in experimental field; among them grasses were four, sedges one and remaining weed flora were from broad leaf category. The predominant weed species were *Digitaria sanguinalis, Cynodon dactylon, Eleusine indica, Echinochloa colona* amaong grasses; *Cyperus rotundus* among the sedges and the broad leaf weeds were *Cleome viscose, Chenopodium album, Euphorbia hirta, Digeria arvensis, Physalis minima* and *Amaranthus viridis*. Similar observation was also reported by Das et al. and Kundu et al. [3,4].

Number of broad leaf weeds/m²

At 30 DAS, broad leaf weed population was highest in weedy check treatment (T_1). The lowest value of broad leaf weed population was recorded in T_3 (total weed free plot). All weed management treatments were significantly better than T_1. At 40 DAS also similar trend was observed (Table 1). All weed management treatments were superior to T_1. The highest suppression of broad leaf weed population was recorded in T_8 (Imazethapyr@ 25 g a.i. ha[-1]), however it was statistically at par with most of the herbicide application treatments [except T_{12} (Fenoxaprop-p-ethyl@ 100 g a.i. ha[-1])]. Most of the herbicides were found effective in controlling broad leaf weed population.

Number of grassy weeds/m²

At 30 DAS, grassy weed population was highest in weedy check treatment (T_1). The lowest number of grassy weed population was

recorded in T_3 (total weed free plot). All weed management treatments were significantly better than T_1. Most of the herbicides were better than the hand weeding treatment (T_2) in managing grassy weeds (Table 1). Similar observation was noticed at 40 DAS. The highest suppression of grassy weed population was recorded in T_3 which was statistically superior to all other treatments. The population of grassy weeds was quite high in weedy check treatment (T_1).The herbicides like Vellore 32 (Pendimethalin 30 EC+Imazethapyr 2 EC), Pendimethalin, Imazethapyr, and Fenoxaprop-p-ethyl were found effective in controlling grassy weed population.

Number of sedge weeds/m²

Sedge weed population was slightly lower than grassy and broad leaf weeds on both the date of observations. At 30 DAS, sedge weed population was maximum in weedy check treatment (T_1). The minimum number of sedge weed population was recorded in T_3 (total weed free plot). All weed management treatments were significantly better than T_1. Most of the herbicides were ineffective than the hand weeding treatment (T_2) in managing sedge weeds (Table 1). More or less similar type of data was recorded at 40 DAS. The highest suppression of sedge weed population was recorded in T_3 which was statistically superior to all other treatments. The population of sedge weeds was quite high in weedy check treatment (T_1). The herbicides like Imazethapyr, Pendimethalin, and Quizalofop-p-ethyl were found effective in controlling sedge weed population. The treatment T_2 (hand weeding at 20 and 40 DAS) was also effective in controlling sedge weed population.

Leaf area index (LAI)

Leaf area index is an important growth factor of a crop. LAI of the crop was determined at 30, 45 and 60 DAS and are presented in Table 2. Among three occasions the more values of LAI were recorded in 45 DAS in comparison with 30 and 60 DAS.

It is evident that the leaf area index at 30 DAS was highest in T_3 (total weed free plot) which was closely followed by the treatments like T_2 (hand weeding at 20 and 40 DAS), T_{14} [Vellore 32 (Pendimethalin

Treatment	No. of broad leaf weeds/m²		No. of grassy weeds/m²		No. of sedge weeds/m²	
	30 DAS	40 DAS	30 DAS	40 DAS	30 DAS	40 DAS
T_1	30.94	40.29	65.40	67.27	22.43	27.93
T_2	15.98	20.89	16.27	17.29	6.44	7.44
T_3	1.48	1.50	1.22	1.50	1.75	2.50
T_4	20.33	21.11	11.88	20.78	12.50	14.49
T_5	16.73	19.49	13.23	17.60	14.00	17.51
T_6	20.94	22.45	14.48	21.44	10.00	15.48
T_7	15.99	20.20	12.61	16.5	10.00	15.32
T_8	13.40	16.93	14.01	18.48	11.50	15.88
T_9	15.45	17.90	12.22	23.32	11.50	12.95
T_{10}	15.88	19.96	13.11	21.01	11.00	15.51
T_{11}	13.04	21.88	15.22	16.04	16.00	16.72
T_{12}	19.34	22.23	13.66	19.51	13.50	16.73
T_{13}	14.85	17.22	13.55	17.44	12.50	19.61
T_{14}	14.44	21.88	13.22	18.44	12.00	17.99
SE m (±)	1.26	1.61	0.72	0.80	1.42	1.55
CD at 5%	3.80	4.85	2.22	2.46	4.29	4.69

Note T_1: No weeding (weedy check); T_2: Hand weeding at 20 & 40 DAS; T_3: Total Weed free (Regular weeding as and when required); T_4:Quizalofop-p-ethyl@37.50 g *a.i.* ha[-1]; T_5:Quizalofop-p-ethyl@75.00 g *a.i.* ha[-1]; T_6: Pendamethalin@1 kg *a.i.* ha[-1]; T_7:Pendamethalin@2 kg *a.i.* ha[-1]; T_8: Imazethapyr @ 25 g *a.i.* ha[-1]; T_9: Imazethapyr@40 g *a.i.* ha[-1]; T_{10}: Imazethapyr@55 g *a.i.* ha[-1]; T_{11}: Fenoxaprop-p-ethyl@50 g *a.i.* ha[-1]; T_{12}: Fenoxaprop-p-ethyl@100 g *a.i.* ha[-1]; T_{13}: Vellore 32 (Pendamethalin 30 EC+Imazethapyr 2 EC)@0.75 kg *a.i.* ha[-1]; T_{14}: Vellore 32 (Pendamethalin 30 EC+Imazethapyr 2 EC)@1.00 kg *a.i.* ha[-1].
*DAS: Days after sowing.

Table 1: Effect of herbicides on weed population of green gram crop. (Pooled data of two years).

Treatment	Leaf area index			Dry mass of aerial plant parts (g/m²)		
	30 DAS	45 DAS	60 DAS	30 DAS	45 DAS	60 DAS
T_1	2.19	2.67	1.56	19.88	73.49	145.40
T_2	3.23	4.15	4.02	26.83	128.51	333.56
T_3	3.32	4.38	4.14	27.95	130.25	360.83
T_4	2.75	3.51	3.28	24.02	105.82	232.50
T_5	2.86	3.54	3.36	24.67	109.84	264.13
T_6	3.02	4.00	3.91	26.45	126.85	306.53
T_7	3.06	4.06	4.03	27.00	127.36	312.25
T_8	2.72	3.49	3.44	24.33	115.66	256.00
T_9	2.97	3.66	3.60	25.21	119.05	300.30
T_{10}	2.91	3.57	3.43	25.09	121.00	285.62
T_{11}	2.65	3.38	3.21	21.86	110.63	216.97
T_{12}	2.70	3.45	3.16	22.04	108.76	225.36
T_{13}	3.08	3.79	3.71	26.05	120.37	300.94
T_{14}	3.14	4.22	4.06	26.13	126.42	316.22
SE m (±)	0.10	0.24	0.16	0.72	3.42	7.16
CD at 5%	0.31	0.74	0.50	2.18	10.27	21.50

Note T_1: No weeding (weedy check); T_2: Hand weeding at 20&40 DAS; T_3: Total Weed free (Regular weeding as and when required); T_4: Quizalofop-p-ethyl@37.50 g *a.i.* ha⁻¹; T_5: Quizalofop-p-ethyl@75.0 g *a.i.*ha⁻¹; T_6: Pendamethalin@1 kg *a.i.* ha⁻¹; T_7: Pendamethalin@2 kg *a.i.* ha⁻¹; T_8: Imazethapyr@25 g *a.i.* ha⁻¹; T_9: Imazethapyr@40 g *a.i.* ha⁻¹; T_{10}: Imazethapyr@55 g *a.i.* ha⁻¹; T_{11}: Fenoxaprop-p-ethyl@50 g *a.i.* ha⁻¹; T_{12}: Fenoxaprop-p-ethyl@100 g *a.i.* ha⁻¹; T_{13}:Vellore 32 (Pendamethalin 30 EC+Imazethapyr 2 EC)@0.75 kg *a.i.* ha⁻; T_{14}: Vellore 32 (Pendamethalin 30 EC+Imazethapyr 2 EC) @1.00 kg *a.i.* ha⁻¹.

Table 2: Effect of herbicides on Leaf area index and Dry mass of aerial plant parts (g/m²) of green gram crop. (Pooled data of two years).

30 EC+Imazethapyr 2 EC)]@1.00 kg a.i. ha⁻¹), T_{13} [Vellore 32 (Pendimethalin 30 EC+Imazethapyr 2 EC)]@0.75 kg a.i. ha⁻¹)), T_7 (Pendimethalin@2 kg a.i. ha⁻¹) and T_6 (Pendimethalin@1 kg a.i. ha⁻¹), there was no significant difference and they were statistically at par . The lowest value of leaf area index was obtained in weedy check (T_1) which was significantly lower than all other treatments. At 45 DAS also, the highest LAI was recorded in T3 (total weed free treatment), it was, however, at par with T_2 (hand weeding at 20 and 40 DAS),T_{14} [Vellore 32 (Pendimethalin 30 EC+Imazethapyr 2 EC)]@1.00 kg a.i. ha⁻¹) T_{13} [Vellore 32 (Pendimethalin 30 EC + Imazethapyr 2 EC)@0.75 kg a.i. ha⁻¹], T_7 (Pendimethalin@2 kg a.i. ha⁻¹), T_6 (Pendimethalin@ 1 kg a.i. ha⁻¹) and T_9 (Imazethapyr@40 g a.i. ha⁻¹). The leaf area index was observed lowest in weedy check which was significantly lower than all other treatments except T_{11} (Fenoxaprop-p-ethyl@50 g a.i. ha⁻¹).

On an average the LAI value was decreased slightly at 60 DAS as the crop reached its maturity stage. At this stage also highest leaf area index was found in T_3 (total weed free treatment), it was, however, at par with T_2 (hand weeding at 20 and 40 DAS), T_{14} [Vellore 32 (Pendimethalin 30 EC+Imazethapyr 2 EC)@1.00 kg a.i. ha⁻¹], T_{13} [Vellore 32 (Pendimethalin 30 EC+Imazethapyr 2 EC)@0.75 kg a.i. ha⁻¹], T7 (Pendimethalin@ 2 kg a.i. ha⁻¹) and T_6 (Pendimethalin@1 kg a.i. ha⁻¹). The lowest value of LAI was recorded in weedy check (T_1) which was significantly lower than all other treatments. The better treatments in this field of discussion were T_3, T_2 and T_{14}. The observations were well consistent with that of Singh et al. Srivastava et al [6,7].

Dry matter accumulation

The real picture of crop growth can be obtained from the data of dry matter accumulation. Dry mass of aerial parts of plants/m² was determined in this experiment determined at 30, 45 and 60 DAS and presented here (Table 2) as dry matter accumulation of green gram crop. Dry biomass of the crop increased gradually with the age of the crop. At 30 DAS, the highest dry matter accumulation was recorded in T_3 (total weed free treatment), however, it was statistically at par with T_2 (hand weeding at 20 and 40 DAS), T_7 (Pendimethalin@ 2 kg a.i. ha⁻¹), T_6 (Pendimethalin@ 1 kg a.i. ha⁻¹), T_{14} [Vellore 32 (Pendimethalin 30 EC+Imazethapyr 2 EC)@ 1.00 kg a.i. ha⁻¹)] and T_{13} [Vellore 32 (Pendimethalin 30 EC+Imazethapyr 2 EC)@0.75 kg a.i. ha⁻¹]. The lowest value (19.88 g/m²) was observed in weedy check (T_1).

Yield components and seed yield

Yield of green gram crop is influenced by a number of yield attributing characters, like number of pods/m², number of seeds/pod and test weight of seeds. Observation on yield attributes or components like numbers of pods/square meter, numbers of seeds/pod, test weight of seed (1000 seed weight) and seed yield were recorded at the time of harvest and have been presented in the Table 3.

Number of pod/m²

In pulse crops number of pods/m² is the most important determinant of grain or seed yield. The number of pods/m2, which normally gives a more reliable or accurate picture and contributing most in determining the yield, is presented here as main yield component (Table 3). The number of pods/m² ranged from 208.0 in T_1 (control treatment) to 443.2 in T_3 (total weed free treatment). In T_3 maximum number of pods/m² was obtained and it was followed by T_2 and T_{14}; treatments T_3 and T_2 were, however, statistically at par with each other. All weed management treatments were significantly better than T_1 (control treatment). Treatments T_{11}, T_{12} and T_4 were statistically at par, but these three treatments produced significantly lower number of pods than other weed management treatments. Among the herbicides Vellore 32 (Pendimethalin 30 EC+Imazethapyr 2 EC) and Pendimethalin produced more number of pods/unit area. The results confirm the findings of Kumar et al. Veeraputhiran and Singh et al [5,7,8].

Number of seeds/pod

Number of seeds/pod is another important yield component of green gram crop. The weed management treatments had no significant effect on number of seeds/pod. The number of seeds/pod varied from 9.60 in T_1 (control treatment) to 10.15 in T_3 treatment (total weed free), however, there was no significant difference and all the treatments were statistically at par (Table 3).

Treatment	No. of pods/m²	No. of seeds /pod	Test weight (g)	Seed yield (Kg ha⁻¹)	B : C ratio
T_1	208.0	9.60	39.89	460	1.24
T_2	409.7	10.10	41.20	970	2.08
T_3	443.2	10.15	39.39	1005	1.83
T_4	281.3	9.80	41.05	758	2.16
T_5	331.6	9.70	40.12	827	2.15
T_6	381.0	9.65	38.95	912	2.53
T_7	377.6	10.00	40.70	910	2.26
T_8	317.4	10.05	39.80	803	2.45
T_9	361.9	9.90	40.02	900	2.69
T_{10}	339.5	9.95	39.40	864	2.55
T_{11}	256.8	9.80	38.16	680	1.99
T_{12}	263.3	9.85	39.60	706	1.93
T_{13}	370.3	10.00	40.80	905	2.75
T_{14}	402.1	10.10	40.45	925	2.75
SE m (±)	12.1	0.19	1.14	30.4	-
CD at 5%	36.4	NS	NS	91.4	-

Note T_1: No weeding (weedy check); T_2: Hand weeding at 20 and 40 DAS; T_3: Total Weed free (Regular weeding as and when required); T_4: Quizalofop-p-ethyl@37.50 g *a.i.* ha⁻¹; T_5: Quizalofop-p-ethyl@75.00 g *a.i.* ha⁻¹; T_6: Pendamethalin@1 kg *a.i.* ha⁻¹; T_7: Pendamethalin@2 kg *a.i.* ha⁻¹; T_8: Imazethapyr@25 g *a.i.* ha⁻¹; T_9: Imazethapyr @40 g *a.i.* ha⁻¹; T_{10}: Imazethapyr @55 g *a.i.* ha⁻¹; T_{11}: Fenoxaprop-p-ethyl@50 g *a.i.* ha⁻¹; T_{12}: Fenoxaprop-p-ethyl @100 g *a.i.* ha⁻¹; T_{13}: Vellore 32 (Pendamethalin 30 EC+Imazethapyr 2 EC)@0.75 kg *a.i.* ha⁻; T_{14}: Vellore 32 (Pendamethalin 30 EC+Imazethapyr 2 EC)@1.00 kg *a.i.* ha⁻¹.

Table 3: Effect of herbicides on Yield components, Seed yield and Benefit - cost ratio (B: C ratio) of green gram crop.

Test weight of seeds

The test weight of seeds (refers to 1000 seed weight) generally does not vary so much. In this experiment also the differences in test weight (Table 3) were at par (statistically non-significant). The test weight of seed was highest in T_2 and it was lowest in T_{11}, however, there was no significant difference.

Seed yield

Seed yield of the crop was distinctly influenced by the weed management treatments. The maximum seed yield was obtained in T_3 (total weed free treatment) followed by T_2 (hand weeding at 20 and 40 DAS) and T_{14} [Vellore 32 (Pendimethalin 30 EC+Imazethapyr 2 EC)@1.00 kg a.i. ha⁻¹)]. The minimum seed yield was obtained in T_1 (control-weedy check). The reduction in yield under the control treatment (i.e. in T_1) may be attributed to reduced growth and number of plants and number of pods per unit area. The average seed yield obtained was significantly more with weed management. Crop performance was not good in the control treatment, thus, the yield per hectare was significantly lower than that obtained in other treatments (Table 3). The reduction in seed yield in weedy treatment as compared to hand weeded and total weed free were 160% and 169%, respectively. Seed yield was reduced by Fenoxaprop-p-ethyl treatments (T_{11} and T_{12}) in comparison with other herbicidal treatments. The results are in conformity with the findings reported by Yadav and Singh, Singh et al, Singh et al, Srivastava, Malik et al, and Veeraputhiran [5,6,8-11].

Benefit-cost ratio

From the data it was clear that the benefit : cost ratio was highest in T_{14} [Vellore 32 (Pendimethalin 30 EC+Imazethapyr 2 EC)@1.00 kg a.i. ha⁻¹)] and that was closely followed by T_{13} [Vellore 32 (Pendimethalin 30 EC+Imazethapyr 2 EC)@0.75 kg a.i. ha⁻¹)], T_9 (Imazethapyr@40 g a.i. ha⁻¹), T_{10} (Imazethapyr@55 g a.i. ha⁻¹), T_6 (Pendimethalin@1 kg a.i. ha⁻¹) and T_8 (Imazethapyr@25 g a.i. ha⁻¹). The treatments like T_3 (total

weed free treatment) and T_2 (hand weeding at 20 and 40 DAS) although produced higher seed yield but the B: C ratios were low due to high cost of cultivation. The weedy check treatment (T_1) had the lowest B: C ratio due to poor yield in this treatment. The tremendous weed infestation in weedy check treatment drastically reduced the yield of the crop. Similar findings were also reported by Kundu et al. and Randhwa et al. [12,13]. The treatments like T_{14}, T_{13}, T_9, T_{10}, T_6 and T_8 are economically viable and the concerned herbicides are effective for weed management of green gram crop.

Conclusion

From the result it can be concluded that all the weed control treatments effectively controlled weeds and significantly reduced their population and dry weight. However, application of Vellore 32 (Pendimethalin 30 EC+Imazethapyr 2 EC)@1.00 kg a.i. ha⁻¹ was found most effective in reducing population and dry mass of weeds and producing maximum yield of green gram.

References

1. Gomez KA, Gomez AA (1984) Statistical Procedures for Agricultural Research. John Wiley and Sons, Inc. pp:307.

2. Jackson ML (1967) Soil chemical analysis.Prentice Hall of India Pvt. Ltd, New Delhi, pp: 183-347 and 387-408.

3. Das NR, Bhattacharya SP, Das AK (1997) Weeds in crops of West Bengal in Summer Season. The world weeds 4: 198.

4. Kundu R, Bera PS, Bramachari K (2009) Effect of different weed management practices in summer mungbean (*Vigna radiate* L.) under new alluvial zone of West Bengal. Journal of Crop and Weed 5: 117-121.

5. Singh G, Khajuria V, Gill R, Lal SB (2001) Effect of weed management practice in summer mung (*Vigna radiate* L.) Biennial Conference of Indian Society of Weed Science.

6. Srivastava M, Kumar N, Verma, P, Kaleem Mohd (2003). Effect of selected herbicides treatment on growth and yield of zaid season black gram (*Phaseolus mungo*). Biennial Conference of Indian Society of Weed Science.

7. Kumar A, Tewari, AN (2004) Crop weed competition studies on summer sown black gram (*Vigna mungo* L.). Indian journal of Weed Science 36: 76-78.

8. Veeraputhiran R (2009) Effect of mechanical weeding on weed infestation and yield of irrigated black gram and green gram. Indian journal of Weed Science, 41: 75-77.

9. Yadav VK, Singh SP (2005) Losses due to weeds and response to pendimethalin and fluchloralin in varieties of summer sown *Vigna radiata*. Annuals of Plant Protection Sciences 13: 454-457.

10. Singh KS, Singh R, Kaleem Mohd (2002) Effect of different herbicides for control weed in green gram (*Vigna radiata* L.). Biennial conference of Indian Society of Weed Science.

11. Malik RS, Yadav A, Malik RK, Singh S (2005) Performance of weed control treatments in mung bean under different sowing methods. Indian Journal of weed science 37: 273-274.

12. Kundu R, Brahmachari K, Bera PS, Kundu CK,Roy choudhury S (2011) Bioefficacy of imazethapyr on the predominant weeds in soybean. Journal of Crop and Weed 7:173-78.

13. Randhwa JS, Deol JS, Sardana Virender, Singh Jaspal (2002) Crop weed competition studies summer green gram (*Vigna radiate* L.). Indian Journal Weed Science. 34: 299-300.

8

Effect of Fertilizer and Rhizobium Inoculation on Growth and Yield of Soyabean Variety (*Glycine max* L. Merrill)

Adeyeye AS[1]*, Togun AO[2], Olaniyan AB[3] and Akanbi WB[3]

[1]*Department of Crop Production and Protection, Federal University Wukari, Taraba State, Nigeria*
[2]*University of Ibadan, Ibadan, Oyo State, Nigeria*
[3]*Ladoke Akintola University of Technology, Ogbomoso, Oyo State, Nigeria*

Abstract

The study assessed the response of three soybean varieties: (TGX 1740 - 2F, TGX 1842 - IE, and TGX 1448-2E) to three fertilizer treatments (4 t/ha compost, 30 kgN/ha urea, 2 t/ha compost+30 kgN/ha urea and a control) with or without Rhizobium Inoculation using strain R.25B+2180A. The experiment was a randomized complete block design with three replications. Data collected on vegetative, reproductive and grain yield parameters were subjected to analysis of variance ANOVA at 5% probability level. Fertilizer effects were significant on all the parameters assessed. In this study, soybean grain yield varies from 0.2 t/ha in plants nourished with 30 kgN/ha to 0.96 t/ha in plants that received 4 t/ha compost. Inoculation of soybean seed with rhizobium significantly improved the grain yield of soybean. Inoculated plants produced grain yield that is 35% higher than non-inoculated ones. The interactive effect of fertilizer types and inoculation was not significant on soybean grain yield and yield parameters. It was concluded that the use of 4 t/ha compost in combination with appropriate rhizobium strain could be a good agronomic practice for the production of high quality grain in soybean.

Keywords: Compost; Fertilizer; Nitrogen; Rhizobium; Soybean

Introduction

Soybean (*Glycine max* (L.) Merrill) is a legume of very high nutritional value especially in Nigeria because of the limited animal protein in human diet [1]. The worldwide recognition of soybean is mainly due to its high nutritional value and high seed protein content of about 38-42% [2]. To improve the yield and quality of soybean seeds, many cultural practices had been reported in different countries of the world, most especially rhizobium inoculation, fertilization with organic and inorganic materials [3,4]. Organic manure in sufficient quantities release nutrients rather slowly and steadily over a long period and also improves the soil fertility status by activating the soil microbial biomass [5,6]. The use of organic manure is limited by the huge quantities required in order to satisfy the nutritional needs of crops in view of its low nutrient content. Such huge quantities are obviously not obtainable and even if they were, transportation and handling costs would still constitute a major constrain, Also the use of inorganic fertilizer is on the decline by the farmers due to high cost and scarcity. The perennial fertilizer scarcity experienced in the country is partly attributable to the high cost. In order to address the bulkiness of organic materials, and the high cost of inorganic fertilizers complementary use of organic and mineral fertilizers has been recommended for long term cropping in the topics [7,8]. High and sustained crop yield can be obtained with judicious and balanced N fertilization combined with organic matter amendment [4,9].

Seed inoculation with Rhizobium strain enhances nitrogen fixation in soybean whether cultivar is promiscuous or not and several researches reported that significant yield increase were obtained by inoculation of soybean with appropriate bacteria before sowing [10-12]. The effectiveness and efficiency of nitrogen fixing process however depend mainly on the status of native soil rhizobium population and their compatibility with the soybean planted in the given soil. Olufajo et al. also reported the responses of five promiscuous nodulating soybeans to some selected Bradyrhizobial strains and concluded that there was a beneficial effect from nodulation of all the above soybean cultivars especially in the soils having low populations in indigenous rhizobia. Also yield and yield components are complex traits, which exhibit polygenic or quantitative inheritance pattern [13]. Quantitative traits are govern by multiple genes whose expression is greatly influenced by exposed external environment and thus it results into scale or rank shift of their performance [13,14]. Reports are scanty on the use of fertilizers with Rhizobium inoculation on the growth characteristics and yield of soybean. Hence the general objective of this study was to develop a technological package of fertilizers application with Rhizobium inoculation for the production of soybean for maximum grain yield in Nigeria.

Materials and Methods

Two years field trial was conducted to assess the effects of fertilizer application and Rhizobium inoculation on the growth and grain yield of soybean varieties. The field was ploughed, harrowed and the layout was done. The total land area for the experiment is 28 m × 12 m=(336 m²). The plot size is 3 m × 2 m with planting spacing of 60 cm × 5 cm. There were 1 m gaps between sub plots, while 2 m gaps were left between replicates. The treatments tested were three soybean varieties: TGX 1740-2F, TGX 1448-2E, and TGX 1842-1E: four types of fertilizers: 0 t/ha compost, 4 t/ha compost, 30 kgN/ha urea and 2 t/ha compost+30 kgN/ha and two levels of inoculations (inoculated (+) and non-inoculated (-). The experiment was a 3 × 4 × 2 factorial fitted into randomized complete block design with three replicates.

The compost was prepared from maize stover and poultry manure combined in ratio 3:1 maize stover: poultry manure on dry weight

---

*Corresponding author: Adeyeye S, Department of Crop Production and Protection, Federal University Wukari, Taraba State, Nigeria
E-mail: solorach2002@yahoo.com

basis [15]. The heap was watered and turned every fortnight, while the $_pH$ and temperature were taken weekly. The compost treatments were applied a week before planting while urea fertilizer in both sole and where it is combined with compost were applied a week after planting. Inoculation of seeds was done using Rhizobium strain R.25B+2180A which is peat base inoculants. It was thoroughly mixed with sugar coated soybean seeds and later air-dried (15 g of inoculums was applied to 1 kg of seeds).

Data collection and analysis

Destructive sampling was done at 2, 4, 8, 10, 12 and 14 weeks after planting, where 3 plants were taken per plot for measurement. The parameters measured were: stem height per plant, Number of leaves, nodes, branches and flowers, dry matter weight of root, stem and leaves, and grain yield. Data collected were subjected to analysis of variance (ANOVA) and the means separated using Duncan Multiple Range Test at 5% level of probability.

Results

Vegetative growth

The effect of variety on plant height, number of leaves, number of nodes and branches were also significant at all sampling times in the field except for number of leaves at 4WAP, plant height at 8 and 10WAP (Figure 1). TG × 1842-IE showed significantly superiority in number of leaves per plant to other varieties studied (Figure 2). The same trend was observed at 8WAP with respect to the number of nodes where variety TG × 1842-IE was also superior to TG × 1740-2F and TG × 1448-2E (Figure 3). Plant height was significantly higher through application of compost at 4 t/ha and the combine application rate of 2 t/ha compost+30 kgN/ha urea at all sampling times compared to the control and 30 kgN/ha rate. Number of leaves showed significant difference at 4 WAP among the fertilizer treatments, but at 6 and 12 WAP, application of 30 kgw/ha and compost rate of 4 t/ha significantly produced more leaves per plant followed by the combine application rate of 2 t/ha compost+30 kgN/ha urea (Figure 2). Number of node per plant was significantly higher in plants that received compost application of 4 t/ha and 2 t/ha compost+30 kgN/ha fertilizer (Figure 3). There was significant difference among fertilizer treatments on the number of branches per plant at 6WAP but at 10WAP, combined application rate of 2 t/ha compost+ 30 kgN/ha urea produced significantly higher branches followed by 4 t/ha compost rate while control had the least (Figure 4). Inoculation had no significant effects on growth parameters studied at all the sampling times except for plant height at 2WAP, number of branches at 6WAP and number of nodes at 2WAP (Figures 1-4).

Dry matter accumulation

Varietal effect was significant on stem dry weight, leaf dry weight, root dry weight and pod dry weight during the growth and development through the sampling period. TG × 1448-2E accumulated higher stem dry matter except at 12 WAP when compared with other varieties (Table 1). However at 10 and 12 WAP, TG × 1442-IE had higher stem dry matter accumulation than other two varieties. Leaf dry weight was significantly higher in both varieties TGX 1842-IE and TGX 1448-2E compared with TGX 1740-2F at 8 and 10WAP. However at 12 WAP there were no significant effects among all the varieties studied (Table 2). Dry matter accumulated in the root of soybean was not significantly different among the varieties at 6, 8 and 10WAP except at 12WAP when TGX 1740-2F produced significantly highest values for root dry weight followed by TGX 1448-2E and least with TGX 1842-IE (Table 3).

Fertilizer treatments resulted in higher root dry matter accumulation compared to the control. (None fertilized plants). Application of 2 t/ha compost+30 kgN/ha urea produced significant higher root dry matter weight although were similar most times with 4 t/ha compost rate (Table 3). This trend was also observed with stem dry matter yield, leaf dry matter yield and pod dry matter yield. Inoculation was not significant on dry matter accumulation yield of soybean plant except for leaf dry weight at 6WAP where inoculated plant had significant higher value when compared to non-inoculated plant (Table 3). Number of flower were not significant at all stages of growth except at 8WAP where TGX1740-2F performed best than other varieties (Table 4). Pod number had significant effect on the varieties at 6, 8,10 and 12WAP and at 6WAP TGX1842-1E produced significantly higher pod number than the other varieties (Table 4). Fertilizer types produced significant effect on the flower and pod number and compost rate of 4 t/ha gave higher flower number followed by the combined rate of 2 t/ha compost+30 kgN/ha urea with the least from 30 kgN/ha urea and the control (Table 4). The highest pod number was produced at 8 and 10WAP from the combined fertilizer of 2 t/ha compost+30 kgN/ha urea (Table 4). Inoculation had no significant effect on the number of flower and pod. However inoculated plants had higher mean values of flower and pod when compared with non-inoculated plants (Table 4).

Seed characteristics and seed yield

The effect of variety on seed characteristics of soybean at maturity showed variety TGX 1842-IE to have significantly higher pod weight while TGX 1740-2F had the significantly least pod weight (Table 5). There was no significant difference among the varieties in terms number of seed per plant, although the variety TGX 1842-IE had the higher mean value compared to other varieties (Table 5). TGX 1740-2F had significantly higher seed weight compared to either TGX 1448-2E or TGX 1842-IE. Also TGX 1740-2F had significantly higher seed weight per plant compared with other varieties. There was no significant difference between varieties TGX 1740-2F and TGX 1842-IE with respect to husk dry weight but are different significantly from variety TGX 1448-2F effect of variety was also significant on grain yield Application of 4 t/ha compost produced highly significant pod weight when compound to other fertilizer treatments. Also, 4 t/ha application rate produced significantly higher number of seed per plant followed by combined application of 2 t/ha compost+30 kgN/ha urea studied (Table 5). The interactive effect of variety and fertilizer types was significant on soybean seed weight, husk weight and grain yield (Table 6). The interaction between variety and inoculation had no significant effect on all the seeds and pods parameters taken. There was a significant interaction effect however on variety and fertilizer types on seed weight, husk weight and grain yield while fertilizer and inoculation interaction was only significant on soybean seed weight and grain yield (Table 6).

Discussion

The results showed that fertilizer types influence the growth, dry matter production and yield of soybean varieties. While inoculation of soybean seed alone had no significant effect. Application of compost and urea fertilizer enhanced development of vegetative parameters such that number of leaves, node number and number of branches were significantly better in fertilized plants then unfertilized plants i.e., control. This suggested reduced in crop productivity under a condition of limited nutrients, most especially N-availability. This is in line with the observation of Smith et al. [9], Makinde et al. [15] Akanbi et al. [16] and Daramola et al. [17]. They reported that availability of

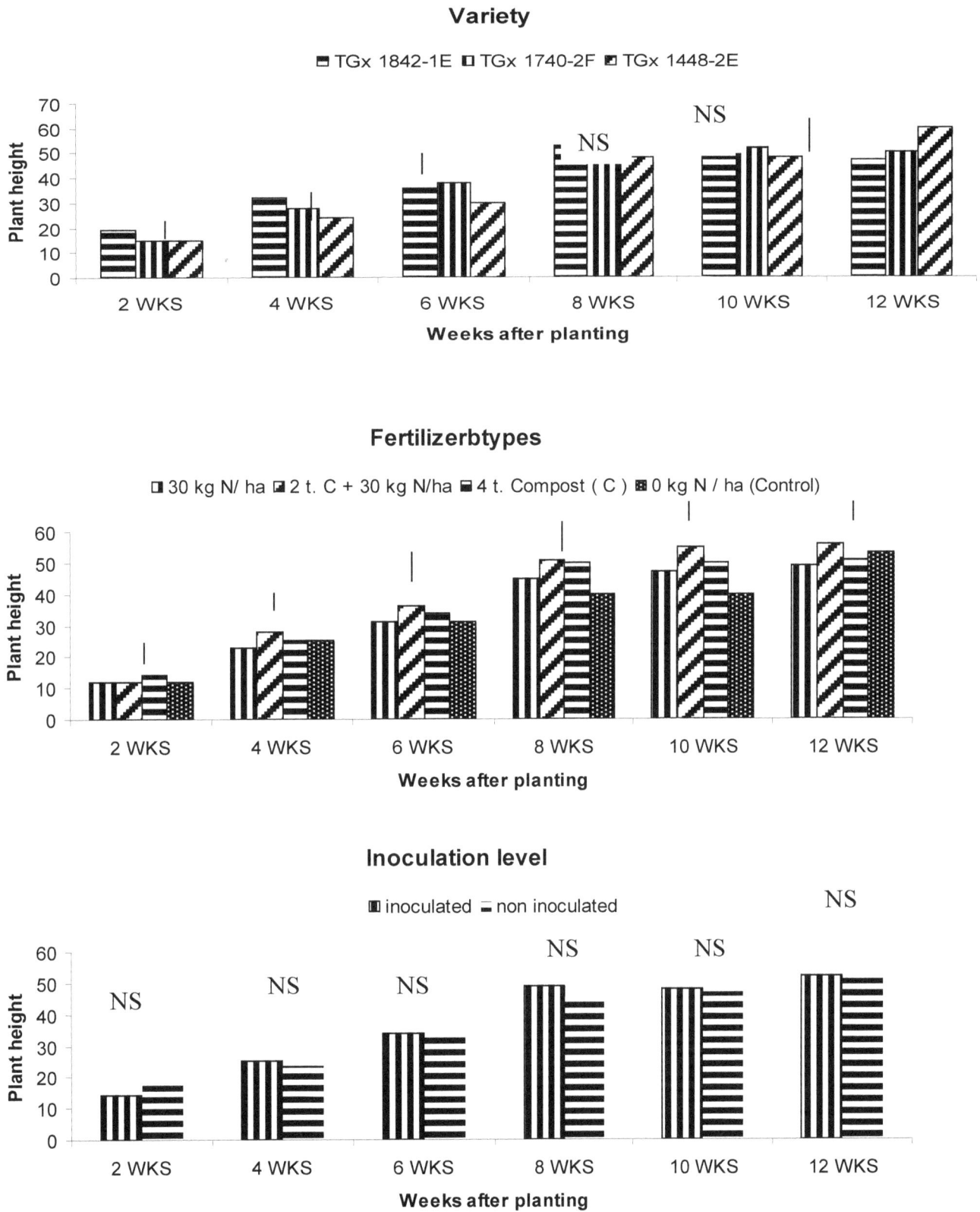

Figure 1: Effect of variety, fertilizers and inoculation on soybean plant height at different growth stages.

Variety

■ TGx 1842-1E ▣ TGx 1740-2F ▣ TGx 1448-2E

Fertilizer Types

▨ 30 kg N/ ha ▨ 2 t. C + 30 kg N/ha ▨ 4 t. Compost (C) ▣ 0 kg N / ha (Control)

NS

Inoculation level

▣ inoculated   ▤ non inoculated

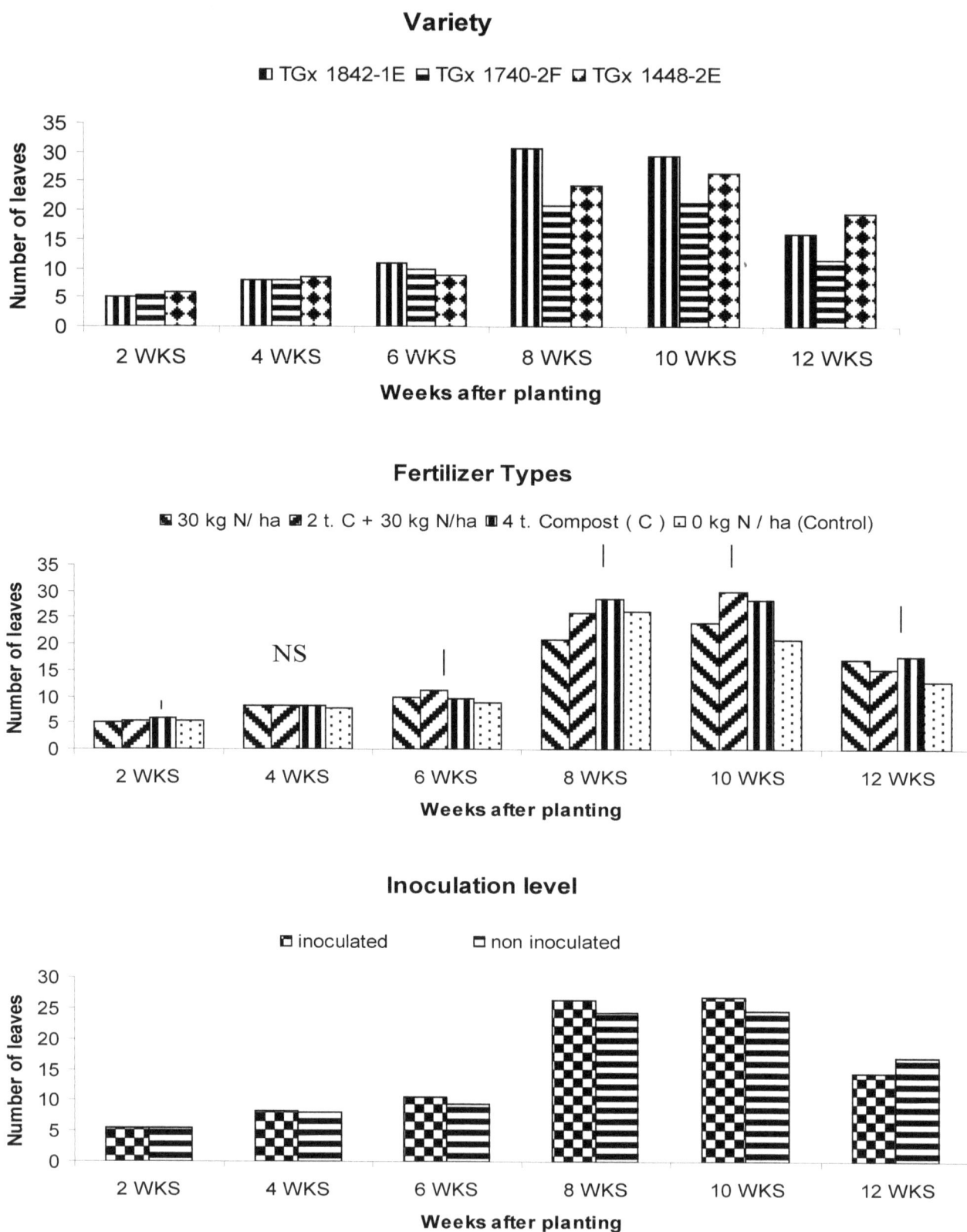

NS=not significant

Figure 2: Effect of variety, fertilizers and inoculation on soybean number of leaves at different growth stages.

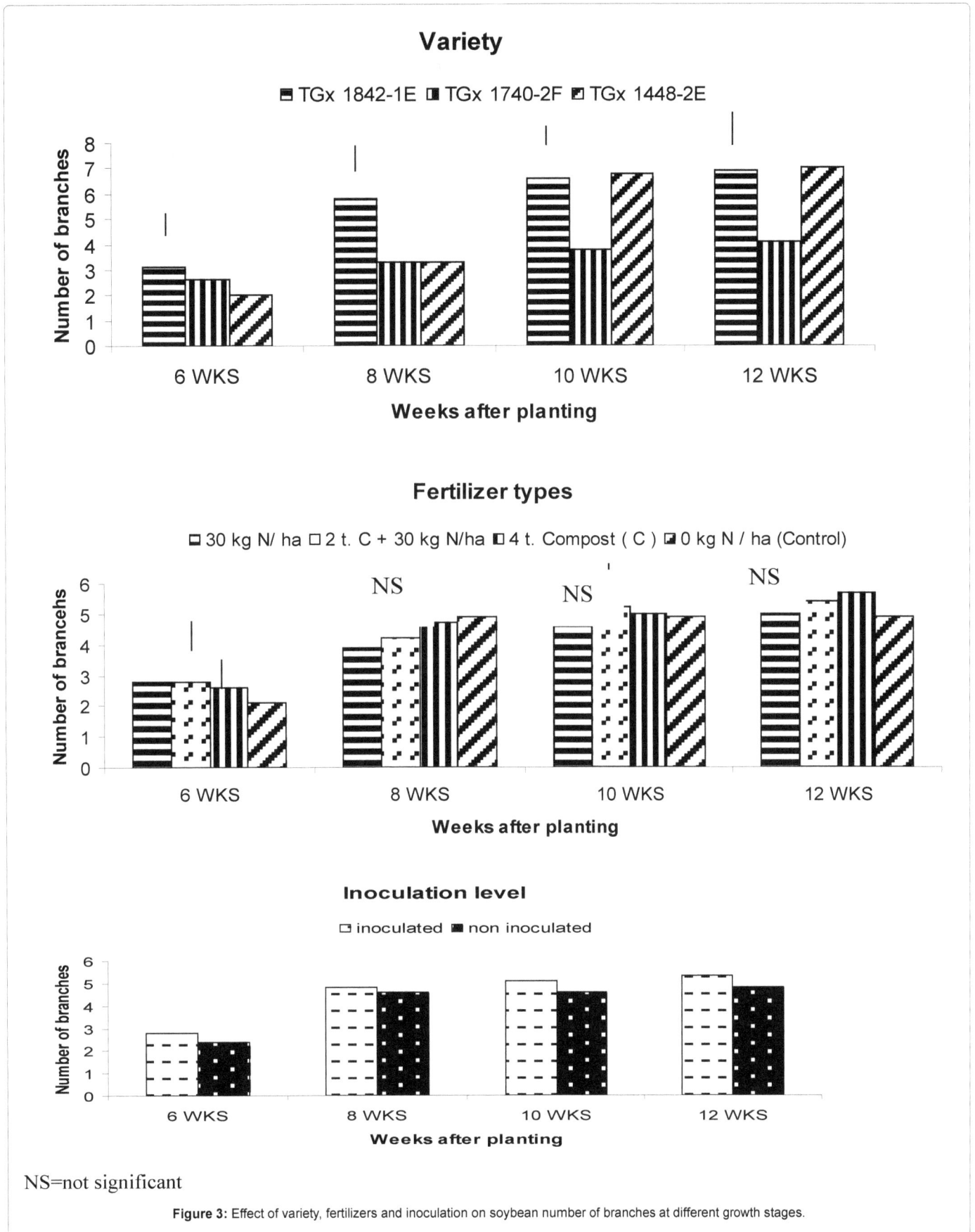

NS=not significant

Figure 3: Effect of variety, fertilizers and inoculation on soybean number of branches at different growth stages.

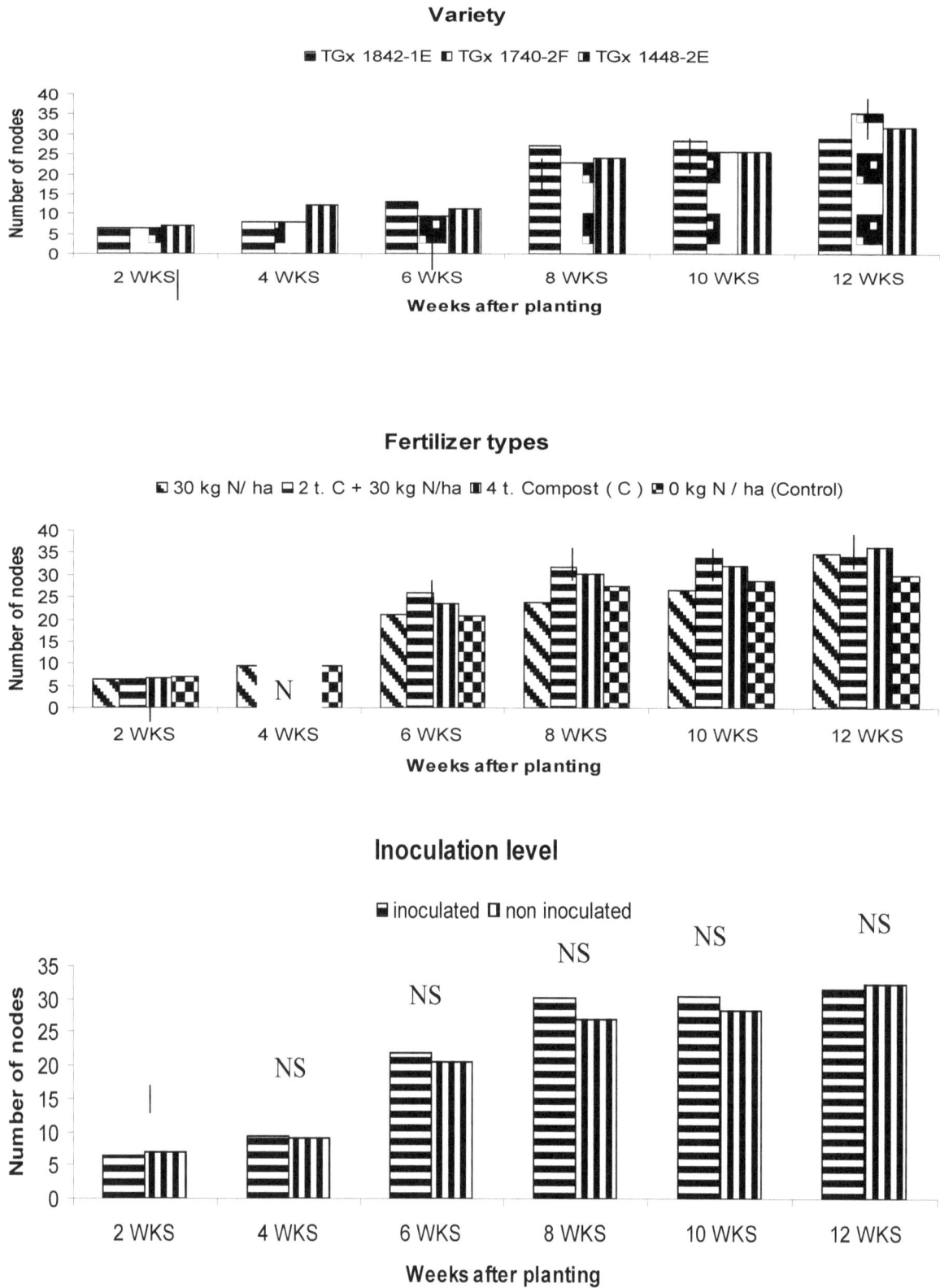

Figure 4: Effect of variety, fertilizers and inoculation on soybean number of nodes at different growth stages.

NS=not significant

Properties	Babcock
pH (H_2O)	5.8
Organic Carbon (%)	0.55
Total N (%)	0.03
Available P (ppm)	0.68
Fe (mg/kg)	9.16
Cu (mg/kg)	1.2
Zn (mg/kg)	5.12
Exchangeable K (c mol/kg)	0.21
Exchangeable Na (c mol/kg)	-
Exchangeable Ca (c mol/kg)	0.8
Exchangeable Mg (c mol/kg)	0.05
Exchangeable Acidity (c mol/kg)	-
E C E C (c mol/kg)	1.06
Base saturation (g/kg)	926
Sand (g/kg)	950
Silt (g/kg)	120
Clay (g/kg)	12
Textural Class	Loam sand

Table 1: Pre cropping soil chemical and physical analysis of experimental soil.

Treatment (Weeks after planting)	Stem dry matter yield (g/plant)						Leaf dry matter yield (g/plant)					
	2	4	6	8	10	12	2	4	6	8	10	12
Variety												
TGx 1842-1E	0.14b	0.7ab	1.6ab	12.9a	4.6a	5.9a	0.1b	1.4ab	2.3b	10.5a	6.9a	3.5a
TGx 1740-2F	0.16ab	0.6b	1.3b	4.5b	2.5b	4.6b	0.4a	1.1b	3.1a	5.9b	3.6b	4.4a
TGx 1448-2E	0.19a	0.8a	1.7a	5.6ab	5.2a	4.1b	0.4a	1.8a	2.0b	8.9a	6.6a	4.2a
Fertilizer Types												
30 kgN/ha	0.14c	0.6b	1.6	4.6b	3.1bc	4.0b	0.3b	1.0b	1.9c	6.3b	3.8b	3.6bc
2 t/ha C+30 kgN/ha	0.2ab	0.8a	1.7a	6.5ab	5.6a	5.7a	0.4ab	1.6a	2.9b	8.7ab	7.1a	4.2b
4 t/ha C	0.2a	0.8ab	1.4a	14.2a	4.8ab	6.4a	0.5a	1.7a	3.1a	11.2a	7.6a	5.9a
0 kgN/ha (control)	0.15c	0.6ab	1.5a	5.6ab	2.8c	3.5b	0.4b	1.4ab	3.6a	7.6b	4.4b	2.5c
Inoculation												
Inoculation	0.15b	0.6a	1.8a	9.6a	4.4a	4.9a	0.4a	1.4a	2.6a	8.7a	5.6a	4.1a
Non inoculation	0.18a	0.8a	1.3b	5.9a	4.4a	4.9a	0.4a	1.4a	2.2b	8.2a	5.8a	3.9a

Note: Values with different letters along column are significantly different using DMRT at 5% probability level.

Table 2: Effect of variety, fertilizer types and inoculation on soybean stem and leaf dry matter yield at different growth periods.

Treatment	Root dry matter yield (g / plant)						Pod dry matter yield (g/plant)	
	2	4	6	8	10	12	8	10
	Weeks after planting							
Variety								
TGx 1842- IE	0.08b	0.5a	0.66a	2.2a	1.9a	1.87b	5.9a	20.1a
TGx 1740-2F	0.13a	0.5a	0.75a	1.8a	1.9a	2.32a	2.7b	14.4b
TGx 1448 – 2E	0.10ab	0.5a	0.80a	1.8a	2.2a	2.13ab	3.9b	13.9b
Fertilizer types								
30 kg N/ha	0.08b	0.4c	0.67b	1.6b	1.7bc	1.73b	2.6c	11.7b
2 t C+30 kg N/ha	0.10ab	0.54a	0.87a	1.9ab	2.9a	2.40a	5.6a	18.0a
4 t c.	0.12a	0.53ab	0.76ab	2.4a	2.2ab	2.51a	5.1ab	18.9a
0 kg N/ha (Control) 0.12a		0.41bc	0.67b	1.8b	1.3c	1.78b	3.4bc	15.9a
Inoculation								
Inoculated	0.09a	0.50a	0.78a	2.0a	2.2a	2.07a	4.1a	15.1a
Non inoculated	0.01b	0.45a	0.70a	1.9a	1.8a	2.14a	4.3a	17.2a

Note: Values with different letters along column are significantly different using DMRT at 5% probability level

Table 3: Main effects of variety, fertilizer type and inoculation on field grown soyabean root and pod dry matter yield at different growth periods.

Treatment	Number of flowers/plant			Number of pods/plant		
	6	8	10	8	10	12
Variety	Weeks after planting					
TGx 1842- IE	13.3a	47.9b	14.3a	35.8a	63.1a	69.9ab
TGx 1740 – 2F	15.4a	71.8a	21.4a	14.3b	34.7b	58.8b
TGx 1448 – 2E	13.8a	65.0a	18.3a	9.9b	60.6a	81.8a
Fertilizer types						
30 kg N /ha	14.6a	46.4c	16.4a	16.1b	45.1b	67.9a
2 t C. + 30 kg N /ha	15.0a	63.9b	21.7a	20.3ab	61.7a	68.4a
4 t C.	15.6a	78.1a	17.3a	23.8a	61.2a	75.0a
0 kg N /ha (Control)	11.4a	38.1bc	16.4a	19.9b	43.1b	66.3a
Inoculation						
Inoculated	14.2a	63.2a	17.4a	21.2a	63.3a	69.5a
Non inoculated	14.3a	60.1a	18.4a	18.7a	62.3a	68.8a

Note: Values with different letters along column are significantly different using DMRT at 5% probability level. C=Compost.

Table 4: Main effects of variety, fertilizer type and inoculation on field grown soybean number of nodules, flowers and pods at different growth periods.

Treatment	Pod weight	Number of seeds	Seed weight	Husk weight
Variety				
TGx 1842-IE	36.2a	255.1a	21.35a	14.6a
TGx 1740–2F	23.6b	130.1b	14.2b	9.5b
TGx 1448–2E	25.5b	159.7b	15.4b	9.9b
Fertilizer type				
30 kg N/ha	21.4b	140.3a	12.3b	8.8b
2 t C+30 kg N/ha	30.3ab	198.3a	17.6ab	12.6ab
4 t C.	34.6a	199.8a	20.7a	13.9a
0 kg N/ha (Control)	27.6ab	188.1a	17.5ab	9.9ab
Inoculation				
Inoculated	30.1a	195.9a	18.2a	11.9a
Non inoculated	26.8a	167.4a	15.9a	10.6a

C=Compost; Note: Values with different letters along column are significantly different using DMRT at 5% probability level.

Table 5: Main effects of variety, fertilizer types and inoculation on soybean pod and seed characteristics.

Source of variation	Pod weight	Number of seeds	Seed weight	Husk weight	Grain yield
Variety (V)	7.31**	1.05ns	12.72**	3.52*	1.92**
Fertilizer (F)	7.11**ns	4.46**	11.33**	7.46**	1.70**
Inoculation (I)	0.01ns	1.98ns	0.03ns	0.16ns	0.01ns
V × F	1.33ns	1.88ns	2.74*	2.18*	0.42*
V × I	0.03ns	0.50ns	0.10ns	0.93ns	0.01ns
F × I	3.18*	2.59ns	3.55*	1.74ns	0.60*
V × F × I	0.58ns	0.70ns	0.94ns	0.52ns	0.20ns

*, **, ns significant at 0.05 and 0.01 probability level; not significant

Table 6: Summary of ANOVA for main and interactive effects of variety, fertilizer types and inoculation on soybean pod and seed characteristics.

adequate nutrients could improve crop growth and yield parameters. Inoculation effects were not significant in most of the growth parameter measured at all stages of growth. The reason for this may be due to sufficient nitrogen being released from the organic matter or biological antagonism from other microorganisms indigenous to the soil used. These could be answered by future research. Dry matter in the different plants increased significantly with compost rate of 4 t/ha followed by the combine application of 2 t/ha compost+30 kgN/ha urea, although were similar most times with 4 t/ha compost. This corroborate earlier assertion that soybean require between 30 kgN/ha to 60 kgN/ha fertilizer for optimum performance [3,18].

Application of compost at 4 t/ha and combine application of 2 t/ha compost+30 kgN/ha urea are superior to other treatments with respect to seed yield. This also suggested that the nutrient use efficiency as a result of applied fertilizer by soybean plants, which improved the synthesis and translocation of photosynthesis from the sources to the sink and significantly increased in number and weight of pods and seeds [19].

Conclusion

From this study, combine application of 2 t/ha compost+30 kgN/ha urea is recommended for the production of soybean in the study area. The treatment was as effective as application of 4 t/ha compost. This reduces the quantity of inorganic fertilizer required whenever the two are needed to be applied together.

References

1. IITA (1990) Annual report for 1989/90. International Institute of Tropical Agriculture, Oyo Road, Ibadan, Nigeria. pp: 53-57.

2. Messina M, Messina V (2010) The role of soy in vegetarian diets. Nutrients 2: 855-888.

3. Chiezey UF, Odunze AC (2009) Soybean response to application of poultry manure and phosphorus fertilizer in the sub-humid savanna of Nigeria. J Ecol Nat Environ 1: 25-31.

4. Manral HS, Saxena SC (2003) Plant growth, yield attributes and grai yield of soybean as affected by the application of inorganic and organic sources of nutrients. ENVIS Bulletin 8: 16-18.

5. Ayuso MA, Pascal JA, Garcia C, Hernandez T (1996) Evaluation of urban wastes for agricultural use. Soil Sci Ptant Nutr 142: 105-111.

6. Belay A, Classens AS, Wehner FC, De Beer JM (2001) Influence of residual manure on selected nutrient elements and microbial composition of soil under long-term crop rotation. S Afr J Plant Soil 18: 1-6.

7. Palm CA, Myers RJK, Nandwa SM (1997) Combined use of organic and inorganic nutrient sources for soil fertility maintenance and replenishment. In: Replenishing Soil Fertility in Africa. SSSA Spec Publ 51, SSSA, Madison, pp: 193-217.

8. Ipimoroti RR, Daniel MA, Obatolu CR (2002) Effect of organo-mineral fertilizer on Tea growth at Kusuku, Mambilla Plateau, Nigeria. Moor J Agric Res 3: 180-183.

9. Makinde EA, Akande MO, Agboola AA (2001) Effect of fertilizer type on performance of melon in a maize-melon intercrop. ASSET Series 1: 151-158.

10. Joshi JM, Nkumbula S, Javaheri F (1986) Seed inoculation response for promiscuous soybean cultivars. Soybean genetics newsletter 13: 206-212.

11. Kim SD, Yoo ID, Hong EH, Shin MK, Choe JH (1988) Effect of Rhizobium inoculant application on Inoculation and Nitrogen fixation in different soil types in soybeans. Research reports of rural development administration, Upland and industrial crops, Korea Republic 30: 9-13.

12. Ibrahim SA, Mahmoud SA (1989) Effect of inoculation on growth, yield and nutrient uptake of some soybean varieties. Egyptian J Soil Sci 29: 133-142.

13. Mahendra D, Todd CW, Richard H, Danien SP, George EB, et al. (2016) Genotype x environment interaction and stability analysis for watermelon fruits yield in the United States. Crop Sci 56: 1645-1661.

14. Mahendra D, Todd CW, Penelope PV, Richard H, Daniel SP, et al. (2016b) Stability of fruit quality traits in diverse watermelon cultivars tested in multiples environment. Horticulture Research 23: 16066.

15. Akanbi WB, Adebooye CO, Togun AO, Ogunrinde JO, Adeyeye AS (2007) Growth herbage and seed yield and quality of Telfairia Occidentals as influenced by cassava peel compost and mineral fertilizer. World J Agric Sci 3: 508-516.

16. Smith SR, Hall JE, Hadley P (1992) Composting sewage, sludge wastes in relation to their Suitability for use as fertilizer materials for vegetable crop production. Acta Hort 302: 202-215.

17. Daramola DS, Adeyeye AS, Lawal D (2006) Effect of application of organic and inorganic nitrogen fertilizers on the growth and dry matter yield of Amaranthus Cruentus. Acta Satech 3: 1-6.

18. Chiezey UF (1992) Perfromance of soyabean (Glycine max (L) Merill) cultivar 'Samsoy-2' as affected by rate and time of application of Nitrogen fertilizer. PhD Thesis, ABU, Zaria.

19. Babatola LA, Olaniyi JO (1997) Effect of NPK fertilizer levels and plant spacing on performance and shelf life of Okra. In: Proc 15th HORTSON conference, NIHORT, Ibadan.

9

Effect of Nitrogen Rates and Irrigation Regimes on Water Use Efficiency of Selected Potato Varieties in Jimma Zone, West Ethiopia

Egata Shunka Tolessa[1]*, Derbew Belew[2], Adugna Debela[2] and Beshir Kedi[2]

[1]Ethiopian Institute of Agricultural Research, Holetta Research Center, Horticulture Research Division, Ethiopia
[2]Jimma University College of Agriculture and Veterinary Medicine, Ethiopia

Abstract

Ethiopia has possibly the greatest potential for potato production. But its contribution to food security is less due to poor agronomic techniques and other factors which require improving the ways of resources use. This experiment was conducted in Jimma University College of Agriculture and Veterinary Medicine greenhouse to study the effect of nitrogen rates and irrigation regimes on water and nitrogen use efficiency of selected potato varieties (Jalenie, Guassa and Degemegn), using three nitrogen rates (130, 110, 90 kg/ha), and three irrigation regimes (full irrigation (100%), 80% and 60% of full irrigation) on clay textured fine top soil filled to poly ethylene pot of 15 liter and 30 cm upper diameter. The experiment was 3 × 3 × 3 factorial with three replications laid down in a Randomized Complete Block Design. Interaction of variety and irrigation significantly affected water use efficiency (WUE). Jalenie variety recorded the highest WUE at 80% irrigation, but was on par with Guassa varieties at 100% irrigation. The lowest WUE was obtained from Degemegn variety at 100% irrigation even though there was no significant difference among the three irrigations. From the results, it can be concluded that irrigation regimes and variety were significantly affected water use efficiency of the potato varieties while the nitrogen rates and interaction between or among factors holding nitrogen combination were not influenced the water use efficiency of the potato varieties significantly. As this is output of greenhouse condition, open field experiment is suggested to be carried out to come up with conclusive results.

Keywords: Variety; WUE (water use efficiency); Irrigation regimes; Nitrogen rates

Introduction

Potato (*Solanum tuberosum* L.) ranks fourth among the world's crop production in volume after wheat, rice and corn [1]. But it is first from Root and Tuber crops followed by cassava, sweet potato and yam [2]. Potato has got production potential of about 327 million tons and 18.6 million hectares worldwide [3]. Potato was introduced to Ethiopia in 1858 (19th century) by a German Botanist Schimper [4]. Since then, farmers in Ethiopian high lands began cultivating the potato tuber as compensation when other crops failed. In Ethiopia, the estimated land under potato cultivation each year is over 160,000 hectares [5]. Based on FAO data, potato production in Ethiopia has increased from 280,000 tons in 1993 to around 525,000 tons in 2007 [6].

Potato is temperate crop that satisfactorily grows and yields well in cool and humid climates [7]. It is a major food crop in many countries being grown from the tropics to the sub-polar. Among African countries, Ethiopia has possibly the greatest potential for potato production as 70% of its arable land mainly in highland areas with altitude greater than 1,500 m above sea level is considered suitable for potato [8]. Since the highlands are also home to higher percent of Ethiopia's population, the potato can play a key role in ensuring national food security if production potentials are exploited well [6].

The ideal growth requirements for potato include high and nearly constant soil matric potential, high soil oxygen diffusion rate, adequate incoming radiation and optimal soil nutrients [9]. Among other environmental conditions, temperature and photoperiod are known to affect the various physiological processes of the potato plant [10]. Optimum temperatures for foliage growth and net photosynthesis are 15-25°C and 20°C for tuberization. At temperature above 29°C tuberization is inhibited, foliage growth is promoted and net photosynthesis and assimilate partitioning to the tubers are reduced [11]. In natural environment plants are subjected to many stresses that have a great impact on growth, development and finally yield of crops. These factors can be biotic and abiotic. Among these factors, drought and nutrients suboptimal use are major abiotic factors that limit crop production [12].

Early studies have shown that water is the most important limiting factor for potato production and it is possible to increase production levels by well-scheduled irrigation programs throughout the growing season for efficient use of water [13]. Most researchers reporting the influence of water stress on potato yield in terms of its effect on aerial parts [14]. In course of improving water and nitrogen use efficiency researchers indicated use of drip irrigation for most crop commodities; mainly for vegetables and fruits [15]. For efficient use of water, supplementing rainfall by irrigation water to satisfy the needs of the crop at each growth stages is important to attain the required yields, especially in periods of limited rainfall. This is a key operation to avoid water shortage and over-irrigation which can reduce yields through reducing soil aeration that in turn reduce uptake(water and nutrient) and increasing nitrogen leaching [15].

Potatoes are generally sensitive, especially to deficiencies and excesses of N [16]. After beginning the tuber bulking phase, potatoes require a higher and steady supply of N. Mid-season N shortage reduces canopy growth and often causes premature senescence, which can reduce yields [17]. Potatoes require relatively high amounts of fertilizer

*Corresponding author: Egata Shunka Tolessa, Ethiopian Institute of Agricultural Research, Holetta Research Center, Horticulture Research Division, PO Box 3002, Addis Abba, Ethiopia
E-mail: egata.shunka@yahoo.com

because of high nutrient demand and a shallow, as well as inefficient rooting system [18,19]. In addition to shallow rooting, many potato cultivars have relatively inefficient nutrient and water use efficiency systems [20]. The consequence of poor efficiency and high water/fertilizer rates in potato is the potential for significant N contamination to surface and groundwater [21,22]. Although not studied as extensively as N in potatoes, high soil P is a potential environmental problem as well [23]. Understanding nitrogen application rates and irrigation regimes that enhance the efficient use of both water and nitrogen, and developing wisdom of efficient use of resource management practices could minimize the potential N losses thereby reducing production cost and increasing farm profit.

Water use efficiency is defined as the tuber yield obtained per unit of water consumed. According to Hassan et al. [24] WUE of potato ranges from 69 to 233 kg ha^{-1} mm^{-1}. Kiziloglu et al. [25] reported change of WUE between 63.4 to 44.1 kg ha^{-1} mm^{-1}. The WUE varied with growing season [26].

Though potato has been under cultivation for 154 years in the country, its production was not widely spread and it contributed little to food security in the country. According to Yilma [8], about 70% of cultivated agricultural land is suitable for potato production. But the production potentials are not exploited well as still it is under produced and utilized. The national average yield is approximately 7.9 tons/ha [27], which is very low compared to the world average of 16.4 tons/ha [2]. The main reason associated to this under production and utilization of potato is lack of high yielding and disease resistant improved potato varieties, problems of pests and disease especially potato late blight [28], are also the causes of underutilization of potato in Ethiopia. Moreover, lack of sufficient quantity of good quality seed, poor agronomic techniques and lack of storage facilities.

In Ethiopia, utilization of irrigation water for potato production is not well known [27]. When irrigated there is excessive and shortage problem as the farmers were using the same amount of water and intervals, regardless of crop species and growth stage [29]. Excessive irrigation of potatoes results in water loss and significantly increases of runoff (soil erosion) from production fields. There is also soil nutrient leaching which leads to contamination of the groundwater due to fertilizers and other chemical products [30]. In addition, it increases production costs, reduce yield by affecting soil aeration, favors the occurrence and severity of diseases and pests. On the other hand, deficient irrigation promotes a reduction of tuber quantity and lower yield due to reduced leaf area and/or reduced photosynthesis per unit leaf area [31]. Optimizing the water and nitrogen supply is an important issue as it varies with many external and crop factors.

In Ethiopia, information about plant water and nitrogen use efficiency is limited. The rates of nitrogen fertilizer used for released potato varieties from Ethiopian research centers are similar. But application of 138 kg N and 20 kg P/ha is found to be the appropriate rate for optimum productivity of Gorebiella variety on the vertisols of Debere Berhan in the central highlands of Ethiopia under rain fed conditions [32] even though the variety is one of the newly released ones, that can be an insight to conduct trials for other varieties to develop optimum rate enhancing economic return. On the other hand, other varieties are cultivated by applying blanket recommendation which is equal to 110 kg N/ha. This blanket application can lead to excessiveness or shortage. When excessive nitrogen is applied crop yield is reduced; cost of production increased and environment is polluted especially soil and ground water is acidified [21]. Shortage of nitrogen application is also reducing yield. Achieving optimum nitrogen rate applications

should be considered as it varying with soil, crop and water available to the crop for optimum return and farm profit.

In addition to this, the information about effect of rates of N-fertilizer application and irrigation regimes on water and nitrogen use efficiency is also scarce. Therefore, the present research was conducted in Jimma University College of Agriculture and Veterinary Medicine in the greenhouse to quantify and compare the water use efficiencies of three potato varieties (Jalenie, Guassa and Degemegn) and also to determine the interaction effect of rates of nitrogen and irrigation regimes on water use efficiency of the three varieties.

Materials and Methods

Area description

The experiment was conducted in Jimma University College of Agriculture and Veterinary Medicine Greenhouse, situated at latitude and longitude of 7°40'N 36°50'E and 7.667°N 36.833°E, respectively in 2011. Jimma is located 354 km southwest of Addis Ababa.

Light condition in greenhouse

The average shading capacity of the greenhouse was actually 26.87%. There were variations in light intensity reaching to inside greenhouse during the growing period depending on the season of the year and absence or presence of cloud during measurement.

Relative humidity and temperature

The average relative humidity of the greenhouse throughout the growth period was 36.81% while the maximum and the minimum values of the relative humidity were 54.3 and 17.7%, respectively. The average dry bulb temperature of the greenhouse throughout the growth period was 26.69°C while the maximum and the minimum values of the dry bulb temperature were 30.70°C and 22.60°C respectively. The fluctuation of relative humidity was highest when compared to the other parameters recorded.

Growing media soil water conditions

The soil medium used for growing the potato varieties was prepared from clay with 8.7 pH, 0.86 g/cm^3 bulk density, 0.5 EC/ds/m as well as 4.3, 7.5 and 0.192% organic carbon, organic matter and nitrogen content, respectively. 12 kg soil was filled to 15 litter pots and the pots were arranged in three blocks.

The field capacity of the soil was 37.82% while the permanent wilting point of the soil was 23.11%. The water holding capacity of the soil was 147.1 mm/m. The water amount below the permanent wilting point was unavailable to plants. The depletion factor for the irrigation were 0.25 for the 55th day after planting 0.3 and 0.5 for 56-90th and beyond 90th days after planting, respectively [33].

Experimental treatment, design and procedures

The plant materials used for the experiment were sprouted tubers of Jalenie, Guassa and Degemegn potato varieties obtained from potato seed multiplying farmers of Bishida District of Jimma zone. Jalenie and Guassa were light green potato varieties with white flower released in 2002 from Holleta Agricultural Research Center and Adet, respectively. Jalenie grows in altitude rage of 1600-2800 masl with 750-1000 mm annual rain fall and has maturity period of 90-120 days after planting while Guassa grows 2000-2800 masl with 1000-1500 mm annual rain fall and matures in 110-115 days after planting. Degemegn variety is deep green none flowering potato variety released from Holleta research center in 2002 and grows in 1600-2800 masl altitude range with 750-1000 mm annual rain fall and matures in 90-120 days after planting. Jalenie and Guassa grow up to 95.24-126.11 cm and 97.54-115.71 cm heights respectively while Degemegn grows up to 93.39- 107.73 cm heights. These varieties were selected due to their wide agro-ecological zone adaptability and suitability to Jimma growing condition.

The experiment was arranged in $3 \times 3 \times 3$ factorial combination with three replications laid down in randomized complete block design. The factors were nitrogen in three rates (130 kg/ha=2.93 g/pot, 110 kg/ha=2.48 g/pot, 90 kg/ha=2.03 g/pot), irrigation in three regimes (full irrigation=100%, 80% and 60% of full irrigation) and three varieties (Jalenie, Guassa and Degemegn).

Soil property test

Soil property test before production was made taking six representative disturbed samples randomly from top 30 cm depth at six positions.

Growing media prepared

The soil media used for growing the potato varieties was prepared from uniform soil and 12 kg filled in each of 243 pots of equal size. Each treatment had three pots. The filled pots were arranged in three blocks where one sprouted tubers of the same size were planted at 10 cm depth after watering the media well. Before planting the tubers, the irrigation scheduling was done using two installed tensiometer at 12 cm and 24 cm depth of the growing media to control irrigation frequency after calculating readily available soil water or irrigation water amount. The irrigation management was carried out between 20 and 50 cent bars [31,34]. But after April 13 near flowering and tuberization stage the crop wilts even though the tensiometer readings were not reached. Due to this reason watering was done before adjusted tensiometer reading was achieved.

Irrigation water amount applied

The amount of water irrigated once was calculated based on field capacity and wilting point concept of the soil in the pots which was determined in laboratory together with soil property tests. The total available soil water was calculated by subtracting permanent wilting point % from field capacity % in volume from which irrigation water amount or readily available soil water was determined by multiplying by 1000 times root depth (m) and available soil water depletion factor.

Effective rooting depth used

30 cm and 60 cm root depth used was obtained from FAO AGL [33] together with P (irrigation depletion fraction or maximum allowable depletion) but the active uptake is confined to the top 30 cm.

Irrigation methods and criteria used

Watering was done manually using watering cane. The lower limit water potential to begin irrigation was determined by applying pre-experimental trial, installing two tensiometers (Reich BSR Jecknik mmbar or kpa 35 cm and 30 cm length) at 12 and 24 cm on one media having 25-50% available soil water depletion [33]. The irrigation was performed at irrigation criteria of 20-25 cent bars for 25% available soil water depletion, 30-35 cent bars for 30% available soil water depletion and 44-50 cent bars for 50% available soil water depletion. The last irrigation was with held 10-15 days before harvest to allow the tubers to harden their skin before harvesting.

Fertilizer application time and method used

The fertilizers used were Urea (CO ($[NH_2]_2$) (46% N) and 90 kg/ha of DAP (46% P_2O_5)) The amount of ertilizers used in this study was applied using band method. Nitrogen fertilizer was applied in two splits. Half of the nitrogen fertilizers and entire phosphorus requirement was applied as basal while the remaining amount was applied at 45 days after planting [32]. The amount of phosphorus requirement was 90 kg/ha. All of the other cultural practices used throughout the growing season were similar to those that were practiced by regular farmers.

Crop evapotranspiration

Crop evapotranspiration was obtained from root zone soil water balance [35,36] using formula [37]: $I+P=ET+Dr+Ro \pm S$, where I=irrigation water applied, P=precipitation, ETC=crop water requirement, Dr=deep percolation, R_o=runoff and S=soil moisture change. Here actually P and R_o=0, as the experiment was conducted in Greenhouse using container or pot. So the net formula for root zone soil water balance applied was $I=ETC+Dr \pm S$ or $ETC=I- Dr \pm S$.

Climatic condition

For understanding and taking a measure for dangerous condition occurrence, the internal greenhouse air temperature, relative humidity and internal and external solar radiation was monitored, measured and recorded.

Harvesting and dry matter preparation

Tuber harvesting was done once at proper physiological maturity (70% leaves withering). Tuber and shoot dry matters were measured after drying sample biomass in oven dry at 65°C until constant weight was achieved.

Data collection

The solar radiation was measured by light meter or LUX meter (TES 1332, BATT 006P9V, NO.:010300137, Made in Taiwan) and recorded while dry bulb and wet bulb temperature, relative humidity were measured using Digital Sling Psychycro meter (AZ8716, REAL S/N: 96788223, Model: 8716, Made in China) and recorded. The result of soil moisture from tensiometer was recorded. Soil samples before and after production was taken.

Tuber and above ground biomass fresh weight (g)

Three pots of the one treatment whole tuber fresh weight was taken at maturity and averaged for representing treatment output per block while four representative shoot were taken from each pot of the treatment, chopped, weighed and averaged for each treatment.

Water use efficiency (WUE)

The ETC was estimated from soil water balance. Water use efficiency was computed using:

WUE=above ground Biomass and tuber weight (g/pot)

ETC (mm) ETC (mm)

Data analysis

Data was subjected to analysis of variance using proc GLM (general linear model) procedure of SAS 9.2 software [38]. The means were compared with Least Significant Difference (LSD) at 5% significance level and correlation analysis was done to investigate relationship of water use and nitrogen use efficiency using the same software.

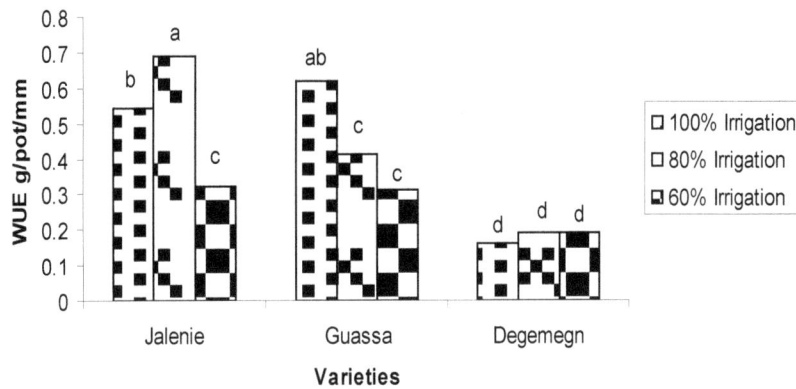

Figure 1: Variety and irrigation interaction effect on water use efficiency of potato.

Treatment	Water amount(mm)	Average tuber weight (g)	WUE (g tuber/pot/mm)	WUE (g above ground fresh weight/pot/mm)
Irrigation				
100%=I1	573.68a**	253.25a**	0.44133a**	0.53741b**
80%=I2	458.95b **	195.81aᵃ*	0.42885a**	0.55924b**
60%=I3	339.5c**	94.21b**	0.27500b**	0.61159a**
Variety				
Jalenie	457.6ns	245.51a **	0.51822a**	0.59599a**
Guassa	457.6ns	216.47a**	0.44704a**	0.57158ab**
Degemeng	457.6ns	81.18b**	0.17993b**	0.54068b**
Nitrogen				
130 kg/ha	457.6ns	170.33ns	0.35730ns	0.56551ns
110 kg/ha	457.6ns	195.03ns	0.40137ns	0.56396ns
90 kg/ha	457.6ns	177.91ns	0.38652ns	0.57878ns
LSD	20	59.371	0.1274	0.0462
C% at α=5%	2.825085	14.25899	19.40995	14.84519

**- means of the same factor followed by the same letter with in the column are not significantly different at 1% level of probability, LSD-Least Significant Difference, CV% - Coefficient of Variance. Ns=none significantly difference at 5% level of probability.

Table 1: Effect of irrigation, variety and nitrogen rates on WUE of potato.

Results and Discussion

Water Use Efficiency (WUE) g/mm: Variety and irrigation interaction significantly affected the WUE calculated from ratio of fresh tuber weight to irrigated water in mm (Figure 1). Jalenie recorded the highest WUE at 80% irrigation, but was not significantly different from Guassa at 100% irrigation. The lowest WUE was obtained from Degemegn at 100% irrigation. However, it was not statistically different from WUE of the same varieties at 80 and 60%. Decreasing the irrigation water by 20% increased the WUE by 14.4% further decreasing to 40% reduced the WUE by 40.33% in Jalenie variety while decreasing the irrigation water by 20% and 40% decreased the WUE of Guassa by 33.6 and 49.6% respectively. In Degemegn variety, decreasing the irrigation water by 20% and 40% had no significant effect on WUE.

Water use efficiency of above ground fresh weight was significantly affected by variety and irrigation (Table 1). Jalenie variety recorded significantly high WUE but was not significantly different from Guassa variety. However, the WUE of Guassa was not significantly different from that of Degemegn variety. Nitrogen rates and interactions holding nitrogen did not affect water use efficiency of above ground fresh weight.

WUE of tuber was increasing with increasing irrigation water from 60-100% (Table 1). A significantly positive correlation coefficient (r=0.225) was observed between WUE and irrigation water amount (Table 2). This may be because when the amount of water irrigated

increased to the field capacity, the potato varieties get better supply that satisfy their needs for better tuber formation that directly involved in increment of the WUE. Significantly strong positive association was also found between WUE and Harvesting index, tuber to shoot ratio, total dry weight, tuber number, tuber fresh and dry weigh (Table 2).

The WUE of tuber in this study was variable with varieties and increased with increased irrigation water amount. These results agree with findings of Darwish et al. [39] which obtained the lowest WUE from 60% of full irrigation while 80%, 100% and 120% irrigation provided maximum WUE, respectively. Steyn et al. [40] reported similar results in similar experiment with irrigation regimes of 100, 80 and 60%. On the other hand, Kirda [41] found a contradictory result from drip irrigation. Onder et al. [7] also reported decreasing WUE with increasing water supply. Similar reports were presented in Kashyap and Panda [42] and Yuan et al. [9]. According to Badr et al. [43] finding, fully irrigated potato increased N uptake and tuber yield which implies that better water use efficiency than water stressed treatment. 80% irrigation was reduced nitrogen losses by 58 to 81% compared to 20% irrigation regime [44], indicating that better watering to the field capacity encourages better utilization of not only water but also other nutrients. David et al. [45] showed different response of potato varieties to fully irrigate and stressed treatments which correlate the two findings as varieties recorded significantly different water use efficiencies resulted in due to different yield development of the varieties under same and different irrigation regimes. Finding

	AWA	TWUE	SFWUE	ATW	ATDW	ATPDW	ATNO.	HI	LAI	APNO.
APNO.	0.25*	0.40**	0.68**	0.42**	0.40**	0.29**	0.53**	0.26ns	0.62**	1
LAI	0.23*	0.34**	0.48**	0.37**	0.40**	0.31**	0.38**	0.32**	1	
HI	0.28**	0.85	-0.20ns	0.84**	0.79**	0.66**	0.76**	1		
ATNO	0.31**	0.83**	0.32**	0.83**	0.79**	0.67**	1			
ATPDW	0.58 **	0.78**	0.60**	0.84**	0.92**	1				
ATDW	0.44**	0.88**	0.20ns	0.91**	1					
ATW	0.42**	0.97**	0.42**	1						
SFWUE	-0.11ns	0.38**	1							
TWUE	0.225**	1								
AWA	1									

*, **: significant correlation at P<0.05 and P<0.01 probability levels, respectively; ns: non- significant; I=Irrigation, N=Nitrogen, AWA=Average water amount, TWUE=water use efficiency from fresh tuber weight, FWWUE=water use efficiency from fresh shoot weight, NUE=Nitrogen utilization efficiency, UPTNUE=Nitrogen uptake Efficiency, ATW=Average Tuber Weight, ATDW=Average tuber dry weight, ATNO=Average tuber number, HI=Harvesting Index, LAI=Leaf area index, APNO=Average stem number

Table 2: Relationship of WUE with yield and yield components.

of Ahmadi et al. [46] indicated non-significant different water use efficiencies between potato varieties subjected to different irrigation water amounts in contrary to this experiment which may be due to growing condition and managements other than irrigation. Decreased water use efficiency was reported with increasing water supply [47]. Maximum potato performance was recorded at full irrigation [48].

Conclusion

Many researchers had reported Variability of WUE with variable variety, irrigation regimes and nitrogen rates. In similar way, interaction of variety and irrigation significantly affected water use efficiency (WUE). Jalenie variety recorded the highest WUE at 80% irrigation, but was on par with Guassa varieties at 100% irrigation. The lowest WUE was obtained from Degemegn variety at 100% irrigation even though there was no significant difference among the three irrigations. From the results, it can be concluded that irrigation regimes and variety were significantly affected water use efficiency of the potato varieties while the nitrogen rates and interaction between or among factors holding nitrogen combination were not influenced the water use efficiency of the potato varieties significantly. Jalane variety is the most water efficient of all followed by Guassa variety. As this is output of greenhouse condition, open field experiment is suggested to be carried out to come up with conclusive results.

Acknowledgements

I express my thanks to Dr. Sentayehu A and Yilikal B for their valuable insight. My thanks are also extended to the staff members of Plant Science as well as other sections of Agarfa College especially to Vice Dean, Akele Molla and Dean of the college, Abera Worku for allowing me to undergo my experiment safely. I also thank staff members of Debre Zeyit Agricultural Research Centre and Jimma University who are working in soil and animal nutrition laboratory as well as horticulture section for extending their support during my laboratory work. Above all, I would like to thank the Almighty God for that He made all things possible for me to finish the study.

References

1. Fabeiro C, de Santa Olalla FM, de Juan JA (2001) Yield and size of deficit irrigated potatoes. Agricultural Water Management 48: 255-266.

2. FAO (2004) FAOSTAT Agricultural data. Provisional 2003 Production and Production Indices Data. Crop primary.

3. FAO (2006) FAOSTAT agriculture. Rome, United Nations Food and Agriculture.

4. Horton D (1987) Potatoes: Production, Marketing and Programs for developing countries. West View Press, London.

5. Ktheisen (2009) International Potato Center: World Potato Atlas, Ethiopia.

6. FAO (2008) The State of Food Agriculture, Rome.

7. Onder S, Caliskan ME, Onder D, Caliskan S (2005) Different irrigation methods and water stress effects on potato yield and yield components. Agric Water Manage 73: 73-86.

8. Yilma S (1991) The potential of true potato seed in potato production in Ethiopia. Acta Hurticultrae 270: 389-394.

9. Yuan BZ, Nishiyama S, Kang Y (2003) Effects of different irrigation regimes on the growth and yield of drip-irrigated potato. Agric Water Manage 63: 153-167.

10. Tsegaw T (2006) Response of potato to paclobutrazol and manipulation of reproductive growth under tropical conditions. Department of Plant Production and Soil Science, University of Pretoria, South Africa.

11. Levy D (1992) Potato in hot climates-could we do more. Proceeding of advanced potato production in the hot climates symposium, Israel. pp: 3-7.

12. Reddy AR, Chaitanya KV, Vivekanandan M (2004) Drought-induced responses of photosynthesis and antioxidant metabolism in higher plants. J Plant Physiol 161: 1189-1202.

13. Panigrahi B, Panda SN, Raghuwanshi NS (2001) Potato water use and yield under furrow irrigation. Irrig Sci 20: 155-163.

14. Lahlou O, Ouattar S, Ledent JF (2003) The effect of drought and cultivar on growth parameters, yield and yield components of potato. Agronomy 23: 257-268.

15. Shirie JM, Tobeh A, Hokmalipour S, Jamaati-e-Somarin SH, Abbasi A, et al. (2006) Potato (Solanum tuberosum) response to drip irrigation regions and plant arrangements during growth stages. pp: 1-2.

16. Biemond H, Vos J (1992) Effects of nitrogen on the development and growth of the potato plant. 2. The partitioning of dry matter, nitrogen and nitrate. Ann Bot 70: 37-45.

17. Westermann DT (2005) Nutritional requirements of potatoes. Am J Potato Res 82: 301-307.

18. Munoz F, Mylavarapu RS, Hutchinson CM (2005) Environmentally responsible potato production systems: A review. J Plant Nutr 28: 1287-1309.

19. Pack JE, Hutchinson CM, Simonne EH (2006) Evaluation of controlled-release fertilizers for northeast Florida chip potato production. J Plant Nutr 29: 1301-1313.

20. Love SL, Novy R, Corsini DL, Bain P (2003) Variety selection and management. Potato Production Systems.

21. Honisch M, Hellmeier C, Weiss K (2002) Response of surface and sub-surface water quality to land use changes. Geoderma 105: 277-298.

22. Madramootoo CA, Wayo KA, Enright P (1992) Nutrient losses through tile drains from potato fields. Appl Eng Agric 8: 639-646.

23. Davenport JR, Milburn PH, Rosen CJ, Thornton RE (2005) Environmental impacts of potato nutrient management. Am J Potato Res 82: 321-328.

24. Hassan AA, Sarkar AA, Ali MH, Karim NN (2002) Effect of deficit irrigation at different growth stages on the yield of potato. Pak J Biol Sci 5: 128-134.

25. Kiziloglu FM, Sahin U, Tune T, Diler S (2006) The effect of deficit irrigation on potato evepotranspiration and tuber yield under cool season and semiarid climatic conditions. J Agron 5: 284-288.

26. Nagaz K, Masmoudi MM, Mechlia NB (2007) Soil salinity and yield of drip irrigated potato under different irrigation regimes with saline water in arid conditions of southern Tunisia. J Agron 6: 324-330.

27. Peter RG, Wachira K, Oscar O, Agajie T, Gebremedhin W, et al. (2009) Improving Potato Production in Kenya, Uganda and Ethiopia: A System Diagnosis. Potato Research 52: 173-205.

28. Gebremedhin W (2001) Country Profile of Potato Production and Utilization Ethiopia. Ethiopian Agricultural Research Organization, Holeta Agricultural Research Center, National Potato Program.

29. Geremew EB (2008) Modelling the soil water balance to improve irrigation management of traditional irrigation schemes in Ethiopia. PhD Thesis, University of Pretoria.

30. Al-Jamal MS, Sammis TW, Ball S (2001) A case study for adopting the nitrate chloride technique to improve irrigation and nitrogen practices in farmer's fields. Appl Eng Agric 17: 601-610.

31. Van Loon CD (1981) The effect of water stress on potato growth, development, and yield. Am Potato J 58: 51-69.

32. Zelalem A, Tekalign T, Nigussie D (2009) Response of potato (Solanum tuberosum L.) to different rates of nitrogen and phosphorus fertilization on vertisols at Debre Berhan, in the central highlands of Ethiopia. African Journal of Plant Science 3: 016-024.

33. FAO (Food and Agriculture Organization) and AGI (Land and water development division) (2002) Onion water management. The AGLW Water Management Group, Land and Water Development Division Rome, Italy.

34. Holder CB, Cary JW (1984) Soil oxygen and moisture in relation to Russet Barbank Potato yield and quality. American Potato Journal 61: 67-75.

35. Tolga E, Chalim OA, Yasim E, Hakan O (2005) Crop water stress index for potato under furrow and drip irrigation systems. Potato Research 48: 49-58.

36. Samuel BSTC, Erika AS, Marc LF, Margaret ST, Stephanie AS, et al. (2009) Soil drying and nitrogen availability modulate carbon and water exchange over a range of annual precipitation totals and grassland vegetation types. Global Change Biology 15: 3018-3030.

37. Waskom RM (1994) Best management practices for irrigation management. Colorado State University Cooperative Extension, pp: 1-12.

38. SAS Institute Inc (2009) SAS 9.2 stored processes developer's guide. Cary, NC, USA.

39. Darwish TM, Atallah TW, Hajhasan S, Haidar A (2006) Nitrogen and water use efficiency of fertigated processing potato. Agric Water Manage 85: 95-104.

40. Steyn JM, Kagabo DM, Annandale JG (2007) Potato growth and yield responses to irrigation regimes in contrasting season of a subtropical region. 8th African Crop Science Conference Proceedings, pp: 1647-1651.

41. Kirda C (2002) Deficit irrigation scheduling based on plant growth stages showing water stress tolerance. Deficit Irrigation Practices, Water Reports 22, FAO, pp: 3-10.

42. Kashyap PS, Panda RK (2003) Effect of irrigation scheduling on potato crop parameters under water stressed conditions. Agric Water Manage 59: 49-66.

43. Badr MA, El-Tohamy WA, Zaghloul AM (2012) Yield and water use efficiency of potato grown under different irrigation and nitrogen levels in an arid region. Agr Water Manage 110: 9-15.

44. Guodong L, Yuncong L, Ashok KA, David MP, James D (2012) Enhancing Nitrogen Use Efficiency of Potato and Cereal Crops by Optimizing Temperature, Moisture, Balanced Nutrients and Oxygen Bioavailability. Journal of Plant Nutrition 35: 428-441.

45. David S, Evelyn RFV, Raymundo G, Felipe DeM, Roland S, et al. (2016) Yield and Physiological Response of Potatoes Indicate Different Strategies to Cope with Drought Stress and Nitrogen Fertilization. Am J Potato Res 93: 288-295.

46. Ahmadi SH, Agharezaee M, Kamgar-Haghighi AA, Sepaskhah AR (2016) Water-saving irrigation strategies affect tuber water relations and nitrogen content of potatoes. International Journal of Plant Production 10: 275-288.

47. Mohammad S, Jaber S (2016) Determination of optimal combination of applied water and nitrogen for potato yield using response surface methodology (RSM). Biosci Biotech Res Comm 9: 46-54.

48. Fathi P, Soltani M (2013) Optimization of water uses efficiency and yield in potato using marginal analysis theory. Journal of Soil and Water Resources Conservation 2: 85-93.

10

Effect of Foliar Nutrition on Growth, Yield Attributes and Seed Yield of Pulse Crops

Uma Maheswari M* and Karthik A

Department of Agronomy, Agricultural College and Research Institute, Coimbatore, Tamil Nadu, India

Abstract

Field experiment was conducted to find out the influence of foliar nutrition on growth and seed yield of pulse crop during Rabi 2013-14. The experiment was laid out in split plot design and replicated thrice. The pulse crop viz., blackgram, greengram, cowpea and horse gram were tried as treatments under main plot. Foliar nutrient sprays viz., 2% DAP, 1% KCL, 1% boron, 1% $MgSO_4$, 1% $ZnSO_4$ and without foliar spray were fitted under sub plot. Foliar spray treatment with the aqueous solution of nutrients was done to the 30 and 45 DAS of pulse crop. Significant increase was recorded in plant height, dry matter production, and number pod plant[-1], number of seed plant[-1], test grain weight, yield and grain yield with foliar application of nutrients. Maximum grain yield was recorded when spread with 2% DAP followed by 1% KCl at flowering and 15 days later is the viable nutrient management package to the pulses for getting higher income through higher productivity.

Keywords: Pulse crop; Foliar nutrition; Di ammonium phosphate; Potash; Seed yield

Introduction

Pulses occupy a unique position in Indian agriculture in virtue of the fact that they provide the rich source of vegetable protein and calories to the average Indian diet. Besides being a rich source of protein, they maintain soil fertility through biological nitrogen fixation in soil and thus play a vital role in furthering sustainable [1]. India is the largest producer and consumer of pulses in the world accounting for 33.6 percent of the world area and 24 percent of the world production of pulses [2]. The area under pulse crop is increasing continuously but productivity is decreasing year by year. The reasons for decreasing productivity are due to decreasing soil fertility especially macro and micronutrients, imbalanced use of fertilizer and occurrences of physiological disorders factors such as inefficient partitioning of assimilates, poor pod setting, excessive flower abscission and lack of nutrients during the critical stages of crop growth leads to nutrient stress, poor growth and productivity were found to be some of the yield barriers of pulse crop [3]. These nutrients are more important because in pulse crop to synchronized flowering altered the source-sink relationship due to rapid translocation of nutrients from leaves to the developing pods. To overcome these constraints, additional nutrition through foliar feeding is play a vital role in pulse production by stimulating root development, nodulation, energy transformation, various metabolic processes and increasing pod setting and thereby increasing the yield [4]. This is one of the most efficient ways of supplying essential nutrients to a growing crop. Considering the above facts the experiment was conducted to incredulous this problem by foliar feeding of nutrients.

Materials and Methods

The field experiment was conducted during Rabi 2013-14 at College of Agricultural Technology, Kullapuram, Theni situated in the Southern agro climatic zone of Tamil Nadu at 10°5' North latitude and 77°5' East longitude at an altitude of 40 m above mean sea level. The soil of the experimental field was sandy clay loam in texture with the available nitrogen 234.26 kg ha[-1], phosphorus 16.76 kg ha[-1], potassium 294.24 kg ha[-1] and organic carbon content 0.30%. The experiment was laid out in split plot design and replicated thrice, assigning pulse crops (Blackgram (M_1), Greengram (M_2), Cowpea (M_3) and Horse gram (M_4) to main plots and foliar nutrients spray (2% DAP (S_1), 1% KCL (S_2), 1% boron (S_3), 1% $MgSO_4$ (S_4), 1% $ZnSO_4$ (S_5) and without foliar spray (S_6) to sub plot treatments. Good viable local variety seeds of green gram, black gram, cowpea and horse gram with local variety having germination of 93 per cent were used at the rate of 20 kg ha[-1]. The seeds were treated with *Rhizobium* culture half an hour before sowing. The treated seeds were sown at 30 cm between rows and 10 cm between seed to seed to maintain optimum plant population. The recommended fertilizer dose of 25 kg N, 50 kg P_2O_5, 25 kg K_2O ha[-1] were applied as basal through urea, single super phosphate and muriate of potash in lines and incorporated at the time of sowing. Foliar application was done at flowering and pod filling stages of crop growth using high volume sprayer with a spray volume of 500 litre ha[-1]. Recommended crop management practices including plant protection remained common to all the treatments. The data collected for pulse crops were statistically analyzed following the procedure given by Gomez and Gomez [5]. Whenever significant difference existed, critical difference was constructed at five per cent probability level. Such of those treatments where the difference are not significant were denoted as NS.

Results and Discussion

Growth characters

The foliar application of 2% DAP produced significantly increased the plant height and number of branches plant[-1] in black gram, greengram, cowpea and horse gram which was statistically on par with spray of 1% KCl and 1% boron. This might be due to enhanced level of nutrient available in the rhizo-ecosystem of the foliar applied nutrients resulting in better plant growth and development. Application of nutrients would have resulted in better vegetative growth as observed by taller plants, more branches and efficient nodulation. This favourable influence of foliar application of nutrients could be ascribed to more and quick access to nutrients by plants at seedling and early development

*Corresponding author: Uma Maheswari M, Department of Agronomy, Agricultural College and Research Institute, Coimbatore-641 003, Tamil Nadu Agricultural University, Tamil Nadu, India, E-mail: umavalarmathi987@gmail.com

stages [6]. The leaf area index and dry matter production was increased with the foliar nutrition of 2% DAP and it was comparable with foliar spray of 1% KCl and 1% boron at flowering and pod setting stage of pulse crop (Table 1). This might be due to increased availability of nutrients to plants leading to maximum plant growth in terms of plant height and leaf area which in turn contributed higher DMP production. During this study we examined that these results also resemble the findings of Hussain et al. [7]. Foliar spray is a well-established tool to complete and to enrich plant nutrition. Foliar feeding can provide the nutrients needed for normal developments of crops in cases where absorption of nutrients from the soil is disturbed. As uptake of nutrients through the foliage is considerably faster than through roots, foliar sprays is also the method of choice when prompt correction of nutrient deficiencies is required.

Yield attributes

Days taken to attain 50% flowering were significantly delayed with increasing levels of foliar nutrients. The possible reason might be that supply of nitrogen and other nutrients are associated with protein synthesis consequently material needed for the formation of fruiting body remains scarce. To fulfil the gap, metabolism remains engaged in cell division, elongation and multiplication for vegetative growth for a longer period and thus there was a delay in the flowering [8]. Foliar spray of 2% DAP recorded the highest values for yield attributing characters viz., number of pods plant^{-1}, number of seeds pod^{-1}, pod length and test grain weight than other foliar spray treatments (Table 2). The increase in yield attributes might be due to supplementation of nutrients at the critical stage without physiological stress. Foliar application of nutrients enhanced the number of floral buds, prevented the floral shedding by maintaining optimum bio-physiological conditions in plants. Adequate

and continuous nutrient availability through soil and foliar nutrition promotes the supply of assimilates to sink or yield container, thus enlarging the size of the yield structure. The findings of Hamayun et al. have also confirmed the results of present study.

Grain yield

The foliar application of 2% DAP produced significantly increased the grain yield in black gram, greengram, cowpea and horse gram which was statistically on par with spray of 1% KCl and 1% boron (Table 2). The impact of the foliar nutrients to meet the nutrient demand of the crop at the critical stage on-site, where they are needed without stress, would have resulted in better growth and development of the crop and ultimately the yield attributing characters and yield on one hand. The balanced growth habit, which induced more flower and fruiting body production with timely supply of nutrients through foliar spray might have reduced shedding of flowers and fruits, which led to a positive source-sink gradient of photosynthates translocation due to growth regulator on the other hand. These favourable effects might have attributed for higher yield of green gram, black gram, cowpea and horse gram under the foliar spray of nutrients and growth regulators. This finding is in line with the results of Manivannan et al. [9].

Conclusion

The present study indicated that foliar application of 2% DAP resulted in higher grain yield in black gram, greengram, cowpea and horse gram. Which was followed by 1% KCl spray. The growth parameters and yield attributes were also found to be higher when 2% DAP is given as spray at flowering and pod filling stages. This may be due to balanced growth habit, which induced more flower and fruiting

Treatment	Plant height (cm)							Number of branches plant^{-1}						
	S_1	S_2	S_3	S_4	S_5	S_6	Mean	S_1	S_2	S_3	S_4	S_5	S_6	Mean
M_1	42.27	42.09	41.67	40.20	40.70	39.03	40.99	46.54	45.79	45.13	44.13	44.16	43.17	44.82
M_2	50.10	47.80	46.77	48.47	49.70	48.33	48.53	53.54	50.87	50.98	52.57	53.24	52.54	52.29
M_3	74.77	69.54	69.63	69.60	68.43	63.63	69.27	76.50	72.83	72.80	72.83	71.87	67.17	72.33
M_4	71.03	65.40	63.97	60.93	62.43	55.30	63.18	74.50	68.54	67.50	64.20	65.80	57.87	66.40
Mean	59.54	56.21	55.51	54.80	55.32	51.58		62.77	59.51	59.10	58.44	58.77	55.19	
	M	S	S at M					M	S	S at M				
SEd	0.97	1.60	3.22					1.14	1.38	2.77				
CD(P=0.05)	2.20	1.70	3.40					2.78	2.80	5.59				
Treatment	Leaf area index (LAI)							Dry matter production (g plant^{-1})						
	S_1	S_2	S_3	S_4	S_5	S_6	Mean	S_1	S_2	S_3	S_4	S_5	S_6	Mean
M_1	0.8	0.6	0.6	0.7	0.6	0.4	0.6	81.57	80.28	73.20	70.87	81.91	64.20	75.34
M_2	0.9	0.7	0.6	0.6	0.5	0.5	0.6	89.91	78.61	74.28	70.17	77.91	64.83	75.95
M_3	1.7	0.9	0.9	0.8	0.9	0.7	1.0	101.50	95.20	91.83	82.20	73.94	66.24	85.15
M_4	1.0	0.8	0.7	0.6	0.7	0.6	0.7	92.83	89.62	82.83	75.50	86.57	67.83	82.53
Mean	1.1	0.8	0.7	0.7	0.7	0.5		91.45	85.93	80.54	74.69	80.08	65.78	
	M	S	S at M					M	S	S at M				
SEd	0.020	0.027	0.054					1.038	1.396	2.793				
CD(P=0.05)	0.076	0.054	0.147					2.542	2.823	5.646				

Main plot : Pulse crops			Sub plot : Foliar nutrition at 30 and 45 DAS		
M_1	:	Black gram	S_1	:	2% DAP
M_2	:	Green gram	S_2	:	1% Kcl
M_3	:	Cowpea	S_3	:	1% Boron
M_4	:	Horse gram	S_4	:	1% Magnesium Sulphate
			S_5	:	1% Zinc Sulphate
			S_6	:	Control (No foliar spray)

Table 1: Effect of foliar nutrition on growth characters of pulse crops.

Treatment	50% flowering (Days)	Number of pods plant^{-1}	Number of seeds pod^{-1}	Test grain weight (g)	Seed yield (kg ha^{-1})
M$_1$	53.60	29.10	20.00	596	16695
M$_2$	43.50	27.60	19.30	584	16065
M$_3$	39.90	25.90	18.90	507	12215
M$_4$	39.70	25.50	18.80	453	10305
SEd	0.25	0.95	0.15	12	-
CD(P=0.05)	0.63	0.23	0.37	29	-
S$_1$	51.30	28.90	19.90	599	17022
S$_2$	45.30	27.40	19.20	529	12749
S$_3$	44.90	27.00	19.00	511	12735
S$_4$	52.40	29.40	20.00	612	16697
S$_5$	27.10	22.60	19.30	423	9900
S$_6$	132	11321	132	132	56
SEd	0.90	0.61	0.10	13	-
CD(P=0.05)	1.84	1.23	0.20	25	-
M at S					
SEd	1.64	1.08	0.23	25	-
CD(P=0.05)	3.36	2.12	0.52	54	-
S at M					
SEd	1.81	1.22	0.20	25	-
CD(P=0.05)	3.69	2.46	0.41	51	-

Main plot : Pulse crops DAS			Sub plot : Foliar nutrition at 30 and 45		
M$_1$	:	Black gram	S$_1$	:	2% DAP
M$_2$	:	Green gram	S$_2$	:	1% Kcl
M$_3$	:	Cowpea	S$_3$	:	1% Boron
M$_4$	:	Horse gram	S$_4$	:	1% Magnesium Sulphate
			S$_5$	:	1% Zinc Sulphate
			S$_6$	:	Control (No foliar spray)

Table 2: Effect of foliar nutrition on yield attributes and seed yield of pulse crops.

body production with timely supply of nutrients through foliar spray might have reduced shedding of flowers and fruits, which led to a positive source-sink gradient of photosynthates translocation due to growth regulator. Hence, foliar application of 2% DAP or 1% KCl will be viable and feasible option in order to get higher yield in pulse crops.

References

1. Balusamy M, Meyyazhagan M (2000) Foliar nutrition to pulse crop. In: Proc Symp on recent advances in pulse up production technology, TNAU, Coimbatore.

2. Pramanik SC (2009) Rainwater management techniques for successful production of pulses in rain fed areas. Indian Farming 58: 15-18.

3. Ali MA, Abbas G, Mohy-ud-Din Q, Ullah K, Abbas G, et al. (2010) Response of mungbean (Vigna radiata L.) to phosphatic fertilizer under arid climate. J Animal & Plant Sci 20: 83-86.

4. Ravisankar N, Chandrasekharan B, Sathiyamoorthi K, Balasubramanian TN (2003) Effect of agronomic practices for multi-blooming in greengram (Vigna radiata L.) (cv. Pusa bold). Madras Agric J 90: 166-169.

5. Gomez KA, Gomez AA (2010) Statistical procedures for agricultural research. 2nd edn. Wiley India Pvt Ltd, India.

6. Nawange DD, Yadav AS, Singh RV (2011) Effect of phosphorus and sulphur application on growth, yield attributes and yield of chickpea (Cicer arietinum L). Legume Res 34: 48-50.

7. Nazir H, Mohammad M, Rehana HK (2011) Response of nitrogen and phosphorus on growth and yield attributes of blackgram (Vigna mungo L.). Res J Agric Sci 2: 334-336.

8. Dixit PM, Elamathi S, Kishor KZ, Neeta C (2008) Effect of foliar application of nutrients and NAA in mungbean. J Food Legume 21: 277-278.

9. Manivannan V, Thanunathan K, Imayavaramban V, Ramanathan N (2003) Growth and growth analysis of rice fallow blackgram as influenced by foliar application of nutrients with and without Rhizobium seed inoculation. Legume Res 26: 296-299.

11

Agronomic and Physicochemical Evaluation of Sweet Potato [*Ipomoea batatas* (L.) Lam.] Collections in Ethiopia

Solomon Ali*, Wassu Mohammed and Beneberu Shimelis

Department of Plant Science, Debre Markos University, Ethiopia

Abstract

The productivity of sweet potato [*Ipomoea batatas (L.)* Lam.] is mainly dependant on the acquisition accessions which posses desirable traits and development of high yielding varieties with desired quality attributes. For this purpose, Haramaya University collected 116 sweet potato accessions from International and National sources to develop varieties for eastern Ethiopia; however, the accessions characterization and documentation were not exhaustively done to support the improvement program. Therefore, this study was conducted during 2012/2013 cropping season to characterize, evaluate, and documenting of agronomic and physicochemical attributes of sweet potato accessions at Haramaya. Augmented design consisting of 114 entries/tests and two checks were used. Varied number of accessions recorded significantly higher values than the mean of the checks for days to physiological maturity, above ground fresh biomass, storage root fresh weight, total storage root yield, marketable storage root yield, reducing sugar, total sugar, and total starch content, pH, dry matter content, total soluble solid, specific gravity and peel content. Tis-9465-7 had the highest storage root fresh weight yield; marketable storage root yield and total storage root yield and Koka-12 and CN-2069-7 exhibited significantly highest values than mean of the checks for days to physiological maturity and above ground fresh biomass, respectively. CN-1752-14, CN-2056-8 and Tis-80/043-1 for reducing sugar, pH and total soluble solid, respectively, exhibited significantly highest values, while CN-1752-15 recorded the highest total sugar and total starch content. Korojo had significantly highest values for specific gravity, dry matter. Tis-82/0602 were exhibited that the lowest in peel content. Elliptic shape (27.19%) and horizontal constriction (45.62%) defect were dominant in the accessions. Most of the accession had white skin color (22.6%) while 21.92% accessions had creamy flesh color.

Keywords: Sweet potatoes; Proximate analysis; Reducing sugar; Total starch content

Introduction

The sweet potato[*Ipomoea batatas* (L.) Lam.] is a dicotyledonous plant which belongs to the family *Convolvulaceae*. It is a tuberous root crop important for food security and cultivated in over 100 developing countries and ranks among the five most important food crops in over than 50 of those countries. Over 95% of the global sweet potato production is in developing countries. In Ethiopia, sweet potato has been cultivated for the last several years and over 95 percent of the crop is produced in the Southwest, eastern and southern parts, where it has remained for many years as one of the major subsistence crops especially in the periods of drought [1,2].

Sweet potato is cultivated in Ethiopia mostly for human consumption. It ranks third after Enset [*Ensete ventricosum* (Wele) Cheesman] and Potato (*Solanum tuberosum L.*) as the most important root crops produced in the countries. Sweet potato covers about 81000 hectares of land in Ethiopia with an average national yield of about <9 t/ha on farm and 25-36 t/ha on research centers [3]. Conservation of genetic diversity within a crop species is the basis of all variety improvement. However, if the improved variety replaces traditional farmers' varieties, as it often does, the result may still be genetic erosion. Therefore, collecting and conserving farmers' varieties is an essential activity as equal to improving and disseminating new varieties. Haramaya University has released and made recommendation for cultivation two sweet potato varieties namely; Barkume andAdu for eastern part of the country. Moreover, there were 114 accession was maintained in Ethiopia and the two released varieties were also maintained for years which were obtained from International and local sources. However, extensive agronomic and physicochemical attributes has not been carried out to identify which accession(s) attributed what and potentially used for which purpose(s). This necessitates studying and documenting the agronomic and physicochemical attributes of these accessions. Therefore, this research was initiated with the objective of characterization, evaluation and documenting of agronomic and physicochemical attributes of sweet potato accessions in Ethiopia.

Materials and Methods

Sweet potato accessions were grown using unreplicated plot under rainfed conditions during the year 2012/2013 main cropping season at Haramaya, Ethiopia research field.

Description of the experimental materials

One hundred fourteen (114) sweet potato accessions and two released varieties (Adu and Barkume) were used in this study. The accessions were collected from eastern Ethiopia, other regions of the country and International Research Canters. The two varieties, Adu and Barkume were released for eastern Ethiopia for cultivation by Haramaya University in 2007 after fulfilling the requirements set by the National Variety Release Committee. The accessions were planted at Haramaya University research field using augmented design in 2012/13 main growing season (Table 1).

*Corresponding author: Solomon Ali, Department of Plant Science, Debre Markos University, Ethiopia, E-mail: ethiotrust@gmail.com

No	Accession	No	Accession	No	Accession	No	Accession
1	Tis-8441-11	30	Tis-8441-4	59	CEMSA	88	CN-1753-16
2	Tis-8441-3	31	Tis-9465-2	60	Bacariso	89	CN-1752-14
3	Tis-82/0602-12	32	Tis-80/043-3	61	Awassa-83	90	CN-2065-18
4	Tis-70357-7	33	Tis-9465-10	62	Nefissie	91	CN-2059-9
5	Tis-9465-7	34	Tis-9068-8	63	CN-2065-5A	92	CN-2065-16
6	Tis-8250-9	35	Tis-70357-5	64	CN-2065-11	93	CN-2065-15
7	Tis-9065-5	36	Tis-9465-8	65	CN-2065-1	94	CN-1753-5
8	Tis-82/0602-2	37	Becale type-3	66	CN-2065-10	95	CN-1775-4
9	Tis-80/043-1	38	Koka-26	67	CN-2065-7	96	CN-1775-3
10	Tis-9068-6	39	Wondogenet	68	CN-2065-8	97	CN-1753-1
11	Tis-82/0602-6	40	Tis-9068-2	69	CN-2065-12	98	CN-1753-7
12	Tis-82/0602-1A	41	Koka-9	70	CN-2065-5B	99	CN-1753-8
13	Tis-70357-4	42	Guracha	71	CN-2065-6	100	CN-1754-6
14	Tis-8250-4	43	Arbaminch	72	CN-2066-4	101	CN-1754-5
15	Tis-9465-1	44	Abadiro	73	CN-2066-2	102	CN-1754-3
16	Tis-9465-8	45	Koka-14	74	CN-1752-8	103	CN-1753-11
17	Tis-9065-1	46	Cuba-1	75	CN-1752-9	104	CN-1753-12
18	Tis-8441-1	47	Koka-12	76	CN-1752-15	105	CN-1753-13
19	Tis-9468-7	48	Becale	77	CN-2059-4	106	CN-1753-14
20	Tis-80/043-2	49	Becale type-1	78	CN-2059-3	107	CN-1753-17
21	Tis-82/062-11	50	Alemaya-local-2	79	CN-2059-20	108	CN-1753-18
22	Tis-8250-7	51	Alemaya-local-3	80	CN-2059-5	109	CN-1754-12
23	Tis-9465-9	52	Becale-type-2	81	CN-2059-8	110	CN-2054-5
24	Tis-9068-3	53	Lesh type—1	82	CN-1752-5	111	CN-2054-7
25	Tis-8250-8A	54	Korojo-1	83	CN-1752-6	112	CN-1754-11
26	Tis-8250-2	55	Becale-B	84	CN-2054-1	113	CN-1753-20
27	Tis-8250-1	56	Korojo	85	CN-2054-2	114	CN-1753-19
28	Tis-70357-2	57	Becale-1B	86	CN-1754-9	115	Adu
29	Tis-82/0602-1B	58	Korojo-2	87	CN-1753-15	116	Barkume

Note: Accessions started with Tis and CN were obtained from Nigeria and Asian Vegetable Center, respectively, and Cuba 1 was obtained from Cuba. The remaining are categorized as Alamaya collection, which were collected from eastern Ethiopia (Abadiro, Alemaya-local-2, Alemaya-local-3,) and other regions of the country such as central Ethiopia (Koka-26, Koka-9, Koka-12, Wondogenet), southern Ethiopia (Arbaminch, Awassa-83) and the last two varieties are released by Haramaya University.

Table 1: List of Sweet potato accessions and cultivars.

Experimental design and procedure

The Accessions were tested in augmented block design with 19 replications. Each replication contained 6 accessions and 2 checks. Each check was appearing once in each block. The checks were replicated 19 times and 114 entries/tests were not replicated. Hundred cm and 30 cm was maintained between rows and plant, respectively. Twelve holes per plot were prepared and one vine cutting was planted in each hole of the ridge and the size of each plot was 3.3 m × 7 m (23.1 m²).

Phenological and growth related traits

The following parameters were recorded from 10 plants in each plot left the two plants grown at both ends of each row/plot as border plant. Days to physiological maturity, Number of branches per plant, Vine length (cm) were determined using a standard procedure. Days to physiological maturity was recorded on plot basis.

Yield and yield components

Root fresh weight (g/plant), Above ground fresh biomass yield (g/plant), Above ground dry biomass (g/plant), Average number of storage roots per plant, Average mass of storage root (g/plant), Marketable storage roots number/plant, Unmarketable storage roots number/plant, Total storage root yield (t/ha), Marketable storage root yield (t/ha) and Unmarketable storage root yield (t/ha):

Storage root physical attributes

The following storage root physical attributes were recorded as per [4] descriptor for the crop, Storage root shape, Storage root defects, Storage root skin color and Storage root flesh color

Chemical attributes of storage roots

Chemical attributes of sweet potato accessions storage roots were measured through the following parameters and procedures. Sugar analysis, reducing sugar, total starch content, pH, total soluble solid, specific gravity, moisture content, peel content and dry matter were determined using a standard format

Statistical analysis

The data were subjected to analysis of variance using the Statistical package for augmented design (SPAD) software [5]. Means that differ significantly were separated using critical difference in each category.

Results and Discussion

Analysis of variance was computed for 22 phenological, growths, yield, yield components, physical and chemical attributes of sweet potato accessions and are presented in Table 2. The result revealed that the presence of highly significant differences (P<0.01) among accessions for reducing sugar, total sugar, total starch content, pH,

Trait	Replication (18)	Mean squares		Among control (1)	Among tests (109)	Tests vs control (1)
		Accession (115)	Error (18)			
DTPM	2	380.71**	2.58	351.4**	261**	463.4**
NB	2.32	4.81 ns	2.85	4ns	4.89ns	7.12ns
VL	336.38	436.66 ns	231.39	151.6ns	436.23ns	81.03ns
AGFBY	161301.68	106538.44 ns	56312.68	221885.81*	97485.46 ns	106516.65 ns
RFW	55646.28	66130.4*	26267.22	351318.6**	44089.43ns	2271536.94**
AGDBY	161301.68	106538.44 ns	56312.68	212975.84*	95178.10 ns	230199.24ns
ANSR	5.35	2.88 ns	3.96	7.5 ns	1.52 ns	6.87 ns
AMSR	6952.7	2158.2 ns	2278.8	21.9 ns	1957.5 ns	10632.8*
MSRN	1.36	0.95 ns	0.8	1.8 ns	0.6 ns	0.29 ns
USRN	2.62	1.15 ns	2.07	3.2 ns	0.78 ns	5.31 ns
TSRY	20.02	24.79*	9.59	53.8 ns	14.23**	234.46**
MSRY	10.4	15.15*	5.79	30.7*	9.09 ns	141.93**
USRY	8.84	4.67 ns	3.67		4.71 ns	1.38 ns
DM	13.01	21.61*	6.16	23.1*	16.83*	16.47 ns
PC	61.77	95.2*	37.86	212**	80.84*	1.68 ns
RS	0.42	2.18**	0.23	2.6**	2.12**	2.22*
TS	1.29	2.72**	0.55	5.7**	2.25**	0.94 ns
TSC	1.16	3.55**	0.49	11.9**	2.02**	5.8*
pH	0.04	2933.79**	0.04	0.05ns	3204.2**	117.35**
TSS	0.75	2.85**	0.9	5.6**	2.27*	0.22ns
MC	1.71	2.71ns	1.95	1.6 ns	2.27ns	14.36**
SG	0.73	15.26**	0.8	0.50 ns	18.63**	2.67 ns

*, ** and ns: Significant at P<0.05, P<0.01 and non significant, respectively.

DTPM: Days to Physiological Maturity; NB/pl: Number of Branches Per Plant; VL: Vine Length; AGFBY: Above Ground Fresh Biomass Yield; RFW: Root Fresh Weight; AGDBY: Above Ground Dry Biomass Yield; ANSR: Average Number of Storage Roots Per Plant; AMSR: Average Mass of Storage Roots; MSRN: Marketable Storage Root Number; USRN: Unmarketable Storage Root Number; TSRY: Total Storage Root Yield; MSRY: Marketable Storage Root Yield; USRY: Unmarketable Storage Root Yield; DM: Dry Matter Content; PC: Peel Content; RS: Reducing Sugar; TS: Total Sugar; TSC: Total Starch Content; TSS: Total Soluble Solid; MC: Moisture Content and SG: Specific Gravity.

Table 2: Mean squares for 22 traits of sweet potato [*Ipomoea batatas* (L.) Lam.] collections on the basis of adjusted means.

total soluble solid, days to physiological maturity and specific gravity while significant differences (P<0.5) was observed for root fresh weight, total storage root yield, marketable storage root yield, dry matter content and peel content. However, non-significant differences among accessions was observed for number of branches per plant, vine length, above ground dry biomass yield, average number of storage roots per plant, average mass of storage roots, marketable storage root number/plant, unmarketable storage root number/plant and moisture content.

As the results are presented in Table 2, there were highly significant (P<0.01) differences between the control (check) varieties for days to physiological maturity, storage root fresh weight, dry peel content, reducing sugar, total sugar, total starch content and total soluble solid while significant (P<0.5) differences were observed for above ground fresh biomass yield, above ground dry biomass yield, marketable storage root yield, dry matter. However, non-significant differences between the check varieties was observed for number of branches per plant, vine length, average number of storage root, marketable storage root number, unmarketable storage root number, average mass of storage root, total storage root yield, pH, moisture content and specific gravity.

Analysis of variance also exhibited highly significant (P<0.01) differences among tests for days to physiological maturity, total storage root yield, reducing sugar, total sugar, total starch content, pH and specific gravity. Likewise significant (P<0.5) differences were exhibited among test entries for dry matter content, peel content and total soluble solid. However, number of branch per plant, vine length, above ground fresh biomass, above ground dry biomass, storage root fresh biomass, average number of storage root, marketable storage root number, unmarketable storage root number, average mass of storage root, marketable storage root yield, unmarketable storage root yield and moisture content were found to be non significant. The result in Table 2 revealed that the presence of highly significant (P<0.01) differences among test versus control for days to physiological maturity, storage root fresh weight, total storage root yield, marketable storage root yield, total starch content, pH and moisture content while significant differences (P<0.5) was observed for average mass storage root, reducing sugar and total sugar. However, non-significant differences among test versus control were observed for number of branch per plant, vine length, above ground fresh biomass, above ground dry biomass, average number of storage root, marketable storage root number, unmarketable storage root number, unmarketable storage root yield, dry matter content, peel content, total soluble solid and specific gravity.

Generally, it was observed significant differences among entries, among test versus control/check varieties, between check varieties of sweet potato studied for considerable number of traits which can be exploited in breeding program or that will allow breeders to select entries for desirable trait(s) that they wish to improve [6]. Describes two basic principles for plant breeding, 'selection for yield' and 'defect elimination'. Therefore, the basic philosophies behind plant breeding programme are to develop cultivars with better yield potential and quality attributes as well as to develop cultivars that have genetic resistance against production hazards that can prevent a cultivar from expressing its yield potential [7]. Based on these philosophies the sweet potato breeding programme may relies on improvement of storage roots yield and improved the quality of the storage roots as per the end use and the observed differences among entries may allow the breeders to use accessions for different objectives.

Variation is the occurrence of difference among individuals due to difference in their genetic composition and/or the environment in which they are raised [8,9]. If the character expression of two individuals could be measured in an environment exactly identical for both, difference in expression would result from genetic control and hence such variation is called genetic variation [8]. The presence of variation in the germplasm for the trait of interest is, therefore, very important. Therefore, information generated in this study on variation of accessions can be utilized by the breeders since the observed variability greatly helps in formulating sound crop breeding and improvement program [10] (Table 2).

Phonological and growth traits

Days to physiological maturity: The data in Table 2 showed that day to physiological maturity among accessions, between check/control varieties and among test versus control was highly significantly (P<0.001). Accessions registered for physiological maturity ranged from 199 to 111 days. Koka-12 and CN-2059-4 accessions were late maturing than most of the accessions and checks, respectively. However, when early maturing is considered as desirable trait, Korojo-2 and Tis-9065-5 accessions were earlier than most of the accessions and check varieties. Sixteen accessions mature earlier than checks.

These accessions had the advantage of earliness than others. Particularly, Korojo-2 (111 days) and Tis- 9065-5 (120 days) significantly matured earlier than checks while Arebaminch, Koka-12, CN-2059-4 were significantly late maturing than both checks.

The observed differences of maturity among accessions as well as checks and entries or test accessions may be mainly attributed to the genetic constitution of the new entries as well as check varieties since all accessions were tested in one location with similar management. This suggestion might be supported by Zhang et al., [11] who reported that physiological maturity is genetically controlled trait in sweet potato. In agreement with this study result Teshome et al. [12] reported that four sweet potato varieties tested at Adamitulu was between 114 and 124 days for days to physiological maturity [13].

Number of branches per plant and vine length: Non- significant differences were observed among accession, control, tests and tests versus control for number of branches per plant (Table 2).

The number of branches per plant was ranged from 20.447 and 1.47. The largest number of branches per plant was recorded on CN-2065-15 and CN-2054-7 than most of the accessions and checks. However, Korojo-2 and Tis-82/0602-6 exhibited the lowest number of branches per plant than the rest of the accessions. The observed differences of branch number among accessions as well as checks and accessions may be mainly attributed by the genetic constitution of the new entries as well as check varieties. This suggestion might be supported by Juo and Mukhtar et al. [14,15] who reported that branch number is genetically controlled trait in sweet potato.

Similar to the result for number of branches per plant, it was observed that non- significant differences for vine length among accession, control, tests and tests versus control (Table 2). The mean of the check varieties for vine length was 58.487 cm while the mean of accessions was 67.729 cm. The length of vine was ranged between 3.62 and 138.862 cm. Nefissie and CN-2065-5A were exhibited the longest vine than most of the accessions including check while Tis-9465-1 (3.62 cm) and Tis-9465-8 (7.112 cm) was recorded the shortest from

the entire accessions. [15] Mukhtar reported that the difference among accession, checks and checks and accession was due to difference in genetic constitution.

Mean performances of accessions for yield and yield components

Above ground fresh and dry biomass yield: Analysis of variance results presented in Table 2 revealed that the presence of significant (P<0.05) difference between checks/among control for above ground fresh and dry biomass but non- significant differences were observed among accessions, tests and test versus control. Though, statistically non- significant differences were observed among accessions, the variation among accessions was too large which ranged from 265 to 5060 g for above ground fresh biomass. The mean of accessions for above ground fresh biomass was 1875.88 g, while the mean of checks was 1365 g. Accession CN-2059-7 and Koka-9 were registered highest above ground fresh biomass than most of the accessions and checks. However, Neffsie and CN-2065-1 accessions were exhibited the lowest above ground fresh biomass yield than others.

The observed differences of above ground fresh biomass among accessions as well as checks and accessions may be mainly attributed by the genetic constitution of the new entries as well as check varieties. This statement is in agreement with Chowdhury and Mukhtar [15,16] who reported that above ground fresh biomass yield is genetically controlled trait in sweet potato. In agreement with this study result [12] reported that the above ground fresh biomass yield of sweet potato varieties were in the range between 429.23 to 2516 g, the mean of accession for above ground dry biomass yield was 952.3 g and mean of checks was 785.5 g. The range for above ground dry biomass was between 210 and 2145 g. Accession Abadiro and Koka-12 were found to be superior for above ground dry biomass yield than most of accessions. On the other hand, CN-1753-8 and Tis-8250-8A were inferior for above ground dry biomass yield since the accessions exhibited the lowest values from all accessions including checks.

Storage root fresh weight and average mass of storage root: Analysis of variance showed that there were significant (P<0.05) differences among accessions and the presence of highly significant (P<0.01) differences between check varieties/among control and control versus tests but differences among tests were not significant for storage root fresh weight (Table 2). The mean storage roots fresh weight of accessions and checks were 484.51 and 486.09 g, respectively. The range for storage root fresh weight was between 74.94 to 854.26 g.Tis-9068-7 and Koka-12 were recorded the highest fresh weight than most of the accessions (mean storage root fresh weight) and checks. Likewise, the adjusted means of CN-1754-6 and Nefissie were the lowest storage roots fresh weight from most of accessions. Janssens [17] reported that the difference among accessions for storage roots weights was due to differences in genetic constitution of genotypes tested [18], ewthwaite, Kenneth and Richardson reported that the weight of fresh storage roots of sweet potatoes were ranged from 210 to 716 g and 280 to 1520 g, respectively, which the results are in agreement with the present finding.

Non- significant differences were observed among accessions and tests for average mass of storage roots but it was evident that the presence of significant (P<0.05) differences for test versus control (Table 2). The mean of the accessions for average mass of storage root was 140.05 g while the mean of checks was 168.5 g. Accessions for average mass of storage roots ranged between 18 to 318 g.Tis-9060-8 and Tis-82/0602-12 were exhibited the highest average mass of storage root of all entries as well as the mean of checks while CN-1753-13 and CN-1753-14 ranked the first from the last. The observed differences of average mass of storage root among accessions as well as checks and accessions could be due to genetic constitution of the new entries as well as check varieties since all accessions were tested in one location with similar management. This suggestion might be supported by Mukhtar et al. [15] who reported that average mass of storage root is genetically controlled trait in sweet potato.

Marketable, unmarketable and total storage root number: Statistically non-significant differences were observed for marketable, unmarketable and total storage root number among tests, accessions and tests versus control (Table 2). However, the range for these parameters was too large. Accessions registered for total marketable storage roots number were 2 to 11, 0.695 to 5.395 and 0.816 to 6.316 marketable and unmarketable root number, respectively. CN-2065-16 and CN-2065-15 had the highest number of marketable storage root where as CN-2065-16 and CN-2065-15 were exhibited the highest number of unmarketable storage root and CN-2065-16 and Becale were recorded the highest number of total storage roots number. on the other hand accession CN-2054-7 and Tis-82/0602-1A were recorded the lowest number of marketable storage roots and Tis-8250-9 and Tis-8441-3 had the lowest unmarketable storage roots number while CN-1753-14 and CN-1753-13 were exhibited the lowest storage roots number from most of accessions. Marketable as well as unmarketable storage roots number was highly controlled by genetic constitution as well as environment [19]. There was a variation in storage roots number between cultivars. The observed differences among accessions in this study might be due to genetic differences since all accession as well as checks receive equal management and treatment.

Marketable, unmarketable and total storage root yield: Highly significant (P<0.01) difference was exhibit between test and control and also there was a significant (P<0.05) difference among accessions but the difference among tests was not significant (Table 2) for marketable storage root yield. The result in Table 2 revealed that non-significant differences among accessions, checks, tests and tests versus control for unmarketable storage root yield. The differences which were exhibited among tests and tests versus control were highly significant (P<0.01), but non-significant differences among accessions and between check varieties/ among control for total storage root yield.

The mean of accessions for marketable, unmarketable and total storage root yield was 8.143 t/ha, 3.9 t/ha and 12.025 t/ha while the mean of checks for marketable unmarketable and total storage roots yield were 11.77, 4.431 and 16.21 t/ha, respectively.

marketable storage root yields of Tis-9465-7 and Tis-82/0602-12 were highest among test accessions including checks where as Unmarketable storage root yield of Koka-12 and Bacale exhibited the highest and total storage root yields of Tis-9465-7 and Koka-12 recorded the highest among test accessions and checks. However, Abadiro and Tis-82/0602-1B had low yield for marketable storage root yield and unmarketable storage root yields of Tis-9465-1 and CN-2065-11 were the first from the last in other word those accessions that had lowest unmarketable yield also had highest yield among test accessions. Total storage root yields of Tis-9465-1 and Nefissie were the lowest.

The range for marketable and unmarketable storage root yields were between 0.512 and 22.088 t/ha, 0.001 and 13.631 t/ha, respectively, and the yields of total storage roots ranged from 2.261 to 28.461 t/ha. Marketable storage root yield of sweet potato was inherited genetically [14]. The current result agree with Mwololo [13] who reported the total

storage root yield of sweet potato collections was ranged between 10.3 and 32.22 t/ha.

Dry matter and moisture content: Analysis of variance showed that significant (P<0.05) differences among accessions, tests and control but there was non-significant difference among tests versus control for dry matter content while for moisture content it was observed highly significant (P<0.01) differences among test versus control but difference among accessions and control and tests were not significant (Table 2).

The mean of accessions were 24.878% and 8.85%, respectively, while the mean of checks were 28.42% and 8.8 %, for dry matter content and moisture content, respectively. Dry matter content was ranged from 13.275% to 40.215 % likewise the range of moisture content was between 1.003% and 16.698%.

CN-1753-18 (40.215%) and CN-1753-11 (39.535%) had the highest dry matter content among test accessions including checks. On the other hand, Abadiro and Tis-9468-7 had the lowest dry matter content. Tis-9468-7 and Tis-82/0602-6 were the first in moisture content. On contrary, Becale-B and Tis-8250-2 found to be the first from the last.

The observed differences of dry matter content and moisture content among accessions as well as checks may be mainly due to genetic constitution of the new entries as well as check varieties since all accessions were tested in one location with similar management. This suggestion might be supported by Dominguaz [20] who reported that dry matter content is genetically controlled trait in sweet potato. Catherine and Scott [21,22] reported that the dry matter percentage of different sweet potato varieties were between 13.4 and 29.2%, 25.23 and 41.11%, respectively. Tsakama, Fred, Bonsi and Loretan [23-26] also found that the dry matter in storage roots were ranged from 12.5 to 30.2, 29 to 39.07, 25 to 42 and 25.5 to 31.7%, respectively. As reported by Chen et al., [27] the dry matter percentages of Xushu18, Sushu2 and Sushu8 were 31.9%, 36.7% and 18.6%, respectively. The dry matter of Beauregard, White Star and skin of White Star variety were 17.54%, 17.89% and 18.97%, respectively [28].

Specific gravity and peel content: As it was presented in Table 2, highly significant (P<0.01) differences was observed among tests for specific gravity and there was no significant differences among control and tests versus control. Differences among control was highly significant (P<0.01) and for accessions and tests were significant (P<0.05), but it was non-significant among accession and tests (Table 2) for peel content. The mean of accessions for specific gravity and peel content were 2.194 and 34.709, respectively while the mean of checks was 1.75 for specific gravity and 37.225 for peel content. The result showed that the range for specific gravity was between 0.046 and 42.334 and for peel content was 14.41 and 78.04. The result showed that specific gravity of CN-2054-1 and CN-1753-16 were the highest values among test entries and checks. Tis-70557-2 and Tis-8250-2 had the smallest among test accessions as well as checks. The difference observed between checks, accessions and checks and accession was due to differences in genetic constitution. This statement might be supported by Ruinard [29].

Accession Korojo and Tis-82/0602-12 registered for peel content was the highest from most of accession including checks. Whereas accession Tis-82/0602-6 and Tis-9465-9 were the least from all new entries as well as checks. The difference observed between checks, accessions and checks and accession may be due to differences in genetic constitution. This statement might be supported by Surayia [26]. Who reported that genetic constitution of each accession contributed for difference of peel content.

Storage root shape, color and defect: Sweet potatoes are nutritious and have numerous health benefits. The orange-fleshed and white-fleshed cultivars are the most familiar to consumers. However, it is also possible to identify other cultivars with varying shape, defect, flesh and skin color. As general remark qualitative traits such as shape, color, defect is genetically controlled traits [30,31].

Storage root shape: Figure 1 showed that the storage root shape of 27.19% of 114 sweet potato accessions was elliptic and 18% of accession was long irregular. Other accessions had also different storage root shape i.e., ovate (15.78%), obovate (9.65%), long elliptic (7.89%), long oblong (7.89%), round elliptic (7.89%), round (4.38%) and oblong (1.75%). The highest number of accessions was exhibited elliptic, however; only six accessions had oblong shape. Most of accessions had similar storage root shape of the checks (Figure 1).

Storage root defect: There was a variation between checks and accession this may be due to difference in genetic makeup of each accession. Storage root orientation was varied with varieties. Defects were different between accessions. The defect of both checks Adu and Berkume was Alligator-like skin. As it is presented in Figure 2, 45.62% of 114 test entries had a defect of Alligator-like skin which was similar to the check varieties. The defect 30.70% and 23.68% of 114 accessions were horizontal constriction and longitudinal grooves, respectively. Alligator-like skin sweet potato accessions were the highest percentage proportion from the entire accessions where as longitudinal grooves was showed the lowest percentage proportion (Figure 2).

Storage root skin color: Sweet potato skin color was different for check varieties which Adu had pink color where as Berkume was creamy. Majority of root skin color of accessions was white which accounts 22.8% followed by pink (21.92%). Variation of skin color was due to genetic difference. Storage root skin color of 21.25%, 14.9%, 8.77%, 4.38%, 3.51% and 2.63% of the accessions were cream, brown, purple red, orange, yellow and red, respectively. Pink color was the highest (22.8%) percentage proportion, however, red account the lowest (2.63%) percentage proportion (Figure 3).

Storage root flesh color: The flesh color of sweet potato accessions was different which may be due to genetic difference of accessions. Adu and Berkume exhibited cream and white flesh color, respectively. The flesh color of most of accessions was different. 37.79% of sweet potato collections exhibited cream color followed by white (36.84%). Dark cream was accounts 22.8% and 1.75% of accessions flesh color was pale yellow. Cream color was the highest flesh color percentage where as pale yellow exhibit lowest proportion (Figure 4).

Chemical attributes of collections

Reducing sugar: Highly significant (P<0.01) differences among accession, tests and control and significant (P<0.05) differences were observed among tests versus control for reducing sugar (Table 2). The mean of accession those registered for reducing sugar was 6.143 mg 100 g^{-1} while mean of check was 6.346 mg 100 g^{-1} and reducing sugar content was ranged from 2.576 to 10.331 mg 100 g^{-1}.CN-1752-14 and CN-1752-9 was exhibited that the highest reducing sugar content from most of accession. Whereas Neffsie and Korojo-1 were registered the lowest reducing sugar which is consider as desirable. The difference here between accession, checks and check and accessions may be due to genetic differences for the trait which this statement was in agreement

with Frankin [32] who reported that reported that reducing sugar of storage root was genetically controlled traits.

Hacineza and Picha [33,34] stated that the total reducing sugar in fresh sweet potato was 6.94 mg 100 g^{-1} and, 7.84 mg 100 g^{-1}, respectively. Walter [35] reported that the concentration of reducing sugar of fresh fry type sweet potato ranges from 5.88 to 6.31 mg g^{-1} similar result were also found by Loretan [26]. This work is in agreement with the findings of Ruinard [29] who reported that the reducing sugar concentration of four varieties was between 2.9 and 5.8 mg 100 g^{-1}.

Total sugar: Analysis of variance in Table 2 showed that there was highly significance difference (P<0.01) among accessions, tests and control but among tests versus control the difference was statistically non-significant for total sugar content. The mean of accessions was 13.305 mg 100 g^{-1} while the mean of checks was 13.603 mg 100 g^{-1}.

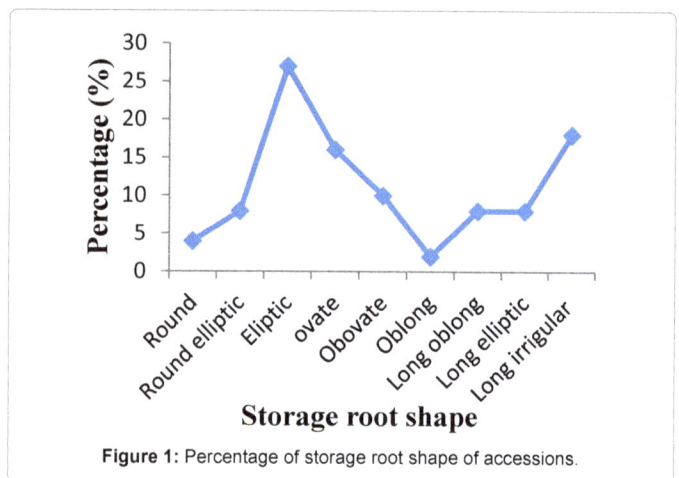

Figure 1: Percentage of storage root shape of accessions.

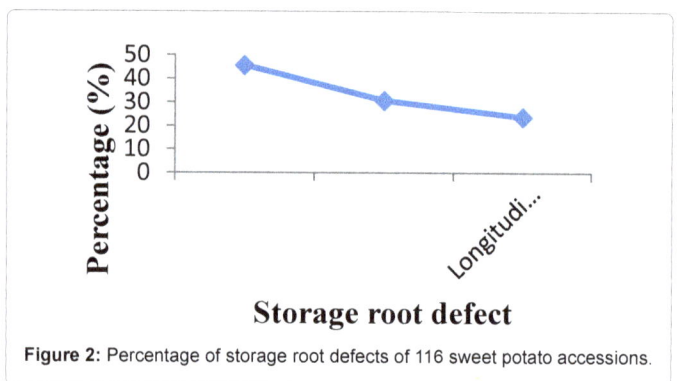

Figure 2: Percentage of storage root defects of 116 sweet potato accessions.

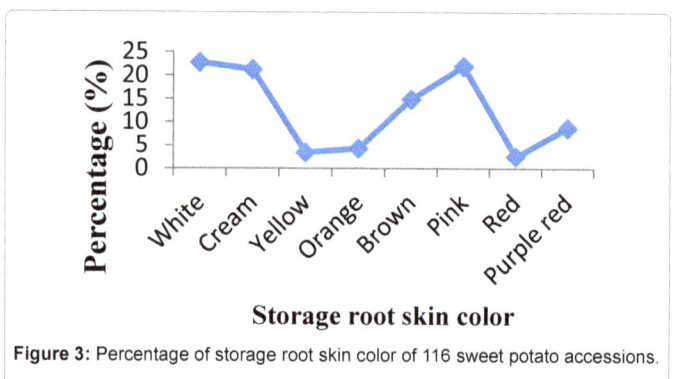

Figure 3: Percentage of storage root skin color of 116 sweet potato accessions.

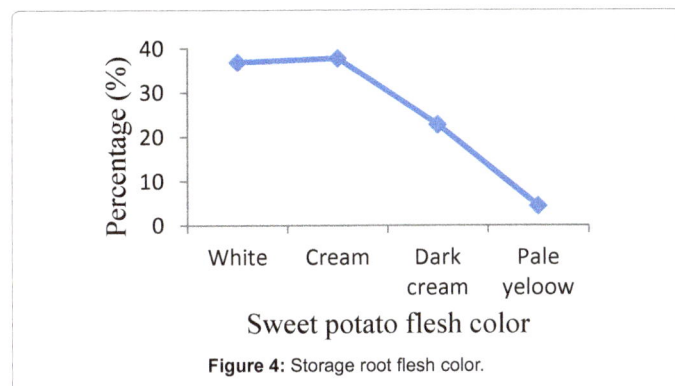

Figure 4: Storage root flesh color.

The range for total sugar concentration was between 9.533 and 17.258 mg 100 g^{-1}. Total sugar concentration of CN-1752-15 and CN-2059-7 was found to be the highest among accessions. However, accession Tis-80/063-3 and Tis-9465-2 were exhibited the lowest. The observed differences of total sugar content among accessions as well as checks and accessions may be mainly attributed to the genetic differences of the entries as well as check varieties since all accessions were tested in one location with similar management. This suggestion might be supported by Frankin [32] who reported that total sugar concentration is genetically controlled trait in sweet potato. Andrade [36] reported that the concentration of total sugar of five sub Saharan Africa sweet potato collection was laid between 1.7 mg 100 g^{-1} to 27 mg 100^{-1} which this result was strongly agree with the present result. According to Onwueme [37] the range of recommended total sugar concentration was between 6.98 to 14.59 g 100 g^{-1} and this result strongly agree with the present finding. Average (11.2 mg 100 g^{-1}) total sugar concentration of four sweet potato varieties was recorded by Hamed [38].

Total starch content: There was highly significant (P<0.01) differences among accessions, tests and control and there was significant (P<0.05) differences among tests versus control for total starch content (Table 2). The mean of accessions for total starch content was 12.569 mg 100 g^{-1} while the mean of check was 12.923 mg 100 g^{-1}.

Total starch content concentration was ranged from 1.167 to 16.402 mg 100 g^{-1}. CN-1752-15 and CN-2059-7 had the highest concentration of total starch content among accessions and checks. WhereasTis-7035-7 and Tis-80/043-3 had the lowest total starch content. There was a difference in total starch concentration between accession, checks and checks and accession. The observed differences may be due to genetic differences among accessions. This suggestion is in agreement with [23].

The present study results agrees with Ruinard [29] report that the total starch content of four sweet potato varieties was laid in the range between 13 mg 100 g^{-1} and 21 mg g^{-1}. Similar results were also reported for Xushu18, Sushu2 and Sushu8 varieties by Chan [27].

pH value and total soluble solid: There was highly significant (P<0.01) differences among accessions, tests and tests versus control but non-significance differences was observed between check varieties/ among control for pH (Table 2). Likewise, highly significant (P<0.01) differences were exhibited among accession and tests and difference among tests were significantly (P<0.05) different for TSS but there was non-significant difference among tests versus control (Table 2). The mean of accessions for pH and TSS was 6.203 and 12.138° brix, respectively, while the mean of checks was 6.06 (pH) and 12.637° brix (TSS). The pH and TSS values were ranged from 5.044 to 7.264 and

7.132 to 7.132° brix, respectively. The pH value of CN-2065-8 and Tis-9465-8 were found to be the highest. However, Tis-82/0602-1A and Becale-1 had the lowest pH values. Kure [39] reported that the pH value of seven sweet potato varieties was ranged from 5.5 to 7.1. Aina and Woolfe et al. [40,41] reported that the range of pH in sweet potato varieties were ranged from 5 to 6.9 and 5.5 to 6.7. These findings are strongly agreed with the present findings.

Summary and Conclusion

Four accessions showed significantly delayed maturity than both checks mean. Koka-12 was found to be the latest maturing accession whereas Korojo was found to be early maturing. CN-2059-7 and CN-1953-1 had the highest and the lowest above ground fresh biomass yield from all entries, respectively. Storage root fresh weights of Tis-9465-7 and CN-1953-7 were the maximum and minimum storage root fresh weight, respectively. Marketable storage root yields of three entries (Tis-9465-7, Tis-82/0602-12 and Tis-70357-7) were significantly highest yield among accessions including checks. Tis-9465-7 was high yielder whereas Tis-82(0602)-1B was the poor yielder. Total storage root yield of Tis-9465-7 was significantly highest while Neffissie storage root yield was the lowest.

The physical attributes of all accessions were evaluated considering shape, defect, skin and flesh color. Accessions had different shape of storage root. The shape of checks (Berkume and Adu) was laid between elliptic to long irregular. Elliptic storage root shape accounts the largest proportion of accession (27.19%) whereas the lowest proportion was oblong (1.75%). About 45.62% of entries storage root defect was horizontal constriction and 23.68% of accessions had longitudinal groove storage root defect. White storage root skin color was dominant which accounts 22.8% of entries where as only few accessions (2.5%) had red skin color. Many accessions (37.79%) had creamy storage root flesh color while only 1.75% of accessions had pale yellow color.

CN-1852-14 and Neffissie had the highest and the lowest content of reducing sugar, respectively. Concentration of total sugar was highest in CN-1752-15 and lowest in Tis-80/043-3. CN-1752-15 and Tis-70357-7 exhibited the highest and the lowest content of total starch concentration, respectively. CN-2065-8 and Tis-80/043-1 had the highest pH and TSS, respectively, whereas Tis-82/0602-1A and CN-1752-8 were found to be posses the lowest pH and TSS, respectively. CN-2054-1, CN-1753-18 and Korojo had the highest specific gravity, dry matter and peel content, respectively. Tis-8250-2, Tis-9468-7 and Tis-82/0602-6 registered the lowest specific gravity, dry matter and peel content, respectively.

References

1. Taboge E, Lema G, Dametew M, Belehu T (1992) Improvement Studies on Ensete and Sweet Potato. Pp. 63-74. In: Horticultural Research and Development in Ethiopia. Proc second. Natl. Hort. Workshop of Ethiopia, 13 Dec., Addis Ababa, Ethiopia.

2. Adhanom N, Tsedeke A, Emana G (1985) Research on Insect Pests of Roots and Tuber Crops: In: Abate T (ed.) A Review of Crop Protection Research in Ethiopia: Proceedings of the first Ethiopian Crop Protection Symposium, 4-7 February, Institute of Agricultural Research, Addis Ababa, Ethiopia.

3. CSA (Central statistical authority) (2011) Ethiopian Agricultural Sample Enumeration, 2009/10; statistical bulletin, 146.Addis Ababa.

4. Huaman Z (1991) IPGRI (International Plant Genetic Resources Institute) Descriptors of sweet potato, Italy. 25-56.

5. Abhishek R, Parsad R, Gupta VK (2010) Statistical package for augmented design (SPAD) I.A.S.R.I, Library Avenue, New delhi.

6. Donald CM (1968) the breeding of Crop ideotype. Euphytica Journal 17: 385-403.

7. Acquaah G (2007) Principles of plant genetics and breeding. Black well publishing. 350 Main Street, Malden, MA 02148-5020, USA. 584.

8. Falconer DS (1990) Introduction to quantitative genetics.3rd ed. John Wiley and Sons, Inc. New York. 450.

9. Allard RW (1960) Principles of plant genetics and breeding. John Willy and Sons, Inc., New York. 663.

10. Welsh RJ (1990) Fundamentals of plant genetics and breeding. John Wiley and Sons, New York

11. Zhang D, Ghislain M, Huaman Z, Golmirzaie A, Hijmans RJ (1998) RAPD variation in sweetpotato [Ipomoea batatas (L.) Lam] varieties from South America and Papua New Guinea. Genetic Resources and Crop Evolution 45: 271-277.

12. Teshome A, Nigussie D, Yibekal A (2012) Sweet Potato Growth Parameters as Affected by Farmyard Manure and Phosphorus Application at Adami Tulu, Central Rift Valley of Ethiopia. Agricultural Science Research Journal 2: 1-12.

13. Mwololo JK, Mburu MKW, Muturi PW (2012) Performance of sweet potato varieties across environments in Kenya. International Journal of Agronomy and Agricultural Research 2: 1-11.

14. Juo ASR (1983) Selected method for soil and plant analysis. Manual Series, NO.1. Ibadan, Nigeria, IITA.

15. Mukhtar AA, Tanimu B, Arunah UL, Babaji BA (2010) Evaluation of the Agronomic Characters of Sweet Potato Varieties Grown at Varying Levels of Organic and Inorganic Fertilizer, World Journal of Agricultural Sciences 6: 370-373.

16. Chowdhury SR, Naskar SK (1993) Screening of drought tolerant traits in sweet potato: role of relative water content. Orissa Journal of Horticulture 21: 1-4.

17. Janssens MJJ (2001) Sweet Potatoes (Ipomoea batatas L) Lam. In: Raemaekers RH (ed.), Crop Production in the tropical. 205-220.

18. Lewthwaite SL, Triggs CM (2010) Sweet potato cultivar response to prolonged drought. The New Zealand Institute for Plant & Food Research Limited, Pukekohe Research Centre, 49 Cronin Road, RD1, Pukekohe, New Zealand. 03-11.

19. Huaman Z, Aguilar C, Ortiz R (1999) Selecting a Peruvian sweet potato core collection on basis of morphological, co-geographical, and disease and pest reaction data. Theoretical and Applied Genetics 98: 840-845.

20. Dominguaz PL (1976) Feeding of Sweet Potato in Monogastrics. Roots, Tuber and Plantains and Bananas in Animal Feeding. 217-233.

21. Catherine G, Fred T, Emmarold M, Alois K (2012) Characterization of Tanzanian elite sweet potato genotypes for sweet potato virus disease (SPVD) resistance and high dry matter content using simple sequence repeat (SSR) markers, African Journal of Biotechnology 11: 9582-9590.

22. Scott GJR, Best R, Rosegrant M, Bokango M (2000) Roots and tubers in the global food system: A vision statement to the year 2020, Lima Peru: International Potato Centre 111.

23. Tsakama M, Mwangwela AM, Manani TA, Mahungu NM (2010) Physicochemical and pasting properties of starch extracted from eleven sweet potato varieties. African Journal of Food Science and Technology 1: 90-9.

24. Fred TEM, Alois (2008) Morphological and agronomical characterization of Sweet potato [Ipomoea batatas (L.) Lam.] germplasm collection from Tanzania. African Journal of Plant Science 2: 077-085.

25. Bonsi C, David PP, Mortley DG, Loretan PA, Hill WA (1994) Selecting Sweet Potato (Ipomoea Batatas) Cultivars for Hydroponic Production. Proceedings of the Tenth Symposium of the International Society for Tropical Root Crops, held in Salvador, Bahia, Brazil.

26. Loretan P, Bonsi CK, Hill WA, Ogbuehi CR, Mortley DG, et al. (1989) Sweet potato growth parameters, yield components and nutritive value for CELSS applications. SAE Tech Paper Ser 891571.

27. Chen Z, Schols HA, Voragen AGJ (2006) Physicochemical Properties of Starches Obtained from Three Varieties of Chinese Sweet Potatoes. Journal of Food Science 68: 431-437.

28. Surayia Z, Sarwar M, Jonathan A, Khan MN, Butt MS (2006) Variation in Physio-Chemical Characteristics of Some Cultivars of Sweet Potato.Pak. J. Bot 38: 283-291.

29. Ruinard J (1976) Notes on Sweet Potato Research in West New Guinea sweet potato symposium, West Irian.

30. Tairo F (2006) Molecular resolution of genetic variability of major sweet potato viruses and improved diagnosis of potyviruses co-infecting sweet potato. Doctoral diss. Dept. of Plant Biology and Forestry Genetics, SLU. Acta Universitatis agriculturae Sueciae 5.

31. Gichuki ST, Barenyiv M, Zhang D, Hermann M, Schmidt J, et al. (2003) Genetic diversity in sweetpotato [Ipomoea batatas (L.) Lam.] In relationship to geographic sources as assessed with RAPD markers. Genetic Resources and Crop Evolution 50: 429-437.

32. Franklin W, Martin (1988) Starch-Sugar Transformation on Sweet and Staple-Type Sweet Potatoes. Tropical Agriculture Research Station, Southern Region Agricultural Research Service, U.S. Department of Agriculture Mayaguez, Puerto-Rico.

33. Hacineaza E, Vasanthakaalam H, Ndirigwe J, Mukwantali C (2010) Comparative study on the β-carotene content and its retention in yellow and orange fleshed Sweet Potato flours. Association for Strengthening Agricultural Research in Eastern and Central Africa 5-8.

34. Picha DH (1985) HPLC determination of sugars in raw and baked sweet potatoes. J. Food Sci 50: 1189-1190.

35. Walter WM Jr, Hoover MW (1986) Preparation, evaluation and analysis of French-fry-type product from sweet potatoes. J Food Sci 51: 967-970.

36. Andrade M, Barker I, Cole D, Dapaah H, Elliott H, et al. (2009) Unleashing the potential of sweet potato in Sub-Saharan Africa: Current challenges and way forward. International Potato Centre (CIP), Lima, Peru. Working Paper: 16-18.

37. Onwueme IC (1991) the tropical tuber crops: Yams, Cassava, Sweet Potato and Cocoyams. John Wiley and Sons, Chichester, UK 189-191.

38. Hamed M, Siliha H, Sandy SK (1973) Preparation and chemical composition of sweet potato flour. Cereal and Bakery Products 50: 133-139.

39. Kure OA, Nwankwo L, Wyasu G (2012) Production and quality evaluation of garri-like product from sweet potatoes. J Nat Prod Plant Resour 2: 318-321.

40. Aina AJ, Falade KO, Akingbala JO, Titus P (2009) Physicochemical properties of twenty-one sweet potato cultivars. Int J Food Sci Techno 44: 1696-1704.

41. Woolfe JA (1992) Sweet potato, an untapped food resource. Cambridge University Press, Cambridge, UK: 643.

12

Effect of Nitrogen and Compost on Sugarcane (*Saccharum Officinarum* L.) at Metahara Sugarcane Plantation

Zeleke Teshome*, Girma Abejehu and Hadush Hagos

Sugar Corporation, Research and Training, P.O. Box 15, Wonji, Ethiopia

Abstract

A study was conducted at Metahara Sugar Estate in Ethiopia with the objectives to determine the optimum rates of nitrogen and compost for sugarcane production, and the effect of compost on soil chemical properties. Four levels of nitrogen (0, 46, 92, and 138 Kg /ha) and five levels of compost (0, 5, 10, 15, and 20 ton/ha) were combined in factorial arrangement. The experiment was laid out in randomized complete block design with three replications. It was conducted on clay soils (Haplic Cambisols). Soil samples were taken before and after treatment application, and analyzed for pH (1:2.5), ECe, organic carbon, total N, available P, and available K. All the cultural practices were executed as per the estate's practice. Analysis of the soil indicated that except for pH value, all the salient soil properties including ECe, organic carbon, total nitrogen, available P, and available K were slightly increased due to compost application. Analysis of variance indicated that interaction effect between nitrogen and compost was significant ($p<0.05$) on stalk girth, stalk weight, cane yield, and sugar yield. Highest cane and sugar yields were obtained when 46 Kg N/ha applied with 15 ton of compost/ha. Net benefit to cost ratio was also found to be the highest, 1.12, for this combination. Therefore, 46 Kg N/ha with 15 ton of compost/ha is recommended for clay soils (Haplic Cambisols).

Keywords: Sugarcane; Compost; Filtercake; Vinasse; Nitrogen; Sugar yield

Introduction

Filter cake and vinasse are the common sugar industry byproducts. Vinasse is a corrosive contaminant liquid residue resulting from alcohol distilleries generated in great amounts which ranges from 12-15 liters per liter of alcohol production [1]. Filter cake is one of sugar processing byproduct mostly used as manure in sugarcane fields as it contains 2.0% P_2O_5, 1.0% nitrogen on dry matter basis, and some organic matter [2]. Vinasse contains high levels of organic matter, potassium, calcium and moderate amounts of nitrogen and phosphorus [1-4], particularly it is rich in potassium. Application of vinasse to the soils for sugarcane fertilization is known to reduce its deposition in waterways and its contaminant effects besides increasing sugarcane yields and reducing fertilizer expenses [5]. In many countries, vinasse is directly applied to sugarcane fields as a substitute for conventional potassium fertilizer products [6,7]. Nevertheless, composting of vinasse with filter cake reduces its deposition and contaminant effects through decomposition of raw organic matter in to soil building humus. In India, vinasse is mixed with filtercake to produce compost. Its application to soil increased the levels of organic matter, K, and P in the soil [2].

In most cases, recycling of vinasse either by direct application or after composting could be considered as an important disposal mechanism especially from environmental point of view. At Metahara Sugar Estate, filter cake together with mineral fertilizers have been used in cane production. Since recently, however, Metahara Sugar Estate is producing large quantity of vinasse from ethanol factory; and the estate is preparing compost for field application by mixing vinasse with filter cake to ensure safe disposal of vinasse. The application of filtercake-vinasse compost to sugarcane fields is used as disposing mechanism in addition to its use as potential source of fertilizer. However, nutrient content of the compost, rate of application, and its effects on sugar yield and soil properties were not yet known.

Therefore, the study was initiated with the following objectives:

- To determine optimum rates of nitrogen and compost for sugarcane production

- To determine the effect of compost on selected soil chemical properties

Material and Methods

Description of the study area

Metahara Sugar Estate is located at about 200 km southeast of the capital city, Addis Ababa. It is situated at 8^0 53' N, 39^0 52' E and an altitude of 950 m.a.s.l. The area has a semi-arid climatic condition. Most of the Estate soils are alkaline with pH normally above 8.0 (the soil may be sodic when pH above 9.5), strongly calcareous with $CaCO_3$ between 6 and 15%, very low or low (<0.2%) in total N, dominated by very high exchangeable Ca and Mg, very low or low (<10 ppm) in available P, very low in organic carbon, CEC between 40 and 75 meq/100 kg soil (some loamy and sandy soils have a CEC <40), and very high levels of exchangeable K but more variable amounts of available K [8]. From the soil fertility point of view, Metahara Sugarcane Plantation soils have been classified into F1 and F2 fertility units.

The experiment was executed on clay soils (Haplic Cambisols) of F1 soil fertility unit covering 63 % of Metahara Sugarcane Plantation. It was conducted for two crop seasons (2010/11-2012/13). The experiment was laid out in factorial randomized complete block design with three

---

*Corresponding author: Zeleke Teshome, Sugar Corporation, Research and Training, P.O. Box 15, Wonji, Ethiopia
E-mail: zeleketesh@yahoo.com

replications. The treatments included four levels of nitrogen (i.e., 0, 46, 92, and 138 Kg/ha) and five levels of vinasse-filtercake compost (i.e., 0, 5, 10, 15, and 20 ton/ha). Urea fertilizer was used as source of nitrogen.

The compost was prepared by the Sugar Estate using windrow composting method at the compost yard. Odour and colour of the compost were considered to determine the maturity before field application. Filtercake-Vinasse compost was applied before planting while nitrogen was applied at 2.5 months of cane age. Size of experimental plot was 52.2 m² (6 furrows of each 6 m long and 1.45 m wide). Test variety used was NCo334. All agronomic practices such as irrigation, molding, weeding etc. were applied uniformly to all experimental plots as per the standard practices followed by the Sugar Estate.

Representative Filtercake-Vinasse compost samples were taken before application. Soil samples were also collected from the study site before treatment application, and after treatment application (5th month). Both soil and compost samples were analyzed for pH (1:2.5), EC (1:2.5), organic carbon, total N, available P and available K following standard analytical procedures [9]. The soil electrical conductivity (EC) measured at 1:2.5 soil: water was converted to electrical conductivity of saturation paste (ECe) by multiplying by a factor of 4.3 [10].

Tiller and stalk population count were done at 3 and 10 months of cane age, respectively. At harvest, 20 millable stalks were sampled from the middle four furrows and the following parameters were determined: stalk length, stalk girth, stalk weight and S% C. Cane yield was calculated from stalk weight and stalk number while sugar yield were calculated from cane yield and S%C.

Statistical analysis

Statistical analysis of the variance was performed for the collected data using SAS software and mean separation was done at 5% probability levels by Duncan Multiple Range Test (DMRT).

Economic analysis

Economic analysis was done by considering the following assumptions and cost components: experimental sugar yield was adjusted down by 15% to reflect actual sugar yield at commercial level [11]. Price of Urea, 600 $USD/ton; haulage cost, 1.94 $ USD/ton of sugar; processing cost, 58 $USD/ton of sugar; compost preparation and transportation cost, 24.45 $USD /ton; production cost of sugar, 259.5 $USD/ton; selling price of sugar, 536 $USD /ton. Net benefit cost ratio was used to compare treatments. For partial budget analysis, factors with significant effects were considered [11]. Finally, net benefit cost ratio (NBCR) was computed using the appropriate relationships [12].

Result and Discussion

Chemical properties of matured compost

The analytical results of compost are presented in Table 1; pH and ECe values of the compost ranged from 6.3-7.1, and 219-258 dS/m, respectively. The values of organic carbon, total nitrogen, available

phosphorus, and available potassium were 17.5%, 1.13%, 0.08%, and 6.33%, respectively. The values of nitrogen, phosphorus, and potassium content of the compost are high as compared to that of Sugar Factory by-products [13]. The C:N of the compost ranged from 14.75 to 19.01 with mean value of 15.48 indicating that the compost was matured; however, the nitrogen content was low in such cases nitrogen from the soil could be immobilized. If the organic materials have a C/N ratio of less than 20, there could be release of mineral nitrogen after application of compost to soil early in the decomposition process. However, the total nitrogen content of the organic substance being added to soil to be considered [14]. The same Author reported that concentrations of nitrogen in the organic manure is between 1.5 and 1.7% are usually sufficient to minimize immobilization of soil nitrogen. The total nitrogen of the compost was 1.13%; this is expected to favor immobilization of soil nitrogen at the start of decomposition. Thus, addition of extra nitrogen is required; otherwise microbes involving in compost decomposition process would compete with sugarcane for nitrogen and this could create shortage of nitrogen required for the crop.

Nutritionally, the compost greatly contributes to the fertility of soil through releasing essential nutrients including macro and micro-nutrients, and activating the living microbes in the soil. Further, it could help in solubilizing fixed elements in the soil. As indicated in Table 1, the mean value of nitrogen was 1.13%; by applying 1 tone of compost per ha, 11.3 kg N/ha can be supplied to the soil if all of the nitrogen is mineralized. However, these nutrients from such organic matters are released very slowly and hence necessitate supplemental fertilizer to avoid nutrient deficiency. In spite of the fact being found small quantities of both macro and micro nutrients in compost, its sole application to soil does not supply the required amount of nutrients to complete its growth and development.

Effect of compost on soil chemical properties

Soil analysis result indicated that the pH was not changed while the ECe values were increased from 2.02 to 2.27 dS/m. However, compared to ECe value of compost (Table 1), the ECe values of soil (Table 2) obtained after application to cane fields can be considered as low or within the normal range for optimal cane growth.

The salient soil properties, namely total nitrogen, organic carbon, available P and available K were not substantially increased (Table 2). This study justify that yield increment for most of compost studies is not only due to supplement of nutrients by compost but also important features of organic materials that activating the soil micro-organism and thereby enhance solubility of fixed plant nutrients in the soil.

Effects of nitrogen and compost on yield attributes and yield of sugarcane

Analysis of variance indicated that except for stalk girth and number, all the parameters including till number, stalk length, stalk weight, cane yield, S%C, and sugar yield were not significant (p<0.05) due to nitrogen rates applied (Table 3).

S/N	Sampling sites	pH (1:1:5)	ECe (dS/m)	Organic carbon, %	Total Nitrogen, %	C:N	Available p, %	Available K, %
1	S1	7.1	258	19	1.26	15.07	0.098	7.11
2	S2	6.6	228	17	0.89	19.01	0.082	6.35
3	S3	6.3	219	16	1.14	14.04	0.085	5.86
4	S4	6.6	237	18	1.22	14.75	0.066	6.01
Mean		6.7	235.4	17.5	1.13	15.48	0.083	6.33

Table 1: Major chemical properties of the compost.

Treatment application	pH (1:2.5)	ECe (dS/m)	OC, %	TN, %	C:N	Avail. P, ppm	Avail. K, ppm
BTA	8.0	2.02	1.60	0.148	10.8	4.19	444
ATA	8.0	2.27	1.69	0.150	11.3	4.77	448

N.B: BTA=before compost application; ATA= after application of compost application.

Table 2: Effect Major soil chemical properties.

Variable	Length (m)	Girth (cm)	Weight (kg)	Tiller number ('ooo ha-1)	Stalk number ('ooo ha-1)	Cane yield (t/ha)	S%C	Sugar Yield (t/ha)
Nitrogen (N)	NS	*	NS	NS	*	NS	NS	NS
Compost (C)	NS	NS	*	*	NS	*	NS	*
N*C	NS	*	*	NS	NS	*	NS	*
CV (%)	7.59	3.50	8.96	6.97	3.83	7.99	5.14	9.29

Table 3: Analysis of variance for the effect s of nitrogen and compost on yield attributes and yield of sugarcane.

N (kg/ha)	Cane Yield, ton/ha				
	Compost, ton/ha				
	0	5	10	15	20
0	262.6 (bcde)	253.5(bcde)	250.9(cde)	274.4(abcde)	275.0 (abcde)
46	287.6 (abcd)	264.5(bcde)	252.4 (cde)	315.6 (a)	264.7(bcde)
92	285.6 (abcde)	263.5 (bcde)	264.3 (bcde)	277.1(abcde)	276.1 (abcde)
138	243.0 (e)	289.3 (abc)	282.7 (abcde)	296.1 (ab)	264.7(bcde)
N (kg/ha)	Sugar Yield, ton/ha				
	Compost, ton/ha				
	0	5	10	15	20
0	34.10(bcd)	33.03(cd)	32.31(d)	36.41 (abcd)	39.44 (abc)
46	36.59 (abcd)	33.99 (bcd)	32.29 (d)	41.35(a)	33.67 (bcd)
92	37.28 (abcd)	35.72 (abcd)	33.11(cd)	35.28 (abcd)	35.28 (abcd)
138	31.61 (d)	38.95(abc)	36.78(abcd)	40.12 (ab)	34.03 (bcd)

Table 4: Cane and sugar yields as affected by the application of nitrogen and compost.

Nitrogen (Kg/ha)	Compost application rate (ton/ha)				
	0	5	10	15	20
0	0.89	0.81	0.70	0.96	1.08
46	0.98	0.82	0.74	1.12	0.77
92	1.00	0.85	0.70	0.83	0.83
138	0.74	0.94	0.80	0.98	0.74

Table 5: Net benefit cost ratio (NBCR) for nitrogen and compost combinations.

Compost application was significant (p<0.05) for tiller number, stalk weight, cane yield and sugar yield, but no significant effect was observed for stalk number, stalk length, stalk girth, and S%C. The interaction effect between nitrogen and compost was significant (P<0.05) for stalk girth, stalk weight, cane yield, and sugar yield while it was non-significant for tiller number, stalk length, stalk number, and S%C (Table 3).

Sole application of nitrogen at higher rate, 138 kg N/ha, gave the lowest cane yield as well as sugar yield; this could be due to the depressing effect of nitrogen when it is applied at higher rates. Combined application of 46 kg N/ha with 15 ton of compost/ha gave maximum cane and sugar yields (Table 4). This is in agreement with the findings of the experiment conducted in India that indicated inorganic fertilizer NP in combination with organic manure gave higher sugar yield than application of NP alone [15]. Similarly, this result is also in agreement with the research conducted at Metahara that gave best cane yield was recorded with the application of both nitrogen and compost than sole application of nitrogen or compost [16].

Economic analysis

Based on net benefit to cost ratio, nitrogen-compost combination (46 Kg N/ha and 15 ton /ha) was relatively with highest NBCR, 1.12,

within one crop cycle (Table 5). This result should not be solely evaluated by its yield outcome or cost of fertilizer reduction, but it should also be evaluated with its potential outlet benefit for excess vinasse disposal.

Conclusion

Soil analysis result indicated that the compost was neutral in reaction with mean pH value of 6.7, and it contained mainly 17.5% organic carbon, 1.13% nitrogen, 0.083% phosphorus, and 6.33% potassium. Except pH, ECe, total nitrogen, organic carbon, available P and available K values were slightly increased due to the application of compost.

Interaction effect between nitrogen and compost was significant (p<0.05) on stalk girth, stalk weight, cane yield, and sugar yield. Highest cane and sugar yields were obtained when 46 Kg N/ha applied with 15 ton of compost/ha. Net benefit to cost ratio was also found to be the highest, 1.12, for this combination. Therefore, based on the present findings the following recommendations are forwarded:

- On clay soils, 15 ton of compost/ha should be applied with 46 Kg nitrogen/ha.

- Compost should be applied before furrowing while nitrogen fertilizer should be applied at 2.0-2.5 of after planting.

- Avoid application of compost on heavy (vertic) clay soils and fields prone to water logging.

- In the future, application rate and method should be studied for ratoon crops.

Acknowledgments

The Authors are grateful to Sugar Corporation, Research and Training, for financing the study and would like to thank those who participated in data collection and summarization. Authors are also indebted to Metahara Sugarcane Plantation Staff for their collaboration while conducting the study.

References

1. Rosseto AJ (1987) Cana-de-acucar, cultivo e utilizacao. Campinas. 435-504.

2. Rao PN (1990) Recent advances in sugarcane, 1st edition. The KCP LTD. India.

3. Melchor H, Garcia GIS, Lopez SP, Espinoza DJL, Estrada LICC, et al. (2008) Vinasse and filtercake compost as nutrient source for sugarcane in a molic gleysol of chips, Mexico. 33: 855-860.

4. Franco A, Marques MO, Melo WJ (2008) Sugarcane grown an Oxisol amended with sewerage sludge and vinasse: Nitrogen contents in soil and plant. J. Agri. sci, 65: 408-414.

5. CIMMYT (International Maize and Wheat Improvement Center) (1988) An Economic Training Manual: From Agronomic Data to Farmer's Recommendations. CIMMYT, Mexico 1-79.

6. Gomez J, Rodriguez O (2000) Effects of vinasse on sugarcane productivity. Revista de la Facultad de Agronomia LUZ. 17: 318-326.

7. Mariano AP, Crivelaro HR, Angelis DF, Bonotto DM (2009) The use of vinasse as an amendment to ex-situ bioremediation of soil and groundwater contaminated with diesel oil. Brazilian Archives of Biology and Technology. 52: 1043-1055.

8. Booker Tate (2009) Re-evaluation of plantation soils at Metahara Sugar Factory. Final Report. Booker Tate Limited, UK.

9. Sahilemedhin S, Taye B (2000) Procedures for soil and plant analysis. Addis Ababa, Ethiopia.

10. Kumar KS, Vijay KS, Verma KS (2002) Influence of use organic manure in combination with organic fertilizer on sugarcane and soil fertility. Indian Sugar 52: 177-182.

11. FAEF (2003) Salinidade dos Solos do Regadio de Cho´kwe`: Projecto Cheias Limpopo. Universidade Eduardo Mondlane, Departamento de Engnharia Rural, Maputo.

12. Camargo OA, Valadares JMAS, Berton RS, Sobrinho JT, Menk JRF (1987) Alteraçao de caracteristicas químicas de um latossolo vermelho-escuro distrofico pela aplicaçao de vinhaça. Boletim Científico do Instituto Agronomico de Campinas 9: 23.

13. HBCNA (2006) Handbook of Cost-Benefit-Analysis - Financial management reference material.

14. Zeleke T, Girma A, Abiy F (2013) Evaluation of composting materials at Metahara Sugar Estate: Advisory note. In: Zeleke Teshome, Abiy Getaneh, Yohannes Zekarias and Fikru W/Mariam (eds). Research and Training Miscellanea. Wonji, Ethiopia.

15. Tisdale SL, Nelson WL, Beaton JD (1985) Soil fertility and fertilizers (4thedn.) Macmillan Publishing Company. New York.

16. Haile G (1993) Response of sugarcane to compost and nitrogen. In:Ambachew Dametie and Girma Abejehu (eds). Review of sugarcane research in Ethiopia: Soils, Irrigation and Mechanization (1964-1998). ESISC Sh Co., Wonji, Ethiopia.

13

Effect of Growth Promoting Substances on Selected Three Ornamental Plants

Okunlola A Ibironke*

Department of Crop, Soil and Pest Management, The Federal University of Technology, PMB 704 Akure, Ondo State, Nigeria

Abstract

An experiment was conducted in the nursery, the department of Crop, Soil and Pest management the Federal University of Technology, Akure, on stimulation of rooting of three ornamentals; *Euphorbia milii, Adenium obesium, and Murraya paniculata*, (Christ thorn, Desert rose and Murraya respectively) using some rooting substances; Indole-3-butyric acid (IBA), Top soil, Coconut water and Tetracycline from July to September, 2013. The experiment was laid out in a Completely Randomized Design (CRD) and replicated four times. Data were collected on number of branches, the number of leaves per cutting, root weight, number of roots and length of roots. The results from the study showed that each of the treatment had significance ($P<0.05$) with respect to a specific plant. Tetracycline was found the best for rooting Christ thorn cuttings. Indole-3-Butyric Acid (IBA) was found the best for rooting Roses cuttings. Coconut water treatment was found the best for rooting Muraya cuttings. The different treatments produced significant variation while there was no significant variation among the three different plant cuttings, but in the interaction between the plant cuttings of the different treatments.

Keywords: Difficult-to-root plant; Root formation; Growth; Development

Introduction

Ornamental plants are essential object of environmental aesthetic beautification and management; they make up the component of urban green spaces, public parks and houses more for relaxation and enjoyment [1,2]. They are grown for the display of aesthetic features including flowers, leaves, scent and overall foliage texture- fruit, stem and bark. They are a valuable tool for the harmonious and practical resolution of many physical site problems, and they provide durable aesthetic satisfaction [3]. Generally, most perennial ornamental plants are multiplied and propagated through asexual means of reproduction such as cuttings, layering or grafting [4]. The cuttings from stems, leaves, roots or terminal buds were the commonly used techniques, due to their ability to retain the characters of the parent and also, for breeding seedless hybrid. Success of rooting ornamental plant cuttings depend on their growth responses, based on nutrient present with the aid of growth promoting substances before planting [5].

Euphorbia milii, Adenium obesium, and *Murraya paniculata* were known as 'difficult to root' ornamental plants, this difficulty led to research on propagating their stem cuttings in different growth promoting substances to observe their responses [6]. Studies have shown that physiological state of the mother plant, the prevailing environmental conditions in the nursery i.e., light, temperature and humidity play important role in rooting and developmental stages of cuttings.

According to McGregor [7] root promoting hormones like cytokinnins, gibberellins, ethylene, abscisic acid, brassinosteroids, jasmonic acids and auxins play major role in the success of rooting the cuttings. Synthetic growth treatments had these phytohormones naturally present in some plants, as active ingredients produced for commercial production. Formation of adventitious root ensure survival of the vegetative stage, this prompted several researchers to investigate the artificial means of initiating roots of stem cuttings, planted for optimum growth [8].

Although, in the production of nursery crops in containers, the selection and preparation of the medium is extremely important and could pay great dividends in terms of plant growth and quality. There is no universal or ideal rooting mix for cuttings [9] an appropriate propagation medium depends on the species, cutting type and propagation system [10]. To this end, different growth treatments and stem cuttings were explored to optimize the rooting of the ornamentals. Thus, the objective for this research was to determine the most effective growth treatment that would facilitate root development and promote rooting of each stem cutting and also to determine the best treatment that enhance vegetative growth in the stem cutting.

Materials and Methods

The experiment was stationed at the nursery of the department of Crop, Soil, and Pest management, Federal University of Technology, Akure with (7°16'N, 5°12'E) located in the rain forest vegetation zone of Nigeria between July and September 2013. The rainfall pattern of Akure is bimodal with a wet season of about seven months occurring during April to October/November and through February to March. The mean daily temperature ranges between 25°C and 37°C.

The materials used were stem cuttings of *Euphorbia milii, Adenium obesium,* and *Murraya paniculata*, (Christ thorn, Desert rose and Murraya respectively) collected from mother plant (stock plant) at Winpool garden, Exotic garden and Lucado horticultural garden within the state, also perforated plastic containers, Top soil, Tetracycline, IBA, Coconut water from matured green fruit and Distilled water. Stem cutting of apical plant cutting with leaf nodes for new root production, within range of 5 cm-10 cm of the different ornamental plants, was partly buried into the soil. Fertile top soil mixed with sandy soil in ratio

*Corresponding author: Okunlola A. Ibironke, Department of Crop, Soil and Pest Management, The Federal University of Technology, PMB 704 Akure, Ondo State, Nigeria, E-mail: okunlolaa1.hort@gmail.com

1:1 was packed into plastic pots, for good aeration and good drainage, this was kept moist and not waterlogged, while the soil was slightly acidic with pH of 5.94. There was partial shade established to prevent cutting from dehydration, especially from hot afternoon sun which can burn off the foliage and therefore prevent diseases. 25 ml of fresh coconut water was added to 100 ml of distilled water, and each replicate of cuttings was treated with 25 parts per ml. 5 capsules equivalent of 5 g Tetracycline was dissolved in 100 ml of distilled water, and each replicate were treated with 25 parts per ml. IBA of 5 g was weighed and dissolved with 100 ml of distilled water, and each replicate was treated with 25 parts per ml. The fourth treatment was distilled water of 100 ml used to treat each replicate in 25 parts per ml as the control for the experiment.

The soil analysis was done using the procedure in the csp laboratory manual booklet. Soil pH was determined. Magnesium was determined with an atomic absorption spectrophotometer. Exchange acidity was determined by Cabonoglu et al. [11] titration method. Soil organic C was determined by the procedure of Walkley and Black using the dichromate wet oxidation method, total N was determined by micro-Kjeldahl digestion method, available P was determined by Bray- 1 extraction followed by molybdenum blue colorimetry. Exchangeable K, Ca and Mg were extracted using 1.0 N ammonium acetate. Thereafter, K was determined using flame photometer and Ca and Mg were determined using the EDTA titration method while sodium (Na) was determined by flame emission photometry. Particle size distribution was determined with a hydrometer.

The experiment was arranged in a completely randomized design (CRD) with four replicates. The data obtained were subjected to analysis of variance (ANOVA) and treatment mean were compared using Duncan's multiple range test (DMRT) at p=0.05 probability level [12]. And every 4 weeks after planting, data on plant height, stem girth, leaf length and number of leaf were taken to assess plant growth from each treatment per replicate. The ornamental plants were harvested after 12 weeks. Yield parameters such as root number, root length, leaf area and net dry weight were recorded. Soil nutrient present were also determined in relation to the initial and present nutrient content.

Results

Initial soil analysis before the experiment showed the nutrient contents of the soil which contains higher percentage of organic matter, phosphorus, nitrogen and potassium and other nutrients in adequate proportions as shown in Table 1. However there were significant differences in the effects of each treatment on the nutrients contents of the soil with coconut water having higher significance and the control showing no significant difference (Table 2). The Table 3 showed the effect of different growth treatment on leaves number of different ornamental plants, Significant (p<0.05) differences were not observed between the plants cuttings used for the experiment. However, it was observed that the plants treated with coconut water had the highest number of leaves followed by Tetracycline, and IBA while the control had the lowest. Table 4 showed the response of ornamental plants within the twelve weeks of planting to number of leaves sprouting. Significant (p<0.05) differences were observed during the 4th and 12th week of the experiment. However, the Murraya plant had the highest number of leaves (Table 5), followed by Christ thorn while Roses had the lowest number of leaves.

The combine effect of different growing media and different ornamental species on leaves number were also analyzed as significant (p<0.05) differences were observed during the 1st and 3rd month of the experiment. However, Murraya plant grown with coconut water had the highest number of leaves while the Roses grown in the control experiment had the lowest number of leaves. The response of the ornamental plants to different growth treatment through number of branches per plant as evaluated in Table 6 Significant (p<0.05) differences were not observed throughout the months of the experiment. However, it was observed that the plants treated with Tetracycline and IBA had highest no of branches while the control had the least. Table 7 showed the response of different ornamental plant to different growth treatment on number of branches. Significant (p<0.05) differences were

Soil analysis	Initial reading before planting
Sodium (C mol/kg)	29.80
Phosphorus (mg/kg)	0.45
Nitrogen (%)	4.60
Potassium (C mol/kg)	39.70
Magnesium (C mol/kg)	8.20
Calcium (C mol/kg)	45.8
Organic matter (%)	26.8
pH	5.94

Table 1: Soil chemical analysis before planting and treatment with growth substances.

Treatments	Species	N	P	K	Ca	Mg	Na	OM	pH
Tetracycline	Murraya	1.60a	0.15a	25.40a	14.40a	5.00a	13.50a	11.70a	6.39a
	Christ thorn	3.20b	0.07b	27.40a	14.00a	5.60a	13.40a	12.50a	6.15a
	Roses	3.50b	0.13a	29.40a	34.80b	4.00a	15.20b	14.10a	5.87a
Coconut water	Murraya	2.30a	0.14a	21.60a	43.40b	4.60a	9.50a	10.10a	6.49a
	Christ thorn	2.70ab	0.12a	26.60b	21.00a	4.60a	12.50c	11.40a	6.25a
	Roses	3.00b	0.09a	24.20b	38.60b	5.40b	11.60b	13.20a	4.87b
IBA	Murraya	3.10c	0.10a	22.80a	20.70a	4.80a	10.60a	11.70a	6.69a
	Christ thorn	1.80a	0.10a	25.00b	24.30b	5.80a	12.30a	12.60a	6.25a
	Roses	2.50b	0.14a	15.00a	44.00c	5.20b	15.00a	13.10a	5.37a
Control	Murraya	2.90a	0.09a	27.00a	4.10a	4.10a	14.60b	11.60a	6.79a
	Christ thorn	2.50a	0.12a	20.40b	4.80a	4.80a	9.20a	12.60a	6.55a
	Roses	2.20a	0.24b	27.20a	5.40a	5.40a	13.90b	13.10a	5.77a

Mean in same column followed by the same letter (s) are not significantly different (p ≤ 0.05) by Duncan's multiple range tests.
Table 2: Effect of the treatments on soil nutrient after planting the stem cuttings.

Treatments	4	8	12
	Weeks after planting		
Tetracycline	4.79a	5.75a	6.96a
Coconut water	4.88a	5.58a	7.04a
IBA	4.54a	5.38a	6.63a
Control	4.38a	5.00a	6.04a

Mean in same column followed by the same letter (s) are not significantly different (p ≤ 0.05) by Duncan's multiple range test.

Table 3: Response of stem cuttings on number of leaves to treatments for 12 weeks.

Treatments	4	8	12
	weeks after planting		
Murraya	5.25b	6.25b	7.38a
Christ thorn	4.94b	5.88ab	6.54a
Roses	4.56ab	5.19ab	5.56a

Mean in same column followed by the same letter (s) are not significantly different (p ≤ 0.05) by Duncan's multiple range test.

Table 4: Stem cuttings by treatment interaction effect on number of leaves for 12 weeks.

Treatments	Species	1	2	3
		Months after planting		
Tetracycline	Murraya	5.25ab	6.75a	7.45ab
	Christ thorn	4.50ab	5.75a	6.75ab
	Roses	4.00ab	5.25a	5.75ab
coconut water	Murraya	6.50b	7.50a	9.50b
	Christ thorn	4.75ab	6.00a	7.50ab
	Roses	3.75ab	4.00a	5.40ab
IBA	Murraya	4.75ab	5.75a	7.75ab
	Christ thorn	5.25ab	5.75a	7.00ab
	Roses	5.50ab	6.00a	7.50ab
Control	Murraya	4.50ab	5.00a	7.50ab
	Christ thorn	5.25ab	6.00a	6.50ab
	Roses	5.00ab	5.50a	5.00a

Mean in same column followed by the same letter (s) are not significantly different (p ≤ 0.05) by Duncan's multiple range test.

Table 5: Effect of the treatments on number of leaves of each stem cutting.

Treatments	1	2	3
	Months after planting		
Tetracycline	2.13a	2.79a	3.68a
Coconut water	2.29a	2.92a	3.47a
IBA	2.33a	2.67a	3.67a
Control	2.17	2.65a	3.25a

Mean in same column followed by the same letter (s) are not significantly different (p ≤ 0.05) by Duncan's multiple range test.

Table 6: Response of stem cuttings on number of branches to treatments for 12 weeks.

Treatments	1	2	3
	Months after planting		
Murraya	2.38a	2.88a	3.46a
Christ thorn	2.06a	2.56a	3.23a
Roses	2.31a	2.94a	3.34a

Mean in same column followed by the same letter (s) are not significantly different (p ≤ 0.05) by Duncan's multiple range test.

Table 7: Stem cuttings by treatment interaction effect on number of branches for 12 weeks.

not observed throughout the 12 weeks of the experiment. However, Muraya plant had the highest no of branches followed by Roses while Christ thorn had the lowest.

The combine effect of different growth treatment and different ornamental species on number of branches (Table 8) revealed significant (p<0.05) differences were observed during the 2nd and 3rd months of the experiment. However, Murraya plant grown with Tetracycline had the highest number of branches while the Christ thorn grown in the control experiment had the lowest. Treatment with coconut water and Murraya cuttings in coconut water and tetracycline had the highest number of roots and root net dry weight respectively at 12 WAP (Tables 9 and 10). The combine effect of treatments on length of roots for Murraya and Christ thorn cuttings showed a significant increase (P<0.05) Murraya cuttings had the longest roots over Christ thorn and rose (Table 11).

Discussion

In this study coconut water has proven effective as an efficient growth stimulant for the propagation of difficult to root plant. The research work of Khayyat et al. [13] in which soil nutrient was studied, after treatment with Coconut water to evaluate the potential of soil to sustain growth and the retained nutrient in plant, It was observed that

Treatments	Species	1	2	3
		Months after planting		
Tetracycline	Murraya	2.25a	3.00ab	3.75ab
	Christ thorn	1.75a	2.75ab	3.75ab
	Roses	1.75a	2.75ab	3.50ab
coconut water	Murraya	2.50a	2.90ab	3.25ab
	Christ thorn	2.25a	2.25ab	3.50ab
	Roses	2.75a	3.50b	4.10b
IBA	Murraya	2.50a	2.75ab	3.75ab
	Christ thorn	2.25a	2.50ab	3.75ab
	Roses	2.50a	2.75ab	3.15ab
Control	Murraya	2.25a	2.75ab	3.25ab
	Christ thorn	2.00a	2.75ab	3.50ab
	Roses	2.25a	2.75ab	3.75ab

Mean in same column followed by the same letter (s) are not significantly different (p ≤ 0.05) by Duncan's multiple range test.

Table 8: Effect of the treatments on number of branches of each stem cutting.

Treatments	Root no	Root length	Leaf area	Net dry weight
Tetracycline	8.00a	14.32a	13.54a	4.23a
Coconut water	8.23a	14.35a	13.54a	4.55a
IBA	7.50a	13.77a	12.15a	4.27a
Control	7.00a	11.30a	11.52a	3.88a

Mean in same column followed by the same letter (s) are not significantly different (p ≤ 0.05) by Duncan's multiple range test.

Table 9: Response of root cuttings to different treatments for 12 weeks.

	Root no	Root length	Leaf area	Net dry weight
Murraya	8.25ab	15.30bc	7.75ab	5.20b
Christ thorn	7.00a	6.33a	15.20bc	3.35a
Roses	5.50a	13.03bc	12.23ab	5.41b

Mean in same column followed by the same letter (s) are not significantly different (p ≤ 0.05) by Duncan's multiple range test.

Table 10: Root cuttings by treatment interaction on parameters taken for 12 weeks.

Treatments	Species	Root no	Root length	Leaf area	Net dry weight (%)
Tetracycline	Murraya	16.00a	21.00bc	12.20ab	5.20a
	Christ thorn	10.00ab	7.00a	16.80d	3.30ab
	Roses	9.00a	19.00c-e	14.40ab	6.12bc
Coconut water	Murraya	17.00cd	16.50c-e	12.50ab	5.90c
	Christ thorn	11.00ab	6.00a	15.00cd	3.10ab
	Roses	7.00a	13.20bc	14.00ab	5.70bc
IBA	Murraya	12.00ab	18.90ab	10.50ab	4.50ab
	Christ thorn	8.00a	7.00a	10.40ab	3.12a
	Roses	6.00a	11.00bc	12.50ab	6.90bc
Control	Murraya	12.00ab	10.20bc	11.80ab	3.70a
	Christ thorn	10.00bc	12.30a	10.60ab	3.00ab
	Roses	6.00a	8.90ab	10.00ab	4.90b

Mean in same column followed by the same letter (s) are not significantly different ($p \leq 0.05$) by Duncan's multiple range test.

Table 11: Effect of the treatments on parameters taken for each root cutting.

Myo- inositol present in coconut has mechanism for sodium uptake, which plays a major role in the transition of ice plant from non-tolerance of soil salinity to successful adaptation of the ice plant to salinity stress, this result into sprouting and survival of the plant. Furthermore, this study of responses of different ornamental plants to growth promoting substances has shown relevant results by data analysis obtained. The results of the soil analysis of initial and after planting with treatments, the mineral nutrient absorbed by the root hairs determined their rate of sprouting and survival through Plant height, number of branches, number of leaves, number of roots, root length, leaf area and dry net weight as represented by the data. Different treatments produced significant variation while there was no significant variation among the three different plant cuttings, but in the interaction between the plant cuttings of the different treatments.

It was observed from results that maximum sprouting of Murraya was seen by treating the cuttings with Coconut water, which was enhanced by the extensive root hairs of the taproot to absorb water and needed nutrient. Also, this was due to the fact that Cytokinin present in Coconut water encourages cell division and growth. From Tables 9-11 [14] more plant root length, number and net dry weight was recorded by Roses treated with Indole-3-butyric acid, which was related with the rooting length of the lateral root induced by IBA to absorb needed nutrient, and also Auxin derivative in IBA promotes apical growth [15]. Cuttings treated with coconut water significantly increase shoot length, shoot girth, number of leaves, wet root   weight, dry root weight and root length. Asma et al. [2] performed an experiment on *In vitro* propagation of kiwifruit (*Actinidia deliciosa*) using coconut water. During the study, it was observed that the root induction was highly effected by the length of shoots and an appropriate length was pre-requisite for the efficient root formation. The use of coconut water also indirectly effected *In vitro* roots induction since during shoot multiplication; the addition of coconut water to the culture media resulted in maximum shoot length (7.2 ± 0.16) and hence facilitating the efficient root formation. This enhanced root formation ultimately resulted in the high survival rate (>95%) of the grown plants.

Results from this experiment have proven that tetracycline is an efficient growth stimulator for Christ thorn as it also has added advantages in inhibiting pathogenic invasion. From recent researches, it was known that Tetracycline prevents and overpowers disease pathogens by moving through all parts of the plant, activating the immune system, thereby all other systems especially the branches are stimulated for optimum performance, even with the short adventitious root [16]. Cuttings treated with Tetracycline, Coconut water and IBA induced maximum sprouting and plant growth. Plant cuttings in the replicates within the twelve weeks of experiment had no significant differences. The treatment worked for different physiological characters independently, growers can choose based on result analyzed for desired treatment. Tetracycline was found the best for rooting Christ thorn cuttings. Indole-3-Butyric Acid (IBA) was found the best for rooting Roses cuttings [17].

Conclusion

Cuttings treated with Tetracycline, Coconut water and IBA induced maximum sprouting and plant growth. Tetracycline was found the best for rooting Christ thorn cuttings. Indole-3-Butyric Acid (IBA) was found the best for rooting Roses cuttings. Coconut water treatment was found the best for rooting Murraya. Tetracycline treatment helped the ornamental plants to fight pathogenic invasion, because all the plants with the treatment remain healthy throughout period of experiment, though Roses had slow growth development compared to others.

References

1. Okunlola AI (2013) The effects of cutting types and length on rooting of Duranta repensin the Nursery. Global journal of human social science, geography, geosciences, environmental and disaster management. Global journal 13.

2. Asma NA, Kashif SK (2008) In-vitro propagation of croton (Codiaeum variegatum). Pakistan journal of botany 40: 99-104.

3. Jean WH, Young K, Liya G, Yan FN, Swee NT (2009) Chemical Composition and Biological properties of coconut (Cocos nucifera L.). Plant Physiol 14: 5144-5163.

4. Rauf A (2011) Phytochemical Analysis and Radical Scavenging Profile of Juices of Citrus sinensis, Citrus anrantifolia and Citrus limonum. Organic Medicinal Chemistry Letter 4: 1-3.

5. Longman KA (2002) Tropical Tree: Rooting cuttings. A practical manual, Blaketon Hall Limited, Exeter, pp: 73-76.

6. Kiran M, Baloch JC, Waseem K, Jilani MS, Qasim K (2007) Effect of Different Growing Media on the Growth and Development of Dahlia (Dahlia pinnata) Under the Agro-climate Condition of Dera ismail khan. Pakistan Journal of Biological Sciences 10: 4140-4143.

7. McGregor AM (2008) Developing the ornamentals industry in the Pacific: An opportunity for income generation. ACIAR, Australia, pp: 7-8.

8. Hanes S, Leonhardt KW (2008) Tropical floriculture: International trends in production and marketing. Department of Tropical Plant and Soil Sciences, University of Hawaii, p: 125.

9. Hall KC (2003) Manual on nursery practices. Forestry Department, Ministry of Agriculture. 173 Constant Spring Road, Kingston, Jamaica, p: 77.

10. Agbo CU, Omaliko CM (2006) Initiation and Growth of Shoots of Geogronema latifolia Benth Stem Cuttings in Different Rooting Media. Africa Journal of Biotechnology 5: 425-428.

11. Cobanoglu F, Gazi B, Kocatas H, Ozen M (2004) Productivity and Quality Parameters on Fig Young Plant in Tube. Korean society of horticultural science.

12. Steel KA (1990) Fig Cultivation, Foundation of Agricultural Research and Promotion. Publishing No: 20, Yalova, Turkey.

13. Khayyat M, Nazari F, Salehi H (2007) Effect of different pot mixtures on Pothos (Epipremnun aureum L. and Andre 'Golden Pothos') growth and development. American-Eurasian J Agric Environ Sci 2: 341-348.

14. Souidan AA, Zayed MM, Dessouky MTA (1995) A study on improving the rooting of Ficus prestige var. decora stem cuttings. J Agri Sci Cairo 40: 821-829.

15. Skoog L, Yildiz H (2000) Fig cultivation in Brazil. In proceedings of advanced course on fig production. Aegean Univ, Izmir.

16. Rowezak MMA (2001) Response of some Ornamental plants to Treatment with Growth substances. MSc Thesis, Faculty Agriculture, Cairo University, Egypt.

17. George SA (2006) Users Guide, Plant nutrient in Coconut water, Release 6.13, SAS Institute Inc., Cary, NC, USA.

14

Determining Optimum Harvest Age of Sugarcane Varieties on the Newly Establishing Sugar Project in the Tropical Areas of Tendaho, Ethiopia

Hadush Hagos[1]*, Luel Mengistu[1] and Yohannes Mequanint[2]

[1]Ethiopian Sugar Corporation, Research and Training Division, Sugarcane Production Research Directorate, Agronomy and Protection Research Team, Wonji Research Center, P.O.Box 15, Wonji, Ethiopia
[2]Ethiopian Sugar Corporation, Research and Training Division, Sugarcane Production Research Directorate, Wonji Research Center, P.O.Box 15, Wonji, Ethiopia

Abstract

Field experiment was conducted to determine the optimum maturity of the major sugarcane varieties (*Saccharium officinarium* L.) with high sucrose content and sugar yield. Six levels of harvest ages (10, 12, 14, 16, 18 and 20 months) and four major varieties N-14, NCO-334, CO-680 and B52-298 which cover 90% of the area were used in a completely randomized block design with 6x4x3 factorial treatment arrangements. All data's were collected at the end of each level of harvest ages. Analysis of variance (ANOVA) showed that harvest age significantly influenced quality parameters (brix, pol, purity and ERS) and yield parameters (plant height, cane yield and sugar yield) ($P<0.001$). The important parameters of maximizing sugar yield and net revenue in relation to harvest date and crop age is expressed by t/ha/month as an index of time value of sugarcane crop. Considering the time value, increase in harvest age showed a negative impact on brix, pol, estimated recoverable sucrose, cane yield and sugar yield in the tropical area of Ethiopia. As a result high sugar yield was recorded at the early harvesting ages 12 and 14 months. However, optimum sugar yield was recorded on 12 months harvest age with economically acceptable marginal rates of return 178.13%. Therefore, adjusting harvest age to 12 months for the major sugarcane varieties was economically recommended to obtain optimum sugar yield with efficient time use at the tropical areas of Tendaho.

Keywords: Sugar project; ERS (Estimated recoverable sucrose); Sugarcane; Marginal rate of return

Introduction

Age of harvest is one of the most important factors affecting sugarcane productivity. Varietal differences in growth and maturity rates must be considered when harvesting decisions are made [1]. In addition to the difference of varietal maturity rates, environmental conditions, management practices, and pest pressure also influence the optimal harvest age of sugarcane along the coast. The climate elements, temperature, solar radiation, relative humidity and total rainfall variables that account for a major variation in harvest age among sugarcane growing countries [2]. Sugarcane varieties developed in South African Sugarcane Research Institute (SASRI) exhibit pronounced differences in their suitability to different harvest ages with faster maturing varieties being more suited to the 12-month cycle, and slower maturing varieties being suited to the 18-month cycle along the coast [3].

Cane maturity is usually determined by monitoring sugar yield parameters such as: Pol % cane, Brix % cane, commercial cane .sugar (CCS) and ton cane per hectare (TCH). However, most researchers focus their evaluation on Pol % cane and its value ranged from 10.49 - 17.86 [4]. In milling operations, the preferred varieties are those with Pol % cane and Brix % cane values nearly equal at maturity, and a Pol value 16 or greater and purity of 80 % or greater are commercially acceptable [5].

Sugarcane varieties differ in their ability to mature at different stages. In Iran for instance, the optimum age to harvest for certain cane varieties depends on whether the cane is early maturing (10-12 months), medium maturing (12 months) or late maturing (14-16 months) [6]. Some sugarcane varieties have relatively high sucrose content in early season and are defined as early maturing while it is the converse in others which are known as late maturing [6]. The crop season is also variable in different countries being 20 - 24 months in Hawaii, 13-19 months in Jamaica, 12-18 months in India, 16 months in Mauritius, 15 months in Queensland (Australia) and 10-14 months in Brazil [7].

Some sugarcane varieties must be harvested before achieving maximum sucrose levels to sustain early-season milling operations. "Early maturing" varieties are preferentially harvested during this time, recognizing that they may not have reached their peak sucrose content, but may have higher sucrose content than other later-maturing varieties [8]. Consequently, lack of maturity status makes it difficult to make informed harvest scheduling decisions and the time of ripening depends on characteristics which are closely related to the length of growing period [8]. The peak sucrose content of sugarcane at harvest time is affected by different growing and plant physiological conditions during the maturation period. Furthermore, the variation among soil on cane fields causes considerable differences in soil moisture holding capacity, degree of drying, and, consequently, the rate at which cane fields ripen [9].

In Ethiopian Sugar Estates usually cane maturity is customarily determined by taking the crop age and appearance as criteria for several years. From Scientific point of view chronological age of sugarcane is not a reliable guide to determine cane maturity alone [10]. Therefore, other factors such as varieties, weather conditions, and

*Corresponding author: Hadush Hagos, Ethiopian Sugar Corporation, Research and Training Division, Sugarcane Research Directorate, Agronomy and Pretction Research Team, Wonji Research Center, P.O.Box 15, Wonji, Ethiopia
E-mail: hadgos@gmail.com

soil type may have more direct bearing on the real maturity of canes than the crop age [11]. The current sugar production of the Ethiopian Sugar Industry covers only 60% of the annual demand for domestic Consumption while the deficient is imported from abroad. In order to make the country self-sufficient in sugar and export the surplus sugar and produce ethanol and other by-products, the Federal government of Ethiopia is working to establish sugarcane plantation on more than 400,000 ha in addition to the vast expansion project of the previously established farms with erection of high crashing capacity 10 new sugar mills.

The importance of determining yield potentials for sugarcane has been noted by many scientists with goals to aim for barriers to be broken. Law of the minimum suggests that there is always some factor limiting yield. Therefore, yield potential need to be defined in terms of the limiting factor [12]. There are many reasons for lower productivity of sugarcane but the most pertinent is improper implementation of sugarcane management practices [13].

However, harvesting many fields without considering crop age are common constraints in sugarcane production in the tropical areas of Ethiopian. Many sugarcane fields in tropical areas of Ethiopia were covered with over-stand cane having an age range of 20-30 month old. This will cause a decline both in yield and quality of sugarcane production due to heavy lodging, and remobilization of accumulated sucrose to supply newly growing side shoots. Similarly, over aged canes deteriorate their sucrose content by heavy lodging and remobilized to supply the unproductive bull shoots (newly growing shoots) [14]. Optimum harvest age of sugarcane varieties was not studied yet in the tropical area of Ethiopia. Considering this drawback, the study was carried out with the objective to determine the optimum harvest age of the major sugarcane varieties with high sucrose content and sugar yield.

Materials and Method

The experiment was conducted in the newly establishing Sugar Project of Tendaho which covers an area of 50,000 ha. Soils of the area were clay, silty-clay-loam in texture with mean maximum temperatures and average annual rainfall of 37.7°C; 220 mm. Six levels of harvest ages (10, 12, 14, 16, 18 and 20 months) and four major varieties N-14, NCO-334, CO-680 and B52-298 which cover 90% of the area were used in a completely randomized block design with 6x4x3 factorial treatment

arrangements. Each plot had six rows with 6 m length and 1.45 m width for each row (6 m x 1.45 m x 6 rows) having an area of 52.2 m^2 for a single plot. The distance between plots was 2.9 m while it was 4.35 m between replications. The harvested plot consisted of four rows with 6 m length and 1.45 m width each (6 m x 1.45 m x 4 rows) with an area of 34.8 m^2.

To investigate the effects of the treatments quality and yield parameters were measured during the study. At harvest, twenty milleable stalks from the middle four rows were randomly sampled for weight measurement and total population of the middle four rows were counted to estimate cane yield [15]. Half of the twenty stalks were used for stalk length measurement and analysis of quality parameters (brix, pol and estimated recoverable sucrose) in the laboratory. Temperature corrected refractometer brix and saccharometer were used to determine the brix and pol percent of the cane [16]. Estimated recoverable sucrose was the combined effect of brix and pol percent's [17]. At the last sugar yield was estimated from cane yield and estimated recoverable sucrose [18]. All cultural practices were executed based on the current practices of Tendaho Sugar Project except harvesting. Economic analysis was done using partial budget analysis procedures [19]. The effect of harvest age on sugar yield of sugarcane varieties was analyzed using the appropriate analytical software (SAS 9). Mean separation was conducted using Duncan's Multiple Test Range (DMTR) at 5% probability level whenever significant differences were detected in the F-test.

Result and Discussion

Stalk height was significantly (P<0.0001) affected by harvest time (Table 1). The stalk height significantly increased with increasing harvest age until 16 months of the four varieties. This result demonstrated that there was a substantial amount of growth in terms of stalk height at the latest harvesting ages for the sugarcane varieties [20]. According to [21], ripening in sugarcane is characterized by rapid accumulation of sugar with a concomitant reduction in vegetative growth and cane elongation. The current study demonstrated that, a significant increase of stalk height from 10 to 16 months increase cane tonnage substantially during harvest time. However, the continued growth on top of cane in terms of stalk height may pose a problem on sugar recovery during processing [22]. This is due to the fact that the juice from the tops of young cane contains starch, ash, soluble polysaccharides and reducing sugars [23]. The stalk height showed significant (P<0.0001) difference

Harvest age (months)	PH (m)	CY (t/ha/month)	SY (t/ha/month)	Brix%	Pol%	ERS%
10	2.43^d	82.27^a	0.73^b	16.06^b	13.37^b	8.91^c
12	2.64^c	91.36^a	1.03^a	18.11^a	16.05^a	11.24^b
14	2.97^b	81.92^a	0.98^a	17.60^a	16.37^a	11.87^a
16	3.16^a	67.72^b	0.76^b	18.12^a	16.12^a	11.32^{ab}
18	3.19^a	53.33^c	0.48^c	16.10^b	13.51^b	9.07^c
20	3.20^a	40.88^d	0.26^d	14.05^c	10.63^c	6.49^d
Varieties						
N-14	2.83^b	68.91^a	0.71^{ab}	17.29^a	14.77^b	10.00^b
NCO-334	2.91^b	66.22^a	0.64^b	15.19^c	13.35^d	9.30^c
CO-680	3.19^a	72.43^a	0.73^{ab}	16.57^b	14.00^c	9.45^c
B52298	2.79^b	70.76^a	0.76^a	17.64^a	15.28^a	10.51^a
Age*Var	ns	ns	ns	***	***	***
CV (%)	7.89	17.25	20.85	5.12	5.31	6.95

Means followed by the same letter within a column are not significantly different. PH, stalk height (P<0.0001); CY: cane yield (P<0.001); SY: sugar yield (P<0.001); Brix%: Percentage of refractometer brix (p<0.001); Pol%: percentage of sacharometer pol (p<0.001); ERS: estimated recoverable sucrose (p<0.001); MAP: months after planting; ns: non-significant

Table 1: Yield and quality parameters of sugarcane varieties cane as influenced by harvest age in the tropical areas of Tendaho.

HA (month)	Sugar yield (t/ha/month)	Gross field benefit ($ USD /ha/month)	Total variable cost ($USD /ha/month)	Net benefit ($USD /ha/month)	Change in net benefit ($USD /ha/month)	MRR (%)
10	0.73	484.43	0	484.43		
12	1.03	683.51	71.58	611.93	127.50	178.13
14	0.98	650.33	143.16	507.17		
16	0.76	504.34	214.74	289.60		
18	0.48	318.53	286.32	32.21		
20	0.26	172.54	357.89	(185.36)		

HA: harvest age; MRR: Marginal rate of return.

Table 2: Partial budget analysis of the sugarcane cane as influenced by harvest age.

with the four sugarcane varieties. CO-680 variety recorded the highest length as compared with N-14, NCo-334 and B52298 varieties (Table 1). This could be attributed to the difference in growth habit among sugarcane varieties during ripening period. In agreement with this result, another research results recorded there is varietals difference in stalk height among the sugarcane varieties [24].

Cane yield and sugar yield were significantly (p<0.001) influenced by harvest age. The profitability of sugar yield within various harvest ages considers a time value. So, analysis for harvest age was computed in terms of t/ha/month (Table 1). The important parameters of maximizing sugar yield and net revenue in relation to harvest date and crop age is expressed by t/ha/month as an index of time value of sugarcane crop [25]. Considering the time value, the result of this study revealed that, highest cane yield and sugar yield were recorded on 12 months harvest age, followed by 14 months harvest age. Significant increase in cane yield was recorded with an increase in harvest age from 10 to 14 months [26]. The major drop in sugar yield with an age restriction of below 12 months might be due to many hectares of crop being forced to be harvested when expected yields are extremely low as well as older crops being disallowed [26]. There is no significant difference in cane yield among the four sugarcane varieties. However, high cane yield was recorded from CO-680, N-14 and B52-298 sugarcane varieties (Table 1). On another hand, significant (P<0.05) difference in sugar yield was recorded among the sugarcane varieties. The current study revealed the high sucrose content on B52298, CO-680 and N-14 was attributed to a significant increase of sugar yield on these sugarcane varieties (Table 1). Because sugar yield is a function of both cane yield and sucrose accumulation [27].

Increased levels of harvest age significantly (p<0.001) influenced all quality parameters. The interaction of harvest age and sugarcane varieties showed highly significant (p<0.001) influence on quality parameters (Table 1). The highest pol and estimated recoverable sucrose were obtained at the 14 months harvest age (Table 1). This might be due to the dilution effect of sugarcane enzymes changing the reducing sugars and non-sucrose materials (fiber) to sucrose or it could be due to positive impact of harvest age on the yield components (plant height and cane yield) which allow accumulation of additional soluble solid or sucrose by on the harvest age. Percent of soluble solids, percent pol and estimated recoverable sucrose significantly increased as age of sugarcane increased until 14 months [25]. Beyond 14 months harvest age all quality parameters showed a declining trend which indicates the reduction of sucrose content due to heavy lodging and remobilization to supply the unproductive bull shoots (newly growing shoots (14). Harvesting either under-aged or over-aged cane with improper time of harvest leads to loss in cane yield, sugar recovery, poor juice quality and problems in milling [28]. Significant difference (p<0.001) of quality parameters was observed among the four sugarcane varieties. The highest Pol and ERS was recorded on N-14 and B52-298 sugarcane varieties (Table 1). This indicates that, those sugarcane varieties have

the probability of high sucrose accumulation if the properly harvested in the proper age.

Economic Analysis

The profitability of sugar yield within various harvest ages considers a time value. So, the partial budget analysis for harvest age was computed in terms of t/ha/month (Table 2). The important parameters of maximizing sugar yield and net revenue in relation to harvest date and crop age is expressed by t/ha/month as an index of time value of sugarcane crop [24].

The partial budget analysis for showed that extending harvest age above 12 months were dominated (Table 2). Marginal rate of return for 12 months harvest age was 178.13%. Increasing harvest age above 12 months lead to increase in additional costs without compensating benefit. The marginal rate of return obtained at 12 months harvest age was above the 100% of the CIMMYT's minimum rate of return required for adoption of agronomic practices. The 178.13% MRR recorded at 12 months harvest age indicated that for every one dollar invested in sugarcane crop it could give a net return of 1.78 USD Dollars. Therefore, 12 months harvest age is more profitable and advisable to sugarcane cane because it gives opportunity to additional profit from investing additional cost.

Conclusions

Harvesting of sugarcane at a proper time i.e., peak maturity, by adopting right technique is necessary to realize maximum sucrose accumulation and sugar production in the tropical area of Tendaho with a least possible field losses under the given growing environment. Improper harvest age is recurrent problems of pre-harvest cultural practices, which severely affect quality and yield of sugarcane cane. All varieties are promising for the environment. However, N-14 and B52298 was recommended to have high percentage of area coverage because of high sucrose accumulation in early ages. The economic analysis indicated that 12 months harvest age gave the highest net benefit of 611.93 $/ha/month with acceptable MRR of 178.13%, respectively. In addition to this, over stand canes affect the growth of consecutive ratoons and creates a suitable environment for pest multiplication. Therefore, adjusting harvest age to 12 months for the major sugarcane varieties was economically recommended to obtain optimum sugar yield with efficient time use at the tropical areas of Tendaho.

Acknowledgements

We would like to thank Ethiopian Sugar Corporation for financing the research and Tendaho Sugar Factory Project for their unreserved material and other facilities support.

References

1. Donaldson RA, Redshaw KAR, Rhodes R, Antwerpen VR (2008) Season effects on productivity of some commercial South African sugarcane cultivars

and trash production. Proceeding South African Sugar Technology Association 81: 528-538.

2. Jorge H, Garcia H, Jorge I, Bernal N (2010) Improving the harvest season based on the maturity in four sugarcane growing regions in Cuba. Pro Int Sugar Cane Technol 27: 56-59.

3. McIntyre RK, Nuss KJ (1998) Evaluation of variety N12 in field trials. Proc S Afr Sug Technol Ass 72: 28-33.

4. Hunsigi G (1993) Production of sugarcane, Theory and Practice. Springer-Verlag, New York, 19-23.

5. Acland JD (1973) An Introduction to the production of field and plantation crops in Kenya, Tanzania and Uganda. East African Crops. Published by Arrangement with the FAO of the United Nations by Longman Group Ltd., Singapore, 192-201.

6. Calderon H, Besosa RA, Luna A (1996) Evaluation of sugarcane varieties suitable for early harvesting under tropical conditions. Proc Int Soc Sugar Cane Technol 22: 239-297.

7. Salisbury FB, Ross CW (1991) Plant physiology. Wadsworth, California, USA.

8. Gilbert R, Shine J, Miller J, Rice R, Rainbolt C (2004) Maturity curves and harvest schedule recommendations for canal point sugarcane varieties at Florida. University of Florida, 1-12.

9. Muchow RC, Wood AW, Spillman MF, Robertson MJ, Thomas MR (1993) Field techniques to quantify the yield-determining processes in sugarcane. Proc Aust Soc Sugar Cane Technol 15: 336-343.

10. UF (University of Florida) (2003) Sugarcane handbook. Extension digital information source (EDIS) Database sugarcane hand book.

11. Liu DL, Bull TA (2001) Simulation of biomass and sugar accumulation in sugarcane using a process-based model. Ecological Modeling. Field Crop Research 78: 181-211.

12. Booker Tate (2009) Re-evaluation of the plantation soils at Metahara Sugar Factory, Ethiopia 1-50.

13. Inman-Bamber NG (1995) Climate and water as constraints to production in the South African Sugar Industry. South African Sugar Association, South Africa, 12: 18-34.

14. Qudsieh HY, Yosuf S, Osman A, Rahman RA (2001) Physico-chemical Changes in Sugarcane and the Extracted Juice at Different Portions of the Stem during Development and Maturation. Journal of Food Chemistry 75: 131-137.

15. Gilbert RA, Shine JM, Miller JD, Rice RW, Rainbolt CR (2006) The Effect of genotype, environment and time of harvest on sugarcane yields in Florida, USA. Field Crops Research 95: 156-170.

16. Hundioto K (2009) Laboratory handout for Ethiopian Sugar Industries. Ethiopian Sugar Development Agency, Research Directorate. Wonji, Ethiopia 1-40.

17. Shukla SK, Yadav RL, Singh PN, Singh I (2009) Potassium nutrition for improving stubble bud sprouting, dry-matter partitioning, nutrient uptake and winter initiated sugarcane (Saccharum spp. hybrid complex) Ratoon Yield. Indian Institute of Sugarcane Research, India. European Journal of Agronomy 30: 27-33.

18. Wagih ME, Ala A, and Musa Y (2004) Evaluation of sugarcane varieties for maturity earliness and selection for efficient sugar accumulation. Sugarcane Agriculture. Sugar Technology North Australia 6: 297-304.

19. CIMMYT (International maize and wheat improvement center) (1988) Economic training manual: from agronomic data to farmer's recommendations. CIMMYT, Mexico 1-79.

20. Inman-Bamber NG (2004) Sugarcane water stress criteria for irrigation and drying off in Australia. Field Crops Research 89: 107-122.

21. Takayoshi T, Makoto M, Hiroshi N (1999) Characteristics of early maturing sugarcane varieties with a high sugar content in relation to growth and invertase activities. Jap J Trop Agri 43: 271-276.

22. Larrahondo JE, Brice CO, Palma M (2010) An assessment of after harvest sucrose losses from sugarcane field to factory. Int J Sugar Cane Technol 27: 233-238.

23. Ambachew D, Bizuneh A (2007) Establishing and update stalk elongation standards for commercial and semi-commercial sugarcane varieties grown at Metahara. Research Report, Metahara Sugar Factory Agriculture Division, Metahara, Ethiopia.

24. Bakker H (1999) Sugarcane cultivation and management. kluwer academic/plenum publisher, New York, USA, 5-10.

25. Rostron H (1972) Effects of age and time of harvest on productivity of irrigated sugarcane. South African Sugar Association Experiment Station, South Africa 142-150.

26. Muchow RC, Higgins AJ, Rudd AV, Ford AW (1998) Optimizing harvest date in sugar production: A Case study for mossman mill region in Australia. Sensitivity to crop age and crop class distribution. Field Crops Research 57: 243-251.

27. Sundara B (2000) Sugarcane cultivation. Vikas Publishing House Pvt Ltd, New Delhi, India.

28. Khandagave R, Patil B (2007) Manipulation of cutting age, varieties and planting time to improve sugar and cane yield. Int Sugar Cane Technl 26: 212-220.

15

Effect of Fodder Radish (*Raphanus sativus* L.) Green Manure on Potato Wilt, Growth and Yield Parameters

Hayfa Jabnoun-Khiareddine[1]*, Rania Aydi Ben Abdallah[1,2], Fakher Ayed[3], Mouna Gueddes-Chahed[1], Ahmed Hajlaoui[1], Samir Ben Salem[1], Wissem Ben Dhia[1] and Mejda Daami-Remadi[1]

[1]*UR13AGR09- Integrated Horticultural Production in the Tunisian Centre- East, Regional Center of Research on Horticulture and Organic Agriculture, University of Sousse, 4042, Chott-Mariem, Tunisia*
[2]*National Agronomic Institute of Tunisia, 1082 Tunis Mahrajène, University of Carthage, Tunisia*
[3]*Technical Center of Organic Agriculture, 4042, Chott-Mariem, Tunisia*

Abstract

Potato is threatened by several soil-borne fungi causing wilt and root rots. In this study, two fodder radish (*Raphanus sativus* L.) (FR) cultivars (cvs. Boss and Defender), used as green manure preceding a potato crop, were evaluated for their suppressive effects against wilt incidence and severity, potato growth and yield as compared to animal manure. The essay was carried out in a completely randomised design with three types of organic amendment and two potato cultivars (cvs. Spunta and Royal). Incidence of potato wilting noted 100 days post planting (DPP) was high, exceeding 70%, for all soil amendments tested. The extent of vascular discoloration varied depending on amendments used where cv. Defender behaved as the control while the highest extent was noted on potato plants grown in cv. Boss amended plots. As compared to animal manure, the application of cvs. Boss and Defender had increased by 48.43 and 41.28% the incidence of vascular discoloration on cv. Spunta, respectively, while on cv. Royal, only cv. Defender had reduced this parameter by 16.32%. Fungal isolations performed from roots and stems revealed the involvement of several soil-borne pathogens in the recorded plant wilting. Soil manuring using cvs. Boss and Defender FR resulted in significant increment in average stem number per plant and aerial part fresh weight by 22.79 and 21.32% and by 34.62 and 27.03%, respectively, as compared to animal manure. At 100 DPP, potato root fresh weight increase by 8.7 and 33.49% was noted on cv. Spunta compared to 30.34 and 23.48% recorded on cv. Royal. Potato tuber yield was improved by 38.28 and 10.7% and by 28.44 and 27.62% in cvs. Spunta and Royal, respectively, relative to animal manure. The use of FR as green manure may be implemented in the integrated management of soil-borne diseases for the enhancement of potato yield.

Keywords: Growth; Organic amendment; Soil-borne fungi; *Solanum tuberosum* L.; Tuber yield; Wilt severity

Introduction

In Tunisia, potato (*Solanum tuberosum* L.) is one of the strategic crops which occupies about 16% of all Tunisian cultivated areas [1-3]. Potato production is concentrated in relatively small diversified farms where potato is grown for several years in the same fields leading to the build-up of many soil-borne pathogens [4]. In fact, soil-borne diseases are persistent, recurrent problems in potato production, resulting in decreased plant growth and vigor, lower tuber quality incited mainly by skin blemishing pathogens and reduced yield [5,6]. Of particular concern throughout major potato producing regions in Tunisia, due to their constant presence, are Fusarium wilt caused mainly by *Fusarium oxysporum* f. sp. *tuberosi* [7,8], Verticillium wilt incited by *Verticillium dahliae* [9], Black dot caused by *Colletotrichum coccodes* [10], and Rhizoctonia canker and black scurf incited by *Rhizoctonia solani* [2,3,11]. In Tunisia, Fusarium wilt has become in last decade as one of the most serious potato diseases responsible for plant stunting, severe leaf yellowing, vascular discoloration and subsequent plant wilting and death [7,8]. Verticillium wilt is a common potato disease causing premature senescence leading to 30-50% yield losses [9,12]. For *C. coccodes*, symptoms include sloughing of the root cortex, brown lesions on roots and blemishes on tubers [13]. This pathogen can cause premature foliage death and yield losses were reported to be as much as 30% [14,15]. *R. solani* affects potato development from emergence to harvest and typical symptoms include death of preemerging sprouts, cankers on underground stem parts and stolons, a diminished root system, and the formation of sclerotia on progeny tubers which leads to lowered tuber yield and quality [16]. These pathogens are often observed as mixed infections and are also associated with other soil-borne bio-aggressors such as nematodes, which lead to exacerbated potato early

dying syndrome. This increased prevalence and severity of soil-borne fungal diseases was due to the absence of resistant potato cultivars since the most grown cultivars exhibited varying degree of susceptibility to all these pathogens [17-21].

Several strategies have been used, in Tunisia and all over the world, to control these potato diseases, such as soil solarisation, long-term rotations, biological control, host resistance, etc. However, serious losses still occur due mainly to pathogen's long-lived resting structures released in the soil such as *Verticillium* microsclerotia, *Fusarium* chlamydospores and *C. coccodes* and *R. solani* sclerotia formed in senescing and dead potato roots, stems, stolons and tubers. Hence, a crucial factor in the management of diseases caused by these pathogens is to reduce their inoculum level below the critical threshold level before a susceptible crop is planted [22].

In the last few years, old practices such as incorporation of green

*Corresponding author: H. Jabnoun-Khiareddine, UR13AGR09- Integrated Horticultural Production in the Tunisian Centre- East, Regional Center of Research on Horticulture and Organic Agriculture, University of Sousse, 4042, Chott-Mariem, Tunisia, E-mail: jkhayfa@yahoo.fr

manure, animal manure, compost, seed meals and other types of organic amendments have been used to control soil-borne pathogens such as *Phytophthora* spp., *Fusarium* spp., *Verticillium* spp., *Rhizoctonia* spp, and *Sclerotinia* spp. [23-27] leading to interesting and promising results [28]. For example, the incidence of Verticillium wilt of potato has been reduced using pea (*Pisum sativum*), oat (*Avena sativa*), broccoli (*Brassica oleracea*), sudan-grass (*Sorghum vulgare*) and corn (*Zea mays)* as green manures [29-31], and addition of animal manures [32,33].

However, some studies indicate that the effectiveness of organic amendments including *Brassica* residues is variable and, in some cases, can even enhance disease severity [34]. This negative effect of organic amendments may be attributed to increased pathogen inoculums' potential, as the decaying material may serve to sustain saprophytic growth of plant pathogens [35], or carried out by the *Brassica* amendments themselves [34]. Furthermore, increases in tuber yield using green manures crops have been inconsistent, despite the beneficial disease suppressive effects [36-38].

Green manures based on *Brassicaceae* species, incorporated into soil when still green or soon after maturity, are shown able to improve the soil's physical, chemical, or biological properties and thereby to increase the succeeding crop's yield, quality, or both [39]. Among these *Brassicas*, fodder radish (*Raphanus sativus* L.) (FR) presents many unique characteristics such as its relatively high tissue phosphor concentration, rapid dry matter accumulation in the fall, and rapid residue decomposition in the spring [40].

FR and other *Brassicas* produce isothiocyanates as a by-product of glucosinolate break-down, and often have a metham sodium like biofumigant action to suppress weeds, nematodes and other soil-borne pests and diseases [28,37,38,41-44]. Furthermore, addition of organic matter increases resistance of the individual plant as a result of uptake of phenols, phenolic and other compounds, such as salicylic acid, which have an antibiotic effect and also work directly on pathogens [45]. Green manures may also affect soil-borne pathogens indirectly by influencing indigenous microbial populations. In fact, incorporation of soil amendments leads to increased soil microbial activity and diversity [46] as well as density of bacteria [47], fluorescent *Pseudomonas* spp. [32,47,48], streptomycetes, and other actinomycetes [47,49] in soil.

The use of *Brassica* spp. and related plants as green manure crops has been receiving increased attention in recent years for their ability to reduce multiple soil-borne potato diseases [50]. Therefore, the current research was established to evaluate the effect of two FR cultivars used as green manure crop, preceding potato growing, for their ability to reduce potato soil-borne diseases in naturally infested soil in Chott-Mariem region, Centre East Tunisia, and their efficacy in increasing potato growth and tuber yield.

Materials and Methods

Plant material

Potato cvs. Spunta and Royal seed tubers were kindly provided by the Technical Center of Potato and Artichoke, Essaïda, Tunisia. Cv. Spunta is most grown, highly susceptible to vascular wilts and low-yielding, whereas cv. Royal is recently introduced in Chott-Mariem region. Before use, tubers were superficially disinfected with a 10% sodium hypochlorite solution during 5 min, rinsed with tap water and air dried. They were kept two weeks under 15-20°C, 60-80% relative humidity and natural room light for pre-germination.

Two FR (oilseed radish) cultivars: cvs. Boss and Defender were kindly

provided by the General Directory of the Protection and the Control of the Agricultural Product Quality, Ministry of Agriculture, Tunisia, for the management of soil-borne bio-aggressors. These FR cultivars were used as green manure crops preceding potato cropping.

Field site, crop sequences and design

Field trial was conducted in 2011-2012 at the experimental farm of the Regional Center of Research on Horticulture and Organic Agriculture in Chott-Mariem region. This site is under conventional farming system and has a history of potato and other vegetables production practices. The soil has a sandy clay texture (Organic matter 76 g/kg at 0-20 cm depth) and has also a long history of potato soil-borne fungal diseases such as Verticillium and Fusarium wilts, Black dot and Rhizoctonia stem canker.

Prior to planting potato, radishes were sown at a rate of 25 kg/ha and grown for approximately nine weeks in the fall (September 12- November 19, 2011). These green manure crops were cut down at flowering stage and mechanically incorporated into soil to a depth of 15-20 cm, between 20 to 23 November by rotovating. The fresh matter (T/ha) of green manures incorporated into the soil was 6 for cv. Boss and 9 for cv. Defender.

The non-radish control treatment consisted of the standard organic amendment, for that site, which was animal manure which is composed of bovine wastes and was applied at a rate of 50 T/ha two weeks before planting potato.

At multi-germ stage, potato seed tubers were planted on December 12, 2012 in radish and animal manure-amended plots consisting of two 40 m long rows. Seed rows were 1 m apart, with two rows per plot, and within row spacing of 0.4 m.

The trial was set up in a completely randomized design with three replications. For each organic amendment and in each replicate or mini plot, thirty potato plants were used per potato cultivar.

Disease incidence and severity

Observational wilt incidence was determined late in the growing season, 100 days post planting (DPP), as the proportion of plants, per plot and per potato cultivar, exhibiting typical early dying symptoms such as leaf chlorosis, wilting or death. For the assessment of disease severity, ten randomly selected plants, with wilt symptoms, were dug from the rows of each organic amendment mini plot. For each plant, stems were longitudinally cut and visually examined for the presence of vascular discoloration. The extent of this vascular discoloration was noted per plant and its incidence represents the percentage of stems per plant showing discolored vessels.

For each organic treatment, plant stems and roots were mixed, surface sterilized in 0.5% NaOCl, rinsed in sterile water, cut into 3-5 mm-fragments, and plated on Potato Dextrose Agar (PDA) medium for the isolation of fungal pathogens involved in wilting symptoms.

The isolation frequency for each pathogen was calculated as the percentage of root or stem fragments showing pathogen growing colonies relative to the total number of stem fragments plated on PDA.

Potato growth and yield parameters

During this essay, cultural practices involving potato fertility, irrigation and pesticide applications were the most commonly used in the region, as recommended by the Technical Center of Potato and Artichoke for potato production guidelines. At 100 DPP, the same ten randomly selected plants, as described above for disease assessment, served for the measurement of roots, aerial parts and tubers' fresh weights together with average stem number per plant.

Statistical analyses

For all parameters measured (disease incidence and severity, stem number per plant, root, aerial part and tuber fresh weight per plant, as well as the isolation frequency per pathogen), statistical analyses were performed following a completely randomised design with three replications where treatments (FR cv. Boss, FR cv. Defender and the animal manure) and potato cultivars (Spunta and Royal) represented the two fixed factors. Thirty potato plants were used per individual treatment. Mean separation was carried out with Fisher's protected LSD or Duncan's Multiple Range test (at P ≤ 0.05). Statistical analyses were performed using SPSS software version 16.

Results

Effect of the organic amendments tested on potato wilt incidence

Visible potato wilt incidence estimated based on the presence of wilt and foliage chlorosis noted at the end of the growing season (100 DPP) was high, exceeding 70%, for all soil amendments tested. In fact, this incidence was not significantly (at P ≤ 0.05) affected neither by soil amendments and potato cultivars nor by their interaction (Table 1). However, potato plants grown following FR cv. Boss green manure showed higher wilt incidence (85%), even statistically insignificant, than following cv. Defender (77%) or using animal manure (73%) as control amendment and this whatever the potato cultivar used. Furthermore, for all soil amendments tested, wilt incidence noted on cv. Spunta plants tended to be higher than that recorded on cv. Royal, even if statistically insignificant.

Effect of the organic amendments tested on potato wilt severity

Data presented in Table 2 showed that wilt severity, estimated via the extent of vascular discoloration, noted on potato stems varied significantly (at P ≤ 0.05) depending only on the organic amendments tested. The highest extent, of 6.15 cm from the collar, was recorded on potato plants grown in cv. Boss-amended plots, which is 58.91% higher than that noted on plants grown following animal manure. However, plants from cv. Defender-treated plots showed wilt severity statistically comparable to that of plants grown in animal-manured plots.

Organic amendments/ Potato cultivars	cv. Spunta	cv. Royal	Average extent per organic amendment[a*]
Folder radish cv. Boss	6.58	5.69	6.15 a
Folder radish cv. Defender	4.59	2.65	3.62 b
Animal manure	1.94	3.10	2.52 b
Average extent per potato cultivar[b*]	4.37 a	3.81 a	

a Mean vascular discoloration extent per organic amendment for potato cultivars combined.
b Mean vascular discoloration extent per potato cultivar for all organic amendments combined.
*For organic amendments and potato cultivars tested, values followed by the same letter are not significantly different according to Duncan's Multiple Range test at P ≤ 0.05. Ten plants were used per individual treatment.

Table 2: Effect of Boss and Defender fodder radish cultivars used as green manure preceding potato growing, compared to animal manure, on the extent of vascular discoloration on two potato cultivars noted 100 days post-planting.

Organic amendments/ Potato cultivars	cv. Spunta	cv. Royal	Average incidence per organic amendment[a*]
Folder radish cv. Boss	58.5	69.23	63.87 a
Folder radish cv. Defender	51.38	48.11	49.75 b
Animal manure	30.16	57.5	43.83 b
Average incidence per potato cultivar[b*]	46.68 b	58.28 a	

LSD (Organic amendments x potato cultivars)=24.21 at P ≤ 0.05.
a Mean stem vascular discoloration incidence per organic amendment for potato cultivars combined.
b Mean stem vascular discoloration incidence per potato cultivar for all organic amendments combined.
*For organic amendments and potato cultivars tested, values followed by the same letter are not significantly different according to Duncan's Multiple Range test at P ≤ 0.05.
The incidence of stem vascular discoloration represents the percentage of stems per plant showing discolored vessels. Ten plants were used per individual treatment.

Table 3: Effect of Boss and Defender fodder radish cultivars used as green manure preceding potato growing, compared to animal manure, on the incidence of stem vascular discoloration (%) on two potato cultivars noted 100 days post-planting.

Potato cv. Spunta plants showed, for all soil amendments tested combined, 12.73% higher vascular discoloration extent, even statistically insignificant, than that noted on cv. Royal plants.

ANOVA analysis revealed that vascular discoloration incidence, noted on potato plants 100 DPP, varied significantly (at P ≤ 0.05) depending on soil treatments and potato cultivars; a significant interaction was noted between both fixed factors. In fact, data given in Table 3 indicated that this incidence ranged from 30.16%, noted on cv. Spunta plants grown in animal manure-amended plots, to 69.23%, recorded on cv. Royal grown following cv. Boss-treated plots. As compared to animal manure, green amendments using cvs. Boss and Defender FR had increased the incidence of vascular discoloration per plant by 48.43 and 41.28% on cv. Spunta, respectively. However, on cv. Royal, soil amendment using cv. Defender had reduced this parameter by 16.32%, as compared to animal manure, contrarily to cv. Boss, which increased this incidence by 16.94%. Overall and whatever the organic amendment tested (combined data of the three organic amendments tested), the incidence of vascular discoloration noted on cv. Spunta plants was 19.9% lesser than that recorded on cv. Royal plants.

Effect of the organic amendments tested on the isolation frequency of soil-borne pathogens

At the end of the growing season (100 DPP), fungal isolations performed from potato plants grown in organically amended plots and exhibiting wilt symptoms revealed the involvement of several soil-

Organic amendments/ Potato cultivars	cv. Spunta	cv. Royal	Average incidence per organic amendment[a*]
Folder radish cv. Boss	94.78	75.67	85.22 a
Folder radish cv. Defender	81.78	71.33	76.55 a
Animal manure	70.56	75.67	73.11 a
Average incidence per potato cultivar[b*]	82.37 a	74.22 a	

a Mean wilt incidence per organic amendment for potato cultivars combined.
b Mean wilt incidence per potato cultivar for all organic amendments combined.
*For organic amendments and potato cultivars tested, values followed by the same letter are not significantly different according to Duncan's Multiple Range test at P ≤ 0.05.
Wilt incidence was determined as the proportion of plants with leaf chlorosis, wilting or death in each plot for each potato cultivar. Thirty plants were used per individual treatment.

Table 1: Effect of Boss and Defender fodder radish cultivars used as green manure preceding potato growing, compared to animal manure, on wilt incidence (%) on two potato cultivars noted 100 days post-planting.

Organic amendments/ Pathogens	cv. Boss	cv. Defender	Animal manure	Average isolation frequency (%) per fungus[a*]
Fusarium oxysporum	35.00	28.33	53.33	38.89 a
F. solani	12.50	11.67	12.83	12.33 b
Verticillium dahliae	10.83	11.67	11.67	11.39 b
Rhizoctonia solani	7.50	15.83	24.17	15.83 b
Colletotrichum coccodes	32.50	43.33	25.83	33.89 a

a Mean isolation frequency per pathogen for all organic amendments and potato cultivars combined.
*For fungal pathogens, values followed by the same letter are not significantly different according to Duncan's Multiple Range test at P ≤ 0.05.

Table 4: Effect of Boss and Defender fodder radish cultivars, used as green manure preceding potato growing, on the isolation frequency of soil-borne pathogens from roots, as compared to animal manure, noted 100 days post-planting.

Organic amendments/ Pathogens	cv. Boss	cv. Defender	Animal manure	Average isolation frequency (%) per fungus[a*]
Fusarium oxysporum	57.50	55.83	47.50	53.61 a
F. solani	10	0.83	4.17	5 b
Verticillium dahliae	10.83	19.17	15	15 b
Rhizoctonia solani	5.83	9.17	15	10 b
Colletotrichum. coccodes	8.33	17.50	12.50	12.78 b

a Mean isolation frequency per pathogen for all organic amendments and potato cultivars combined.
*For fungal pathogens, values followed by the same letter are not significantly different according to Duncan's Multiple Range test at P ≤ 0.05.

Table 5: Effect of Boss and Defender fodder radish cultivars, used as green manure preceding potato growing, on the isolation frequency of soil-borne pathogens from stems, as compared to animal manure, noted 100 days post-planting.

borne fungal pathogens in the observed signs. In fact, the isolation frequency of these pathogens was not significantly influenced nor by potato cultivars and organic amendments, nor by their interaction (at P ≤ 0.05). As shown in Tables 4 and 5, *F. oxysporum* and *C. coccodes* were most frequently isolated from roots of wilting plants, followed by *R. solani*, *F. solani*, and *V. dahliae* whereas the predominant fungi recovered from stems showing vascular discoloration and cankers were *F. oxysporum*, *V. dahliae*, *C. coccodes*, *R. solani*, and to a lesser extent *F. solani*.

Effect of the organic amendments tested on potato growth and production parameters

Average stem number: Results illustrated in Table 6 indicated that the average stem number noted on potato plants at the end of the growing season (100 DPP) varied depending on organic amendments

Organic amendments/ Potato cultivars	cv. Spunta	cv. Royal	Average stem number per organic amendment[a*]
Fodder radish cv. Boss	3.77	3.40	3.58 a
Fodder radish cv. Defender	3.57	3.47	3.52 a
Animal manure	3.10	2.43	2.77 b
Average stem number per potato cultivar[b*]	3.48 a	3.10 a	

a Mean stem number per organic amendment for potato cultivars combined.
b Mean stem number per potato cultivar for all organic amendments combined.
*For organic amendments and potato cultivars tested, values followed by the same letter are not significantly different according to Duncan's Multiple Range test at P ≤ 0.05. Ten plants were used per individual treatment.

Table 6: Effect of Boss and Defender fodder radish cultivars, used as green manure preceding potato growing, as compared to animal manure, on mean stem number per plant on two potato cultivars tested noted 100 days post-planting.

only. In fact, soil treatment using FR cvs. Boss and Defender led to significant increase in average stem number per plant by 22.79 and 21.32%, respectively, as compared to animal manure.

Root fresh weight: Potato root fresh weight was significantly affected by the organic amendments tested but was not influenced with potato cultivars used; however, a significant interaction (at P ≤ 0.05) was noted between both fixed factors (Table 7). In fact, as indicated in this table, FR cvs. Defender and Boss had increased this parameter by 8.7 and 33.49% on cv. Spunta compared to 30.34 and 23.48% recorded on cv. Royal, respectively, as compared to animal manure.

Aerial part fresh weight: Data given in Table 8 revealed that only the organic amendments tested had significantly impacted the aerial

Organic amendments/ Potato cultivars	cv. Spunta	cv. Royal	Average fresh weight per organic amendment[a*]
Fodder radish cv. Boss	27.87	23.70	25.78 a
Fodder radish cv. Defender	20.30	26.03	23.17 a
Animal manure	18.53	18.13	18.33 b
Average fresh weight per potato cultivar[b*]	22.23 a	22.62 a	

LSD (Organic amendments × potato cultivars)=6.94 g at P ≤ 0.05.
a Mean root fresh weight per organic amendment for potato cultivars combined.
b Mean root fresh weight per potato cultivar for all organic amendments combined.
*For organic amendments and potato cultivars tested, values followed by the same letter are not significantly different according to Duncan's Multiple Range test at P ≤ 0.05. Ten plants were used per individual treatment.

Table 7: Effect of Boss and Defender fodder radish cultivars, used as green manure preceding potato growing, as compared to animal manure, on root fresh weight (g) of two potato cultivars noted 100 days post-planting.

Organic amendments/ Potato cultivars	cv. Spunta	cv. Royal	Average fresh weight per organic amendment[a*]
Fodder radish cv. Boss	126.90	148.50	137.70 a
Fodder radish cv. Defender	108.60	138.13	123.37 a
Animal manure	92.03	88.00	90.02 b
Average fresh weight per potato cultivar[b*]	109.18 a	124.88 a	

a Mean aerial part fresh weight per organic amendment for potato cultivars combined.
b Mean aerial part fresh weight per potato cultivars for all organic amendments combined.
*For organic amendments and potato cultivars tested, values followed by the same letter are not significantly different according to Duncan's Multiple Range test at P ≤ 0.05. Ten plants were used per individual treatment.

Table 8: Effect of Boss and Defender fodder radish cultivars, used as green manure preceding potato growing, as compared to animal manure, on aerial part fresh weight (g) of two potato cultivars noted 100 days post-planting.

Organic amendments/ Potato cultivars	cv. Spunta	cv. Royal	Average fresh weight per organic amendment[a*]
Fodder radish cv. Boss	989.90	727.53	858.72 a
Fodder radish cv. Defender	684.10	719.33	701.72 b
Animal manure	610.93	520.60	565.77 c
Average fresh weight per potato cultivar[b*]	761.64 a	655.82 b	

LSD (Organic amendments x potato cultivars)=184.58 g at P ≤ 0.05.
a Mean tuber fresh weight per organic amendment for potato cultivars combined.
b Mean tuber fresh weight per potato cultivar for all organic amendments combined.
*For organic amendments and potato cultivars tested, values followed by the same letter are not significantly different according to Duncan's Multiple Range test at P ≤ 0.05. Ten plants were used per individual treatment.

Table 9: Effect of Boss and Defender fodder radish cultivars, used as green manure preceding potato growing, as compared to animal manure, on tuber fresh weight (g) of two potato cultivars noted 100 days post-planting.

part fresh weight noted on potato plants at the end of the growing season (100 DPP). In fact, green manuring using FR cvs. Boss and Defender had significantly enhanced potato aerial part fresh weight by 34.62 and 27.03%, respectively, relative to animal manure. This parameter was 12.57% higher on potato cv. Royal plants, even statistically insignificant, than that noted on cv. Spunta plants.

Tuber fresh weight: As shown in Table 9, tuber fresh weight per plant varied significantly (at $P \leq 0.05$) depending on amendments used and on potato cultivars tested; a significant interaction was also noted between both fixed factors. In fact, tuber yield per plant, recorded at the end of the growing season (100 DPP), ranged between 990 g, recorded on cv. Spunta plants grown in cv. Boss-amended plots, and 520.6 g, noted on cv. Royal plants grown following animal manure. In fact, soil treatment using FR cvs. Boss and Defender resulted in 38.28% and 10.7% higher tuber fresh weight on cv. Spunta compared to 28.44% and 27.62% increase recorded on cv. Royal, respectively, as compared to animal manure. Furthermore and whatever the potato cultivar used, manuring with FR cvs. Boss and Defender resulted in 34.11 and 19.37% higher tuber yield per plant than animal manure. Furthermore, cv. Spunta yielded significantly 13.89% higher tuber fresh weight than cv. Royal for all amendments combined.

Discussion

In the present study, two cultivars of FR, grown as green manures preceding potato growing, were assessed, in comparison with animal manure, for their ability to reduce the incidence of potato soil-borne fungal pathogens and to ultimately increase crop growth and yield.

Green manuring with FR cvs. Boss and Defender, before planting potato as main crop, resulted at the end of the growing season in high wilt incidence levels in the treated plots as well as in animal manured ones. However, wilt severity, estimated via vascular discoloration extent, noted following cv. Defender-amendment was comparable to that recorded on plants grown in animal manured plots. In contrast, 58.91% higher vascular discoloration extent, relative to animal manure, was noted on plants from cv. Boss-amended plots.

These results indicated that single application of FR cvs. Boss and Defender did neither reduce wilt incidence nor wilt severity better than animal manure under our field conditions. In fact, the site chosen in this investigation is known to be highly infested by soil-borne and vascular pathogens causing yearly severe losses in terms of quantity and quality of potato tubers (unpublished data). Therefore, the observed wilt incidence noted on potato plants is the coupled effect of the root infecting pathogens, *C. coccodes, R. solani* and *F. solani*, together with the vascular ones namely *F. oxysporum* f. sp. *tuberosi* and *V. dahliae*. In fact, our results are consistent with many other findings showing that manuring soil with *Brassicaceae* species have not shown always efficacy in reducing disease severity and soil-borne inoculums in the soil. In this sense, Neubauer et al. [51] find that soil amendments using *R. sativus* cultivars such as cv. Defender are less effective than *B. juncea* in reducing viable microsclerotia. Furthermore, the incorporation of broccoli (*B. oleracea*) does not significantly reduce inocula of *F. oxysporum* f. sp. *asparagi, R. solani, V. dahliae*, and *Fusarium* spp. [22,52]. In this regard, Hartz et al. [53] mention that soil populations of *V. dahliae* and *Fusarium* spp. are unaffected by *Brassica* spp.-based green manures and there are no evidence of soil-borne disease suppression on subsequent tomato crops. Results from the current study are also in line with those of Davis et al. [54] revealing that two consecutive years of sudangrass, oat, or rye green manures does not reduce inoculum of *V. dahliae*. In the same sense, Geary et al. [55] find that oilseed radish, mustard, and canola exhibit limited effects on the severity of onion pink root caused by *Phoma terrestris*, and are not a viable option for this disease control. Moreover, Molina et al. [56] mention that soil treatments with green manures based on fall rye (*Secale cereale* L.) and sorghum-sudangrass hybrid (*Sorghum bicolor* L. Moench, 'Super Su 22') increase inoculum density as well as potato early dying incidence.

This lack of effectiveness of green manure-based treatments in decreasing wilt incidence and severity may be attributed to several factors such as the limited number of green manure cycles or insufficient physical disruption of plant tissues during plow down that lead to low release of toxic compounds, mainly glucosinolates, as previously suggested [12,57]. These factors are relatively important for long-term survival structures of plant pathogens in soil, particularly, microsclerotia of *V. dahliae* [12]. In fact, glucosinolate concentrations and the resulting production of different forms of isothiocyanates vary greatly among *Brassica* species and even among cultivars within each species [58], and are also affected by environmental conditions and plant development [59,60]. This could explain in parts the lack of efficacy in reducing potato diseases in the present work, as FR is known to have moderate glucosinolate content [61] as compared to other *Brassicaceae* species. In addition, in the field, the expected isothiocyanates concentrations are once more lower, because of a poor release efficiency of these compounds due to incomplete tissue maceration with a mulching implement.

Other studies indicate that the negative effect of these amendments may result from increased pathogens' inoculum potential when the substrate serves to sustain saprophytic growth of plant pathogens [35] or from the increase of inoculum potential of pathogens carried out by the *Brassica*-based amendments themselves [34].

Nevertheless, the effectiveness of organic amendments including Brassicaceous residues has been shown to be variable [62]. In fact, the adoption of a green manure rotation crop has been associated with significant decreases in severity and incidence of Verticillium wilt, black scurf and stem canker [37,38,63], as well as common scab [37,38]. Larkin and Griffin [37] report that *Brassicaceae* crops, including radish, canola, rapeseed, turnip, yellow mustard, and Indian mustard are able to reduce inoculum levels of *R. solani* by 20-56% in greenhouse tests and to decrease subsequent potato seedling disease by 40-83%. Furthermore, oilseed radish incorporated earlier as a green manure has been observed to reduce Rhizoctonia Root and Crown Rot in sugarbeet and to decrease population of *V. dahliae* to a greater degree [64,65]. In addition, fodder and oilseed radishes have also shown efficacy in suppressing many soil-borne bio-agressors such as nematodes and bacteria. It was found that two new oil radish varieties Defender and Comet have significantly lowered population of *Heterodera schachtii* by 95%. Hafez and Sundararaj [66] consider that cv. Defender is the most economically and highly suitable cultivar for the management of sugar beet cyst nematode. Moreover, significant reduction and long-term elimination of bacterial wilt caused by *Ralstonia solanacearum* from the soil was achieved through incorporating especially radish or mustard plants in large amounts into the soil immediately before planting tomatoes which led to 50-70% lower bacterial wilt incidence [67].

In this regard, Davis et al. [68] mention that a single green manure of sweet corn followed by two consecutive potato cropping years, has suppressed by 60-70% Verticillium wilt and increased yields and this effect has occurred even though soil-borne *V. dahliae* inoculum levels have augmented by more than 4-fold from 45 to 182 cfu/g of soil. They also found that, although these treatments show no direct effect on *V. dahliae* soil populations, the colonization of *V. dahliae* on potato feeder-roots and in potato tissue of stem apices are reduced.

These inconsistencies in reducing microbial populations using *Brassicaceae*-based amendments have been attributed to the use of different species, physical environments and target organisms [52], soil temperature [69], and amount of crop residue incorporated [22]. In general, the mechanisms involved in disease control are multiple and can vary with each pathosystem [24].

In the current study, manuring soil with FR cvs. Boss and Defender, at a rate of 25 t/ha, had significantly enhanced all potato growth and yield parameters as compared to animal manure. In fact, potato average stem number per plant and aerial part fresh weight were respectively increased by 22.79 and 21.32% and by 34.62 and 27.03%. Furthermore, roots fresh weight were augmented by 8.7 and 33.49% on cv. Spunta compared to 23.48 and 30.34% recorded on cv. Royal, following FR cvs. Boss and Defender. In this sense, several studies indicate the positive effect on plant's biomass produced in forage radish-amended soils such as for *Lactuca sativa*, *Lupinus nanus*, and *Beta vulgaris* [64,70,71]. In this regard, Subbarao and Hubbard [72] find that the incorporation of broccoli residue results in consistently taller cauliflower plants with greater root and shoot biomass.

Moreover, in the present study, potato growth improvement achieved following single application of cvs. Boss and Defender as green manures, resulted in enhancement of tuber yield per plant by 38.28 and 10.7% noted in cv. Spunta and by 28.44 and 27.62% recorded on cv. Royal, respectively, as compared to animal manure. This increase in potato tuber yield using Brassicaceous species green manures was reported in several previous studies [30,38,54]. In this regard, Lehrsch and Gallian [64] find that planting radish as a green manure in the fall, prior to planting potato the following spring, leads to yield improvement in the grown potato as well as in the subsequent sugar beet crops. It has been proved that amongst six varieties of green manure crops, maximum beet yield was obtained from FR cv. Defender planted plots [67]. Our results are also in line with those of Lazarovits [73] showing that incorporating millet over 3 years at two potato field sites, lead to 20% increased potato yields at one site and 50% at the other, but little decrease in Verticillium wilt incidence was observed.

Interesting results from the present study showed that even though disease reductions are minimal, yield increases due to incorporation of radish biomass are significant and this could be explained by the improved nutrient availability as mentioned in several studies [45,74,75]. In this sense, Lehrsch and Gallian [64] relate the increase in the yield and quality of sugar beet, after fall-incorporated radish biomass, to soil physical and hydraulic properties improvement. In this regard, Schomberg et al. [76] report that oilseed radish grows rapidly in the fall and spring and can scavenge significant quantities of N. Furthermore, as oilseed radish produced a significant amount of biomass in the fall and early spring, it has been suggested that it could be useful in rotations where earlier planting dates are desired and for preventing leaching of residual N [76]. In the same sense, Talgre et al. [77] find that among all the *Brassicaceae* tested, the most effective ones were FR and white mustard which produce the highest biomass and therefore drove more nutrients into the soil. Many studies have also related the disease suppressive effect of green manure to improved soil fertility (mainly NPK contents) [12,78] and soil quality [31]. In fact, green manures and organic amendments are commonly used in crop production systems to increase soil nutrient availability, organic nitrogen and organic matter, which are associated with higher tuber yields [19]. Furthermore, the efficacy of organic amendments is soil-specific and very much rate-dependent. In this sense, Ochiai et al. [31] find that the quantity and not the chemistry of organic inputs is the critical factor for disease suppression.

In the present study, the increase in potato growth and yield seem to be more pronounced using cv. Boss than cv. Defender even though this last cultivar have been incorporated at the rate of 9 kg/ha compared to 6 kg/ha for the former. A possible explanation may involve the beneficial microbial population composition which may be more stimulated by residues' allelochemicals from cv. Boss than those from cv. Defender. Furthermore, cv. Spunta yielded 13.89% significantly higher tuber fresh weight than cv. Royal, for all amendments combined which demonstrated the usefulness of these soil treatments for tuber yield increase in the most grown cultivar in the region even under high soil-borne inoculums levels.

Conclusion

Single application of FR cvs. Boss and Defender resulted in a significant growth and yield improvement of two potato cultivars, Spunta and Royal, comparatively to animal manure which is the most used organic amendment in the region, but did not result in significant decrease in wilt diseases' incidence noted at the end of the growing season. Therefore, additional research is needed to verify if multiple seasons together with higher rates of green manures could be associated with significant decreases in wilt incidence and severity on potato and to identify which mechanisms are related with yield improvement. Further studies are also required to determine the effect of these green manures in reducing the pathogens' inoculum densities and enhancing the activity of beneficial micro-organisms in the soil thus improving its fertility.

Furthermore, green manures, like other types of cultural practices, are best implemented as an important component of an integrated disease management program and not as the only control means for soil-borne diseases. Thus, combining green manures with other cultural, biological or chemical approaches can substantially increase disease control and help in achieving greater sustainability and improving soil fertility and subsequently crop production.

References

1. Anonymous (2015) Groupement Interprofessionnel des légumes, filières des légumes.

2. Djébali N, Tarhouni B (2010) Field study of the relative susceptibility of eleven potato (Solanum tuberosum L.) varieties and the efficacy of two fungicides against Rhizoctonia solani attack. Crop Prot 29: 998-1002.

3. Mrabet M, Elkahoui S, Tarhouni B, Djebali N (2015) Potato seed dressing with Pseudomonas aeruginosa strain RZ9 enhances yield and reduces black scurf. Phytopathol Medit 54: 265-274.

4. Triki MA, Priou S, El Mahjoub M (2001) Effects of soil solarization on soil-borne populations of Pythium aphanidermatum and Fusarium solani and on the potato crop in Tunisia. Potato Res 44: 271-279.

5. Larkin RP (2008) Relative effects of biological amendments and crop rotations on soil microbial communities and soilborne diseases of potato. Soil Biol Biochem 40: 1341-1351.

6. Larkin RP, Tavantzis S (2013) Use of biocontrol organisms and compost amendments for improved control of soilborne diseases and increased potato production. Am J Potato Res 90: 261-270.

7. Daami-Remadi M, El Mahjoub M (2004) Emergence en Tunisie de Fusarium oxysporum f. sp. tuberosi agent de flétrissure vasculaire des plantes et de pourriture sèche des tubercules de pomme de terre. EPPO Bull 34: 407-411.

8. Ayed F, Daami-Remadi M, Jabnoun-Khiareddine H, El Mahjoub M (2006) Potato Vascular Fusarium wilt in Tunisia: Incidence and biocontrol by Trichoderma spp. Plant Pathol J 5: 92-98.

9. Daami-Remadi M, Jabnoun-Khiareddine H, Ayed F, El Mahjoub M (2011) Comparative aggressiveness of Verticillium dahliae, V. albo-atrum and V. tricorpus on potato as measured by their effects on wilt severity, plant growth and subsequent yield loss. Funct Plant Sci Biotechnol 5 (Special Issue): 1-8.

10. Daami-Remadi M, Bouallègue R, Jabnoun-Khiareddine H, El Mahjoub M

(2010) Comparative aggressiveness of Tunisian Colletotrichum coccodes isolates on potato assessed via black dot severity, plant growth and yield loss. Pest Technol 4: 45-53.

11. Daami-Remadi M, Zammouri S, El Mahjoub M (2008) Effect of the level of seed tuber infection by Rhizoctonia solani at planting on potato growth and disease severity. Afr J Plant Sci Biotechnol 2: 34-38.

12. Rowe RC, Powelson MR (2002) Potato Early Dying: Management challenges in a changing production environment. Plant Dis 86: 1184-1193.

13. Stevenson WR, Loria R, Franc GD, Weingartner DP (2001) Compendium of Potato Diseases. APS Press, St. Paul, MN, USA, p: 106.

14. Tsror L, Erlich O, Hazanovsky M (1999) Effect of Colletotrichum coccodes on potato yield, tuber quality, and stem colonization during spring and autumn. Plant Dis 83: 561-565.

15. Tsror L, Johnson DA (2000) Colletotrichum coccodes on potato. In: Prusky D, Freeman S, Dickman MB (Eds) Colletotrichum – Host Specificity, Pathology and Host – Pathogen Interaction, APS Press, pp: 362-373.

16. El Bakali AM, Martín MP (2006) Black scurf in potato. Mycologist 20: 130-132.

17. Daami-Remadi M, Zammouri S, El Mahjoub M (2008) Relative susceptibility of nine potato (Solanum tuberosum L.) cultivars to artificial and natural infection by Rhizoctonia solani as measured by stem canker severity, black scurf and plant growth. Afr J Plant Sci Biotechnol 2: 57-66.

18. Daami-Remadi M, Jabnoun-Khiareddine H, Ayed F, El Mahjoub M (2010) Comparative susceptibility of potato cultivars to Verticillium wilt assessed via wilt severity and subsequent yield reduction. Int J Plant Breed 4: 55-62.

19. Davis JR, Huisman OC, Everson DO, Schneider AT (2001) Verticillium wilt of potato: A model of key factors related to disease severity and tuber yield in south-eastern Idaho. Am J Potato Res 78: 291-300.

20. Ayed F, Daami-Remadi M, Jabnoun-Khiareddine H, El Mahjoub M (2006) Effect of potato cultivars on incidence of Fusarium oxysporum f. sp. tuberosi and its transmission to progeny tubers. J Agron 5: 430-434.

21. Daami-Remadi M, Jabnoun-Khiareddine H, Sdiri A, El Mahjoub M (2012) Comparative reaction of potato cultivars to Sclerotium rolfsii assessed by stem rot and tuber decay severity. In Daami-Remadi M. (Ed) Potato Pathology. Pest Technol 6 (Special Issue 2): 54-59.

22. Blok WJ, Lamers JG, Termorshuizen AJ, Bollen GJ (2000) Control of soilborne plant pathogens by incorporating fresh organic amendments followed by tarping. Phytopathology 90: 253-259.

23. Bailey KL, Lazarovits G (2003) Suppressing soil-borne diseases with residue management and organic amendments. Soil Tillage Res 72: 169-180.

24. Janvier C, Villeneuve F, Alabouvette C, Edel-Hermann V, Mateille T, et al. (2007) Soil health through soil disease suppression: Which strategy from descriptors to indicators? Soil Biol Biochem 39: 1-23.

25. Hoitink H, Boehm M (1999) Biocontrol within the Context of Soil Microbial Communities: A Substrate-Dependent Phenomenon. Annu Rev Phytopathol 37: 427-446.

26. Noble R, Coventry R (2005) Suppression of soil-borne plant diseases with composts: a review. Biocontrol Sci Technol 15: 3-20.

27. Bonanomi G, Antignani V, Pane C, Scala F (2007) Suppression of soilborne fungal diseases with organic amendments. J Plant Pathol 89: 311-340.

28. Larkin RP, Honeycutt CW, Olanya OM (2011) Management of Verticillium wilt of potato with disease-suppressive green manures and as affected by previous cropping history. Plant Dis 95: 568-576.

29. Davis JR, Huisman OC, Everson DO, Sorensen LH, Schneider AT (1999) Control of Verticillium wilt of the Russet Burbank potato with corn and barley. Am J Potato Res 76: 367.

30. Wiggins BE, Kinkel LL (2005) Green manures and crop sequences influence potato diseases and pathogen inhibitory activity of indigenous streptomycetes. Phytopathology 95: 178-185.

31. Ochiai N, Powelson ML, Crowe FJ, Dick RP (2008) Green manure effects on soil quality in relation to suppression of Verticillium wilt of potatoes. Biol Fertil Soils 44: 1013-1023.

32. Conn KL, Lazarovits G (1999) Impact of animal manures on Verticillium wilt, potato scab, and soil microbial populations. Can J Plant Pathol 21: 81-92.

33. Tenuta M, Conn KL, Lazarovits G (2002) Volatile Fatty Acids in Liquid Swine Manure Can Kill Microsclerotia of Verticillium dahliae. Phytopathology 92: 548-552.

34. Lu P, Gilardi G, Gullino ML, Garibaldi A (2010) Biofumigation with Brassica plants and its effect on the inoculum potential of Fusarium yellows of Brassica crops. Eur J Plant Pathol 126: 387-402.

35. Manici LM, Caputo F, Babini V (2004) Effect of green manure on Pythium spp. population and microbial communities in intensive cropping systems. Plant Soil 263: 133-142.

36. Campiglia E, Paolini R, Colla G, Mancinelli R (2009) the effects of cover cropping on yield and weed control of potato in a transitional system. Field Crop Res 112: 16-23.

37. Larkin RP, Griffin TS (2007) Control of soilborne diseases of potato using Brassica green manures. Crop Prot 26: 1067-1077.

38. Larkin RP, Griffin TS, Honeycutt CW (2010) Rotation and cover crop effects on soilborne potato diseases, tuber yield, and soil microbial communities. Plant Dis 94: 1491-502.

39. Cherr CM, Scholberg S, McSorley R (2006) Green manure approaches to crop production: A synthesis. Agron J 98: 302-319.

40. White CM, RR Weil (2011) Forage radish cover crops increase soil test phosphorus surrounding radish taproot holes. Soil Fert Plant Nutr 75: 121-130.

41. Charron CS, Sams CE (1999) Inhibition of Pythium ultimum and Rhizoctonia solani by shredded leaves of brassica species. J Am Soc Hortic Sci 124: 462-467.

42. Deacon JW, Mitchell RT (1985) Toxicity of oat roots, oat root extracts, and saponins to zoospores of Pythium spp. and other fungi. Trans Br Mycol Soc 84: 479-487.

43. Mcsorley R, Ozores-Hampton M, Stansly PA, Conner JM (1999) Nematode management, soil fertility, and yield in organic vegetable production. Nematropica 29: 205-213.

44. Melakeberhan H, Mennan S, Ngouajio M, Dudek T (2008) Effect of Meloidogyne hapla on multi-purpose use of oilseed radish (Raphanus sativus). Nematology 10: 375-379.

45. Lampkin NH, Foster C, Padel S, Midmore P (1999) the policy and regulatory environment for organic farming in Europe. Organic farming: Economics and Policy 1 & 2. University of Hohenheim, Stuttgart.

46. Lupwayi NZ, Rice WA, Clayton GW (1998) Soil microbial diversity and community structure under wheat as influenced by tillage and crop rotation. Soil Biol Biochem 30: 1733-1741.

47. Mazzola M, Granatstein DM, Elfving DC, Mullinix K (2001) Suppression of Specific Apple Root Pathogens by Brassica napus Seed Meal Amendment Regardless of Glucosinolate Content. Phytopathology 91: 673-679.

48. Bulluck LR, Ristaino JB (2002) Effect of synthetic and organic soil fertility amendments on southern blight, soil microbial communities, and yield of processing tomatoes. Phytopathology 92: 181-189.

49. Kinkel LL, Stromberg KD, Flor JM, Wiggins E (2001) Green manures influence pathogen inhibitory potential of indigenous antagonist communities in soil. Phytopathology 91: S49.

50. Larkin RP (2013) Green manures and plant disease management. CAB Reviews: Perspectives in Agriculture, Veterinary Science, Nutrition and Natural Resources. CAB Reviews 8-037: 1-10.

51. Neubauer C, Heitmann B, Müller C (2014) Biofumigation potential of Brassicaceae cultivars to Verticillium dahliae. Eur J Plant Pathol 140: 341-352.

52. Zasada IA, Ferris H, Elmore CL, Roncoroni JA, MacDonald JD et al. (2003) Field application of Brassicaceous amendments for control of soilborne pests and pathogens. Online. Plant Health Progress.

53. Hartz TK, Johnstone PR, Miyao EM, Davis RM (2005) Mustard cover crops are ineffective in suppressing soilborne disease or improving processing tomato yield. HortScience 40: 2016-2019.

54. Davis JR, Huisman OC, Westermann DT, Everson DO, Sorensen LH et al. (1996) Effects of green manures on Verticillium wilt of potato. Phytopathology 86: 444-453.

55. Geary B, Ransom C, Brown B, Atkinson D, Hafez S (2008) Weed, disease, and nematode management in onions with biofumigants and metam sodium. HortTechnol 18: 569-574.

56. Molina OI, Tenuta M, El Hadrami A, Buckley K, Cavers C et al. (2014) Potato early dying and yield responses to compost, green manures, seed meal and chemical treatments. Am J Potato Res 91: 414-428.

57. Morra MJ, Kirkegaard JA (2002) Isothiocyanate release from soil-incorporated Brassica tissues. Soil Biol Biochem 34: 1683-1690.

58. Kirkegaard JA, Sarwar M (1998) Biofumigation potential of brassicas I. Variation in glucosinolate profiles of diverse field-grown brassicas. Plant Soil 201: 71-89.

59. Kirkegaard JA, Wong PTW, Desmarchelier JM (1996) In vitro suppression of fungal root pathogens of cereals by Brassica tissues. Plant Pathol 45: 593-603.

60. Sarwar M, Kirkegaard JA (1998) Biofumigation potential of brassicas. II. Effect of environment and ontogeny on glucosinolate production and implications for screening. Plant Soil 201: 91-101.

61. Kruger DHM, Fourie JC, Malan AP (2013) Cover crops with biofumigation properties for the suppression of plant-parasitic nematodes: A review. S Afr J Enol Vitic 34: 287-295.

62. Colla P, Gilardi G, Gullino ML (2012) A review and critical analysis of the European situation of soilborne disease management in the vegetable sector. Phytoparasitica 40: 515- 523.

63. Larkin RP, Honeycutt CW (2006) Effects of different 3-year cropping systems on soil microbial communities and rhizoctonia diseases of potato. Phytopathology 96: 68-79.

64. Lehrsch GA, Gallian JJ (2010) Oilseed radish effects on soil structure and soil water relations. J Sugar Beet Res 47: 1-21.

65. Harding RB, Wicks TJ (2001) In vitro suppression of soil-borne potato pathogens by volatiles released from Brassica residues. In: Porter IJ (ed) Proceedings of the 2nd Australasian soil-borne diseases symposium, Lorne, Australia, pp: 148-149.

66. Hafez SL, Sundararaj P (2004) Biological and chemical management strategies in the sugar beet cyst nematode management. In: Proceedings of the Winter Commodity Schools – 2004, University of Idaho, pp: 243-248.

67. Reddy PP (2013) Biofumigation. In Recent advances in crop protection. Springer, pp: 37-60.

68. Davis JR, Huisman OC, Everson DO, Nolte P, Sorenson LH, et al. (2010) The suppression of Verticillium wilt of potato using corn as a green manure. Am J Potato Res 87: 195-208.

69. Gamliel A, Stapleton JJ (1993) Characterization of antifungal volatile compounds evolved from solarized soil amended with cabbage residues. Phytopathology 83: 899-905.

70. Lawley YE, Teasdale JR, Weil RR (2012) the mechanism for weed suppression by a forage radish cover crop. Agron J 104: 1-10.

71. Pearse IS, Bastow JL, Tsang A (2014) Radish introduction affects soil biota and has a positive impact on the growth of a native plant. Oecologia 174: 471-478.

72. Subbarao KV, Hubbard JC (1996) Interactive effects of broccoli residue and temperature on Verticillium dahliae microsclerotia in soil and on wilt in cauliflower. Phytopathology 86: 1303-1310.

73. Lazarovits G (2010) Managing soilborne disease of potatoes using ecologically based approaches. Am J Potato Res 87: 401-411.

74. Thorup-Kristensen K (2001) Are differences in root growth of nitrogen catch crops important for their ability to reduce soil nitrate-N content, and how can this be measured? Plant Soil 230: 185-195.

75. Nitta T (1991) Diversity of root fungal floras: its implications for soilborne diseases and crop growth. J Agric Res 25: 6-11.

76. Schomberg HH, Endale DM, Calegari A, Peixoto R, MiyazawaM et al. (2006) Influence of cover crops on potential nitrogen availability to succeeding crops in a Southern Piedmont soil. Biol Fertil Soils 42: 299-307.

77. Talgre L, Lauringson E, Makke A, Lauk R (2011) Biomass production and nutrient binding of catch crops. Zemdirbyste-Agriculture 98: 251-258.

78. Lambert DH, ML Powelson, WR Stevenson (2005) Nutritional interactions influencing diseases of potato. Am J Potato Res 82: 309-319.

16

Distribution of Stem Rust (*Puccinia graminis* f. sp. *tritici*) Races in Ethiopia

Endale Hailu[1]*, Getaneh Woldaeb[1], Worku Danbali[2], Wubishet Alemu[3] and Teklay Abebe[4]

[1]*Ambo Plant protection Research center, 37Ambo, Ethiopia*
[2]*Kulumisa Agricultural research center, 489, Asella, Ethiopia*
[3]*Sinana Agricultural Research Center, 208, Bale-Robe, Ethiopia*
[4]*Alamata Agricultural research center, 56, Alamata, Ethiopia*

Abstract

Wheat is one of the most important cereal crops of Ethiopia. Stem rust caused by *Puccinia graminis* f. sp. *tritici* is amongst the biotic factors which can cause up to 100% yield loss if susceptible cultivar grown and epidemic occurs. The highland of Ethiopia is considered as a hot spot for the development of stem rust diversity. This study was carried out to determine virulence diversity and race distribution of *P. graminis f. sp. tritici* in Ethiopia. One hundred wheat stem rust samples were collected in 2013 cropping season in the Oromia, Amhara and Tigray region. Of sample collected, 66 were viable and analyzed on to the 20 stem rust differentials lines. A total of 9 races were identified, which includes TTKSK, TTKTF, TTKTK, JRCQC, TKTTF, TTKSC, TRTTF, SRKSC and RRKSF. Race TTKSK was predominant and widely distributed in the country with 52% frequency except in Tigray region. The most virulent and new race, TKTTF which causes localized stem rust epidemic in Bale and Arsi was predominantly distributed in oromia region with 36.4% frequency value. Most of the genes possessed by the differentials were ineffective against one or more of the tested isolates except Sr24. Only stem rust resistance gene 24 was found to confer resistance to most of the races prevalent in Ethiopia. These, this gene could be used in combination with other genes through gene pyramiding in breeding for resistance to stem rust in Ethiopia.

Keywords: Stem rust race; *Puccinia graminis f. sp. Tritici*; Stem rust resistance genes

Introduction

Bread wheat (*Triticum aestivum* L. em. *thell*) is the world's leading cereal grain where more than one-third of the population of the world uses as a staple food It is one of the most important cereal crops of Ethiopia [1,2]. It ranked fourth in land coverage and total production after tef, maize and sorghum [3]. Wheat is produced across a wide range of agro ecological and crop management regime. The most suitable area for wheat production falls between 1900-2700 m.a.sl [1]. Despite the large area under wheat in Ethiopia the national average yield is 2.11 t/ha (CSA, 2013), which is far below the average of African and world yield productivity. The low productivity is attributed to a number of factors including biotic (diseases, insect pest and weeds) and a biotic (moisture, soil fertility, etc) and adoption of new agricultural technologies [4]. Among these factors, diseases play a significant role in yield reduction.

Wheat is susceptible to many diseases including the highly destructive ones like rusts (*Puccinia spp.*), Septoria leaf blotches (*Septoria tritici*), Fusarium head blight (*Fusarium graminearum*), tan spot (*Pyrenophora tritici repentis*), smut (*Ustilago tritici*) and powedery mildew (*Erysiphe graminis* f. sp. *tritici*) [5]. Over 30 diseases have been reported on wheat in Ethiopia [6]. Of these, fungal diseases like rusts (stem, stripe and leaf rust), Fussarium head blight (FHB), Septoria blotch, *Helmenthosporium spp.*, and tan spot are the dominant ones that were reported over time [6-9].

Among these rusts is the most important disease of wheat worldwide, in spite of great progress made in their control in many countries [5]. Rusts are the major disease of wheat since no other wheat disease could result in greater loss over large area in a given year [10]. Rusts can cause up to 60 percent of yield loss for leaf or stripe (yellow) rust and 100 percent loss for stem rust. The persistence of rust as a significant disease in wheat can be attributed to specific characteristics of the rust fungi. These characteristics include a capacity to produce a large number of spores which can be wind-disseminated over long distances and infect wheat under favorable environmental conditions and the ability to change genetically, thereby producing new races with increased aggressiveness on resistant wheat cultivars.

The high virulence diversity and evolution rate of the pathogen makes a considerable proportion wheat germplasm at risk. According to Leppik, the highland of Ethiopia is considered as a hot spot for the development of stem rust diversity. Furthermore, studies that were carried out in Ethiopia showed that most previously identified races were virulent on most of varieties grown in the country and are among the most virulent in the world [11]. The past study indicated that race surveys help to generate information regarding the virulence of races and their frequency and distribution patterns across regions and over time . In addition, race survey is important to study evolvement of new races and determine virulence shifts in a population. This information help to detected new race before build inoculums and cause epidemic. Therefore, the currently study was proposed to determine the distribution and dominant virulent races of *Puccinia graminis* f. sp. *tritici* in Ethiopia in 2013.

Materials and Methods

Race analysis was conducted only for stem rust. Stem rust samples were collected from Oromia, Amhara and Tigray regions. A total 100 samples were collected and analyzed. Stems of wheat infected with stem rust were cut into small pieces of 5 to 10 cm in length using scissors and

*Corresponding author: Endale Hailu, Ambo Plant protection Research center, 37Ambo, Ethiopia, E-mail: endalehailui@gmail.com

placed in paper bags after the leaf sheath was separated from the stem in order to keep the leaf sheath dry.

Spores were collected from samples using atomizer collector in capsule and suspension prepared by mixing spores with lightweight mineral oil (Soltrol). The prepared spore suspensions were inoculated using atomized inoculators on seven days old seedlings of the universally rust susceptible variety "Morocco" which does not carry known stem rust resistance to get enough amount of spore to inoculate on stem rust differentials. Greenhouse inoculations were done using the methods and procedures developed by Stakman et al. [10]. The mono-pustule was further multiplied to get enough spores for the differentials. The plants were then moistened with fine droplets of distilled water produced with an atomizer and placed in dew chamber for 18 h dark at 18 to 22°C followed by exposure to light for 3 to 4 h to provide condition for infection and seedlings were allowed to dry their dew for about 2 h. Then, the seedlings were transferred from the dew chamber to glass compartments in the greenhouse where conditions was regulated at 12 h photoperiod, at temperature of 18 to 25°C and relative humidity (RH) of 60 to 70%.

After two weeks of inoculation, the spores of each single pustule were collected in separate capsule and inoculated on the twenty standard differential sets. Five seeds of the twenty wheat stem rust differentials with known resistance genes (Sr5, Sr6, Sr7b, Sr8a, Sr9a, Sr9b, Sr9d, Sr9e, Sr9g, Sr10, Sr11, Sr31, Sr17, Sr21, Sr30, Sr36, Sr38, Sr24, SrTmp, and SrMcN) and one susceptible variety Morocco were grown in 3 cm diameter pots separately in greenhouse. The single pustule derived spores was suspended in soltrol inoculated onto seven-day-old seedlings using atomizers and/or an air pump. After inoculation, the formal procedure was repeated in dew chamber room. Upon removal from the dew chamber, plants were placed in separate glass compartments in a greenhouse to avoid contamination and produce infection.

Stem rust infection types were scored after 14 days of inoculations based on Stakman et al. [10]. Zero to four scales was used in which 0-2 stands for low infection where as 3-4 for high infection. Five latter race code nomenclatures were done based on Roelfs, Martens and Jin et al. [12,13] (Table 1).

Result and Discussions

Stem rust race in Ethiopia

Of 100 stem rust samples collected, 34 did not yield viable isolates at the time of inoculation in the laboratory. Hence, 66 isolates were used for the final race analysis. From 66 isolates studied, 9 races were identified. 57 *Puccinia graminis* f. *tritici* isolates collected from Oromia region were assigned to 7 races. Similarly, the 8 and 1 isolates collected from Amhara and Tigray region belongs to 2 and 1 race, respectively (Table 2). The highly virulent race called Ug99 (TTKSK) was the most abundant and widely distributed race across the country with a frequency of 52%. The second abundant and virulent race was TKTTF (Digelu race) with frequency 36.4%. These two races accounted for almost 88.4% of the stem rust population. The remaining 8 races composed of the rest of the population (11.6%). Of these, the least abundant races were TTKTF, TTKTK, JRCQC, TTKSC, TRTTF, SRKSC and RRKSF, which were detected only at single location.The identification of 9 races from 66 samples is a clear indication of high virulence diversity within the *Puccinia graminis* f. *tritici* population in Ethiopia. Admassu and Fekadu [14] reported that there is high *Puccinia graminis* f. sp. *tritici* population variability in Ethiopia.

Of the 57 isolates studied in Oromia region, race TTKSK (Ug99) was pre-dominant with frequency of 47% followed by race TKTTF (Digelu race) with 42%. In Oromia, the five races TTKTF, TTKTK, JRCQC, TTKSC and TRTTF were the least abundant, each with frequency of less than 5%. In Amhara region, two races TTKSK (Ug99) and SRKSC were identified in which Ug99 was again the most dominant (87.5%) race. Where as in Tigray, only one race RRKSF identified this may be due to viability of samples collected from the region.

Most of the races were virulent to one or more of the resistance genes (Table 2). For instance, the differential host carrying the resistance gene 5, 21, 6, 9g, 17, 9a, 9d and McNair were susceptible to all of the races. Similarly, four differential hosts carrying the resistance gene: 9e, 11,9b and 10 were susceptible to more than 88.8% of the races.

Puccinia graminis f. sp. tritici -code	Set 1	5	21	9e	7b
	Set 2	11	6	8a	9g
	Set 3	36	9b	30	17
	Set 4	9a	9d	10	Tmp
	Set 5	24	31	38	McN
B		Low	Low	Low	Low
C		Low	Low	Low	High
D		Low	Low	High	Low
F		Low	Low	High	High
G		Low	High	Low	Low
H		Low	High	Low	High
J		Low	High	High	Low
K		Low	High	High	High
L		High	Low	Low	Low
M		High	Low	Low	High
N		High	Low	High	Low
P		High	Low	High	High
Q		High	High	Low	Low
R		High	High	Low	High
S		High	High	High	Low
T		High	High	High	High

Source: Roelfs and Martens [12]; Jin et al. [13]; *Low: Infection types 0, ;, 1, and 2 and combinations of these values. **High: Infection types 3 and 4 and a combination of these values.

Table 1: Nomenclature of *Puccinia graminis* f. sp. *tritici* based on 20 differential wheat hosts.

Race	Virulence spectrum (ineffective Sr resistance genes)	No	%
Oromia			
TTKSK	5,21,9e,7b,11,6,8a,9g,9b,30,17,9a,9d,10, 31, 38,MCN	27	47
TTKTF	5,21,9e,7b,11,6,8a,9g,9b,30,17,9a,9d,10,TMP,38,MCN	1	2
TTKTK	5,21,9e,7b,11,6,8a,9g,9b,30,17,9a,9d,10,TMP,31,38,MCN	1	2
JRCQC	21,9e,11,6,9g,17,9a,9d,MCN	1	2
TKTTF	5,21,9e,7b,6,8a,9g,36,9b,30,17,9a,9d,10,TMP,38,MCN	24	42
TTKSC	5,21,9e,7b,11,6,8a,9g,9b,30,17,9a,9d,10,MCN	1	2
TRTTF	5,21,9e,7b,11,6,9g,36,9b,30,17,9a,9d,10,TMP,38,MCN	1	2
	Total	57	100
Amahara			
TTKSK	5,21,9e,7b,11,6,8a,9g,9b,30,17,9a,9d,10, 31, 38,MCN	7	87.5
SRKSC	5,21,9e,11,6,9g,9b,30,17,9a,9d,10,MCN	1	12.5
Total		**8**	**100**
Tigray			
RRKSF	5,21,7b,11,6,9g,9b,30,17,9a,9d,10,38,MCN	1	100
G total		**66**	**100**

Table 2: Races of *P.graminis f.sp.tritici* identified and their virulence spectrum in Amhara, Oromiya and Tigray regions of Ethiopia in 2013.

Sr7b and Sr30 were susceptible to seven races where as Sr38, Sr8a and SrTmp were susceptible to six, five and four races respectively. Sr36 were susceptible only to TKTTF race and conform resistant to the rest eight races. On contrary, Sr24 resistance gene was found to be effective to all races detected in this study and hence can be considered as source of resistance.

Only three of the differential lines carrying resistance gene Sr36, SrTmp and Sr24, were effective against the most dominate race TTKSK (Ug99) whereas only Sr11, Sr24 and Sr31 were effective against the most virulent race TKTT (Table 2).

In general out of nine races identified the most dominant and virulent race were TTKSK and TKTTF. Most of the genes were ineffective except Sr36, SrTmp and Sr24 against TTKSK race. The discovery of the race Ug99 with Virulence to Sr31 in Uganda in 1999. Pretorius et al. [15] represented a real threat to wheat production in the world, including Ethiopia where stem rust epidemics had not occurred since the resistant cultivar lost its resistance in 1993. In Ethiopia Ug99 was first detected in 2003 at six dispersed sites. In this study also this race is widely distributed in the central part of the country. Previous study also indicated that Ug99 were predominantly distributed in the southern and central parts of the country than in northern west of Ethiopia [16].

Similarly, the second most dominant race TKTTF which is known as Digelu race was virulent on 17 stem rust resistant gene and widely distributed in the central and southern part of the country. This race is for the first time reported in Ethiopia and cause localized stem rust endemics in Bale and Arsi zone of Oromia region in 2013. Stem rust resistance gene in Digelu variety SrTmp gene was broken and most farmers grow Digelu variety were highly affected. The discovery of new race in the present study with earlier report [16,17] revealed some differences. The Stem rust resistance gene Tmp became in effective in Ethiopia due to the evolvement of new virulent race TKTTF.

Virulence frequency of *P. graminis f. sp. tritici* isolates on Stem rust resistant genes

The Stem rust differential host carrying the resistance gene 21, 6, 17, 9a, 9d and MCN were found to be ineffective against all stem rust races detected with frequency value 100%. Similarly, six differential stem rust differentials carrying resistance genes Sr9d, Sr21, Sr6, Sr10, Sr9g, and Sr9b were found to be ineffective against most of stem rust races detected, with virulence frequencies of 65.6, 78.1, 75, 81.2, 87.5, and 93.8% respectively (Table 3). In general, about 75% of the Stem rust resistance genes were ineffective to more than 60% of the isolates. McNair 701 (SrMcN) was ineffective to 96.9% of the isolates tested (Table 3).

On contrary, Stem rust resistance gene 24 was effective against all races detected. Of the 20 stem rust resistance genes, the differential hosts carrying SrTmp, Sr31, and Sr17 were was resistant to 87.5, 75.0, and 78.1% of the isolates tested, respectively. Correspondingly, gene Sr38 was effective against 62.5% of the isolates analyzed followed by Sr36 which was effective against 59.4% (Table 3).

Stem rust resistance gene 24 was effective against most of the isolates tested in Ethiopia. [16] Admassu also indicated that no virulent race detected against Sr24 gene in Ethiopia. Use of this gene for breeding in Ethiopia is permanent [18] Countries like Ethiopia in which stem and yellow rust severely occur every year and the majority of wheat grown by subsistence farmers, for whom use of chemical fungicide against

Stem rust resistance gene (Sr gene)	Virulence frequency (%)	Stem rust resistance gene (Sr gene)	Virulence frequency (%)
5	96.97	30	98.5
21	100	17	100
9e	98.5	9a	100
7b	96.97	9d	100
11	63.64	10	98.5
6	100	Tmp	40.91
8a	93.94	24	0
9g	98.5	31	54.55
36	62.12	38	95.5
9b	98.5	McN	100

Table 3: Virulence frequency of *P. graminis f. sp. tritici isolates* (32 isolates) on 20 Stem rust resistance genes.

stem rust is not economical, incessant supply of resistance varieties unquestionably needed to avoid wheat rust epidemics.

Acknowledgment

We would like to offer a great thanks to wheat rust research team of Ambo plant protection research Center for their valuable encouragement and technical support during the whole period of the study. Without the support of some individuals and institutions the successful completion of this experiment would have not been realized. Durable Rust Resistance in Wheat (DRRW) project and International maize and wheat improvement center (CIMMYT) is duly acknowledged for fully funding this work.

References

1. Hailu G, Tanner DG, Mengistu H (1991) Wheat research in Ethiopia: A Historical perspective, IARI and CIMMYT, Addis Ababa 392.

2. Bekele, Verkuiji HM ,Wangi W, Tanner D (2000) Adoption of improved wheat techenologies in Adaba and Dodola woredas of the Bale high lands, Ethiopia, Mexico, D.F. CIMMYT and EARO.

3. CSA (central statistics Authority) (2013) Report on area and crop production forecast for major grain crops. Addis Ababa, Ethiopia: statistical bulletin.

4. Zegeye T, Taye G, Tanner D, Verkuijl H, Agidie A, et al. (2001) Adoption of improved bread wheat varieties and inorganic fertilizer by small-scale farmers in yelmana Densa and Farta districts of North western Ethiopia. EARO and CIMMYT. Mexico city, Mexico.

5. Prescott JM, Burnett PA, LeSaari EE, Ranson J, Bowman J, et al. (1986) Wheat diseases and pests: a guide for field identification. CIMMYT, Mixico city. DF. Mexico 135.

6. Bekele E (1985) A review of research on diseases of berley, tef and wheat in Ethiopia. In: Tsedeke Abate (ed.), A review of crop protection research in Ethiopia. Institute of Agricultural Research (IAR), Ethiopia 79-107.

7. Yirgu D (1967) Plant diseases of economic importance in Ethiopia. Haile Selassie I university, college of Agriculture, Experimental station bulletin no.50, Addis Ababa, Ethiopia 30.

8. Badebo A (2002) Breeding Bread Wheat with Multiple Disease resistance and high yielding for the Ethiopian highlands: broadening Genetic basis of yellow rust and tan spot resistance. Gottingen, Germany: Gottingen University, PHD thesis.

9. CIMMYT (2005) Sounding the alarm on global stem rust: an assessment of race Ug99 in Kenya and Ethiopia and potential for impact in neighboring countries and beyond. Mexico city. Mexico.

10. Stakman EC, Stewart DM, Loegering WQ (1962) Identification of physiologic races of Puccinia graminis var. tritici.' USDA ARS, E716. United States Government Printing Office: Washington, DC 5-50.

11. van Ginkel M, Getinet G, Tesfaye T (1989) Stripe, stem and leaf rust races in major wheat producing areas in Ethiopia. IAR Newslett. Agric Res 3: 6-8.

12. Roelfs AP, Martens JW (1988) An international system of nomenclature for P. graminis f. sp. tritici. Phytopatholog 78: 526-533.

13. Jin Y, Szabo U, Pretorius ZA, Singh RP, Ward R (2008) Detection of virulence

to resistance gene Sr24 within race TTKS of Puceinia graminis f. sp. tritici. Plant Dis 92: 923-926.

14. Admassu B, Fekadu E (2005) Physiological races and virulence diversity of Puccinia graminis f.sp. tritici on wheat in Ethiopi. Phytopathologia Mediterranea 44: 313-318.

15. Pretorius ZA, Singh RP, Wagoire WW, Payne TS (2000) Detection of virulence to wheat stem rust resistance gene Sr31 in Puccinia graminis f. sp. tritici in Uganda. Phytopathology 84: 526-533.

16. Admassu B (2010) Genetic and virulence diversity of Puccinia graminis f. sp.

tritici population in Ethiopia and stem rust resistance genes in wheat. Gottingen, Germany: Gottingen University, PHD thesis.

17. Teklay A, Woubit D, Getanh W (2013) Physiological races and virulence diversity of Puccinia graminis pers. f. sp. Tritici eriks. & e. Henn. On wheat in Tigray region of Ethiopia. ESci J. Plant Pathol 02: 01-07.

18. Teklay A, Getaneh W, Woubit D (2012) Analysis of pathogen virulence of wheat stem rust and cultivar reaction to virulent races in Tigray, Ethiopia. African Journal of Plant Science 6: 244-250.

17

Effects of Nitrogen and Phosphorus on the Growth Performance of Maize (*Zea mays*) in Selected Soils of Delta State, Nigeria

Umeri C*, Moseri H and Onyemekonwu RC

Department Agricultural Science Education, College of Education, Agbor, Delta State, Nigeria

Abstract

The study was carried out in Delta State, Nigeria. Soil samples were collected from nine locations within Delta State namely Agbor, Asaba and Ubulu-uku (Delta North), Abraka, Oghara-Eki, Sapele (Delta Central), Oleh, Ozoro, Patani (Delta South). Surface soils (0-15 cm) and sub-surface soils (15-30 cm) depth. These were analyzed for their physical and chemical properties. In one of the locations (Agbor) found to be deficient in nitrogen (N) and phosphorus (P) field trials on the effects of these nutrients on the performance of maize (*Zea mays*) were carried out. The variety ACR-89DMRESR-W was used. The design was a 4 × 4 factorial scheme fitted into a randomized complete block design given sixteen treatments combinations with three replicates. The following treatments combinations were applied N_0P_{20}kg/ha, N_0P_{40}kg/ha, N_0P_{60}kg/ha, $N_{20}P_0$kg/ha, $N_{20}P_{20}$kg/ha, $N_{20}P_{40}$kg/ha, $N_{20}P_{60}$kg/ha, $N_{40}P_0$kg/ha, $N_{40}P_{20}$kg/ha, $N_{40}P_{40}$kg/ha, $N_{40}P_{60}$kg/ha, $N_{60}P_0$kg/ha, $N_{60}P_{20}$kg/ha, $N_{60}P_{40}$kg/ha, and $N_{60}P_{60}$kg/ha. The parameter measured were plant height and leaf number at 3, 6 and 9 weeks after planting (WAP), respectively. Combined application of 40 kgN/ha+40 kgP/ha significantly increased maize plant height and leaf number among all the treatments. Therefore combined application of 40 kgN/ha+40 kgP/ha is recommended for optimum growth of maize in the study area.

Keywords: Nitrogen; Phosphorus; Growth performance; Nigeria; Maize; Delta state

Introduction

Maize (*Zea mays L*) is a cereal crop which belongs to the family poaceae [1]. It is an important food crop in Nigeria. It forms a major part of cereal crops consumed by man [2] and serve as a source of dietary carbohydrates [3]. It is used for livestock feed and it is the cheapest and palatable livestock feed for animals such as pig, cattle, sheep, and poultry [4]. It is also a source of raw materials for the production of corn sugar, corn starch, corn syrup and corn oil [5].

Nitrogen is a vital plant nutrient and a major yield determining factor for maize production [6]. Its availability in sufficient quantity throughout the growing season is essential for optimum growth of maize. Most farmers in developing countries usually rely on natural soil fertility for crop production.

An application of urea and triple superphosphate (TSP) fertilizers in combination with farmyard manure was found to enhance the effectiveness of N and P fertilizers [7]. Opening up of a long fallow land may provide adequate nutrients to crops; however, cropping such land is only successful within a few years after its opening. Thereafter, subsequent cropping requires fertilizer input most importantly nitrogen to maintain good yields. Studies conducted by Stewart et al. [8] and Niehues et al. [9] revealed that starter nitrogen was able to stimulate the early growth and yield of maize.

Phosphorus is closely concerned with many growth processes in crop plants. It is involved in many biochemical reactions and concerned with the metabolism of carbohydrates, fats and protein and play roles in the breakdown of carbohydrates; phosphorus (P) is another limiting nutrient in maize production. According to Rehman et al. [10] nutrient P affects leaf growth and senescence dynamics in maize. Various factors could be responsible for P availability to crop plants. These include the form of native soil P, the type of P applied to the soil and reaction.

Some of the problems associated with the soil for cultivation of maize in Delta State Nigeria are deficiencies of Nitrogen and Phosphorus, leaching, continuous cropping, oil spillage and exploration [11]. For maize to reach full production capacity there is need to address nutrient deficiency and response to N and P fertilization in Delta State. Thus the objective was to determine the effect of N and P on maize plant height and leave number.

Materials and Methods

The study was carried out in the rain forest belt of Delta State which lies between 5°N and 8°N and longitude 5°W and 7°E of the equator. The soils have loose brownish top soil over a great depth of large differentiated, non-molten, non gravelly, porous sub soil with coarse sand ass the predominant fraction and clay content is up to 35%. The characteristic management and potentials of all these soils for maize cultivation have been reviewed by Omoti et al. [12]. There are two distinct seasons usually the dry season and the rainy season. Temperature is very high during the day and with cool night [13].

The samples were collected from nine locations within Delta State namely Agbor, Asaba and Ubuly-uku (Delta North), Abraka, Oghara-Eki, Sapele (Delta Central), Oleh, Ozoro, Patani (Delta South). These were chosen to reflect the differences in soil and vegetation characteristics. Surface soils (0-15 cm) and sub surface soils (15-30 cm) were sampled with a tabular sampling augur. Representative soil samples were taken and then bulked for each depth and location.

Based on the analysis, soil obtained (0-15 cm) from Agbor was found to be deficient in Nitrogen and Phosphorus against the established critical values. Consequently, the field trial was established in this location.

*Corresponding author: Umeri C, Department Agricultural Science Education, College of Education, Agbor, Delta State, Nigeria
E-mail: chukwukaraymond@gmail.com

The design was a 4 × 4 factorial scheme fitted into a randomized complete block design giving sixteen treatment combinations with three replicates. The following treatment combinations were applied N_0P_{20}kg/ha, N_0P_{40}kg/ha, N_0P_{60}kg/ha, $N_{20}P_0$kg/ha, $N_{20}P_{20}$kg/ha, $N_{20}P_{40}$kg/ha, $N_{20}P_{60}$kg/ha, $N_{40}P_0$kg/ha, $N_{40}P_{20}$kg/ha, $N_{40}P_{40}$kg/ha, $N_{40}P_{60}$kg/ha, $N_{60}P_0$kg/ha, $N_{60}P_{20}$kg/ha, $N_{60}P_{40}$kg/ha, and $N_{60}P_{60}$kg/ha. Each plot measured 2.1 × 1 m^2 with alley of 1 m between plot and replicates. There were a total of 48 plots (16 × 3). The experimental area used was 329.6 m^2.

Maize seeds (ACR-89 DMERSR-W) obtained from International Institute of Tropical Agriculture (IITA) were sown on the 13th of April 2014 at 2 seeds per hole and later thinned to one plant per stand. The spacing was 70 cm × 25 cm giving a plant population of 20 plants per plot. Reading was done at interval 3, 4 and 10 weeks after planting (WAP) respectively.

Plant height readings were taken at 3, 6 and 9 weeks after planting, a tape rule was used to measure the height of the plant from the soil surface to the apex. The mean value was recorded in centimeters. Number of leaves were counted per plant, per plot at 3, 6 and 9 weeks and recorded respectively.

Plant height and number of leaves determined were subjected to appropriate statistical analysis; ANOVA and correlation coefficient.

Results and Discussion

Plant height

Mean plant height at 3 weeks after planting (WAP) ranged from 11.60-12.90 cm and 10.60 cm-13.60 cm when treated only to Nitrogen and Phosphorus fertilizers respectively (Table 1). The mean plant height value was highest when 40 kg/N/ha and 40 kg P/ha were applied and lowest in the control treatment. Application of fertilizer had no significant effect on plant height.

Treatments					
Nitrogen (Kg/ha)	Phosphorus				Mean
	0	20	40	60	
	Cm				
3 weeks					
0	9.83	11.42	13.00	13.88	12.0a
20	9.37	12.50	13.17	11.50	11.6a
40	12.48	10.78	15.27	12.87	12.9a
60	10.7	11.12	12.82	14.75	12.4a
Mean	10.6a	11.5a	13.6a	13.3a	
6 weeks					
0	56.50	59.50	68.50	67.83	63.1b
20	61.83	70.33	69.00	65.83	66.7b
40	69.17	65.50	78.50	73.33	71.6b
60	61.33	72.67	81.33	87.00	75.8ab
Mean	62.2b	67.0ab	74.3a	73.5a	
LSD (N=8.76) P=8.76)					
9 weeks					
0	120.5	122.0	143.7	140.3	131.6b
20	156.7	167.0	156.0	151.3	157.8a
40	162.2	166.3	184.5	179.2	173.1a
60	154.7	167.2	167.0	192.3	170.3a
Mean	148.5a	155.6a	162.8a	165.8a	
LSD (N=20.7)					

Figures in the column and rows for each week followed by the same letter are not significantly different at 5% level.

Table 1: Effect of Nitrogen and Phosphorus fertilizer on mean plant height of maize.

Treatments					
Nitrogen (Kg/ha)	Phosphorus				Mean
	0	20	40	60	
	Cm				
3 weeks					
0	6.17	6.33	6.00	6.00	6.1a
20	6.00	6.17	6.17	6.00	6.1a
40	6.17	6.00	6.83	6.33	6.3a
60	6.33	6.00	6.50	6.67	6.4a
Mean	6.2a	6.1a	6.4a	6.3a	
6 weeks					
0	10.0	10.2	11.0	11.3	10.63b
20	11.7	12.2	11.8	10.3	11.50a
40	12.0	11.7	13.5	13.0	12.55a
60	11.5	11.5	12.2	13.3	12.13a
Mean	11.30a	11.40a	12.13a	11.98a	
LSD (N=1.26)					
9 weeks					
0	14.8	14.3	15.3	14.3	14.68b
20	16.5	16.3	16.0	16.0	16.20a
40	14.8	15.3	15.8	15.7	15.28b
60	16.7	15.8	15.8	17.2	16.38a
Mean	15.7a	15.4a	15.6a	15.8a	16.38a
LSD (N=0.84)					

Figures in the column and rows for each week followed by the same letter are not significantly different at 5% level.

Table 2: Effect of Nitrogen and Phosphorus fertilizer on the number of leaves.

At 6 WAP, the mean plant height ranged from 63.1 to 75.8 and 62.2 to 74.3 cm when treated with Nitrogen and phosphorus fertilizer respectively. The application of P at 40 kg/ha significantly increased plant in comparison to the control but was not significantly different from other rates. The highest height was obtained at 40 kg/N/ha and 40 kg P/ha combined at three weeks after planting and 60 kgP and 40 kgN at six weeks after planting.

At 9 WAP, plant height was not significantly improved by phosphorus fertilization but was with nitrogen fertilizer. Mean plant height ranged from 148.5 to 165.8 cm and 131.6 to 173.1 cm when treated with phosphorus and nitrogen fertilizers respectively. Mean plant height increased with increasing levels of application of phosphorus and nitrogen fertilizers which is in line with studies conducted by Stewart et al. [8] and Niehues et al. [9] revealed that starter nitrogen was able to stimulate the early growth and yield of maize.

Number of Leaves

The mean number of leaves at 3WAP ranged from 6.1 to 6.4 when treated with phosphorus and nitrogen fertilizers respectively (Table 2). The mean values were not significantly different for both fertilizers rates of application.

At 6 WAP, mean number of leaves ranged from 11.30 to 12.13 and 10.63 to 12.55 when treated with phosphorus and nitrogen fertilizers respectively. The values obtained from treatments which received fertilizers were higher than the control.

At 9 WAP, mean number of leaves varied from 14.68 to 16.38 and 15.4 to 15.8 as a result of phosphorus and nitrogen fertilization. The values did not follow definite pattern with increasing fertilizer rates. Nitrogen fertilizer significantly increased mean number of leaves. The rates of 20 kgN/ha and 60 kgN/ha were significantly different from control which is in agreement with Rehman et al. [10] Who stated that nutrient P affects leaf growth and senescence dynamics in maize and Duncan [7] who reported that an application of urea and triple superphosphate (TSP) fertilizers in combination with farmyard manure was found to enhance the effectiveness of N and P fertilizers in maize production.

Conclusion and Recommendations

Nitrogen and phosphorus fertilization are important in the management of soil of the study area due to deficiency of both nutrients. The study revealed that maize growth was significantly enhanced by the application of Nitrogen fertilizer at the rate of 40 to 60 kg/ha compared to other rates of application. Maize response to applied phosphorus was not significantly different. However, the combined application of 40 kgN/ha plus 40 kgP/ha performed better in enhancing growth of maize. It is therefore recommended that the application of 40 kgN/ha plus 40 kgP/ha will effectively enhance maize growth since nitrogen has been reported to favour vegetative growth especially at the initial stage of growth [14].

References

1. Downswell CR, Paliwal RL, Cantrell RP (1996) Maize in third world. West view Press in co-operation with Winrock International Institute for Agricultural development.

2. Onwueme IC, Sinha TD (1991) Field crop production in Tropical Africa: Principles and practice. CTA, Wageningen, PAYSBAS.

3. Wudiri BB, Fatobi TO (1992) Recent development in cereal production media forum for Agriculture. International Institute for Tropical Agriculture in Nigeria, pp: 13-32.

4. Ekpeyong TE (1985) Proximate and mineral values of Nigerian maize varieties. Nigerian Agricultural Journal 20: 197.

5. Anochili BC (1984) Food crop production. Tropical Agriculture Handbook, Macmillan Publishers, pp: 18-22.

6. Shanti KVP, Reddy MS, Sharma RS (1997) Response of maize (Zea mays) Hybrid and composite to different levels of Nitrogen. Indian Journal of Agricultural Science 67: 424-426.

7. Duncan WG (2002) A theory to explain the relationship between corn population and grain yield. Crop Sci 24: 1141-1145.

8. Stewart WM, Dibb DW, Johnston AE, Smyth TJ (2005) The contribution of commercial fertilizer nutrients to food production. Agron J 97: 1-6.

9. Niehues BJ, Lamond RE, Godsey CB, Olsen CJ (2004) Starter nitrogen fertilizer management for continuous no-till corn production. Agron J 96: 1412-1418.

10. Rehman A, Saleem MF, Safdar ME, Hussain S, Akhtar N (2011) Grain quality, nutrient use efficiency and bio-economics of maize under different sowing methods and npk levels. Chilean J Agric Res 71: 586-593.

11. Corliss J (1991) Conserving crop land for the future agricultural research marsh. pp: 15-20.

12. Omoti U, Onwubuya I, Nnabuchi SE (1986) Soil of the Nigeria oil palm belt -Their characteristics and management for oil palm cultivation. Paper presented at the International conference on oil palm, Port-Harcourt, Nigeria.

13. Iloje SI (2003) A general Geography of Nigeria. Heinemann Books, Ibadan.

14. Udoh JD, Ndon AB, Asuquo PE, Ndaeyo UN (2005) Crop production techniques for the tropics. Concept Publications, Lagos, pp: 103-108.

18

In Vitro Propagation of Selected Sugarcane (*Saccharum officinarum* L.) Varieties (C 86-56 and C 90-501) through Apical Meristem

Mulugeta Hailu*

Department of Biotechnology, College of Dry Land Agriculture and Natural Resources, Mekelle University, Mekelle, Ethiopia

Abstract

Sugarcane (*Saccharum officinarum* L.) is monocotyledonous crop plant that mostly propagates through conventional methods. However, conventional propagation lacks rapid multiplication procedures to commercialize newly released varieties within a short period of time. Hence, the objective of this work was to optimize *in vitro* micro propagation protocol for two sugarcane varieties (C 86-56 and C 90-501) through apical meristem explants. The two varieties were cultured on MS medium supplemented with different concentrations of growth regulators on shoot initiation, multiplication, rooting and acclimatization stages. Analysis of variance (ANOVA) revealed that the two varieties showed statistically significant difference in their response to the various hormonal treatments with regard to the parameters measured. For initiation stage vars. C 86-56 and C 90-501 performed best on 1.0 mg/l and 1.5 mg/l of BAP, respectively. On the other hand, multiplication stage was best in MS media enriched with 1.5 mg/l BAP+1.0 mg/l NAA and 1.0 mg/l BAP+1.0 NAA as manifested in terms of mean number of shoots and mean shoot length for vars. C 90-56 and C 90-501, respectively. With regard to root induction, best rooting response in terms of mean root number and mean root length was achieved best in 1/2 MS media enriched with 1.5 mg/l NAA+0.5 mg/l BAP and 1.0 mg/l NAA+0.5 mg/l BAP for vars. C 86-56 and C 90-501, respectively. Survival rate during acclimatization was best on coco peat media for both varieties C 86-56 and C 90-501 survived 90% and 70%, respectively. Lastly, factors causing low acclimatization, tissue dying, contamination and phenol exudation in the study should be further investigated.

Keywords: Acclimatization; Apical meristem; *In vitro*; Growth medium; Rooting

Introduction

Sugarcane (*Saccharum officinarum* L.) is a monocotyledonous crop plant that belongs to the family of Poaceae. It is a clonally propagated crop from which multiple annual cuttings of stalks are typically obtained from each planting. This crop is especially vulnerable to diseases and propagation from cuttings facilitates the spread of pathogens and may results in epidemics. Sugarcane stalks can be infected by various pathogens without exhibiting any symptoms, and therefore there is a high risk of disease transfer during exchange and transport of sugarcane cuttings. Its growth is closely related to temperature [1]. Sugarcane, being a vegetatively propagated crop, has a low seed multiplication rate, which means one hectare of seed cane suffices six to eight hectare of commercial plantation. Lack of flowering potential and multiplication procedures has long been a serious problem in sugarcane breeding programs [2].

However, Propagation of sugarcane conventionally gives low availability of adequate quantity, highly susceptible to disease. On the other hand, availability of adequate amount of quality and disease free planting materials within a short time is the major limiting factor to attain large scale sugarcane production using the conventional method of propagation and the yield of the existing few and old commercial sugar cane varieties is declining and some productive sugarcane varieties also obsolete due to lack of alternative technologies for disease cleansing and rejuvenation. Besides, it is also difficult to understand the different types of sugarcane varieties in response to resistance to different contaminants during traditional propagation. Moreover, due to shortage of sufficient quantity planting material, commercialization of introduced and adapted improved varieties of sugarcane takes several years using the conventional route of planting material multiplication.

To solve the multitude problems of the conventional propagation method the sugar industry utilizes the advantage of micro propagation technology, which is characterized by rapid multiplication to obtain disease free sugarcane varieties. The nutritional requirement for *in vitro* propagation protocol of sugarcane should be according to genotype and explants used [3]; varieties (genotypes) of the same species respond differently to media [4], besides rapid clonal propagation of sugarcane planting materials depends on the genotype and the plant growth regulators combinations used and needs to develop plant growth regulators combinations for each genotype. Similarly, plant growth regulators requirements for *in vitro* propagation responses vary from cultivar to cultivar in sugarcane [5]. The nutritional requirement for every sugarcane variety is specific and exact [6]. In addition, an efficient protocol is needed for any new variety or clone to get rapid shoot initiation, shoot multiplication, root induction and elongation [7]. So it is recommended that an efficient protocol is needed for every new variety or clone of sugarcane to get rapid callus induction, shoot initiation, shoot multiplication and root induction and elongation [8]. Therefore, this study was carried out to develop or optimize *in vitro* protocol for mass propagation of two sugarcane varieties (C 86-56 and C 90-501) through apical meristem.

Objective of the study

The general objective of this study was to develop optimization protocol for *in vitro* regeneration of two sugarcane varieties namely C 86-56 and C 90-501 from shoot apical meristem explants.

---

*Corresponding author: Hailu M, Department of Biotechnology, College of Dry Land Agriculture and Natural Resources, Mekelle University, Mekelle, 231, Ethiopia, E-mail: mulugetahailu16@gmail.com

Specific objectives were

- To determine appropriate concentrations of BAP hormone on initiation culture of apical meristem.To determine appropriate concentration of BAP and NAA hormones for shoot induction and multiplication.To determine appropriate concentration of NAA and BAP hormones for root growth. To evaluate survival rate of plantlets under greenhouse condition.

Materials and Methods

Description of the study area

The study was conducted at Tigray Biotechnology center; plant tissue culture laboratory which is located at Mekelle town, specifically at Ellala near to Tigray Agricultural Research Center that is located at latitude of 13°29'N, longitude of 39°28'E and altitude of 2076 meters above sea level.

Plant material and explant preparation

Mother plants of the two varieties namely C 86-56 and C 90-501 that were used as a source of explants were raised from stem cuttings (setts) obtained from Welkayt sugar factory. According to Dereje [9] those two Cuban varieties were imported in 2006 and passed through agronomic performance evaluation. They were among the selected ones to be commercialized (Figure 1). Before planting, the setts were treated with hot water at 50°C for 2 hours. Explant preparation were made following the method employed by Belay [10]. The actively growing shoot tips with apical mersitem were collected from three months old mother plants to serve as explants. Shoot tips were cut from mother plants at the base with some nodes. After trimming of the leaves, the shoot tips were taken to the laboratory for surface sterilization and explant preparation. Trimmed shoot tips were washed thoroughly under running tap water, outer leaf sheath were removed and cut into about 10 cm length. Thereafter, the shoot tips further washed three times each for 15 minutes with tap water containing liquid soap solution and three drops of Tween-20. Then, explants were taken to laminar airflow chamber, immersed in 0.3% (w/v) Kocide solution for 30 minutes followed by three times washing each for five minutes with sterile distilled water. The shoot tips were then be rinsed in 70% alcohol for one minute and washed with sterile distilled water three times each for five minutes. Finally, the explants were treated with 10% (v/v) sodium hypochlorite solution (4% active chlorine) for 20 minutes (Figure 2). After discarding the sodium hypochlorite solution,

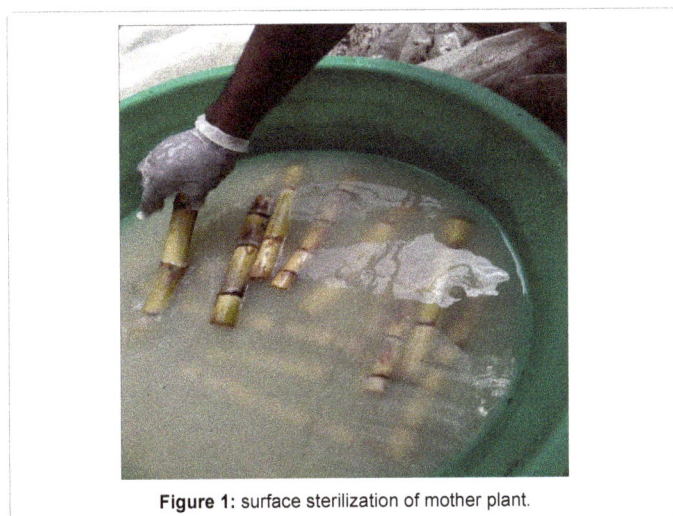

Figure 2: Planting of sterilized mother plant on coco peat soil.

the explants were washed with sterile distilled water three times each for five minutes and the surface sterilized explants were excised and sized to 2.5 cm long for culturing.

Culture media preparation

Full strength Murashige and Skoog (MS) basal medium Murashige and Skoog [11] were used as a culture medium. MS basal medium consisted of 30 g/l sucrose for initiation of apical meristem and shoot initiation, whereas 60 g/l sucrose for rooting. The pH of the medium was adjusted to 5.8 using 1 N KOH and 1 N HCl before being gelled with 5.0 g/l agar and autoclaved at 121°C, 15 psi for 20 minutes (Figure 3). While molten, the medium (40 ml) was dispensed into glass culture jar for culturing and stored under aseptic condition at +4°C until use for shoot initiation.

Initiation of apical meristem

For shoot initiation, the sterilized shoot tips were aseptically transferred to MS-medium prepared as indicated above with supplementation of PGR (BAP) at a concentration of 0.5, 1.0, 1.5 and 2.0 mg/l. MS medium without PGR were used as control. The treatments of initiation for both varieties were as follows

T_1=MS+30 g/l sucrose+5.0 g/l agar+0.0 mg/l BAP

T_2=MS+30 g/l sucrose+5.0 g/l agar+0.5 mg/l BAP

T_3=MS+30 g/l sucrose+5.0 g/l agar+1.0 mg/l BAP

T_4=MS+30 g/l sucrose+5.0 g/l agar+1.5 mg/l BAP

T_5=MS+30 g/l sucrose+5.0 g/l agar+2.0 mg/l BAP

The explants were maintained in dark for 8 hour and light for 16 hour duration. The experiment was laid out in a completely randomized design (CRD) with three replicates.

Shoot induction and regeneration

For shoot initiation, MS basal medium supplemented with BAP in a concentration of 1.0, 1.5, 2.0 and 2.5 mg/l combined with 0.5 mg/l of NAA was used. MS basal medium without PGR, i.e., BAP and NAA were used as a control. Details of the treatments for both varieties were as follows:

T_1=MS+30 g/l sucrose+5.0 g/l agar+0.0 mg/l BAP+0.0 mg/l NAA

T_2=MS+30 g/l sucrose+5.0 g/l agar+0.5 mg/l BAP+1.0 mg/l NAA

Figure 1: surface sterilization of mother plant.

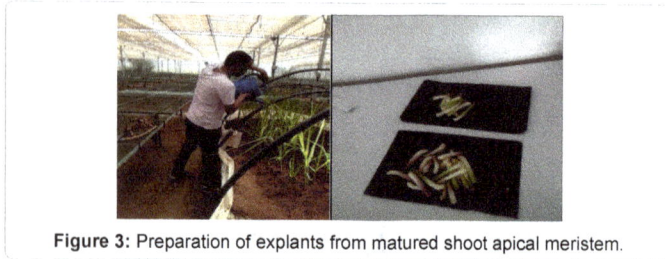

Figure 3: Preparation of explants from matured shoot apical meristem.

T_3=MS+30 g/l sucrose+5.0 g/l agar+1.0 mg/l BAP+1.0 mg/l NAA

T_4=MS+30 g/l sucrose+5.0 g/l agar+1.5 mg/l BAP+1.0 mg/l NAA

T_5=MS+30 g/l sucrose+5.0 g/l agar+2.0 mg/l BAP+1.0 mg/l NAA

Then after, cultures were maintained in a growth room at a temperature of 25 ± 2°C under 16/8 hours light/dark photoperiod adjusted with fluorescent light having 2500 lux light intensity (Figure 4). The incubation chamber had relative humidity of 75-80%. Shoots were allowed to grow 2 to 4 cm and then transferred to rooting media. The experiment was laid out in a completely randomized design (CRD) with a two factors treatment combinations each with three replicates (Figure 5).

Root induction

Well grown 3-5 cm long shoots were aseptically transferred to 1/2 strength MS basal medium containing 0.0, 0.5, 1.0, 1.5, and 2.0 mg/l of NAA. The treatments of rooting for both varieties were as follows:

T_1=1/2 MS+30 g/l sucrose+5.0 g/l agar+0.0 mg/l NAA+0.0 mg/l BAP

T_2=1/2 MS+30 g/l sucrose+5.0 g/l agar+0.5 mg/l NAA+0.5 mg/l BAP

T_3=1/2 MS+30 g/l sucrose+5.0 g/l agar+1.0 mg/l NAA+0.5 mg/l BAP

T_4=1/2 MS+30 g/l sucrose+5.0 g/l agar+1.5 mg/l NAA+0.5 mg/l BAP

T_5=1/2 MS+30 g/l sucrose+5.0 g/l agar+2.0 mg/l NAA+0.5 mg/l BAP

All the cultures were incubated at 25 ± 2°C under 16/8 hours light/dark photoperiod adjusted with fluorescent light having 2500 lux light intensity (Figure 6). The relative humidity of the growth chamber was 75-80%. The experiment was laid out in a completely randomized design (CRD) with NAA factor treatment each with three replicates.

Acclimatization

After four weeks of culture in a rooting media, well rooted *in vitro* plantlets were taken out gently from each PGR treatment bottle and washed under tap water to remove traces of agar that prevent the absorption of nutrients from the acclimatization culture substrates by roots. After this, plantlets were transferred to polystyrene tray that contains three different substrates, namely, coco peat, that is a multipurpose growing medium made up of coconut husk; garden soil: sand: compost in the ratio of 1:1:1 and garden soil and cow dung in the ratio of 1:1. Then, the polystyrene tray was arranged in completely randomized design in computerized green house with relative humidity (RH) gradually reduced from 90 to 60% and temperature of 30 to 31°C for two weeks for primary acclimatization (Figure 7). All the plantlets that survived the primary acclimatization were put in a direct sunlight and nursery shades were provided with adequate amount of water for secondary acclimatization and their performance were monitored for three weeks. Finally, plantlets that survived secondary acclimatization were transplanted to open field.

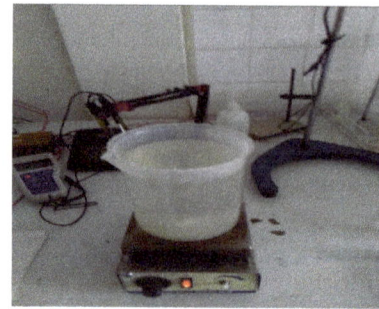

Figure 4: Preparation of Murashige and skoog medium.

Figure 5: Explants on initiation medium.

Figure 6: Sub culturing to shooting stage.

Figure 7: Best cultures of C 86-56 variety on shooting stage (MS+1.5 mg/l BAP+1.0 NAA media).

Data collected

The following *in vitro* plant growth variables were recorded from this experiment and served as sources of quantitative data.

Percent of initiated culture: Percent of culture formed from the apical meristem explants after three weeks.

Average number of days to shoot emergence: Number of days needed by the explants to induce shoots from the first day of culturing in the shooting.

Mean number of shoots: Is the average number of dissectible shoots regenerated and induced from each cultured explants in each type of treatment.

Average number of days to root emergence: Number of days needed by the shoots to induce roots from the first day of culturing in the rooting media.

Mean number of roots: Is the average number of dissectible roots regenerated from each cultured shoot in each type of treatment.

Mean length of shoot: Is the average length of Shoots developed from the base of the shoot to the shoot apex before transferring to rooting media from each cultured explants. Lengths of the shoot were measured using sterilized ruler.

Mean length of root: Is the average length of roots developed from each cultured explants. Length was measured on the 30th day of transferring the shoot to the rooting media. Root length measurement was taken from the point that the root attached to the shoot to root tip.

Survival rate: Is the competence or the ability of the *in vitro* derived plantlets to endure in the *in vivo* condition for acclimatization. Data on this parameter was taken after one month after root transferred to poly house. Accordingly, the survival rate was calculated after three weeks as the ratio of plantlets survived to the total number of plantlets transferred to the poly house and expressed by percentage.

Data analysis

Data were analyzed on the effect of treatments using SAS version 9.1 and means were compared using fisher's Least Significant Difference (LSD) test at $p < 0.0$.

Results and Discussion

Percent of initiated culture

Initiation culture from the apical meristem explants was observed within two weeks after inoculation of the explants on MS medium containing five different concentrations of BAP (0.0, 0.5, 1.0, 1.5, and 2.0 mg/l). The results showed that shoot apical meristem culture initiation or establishment responses in the sugarcane varieties was dependent on the effect of sugarcane varieties (genotype) and BAP (Figure 8). Among the different of BAP tested, sugarcane variety C 86-56 gave the highest initiation culture responses (73.33%) on MS medium containing 0.5 mg/l BAP while C 90-501 gave highest initiation culture responses (76.667%) on MS medium supplemented with 1.5 mg/l BAP as shown in Table 1. This indicated that initiation response in these two varieties is different with respect to the amount of BAP to be used. In line with this, Variation of initiation culture response to different concentration of hormones with variety of sugarcane was reported by Dereje [9]; Tilahun [12]. Control showed no response for initiation in which all explants cultured on control (0.0 mg/l BAP) dried out after explanation.

Shoot regeneration and multiplication

Number of days to shoot emergence: Shoot initiation was observed in PGRs treated cultures at all concentrations. However, no shoot initiation was observed in PGRs free (control) treatments. Previously, Dereje [9] reported that shoot-tip (apical meristem) explants of sugarcane variety C86-12 cultured on hormone free MS medium had slow shooting response. The fact that no shooting response/delayed shooting response observed in PGRs free medium show that the available endogenous hormones may not be sufficient to induce shooting (Figure 9). The result also showed that the number of days to shoot emergence was found to be influenced by different concentrations of BAP combined with constant concentration of NAA. Relatively, C 86-56 variety had faster response at all PGRs concentrations than C

Figure 8: Best cultures of C 90-501 variety on shooting stage (MS+1.0 mg/l BAP+1.0 NAA media).

Treatments	Varieties	
	C 86-56	C 90-501
T_1(0.0 mg/l BAP)	0.000^d	0.000^d
T_2(0.5 mg/l BAP)	73.33^a	66.667^{ab}
T_3(1.0 mg/l BAP)	60.00^{ab}	60.000^c
T_4(1.5 mg/l BAP)	53.33^{bc}	76.667^a
T_5(2.0 mg/l BAP)	43.33^c	50.000^c
Mean	4.600	5.0667
CV	19.4440	16.9016
LSD	1.6272	1.5579

Means followed by the same letter are not significantly different at 5% significance level, CV=Coefficient of variance, LSD=Least significant different.

Table 1: Effect of different concentrations of BAP on the percent of initiated culture from apical meristem explants of C 86-56 and C 90-501 sugar cane varieties.

Figure 9: Best culture of C 90-501 variety on rooting stage (1/2 MS+1.0 mg/l NAA media+0.5 mg/l BAP).

90-501 variety in terms of shoot emergency (Table 2). In both varieties, shoot emergence was found to be earlier at lower concentration, and number of days to shoot emergence increased with increasing PGRs concentrations (Table 2). This finding was in line with that of Sowal [13] who reported the effectiveness of low concentration of BAP to result in rapid shoot multiplication.

Mean number of shoots: Number of shoot/explant was significantly higher in PGRs treated explants than PGRs free cultured explants. Number of shoot /explants was also significantly varied between PGRs treatments with highest number (10.667) counted at 1.5 mg/l of BAP combined with 1.0 mg/l of NAA for variety C 86-56 and 9.333 shoot/explants for variety C 90-501 at 1.0 mg/l of BAP combined with 1.0 mg/l of NAA (Table 2). Previously, Tarique [14] reported that 1.0 mg/l BAP+0.5 mg/l NAA and 1.0 mg/l BAP+0.5 mg/l IBA showed the best result for induction and multiplication of shoots for sugar cane varieties

Sugarcane varieties	Hormones (mg/l)		No of days to shoot mergence	No of shoots per expt.	Shoot length(cm)
	BAP	NAA			
C 86-56	0.0	0.0	0.000[d]	0.000[c]	0.000[d]
	0.5	1.0	13.333[c]	8.333[b]	3.4667[bc]
	1.0	1.0	14.00[c]	8.667[ab]	5.0333[b]
	1.5	1.0	16.33[b]	10.667[a]	8.5333[a]
	2.0	1.0	19.00[a]	9.333[ab]	5.8333[b]
Mean			12.5333	7.4000	4.5733
CV			9.2130	16.3657	17.9602
LSD			2.1007	2.2032	1.4943
C 90-501	0.0	0.0	0.000[c]	0.000[d]	0.0000[d]
	0.5	1.0	12.333[b]	7.333[b]	3.2667[b]
	1.0	1.0	17.00[ab]	9.333[a]	5.6333[a]
	1.5	1.0	18.333[a]	8.667[ab]	4.4333[b]
	2.0	1.0	18.000[a]	8.333[b]	3.1667[c]
Mean			13.1333	6.7333	3.333
CV			10.0246	15.3385	11.037
LSD			2.3952	1.8789	0.663

Means followed by the same letter within a column are not significantly different at 5% significance level, No=Number, Expt. =Explant, Wt. =Weight, CV=Coefficient of variance, LSD=Least significant different.

Table 2: Different shoot parameters measured for apical meristem explants treated with different concentrations of BAP combined with 1.0 mg/l of NAA. Values are mean ± SE, n=3.

of B52-298 and NCO-334, respectively (Figure 10). This shows that different varieties of sugar cane respond differently to different types and concentrations of PGRs, suggesting unique optimization for better performance. Moreover, Genotype specific response to number of shoot regeneration was reported by Gandonou [15] and Behara [8].

Shoot number /explant appeared to increase with increasing concentration of BAP up to 2.0 mg/l. It has been reported that a high level of cytokinin in combination with a low auxin level was essential for the differentiation of adventitious shoots in sugarcane Belay [10]. However, it was observed that mean number of shoots per explants was found to decline with further increase in the concentration of BAP beyond optimum (2.0 mg/l BAP) for both varieties. This finding agrees with Khalafalla [16] who reported that BAP at the concentration of 5.0 mg/l gives low number of shoot per explants and concluded that shoot number decreases as BAP concentration increases beyond optimum. Increasing trend in shoot number per explants up to optimum level is due to the fact that cytokinin (BAP) stimulates protein synthesis and participates in cell cycle control in a cell division George [17].

Effect of growth regulators on shoot length: Average shoot length was significantly higher in PGRs treated explants than explants culture on PGRs free media. Shoot length also showed significant difference between PGRs treatments. Shoots cultured on MS media containing 1.5 mg/l BAP and 1.0 mg/l NAA showed significantly higher mean shoot length (8.533) compared to all other treatments for var. C 86-56. For var. C 90-501 highest mean shoot length (5.633) measured was on MS medium containing 1.0 mg/l BAP and 1.0 mg/l NAA (Table 2). Similar to this result, Dereje [9] reported maximum shoot length 8.4 ± 0.008 for Cuban sugar cane variety C 90-501 when cultured on BAP (1.5 mg/l)+kin (0.5 mg/l). Behera [8] also reported that maximum shoot length of 6.2 ± 0.37 and 4.0 ± 0.61 under BAP (2.0 mg/l)+IBA (0.5 mg/l) and BAP (2.0 mg/l)+IBA (1.0 mg/l) for two sugarcane varieties namely B52-298 and NCO-334, respectively (Figure 11). On the other hand, it was observed that shoot length was found to decline with the increase in the concentration of BAP beyond optimum (2.0 mg/l BAP). This findings agrees with that of Bhatia who explained that increasing the concentration of the PGRs over optimum supplements may lead to negative effects on the morphology of the *in vitro* shoots.

Figure 10: Acclimatization of plantlets on different types of medium.

Figure 11: Screening of survival plantlets in green house.

Effect of growth regulator (NAA) on root induction

Number of days to root emergence: Root initiation was observed in PGRs treated shoots at all concentrations. However, no root initiation was observed in PGRs free (control) treatment. Similar result has been reported by Rashid. The number of days to root emergence was also found to be influenced by different concentrations of NAA. The two varieties showed variation in responding to the different concentrations of NAA in terms of time taken for root emergence. Variety C 86-56 took 13.0 and 18.33 days for root emergence at 0.5 mg/l NAA and 2.0 mg/l containing MS medium, respectively. Whereas C 90-501 formed root earlier (13.67 days) at the highest (2.0 mg/l) of NAA and root formation was delayed at the lowest (1.0 mg/l) of NAA.

Number of roots per shoots: Number of root/shoot was significantly higher in PGR treated shoot than PGR free cultured shoots. Number of roots was also significantly varied between PGR treatments with highest number (12.667) counted at 1.5 mg/l of NAA for variety C 86-56 and 9.0 for variety C 90-501 at ½ MS medium supplemented with 1.0 NAA (Table 3). This result can be complemented by a number of previous studies. For example, Behera [18] found highest number of roots per micro shoots (13.4 ± 1.5) on ½ MS medium supplemented 2.5 mg/l NAA for sugar cane varieties B52-298. The above result contradicts to Dereje [19] reported that maximum root/shoot (17.8) on ½ MS medium supplemented with 5.0 mg/l NAA for C 86-12 sugar cane variety.

Effect of growth hormones on root length: Root length was significantly ($p<0.05$) affected by different concentrations of NAA supplemented to ½ MS medium for both varieties (Table 3). Variety C86-56 produced maximum root length (5.667 cm) on half strength MS media containing 1.5 NAA and 0.5 mg/l BAP. But, variety C 90-501 produced maximum root length (4.7667 cm) on half strength MS media containing 1.0 mg/l NAA and 0.5 mg/l BAP. In line with this, Belay [20] reported root length of 3.2 ± 0.25 cm when grown on ½ strength MS medium supplemented with 1.0 mg/l NAA alone for N14 sugarcane variety. Mangrio [21] obtained average root length of 2.5 cm on ½ MS media supplemented with 3.0 mg/l NAA for sugarcane Variety NCO-334. The effect of variations in the concentrations and combination of the same hormone in most of the cited literatures and in the present work is almost entirely due to variation in the varieties of sugarcane tested by different researchers. That is why it is of paramount importance to optimize genotype specific *in vitro* propagation protocols for every variety.

Survival rate in green house during acclimatization

In vitro induced shoots are very delicate and cannot resist sudden environmental changes that may damage the plantlets unless they are gradually adapted to the new environment. Thus, acclimatization is essential to enable the rooted plantlets to adapt the natural environment in *ex vitro* conditions at controlled temperature and humidity of greenhouse conditions. In the acclimatization stage of this experiment, a total of 150 and 90 well rooted plantlets for variety C 86-56 and C 90-

501, respectively were transferred to greenhouse containing substrates namely, coco peat alone, that is a multipurpose growing medium made up of coconut husk; garden soil: sand: compost in the ration of 1:1:1 and garden soil and cow dung in the ration of 1:1. Then, the polystyrene tray was arranged in greenhouse with relative humidity (RH) gradually reducing from 90 to 60% and temperature of 30 to 31°C for two weeks for primary acclimatization (Figure 12). Generally the acclimatization phase of this experiment revealed that there was a difference in survival rate due to substrate nature and varietal difference. Both varieties had the highest survival value when grown on coco peat alone. On this media substrate, survival rate was 90% and 70% plantlet for C 86-56 and C 90-501 sugar cane varieties, respectively as shown (Table 4).

Summary and Conclusion

Sugarcane, as a globally important industrial crop, mainly is a source of sugar, ethanol, and other important by products. Hence, due consideration to the use of advanced technologies for sugarcane production is mandatory to obtain the unfolded benefits tapped from the crop. On the other hand, in Ethiopia, sugar industry is increasing at an alarming rate and is expected to play a significant role in poverty reduction. Thus, multiplication of sufficient quality of seeding material is needed more than ever before. However, in sugarcane seeding material multiplication usually takes up to 10 years following conventional method; besides the method allows continuation of diseases over vegetative cycles, which leads to drastic yield and quality reduction. To get out of the situation, *in vitro* propagation that enables rapid and large scale production of disease free planting material as being exercised with different crops is a prerequisite.

Based on this fact plant regeneration protocol was optimized in this study through direct Organogenesis for two commercially important Cuban originated sugarcane varieties (C 86-56 and C 90-501) using apical meristem explants. Accordingly, the information below was obtained. For initiation stage of apical meristem explants and more initiated culture var. C 86-56 best perform on 0.5 mg/l BAP mg/l, while var. C 90-501 best perform on 1.5 mg/l of BAP. Shoot parameters were also highly influenced by varieties and the type and combinations of various growth regulators. The effect of varieties and

Sugarcane Varieties	Hormone(mg/l)		No. of days to root emergence	No. of roots per shoot	Root Length(cm)
	NAA	BAP			
C 86-56	0.0	0.0	0.000[d]	0.000[c]	0.00[c]
	0.5	0.5	13.00[c]	7.667[b]	3.767[b]
	1.0	0.5	15.333[b]	9.000[b]	3.83[b]
	1.5	0.5	18.000[a]	12.667[a]	5.067[a]
	2.0	0.5	18.333[a]	12.000[a]	3.700[b]
Mean			12.8667	8.2667	3.183
CV			9.4124	17.668	22.542
LSD			2.2032	2.6572	1.343
C 90-501	0.0	0.0	0.000[b]	0.000[d]	0.000[c]
	0.5	0.5	16.00[ab]	9.000[a]	4.7667[a]
	1.0	0.5	17.00[a]	9.000[a]	4.7667[a]
	1.5	0.5	16.33[a]	7.000[b]	3.33[b]
	2.0	0.5	13.67[a]	5.667[bc]	3.633[ab]
Mean			12.337	5.333	3.2466
CV			21.7643	15.3093	21.0710
LSD			5.0154	1.4854	1.2446

Means followed by the same letter within a column are not significantly different at 5% significance level. No=number, Wt. =weight, CV=Coefficient of variance, LSD=Least significant different.

Table 3: Effect of different concentrations of NAA combined with 0.5 mg/l BAP on rooting responses of C 86-56 and C 90-501 sugar cane varieties. Values are means and n=3.

Types of sugarcane varieties	Types of culture medium	Total No of plantlet transferred	Survived plantlets	Died plantlets	Percent of survived plantlets	Percent of died plantlets
C 86-56	Coco peat only	50	45	5	90%	10%
	Garden soil:sand:compost (1:1:1)	50	36	14	72%	18%
	Garden soil: cow dung(1:1)	50	28	22	56%	44%
C 90-501	Coco peat only	30	21	9	70%	30%
	Garden soil:sand:compost (1:1:1)	30	18	12	60%	40%
	Garden soil: cow dung(1:1)	30	16	14	53.33%	46.66%

No=Number.

Table 4: Effect of different medium substrates on the survival of *in vitro* regenerated plantlets of the two varieties during acclimatization stage in green house.

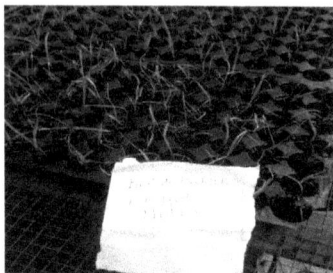

Figure 12: Best plantlets in green house on coco peat medium (survived and died plantlets).

hormones combinations was highly significant (p ≤ 0.05) on average values of shoot parameters. The highest number of initiated explants cultures regenerated more number of shoots and optimum shoot length were observed *in var.* C 86-56 on full MS media supplemented with 1.5 mg/l BAP+1.0 mg/l NAA, whereas maximum percentage of initiated explants cultures regenerated more number of shoots and optimum shoot length were recorded on 1.0 mg/l BAP+1.0 mg/l NAA for C 90-501 variety. Roots regenerated 10 to 15 days after micro shoots were transferred to root induction media for both varieties. Highly significant (p ≤ 0.05) differences were observed among treatments and varieties were also highly significant (p ≤ 0.05). The highest percentage of shoots regenerated number of roots and root length were recorded for *var.* C 86-56 on half strength MS media containing 1.5 mg/l NAA combined with 0.5 mg/l BAP., whereas for *var.* C 90-501 maximum percentage of number of root and root length was recorded on half strength MS media containing 1.0 mg/l NAA combined with 0.5 mg/l BAP. As to the acclimatization response, relatively highest survivability percentages were recorded on coco peat media substrate as compared to the other medium substrates used. Besides from the two varieties C 86-56 survived higher percentage (90%) than C 90-501 (70%) on coco peat media.

Lastly, in this work, the results clearly indicated the importance of evaluating individual variety to optimize a given tissue culture protocol. In other words, genotypic specificity was highly reflected in all of the parameters tested. Genotypic specificity has been reviewed in the literature review part of this paper and many researchers have imposed the evaluation of individual variety to recommend a tissue culture protocol.

Recommendation

Based on the above findings, the following recommendations are made for further investigation of *in vitro* culture of the two varieties.

- Further studies are needed using other hormones such as Kn,2,4-D, IAA, IBA with different concentration and interaction effects for observing their ability to induce shoots and roots for the reproducibility of the protocol optimization through direct or in direct organogenesis.

- Factors causing contamination, low acclimatization, tissue dying, and phenol exudation in the study should be further investigated.

- It is necessary to study the performance and genetic stability of the *in vitro* regenerated seedlings after transplanting in the field necessary.

- To enhance the acclimatization rate of *in vitro* developed plantlets in the glasshouse, various methods have to be manipulated.

- To sum up, the present study has developed protocol optimization for *in vitro* micro propagation of new Cuban origin *Saccharum officinarum* L. varieties (C 86-56 and C 90-501) using apical meristem explants through direct organogenesis. Hence, it is beneficial to use this developed *in vitro* micro propagation protocol as a best road map to large scale propagation to generate large number of seedlings in short period of time.

References

1. Schenck S, Lehrer AT (2000) Factors affecting the transmission and spread of Sugarcane yellow leaf virus. Plant Disease 84: 1085-1088.

2. Jalaja NC (2001) Micro propagation of sugarcane varieties for quality seed production. Manual on Sugarcane Production Technology, Sugarcane Breeding Institute, Coimbatore, pp: 48-52.

3. Soodi N, Gupta PK, Srivastava RK, Gosal SS (2006) Comparative studies on field performance of micro propagated and conventionally propagated sugarcane plants. Plant tissue culture & Biotech 16: 25-29.

4. Roy D (2000) Plant breeding: Analysis and Exploitation of Variation. Alpha Science International Limited.

5. Raman A, Singh SB (2005) Comparative performance of micro propagated and conventionally raised crops of sugarcane. Sugar tech 7: 93-95.

6. Geetha S, Padmanabhan D, (2001) Effect of hormones on direct somatic embryogenesis in sugarcane. Sugar Tech 3: 120-121.

7. Jai G, Sudhir K, Shekhar M (2014) The standardization of protocol for large scale production of sugarcane through micro propagation. International Journal of Plant, Animal and Environmental Sciences 4: 135-143.

8. Behera KK, Sahoo S (2009) Rapid in vitro micropropagation of sugarcane (Saccharum officinarum L. cv-Nayana) through callus culture. Nature and Science 7: 1-10.

9. Dereje S, Kasahun B, Tilaye F (2014) Interaction Effects of 6-Benzylaminopurine and Kinetin on In vitro Shoot Multiplication of Two Sugarcane (Saccharum officinarum L.) Genotypes. Advances in Crop Science and Technology 2: 143.

10. Belay T, Mulugeta D, Derbew B (2014) Effects of 6-Benzyl amino purine Kinetin on in vitro shoot multiplication of sugarcane (Saccharum officinarum L.) varieties. Advances in Crop Science Technology 2: 129.

11. Murashige T, Skoog S (1962) A revised medium for rapid growth and bioassays with tobacco cultures. Physiology Plant arum 15: 473-497.

12. Tilahun M, Mulugeta D, Manju S, Tadesse N (2014) Protocol optimization for in vitro mass propagation of two sugarcane (Saccharum officinarum L.) African Journal of Biotechnology 13: 1358-1368.

13. Sowal MT, Drozdowska L, Szota M (2002) Effect of cytokinins on in vitro morphogenesis and ploidy of sweet potato. Plant Tissue Culture & Biotech 5: 1-8.

14. Tarique HM, Mannan MA, Bhuiyan M, Rahaman MM (2010) Micro propagation of sugarcane through leaf sheath culture. International Journal of Sustainable Crop Production 5: 13-15.

15. Gandonou C, Errabii T, Abrini J, Idaomar M, Chibi F, et al. (2005) Effect of genotype on callus induction and plant regeneration from leaf explants of sugarcane (Saccharum spp.). African Journal of Biotechnology 4: 1250-1255.

16. Khalafalla MM, Abdellatef E, Mohamed Ahmed MM, Osman MG (2007) Micro-propagation of sweet potato (I.batatas) grows regularly. International Journal for Sustainable Crop Production 2: 1-8.

17. George EF, Michael AH, De Klerk GJ (2008) Plant Propagation by Tissue Culture. 3rd edn. Exegetics Ltd, p: 709.

18. Behera SK, Masson A, Sakuma H, Yamagata T (2006) A CGCM study on the interaction between IOD and ENSO. J Clim 19: 1608-1705.

19. Dereje S, Kasahun B, Tilaye F (2015) Interaction Effect of Indole-3-Butyric Acid and α-Naphthalene Acetic Acid on In Vitro Rooting of Two Sugarcane (Saccharum officinarum L.). Advances in Crop Science and Technology S1: 001.

20. Belay T (2016) Effects of Naphthalene Acetic Acid (NAA) and Indole -3- Butyric Acid (IBA) on in Vitro Rooting of Sugarcane (Saccharum officinarum L.). Biotechnology and Biomaterial 6: 1.

21. Mangrio G, Sughra A, Shereen N, Rind B, Dahot M (2016) In vitro regenerability of different sugarcane (saccharum officinarum L.) varieties through shoot tip culture. Pak J Biotechnology 11: 13-23.

19

Ex Vitro Rooting of Sugarcane (*Saccharum officinarum* L.) Plantlets Derived from Tissue Culture

Melaku Tesfa[1]*, Belayneh Admassu[2] and Kassahun Bantte[3]

[1]*Ethiopian Sugar Corporation Research and Training Division, Biotechnology Research Team, Wonji Research Center, Wonji, Ethiopia*
[2]*National Agricultural Biotechnology Research Program, Holetta Agricultural Research Center, Ethiopia*
[3]*Jimma University College of Agriculture and Veterinary Medicine, Jimma, Ethiopia*

Abstract

A study was conducted at Holetta National Agricultural Biotechnology Laboratory with the objective to determine the effect of different concentrations of NAA on *ex vitro* root development of sugarcane microshoots. Five levels of NAA (0, 10, 20, 30 and 40 mg/l) and two levels of genotypes were combined in factorial arrangement. The basal end of the shoots was dipped in NAA solution overnight before the shoots were transferred into a plastic tray containing a mixed growing medium in green house. The results showed that interaction effect of genotype and NAA was highly significant ($p < 0.0001$) on rooting percentage, number of roots per shoot, root length. In genotype N52, best root formation was found on the shoots treated with 20 mg/l NAA by which rooting percentage was 76 ± 5.48 with 5.88 ± 0.04 cm root length and 8.06 ± 0.13 number of roots per plantlets. While in genotype N53 maximum, root formation was recorded on the shoots dipped in 30 mg/l NAA by which rooting percentage was 70 ± 7.07 with 5.42 ± 0.11 cm root length and 4.52 ± 0.19 number of roots per plantlets. Shoots rooted through this method exhibited 100 % survival in both genotypes.

Keywords: Microshoot; *Ex vitro*; NAA; *In vitro*; Medium

Introduction

Sugarcane (*Saccharum officinarum* L.) is a monocotyledonous crop that is grown in the tropical and subtropical regions of the world, and almost cultivated on over 23.8 million ha for its sucrose rich stalk [1]. In Ethiopia, it is cultivated commercially on around 96,000 ha with annual sugar production of 370,000 ton and is a solely raw material for sugar production in the country. Recently, in Ethiopia, sugarcane production has gained attention, allied with its important potential for an environment-friendly bio-fuel (ethanol) production, in creating job opportunity to the nation and generating huge electric power [2].

Sugarcane, since commercially propagated vegetatively by stem cutting, has a low seed multiplication rate (1:10) which resulted in slow seed production of newly released improved varieties. Furthermore, the seed builds up diseases and pests during several cycles of field production, which leads to further yield and quality declines over years [3]. Thus, unavailability of disease-free, true to type planting material is a major limitation in improving sugarcane productivity.

Recently, tissue culture technology plays a leading role in rapid multiplication of disease-free and quality planting material of sugarcane [4]. Accordingly, Ethiopian Sugar Corporation has established its own tissue culture laboratory at Metahara, Kuraz, Tendaho and Fincha sugar factories/projects to produce about 55 million disease free plantlets per year [2]. To do so, *in vitro* propagation protocols have been developed using shoot tip and callus explant for several sugarcane cultivars of Ethiopian Sugar Estates [5-8].

In vitro propagation involves four crucial steps namely, initiation, multiplication and rooting of microshoots and acclimatization of plantlets. However, *in vitro* rooting process is an expensive, labour consuming process and can even double the final price of micropropagated plants. Earlier report shown that *in vitro* rooting may account for 40% total cost of the intensive manipulation needed during *in vitro* propagation [9]. In addition, roots of plantlets raised *in vitro* are generally very fragile and not have root hairs [10]. Therefore, during early acclimatization period, the roots do not function properly to support the plantlets to absorb water and nutrients from the potting medium. *Ex vitro* rooting is more advantageous than *in vitro* rooting in reducing cost of labour, chemicals and equipments, and the time of establishment from laboratory to soil revealed that *ex vitro* rooting reduced more than 50% cost of sugarcane plantlet raised by conventional micropropagation [11-13]. Besides, the plantlets produced after *ex vitro* rooting have better developed root system than the ones produced after *in vitro* rooting [14,15]. Furthermore, rooting and acclimatization phase can be carried out simultaneously; hence, it is more time efficient.

Ex vitro rooting has been applied in micropropagation of various plants species [11,12,16], but there are very limited reports yet on *ex vitro* rooting of sugarcane plantlets. Most reports of *ex vitro* rooting of plant species have involved treatment with exogenous auxin such as indole-3-acetic acid (IAA), indole-3-butyric acid (IBA), and 1-naphthalene-acetic acid (NAA) [12,13,16]. Auxin is applied singly or in a combination at different concentrations to improve rooting frequency of plantlets in the acclimatization period. Therefore, this study was aimed to determine the effect of different concentrations of auxin, NAA on *ex vitro* rooting of two elite sugarcane genotypes.

Materials and methods

The study was undertaken at the National Agricultural Biotechnology Laboratory of the Ethiopian Institute of Agricultural Research, in Holetta. The study was undertaken with two elite sugarcane genotypes

*Corresponding author: Melaku Tesfa, Ethiopian Sugar Corporation Research and Training Division, Biotechnology Research Team, Wonji Research Center, Wonji, Ethiopia, E-mail: melakutesfa2015@gmail.com

viz. N-52 and N-53 obtained from Ethiopian Sugar Corporation, Research and Training Division. The genotypes were selected based on their higher yield performance and sugar quality. In order to carry out explant sterilization and preparation, actively growing shoot tops were excised from 5-months-old screen house grown healthy mother plants of genotype N52 and N53. The trimmed shoot tops (segment) were washed carefully under running tap water for 30 minutes, and reduced to 10 cm length by cutting off at the two ends. Then washed thoroughly for 30 minutes with tap water containing a drop of liquid detergent solution and two drops of tween-20 and rinsed three times with double distilled water. Subsequently, the explant was taken to a laminar air flow cabinet and immersed in 0.1% (w/v) Bavistin® DF 50 % (Carbendizem) fungicide solution, ascorbic acid (0.2% w/v) and citric acid (0.4% w/v) for 30 minutes followed by three times rinsing each for five minutes with sterile double distilled water. The shoot tips were washed again with 70% ethanol for one minute and rinsed with sterile double distilled water three times each for five minute to remove residual ethanol from the shoot tip surface. Finally, surface sterilized with 50 % (v/v) aqueous solution of Sodium hypochlorite (5.25% w/v active chlorine) containing a few drops of tween-20 for 25 minutes. After pouring out sodium hypochlorite solution, the explants were rinsed with sterile double distilled water three times each for five minutes to remove all the trace of the sterilant.

Culture were initiated on [17] medium fortified with BAP, Kinetin and NAA (0.5 mg/l each) [18] and 2% sucrose (w/v) and solidified using agar (agar agar, type I) (0.45 %; w/v) for induction of shoots. The pH of the medium was adjusted to 5.8 followed by autoclaving at 121°C at 105 Kpa pressure for 20 minutes. After 30 days of inoculation the regenerated shoots were transferred to multiplication medium supplemented with 2 mg/l BAP + 0.5 mg/l Kinetin (N52) and 1.5 mg/l BAP and 0.5 mg/l Kinetin (N53) [19]. After 30 days of incubation, shoots were maintained on plant growth regulators (PGRs) free MS medium with 2 g/l activated charcoal for two weeks before transferring rooting stage in order to avoid the carry over effect of hormones from the multiplication media on ex vitro rooting of the plantlets.

After 45 days, healthy micro-shoots having 4 cm heights were employed for ex vitro rooting study. The clumps of in vitro shoots were separated to obtain single micro shoot. The basal portion of these rootless microshoots was dipped in distilled aqueous solution containing auxin, NAA at different concentrations i.e. 0, 10, 20, 30, & 40 mg/l overnight to induce rooting under ex vitro condition. The experiment was arranged in completely randomized design (CRD) with five replications and each treatment had 50 microshoots. After treated with auxins, the shoots were transferred to polystyrene trays containing autoclaved mixture of river sand and forest soil in 2:1 ratio. Subsequently, maintained in greenhouse, which uses Fan-Pad evaporative cooling system providing 25–30°C temperature. During experimenting, high humidity level (80 %-85 %) was maintained by covering the tray with moisten polyethylene sheet and red shade cloth and then sprinkled with water three times a day as necessary and sprayed with quarter strength MS basal medium at weekly interval.

After 4 weeks, the plantlets were carefully removed from the soil mix and data on number of rooted shoots, total number of primary roots and root length were recorded. All microshoots that remain green were considered living and used in calculating rooting percentage. Successfully rooted plantlets were subsequently transferred in medium polyethene bags (15 cm × 20 cm) containing mixture of sand, farm yard manure and soil in 1:1:1 ratio for further hardening and data on survival rate of the plantlets was recorded 4 weeks after transplanting. The collected data were subjected to analysis of variance (F test) using SAS program (Version 9.2). The differences among treatment means were determined by REGQ multiple range test at P<0.05.

Results and Discussion

Statistical analysis of variance showed that the main effect of genotype and NAA and the interaction effect of genotype and NAA highly significant (p<0.0001) on rooting percentage, number of roots per shoot, root length of the two sugarcane genotype (Table 1). The present result also showed that rooting was induced ex vitro over the entire range of NAA concentration tested including the control shoots in both sugarcane genotypes (Table 2).

In the control treatment reduced rooting frequency of 36% and 28% were obtained in genotypes N52 and N53, respectively (Figure 1A and Figure 1B) However, in NAA treated microshoots than 50% of the shoot developed roots regardless of the NAA concentration (Table 2). Shekafandeh [12] observed increased rooting frequency and number of roots from zero percent in untreated shoots to 91.7% and 3.3 roots per shoot, respectively, when the basal end of the shoots were dipped in a solution of 1.5 mg/l IAA and 0.3 mg/l IBA for 24 h before culturing in soil mixture in Myrtle (Myrtus communis L.) plant. Similar results were also reported by Sumaryono and Riyadi [20] in oil palm (Elaeis

Source of variation	DF	Rooting Percentage	Root length (cm)	Number of roots per shoot
		MS	MS	MS
Genotype	1	968***	7.76***	59.19***
NAA	4	2307***	4.73***	25.56***
Genotype*NAA	4	163***	2.08***	4.46 ***
CV %		10.38	5.60	2.72

*** = Very highly significant at P≤0.0001; DF = Degree of freedom; NAA = α-naphthalene acetic acid; MS = Mean square; CV= Coefficient of variation

Table 1: Effect of different concentration of NAA on ex vitro rooting.

	Genotypes					
Treatment	N52			N53		
NAA mg/l	Rooting Percentage	Root length (cm)	Number of root per shoot	Rooting Percentage	Root length (cm)	Number of root per shoot
0	36^d ± 5.48	4.44^{ed} ± 0.30	2.14^h ± 0.13	28^d ± 0.47	2.58^f ± 0.54	1.74^i ± 0.05
10	68^{ab} ± 4.47	4.64^{cd} ± 0.29	5.42^d ± 0.08	50^c ± 7.07	4.32^{ed} ± 0.24	2.56^g ± 0.13
20	76^a ± 5.48	5.88^a ± 0.04	8.06^a ± 0.13	60^{bc} ± 7.07	4.34^{ed} ± 0.21	4.08^f ± 0.08
30	70^{ab} ± 7.07	5.04^{bc} ± 0.05	6.36^b ± 0.11	70^{ab} ± 7.07	5.42^b ± 0.11	4.52^e ± 0.19
40	56^c ± 5.48	4.68^{cd} ± 0.24	5.68^c ± 0.13	54^c ± 5.48	4.08^e ± 0.19	3.88^f ± 0.13
CV%	10.38	5.60	2.72	10.38	5.60	2.72

*NAA =α-naphthalene acetic acid. Values in the same column and variables with different letters are significantly different from each other according to REGWQ at P<0.05.

Table 2: The effect of NAA on rooting percentage, root length and number of roots per shoot.

Figure 1: Acclimatized plantlets. (A) Genotype N52 (B) Genotype N53.

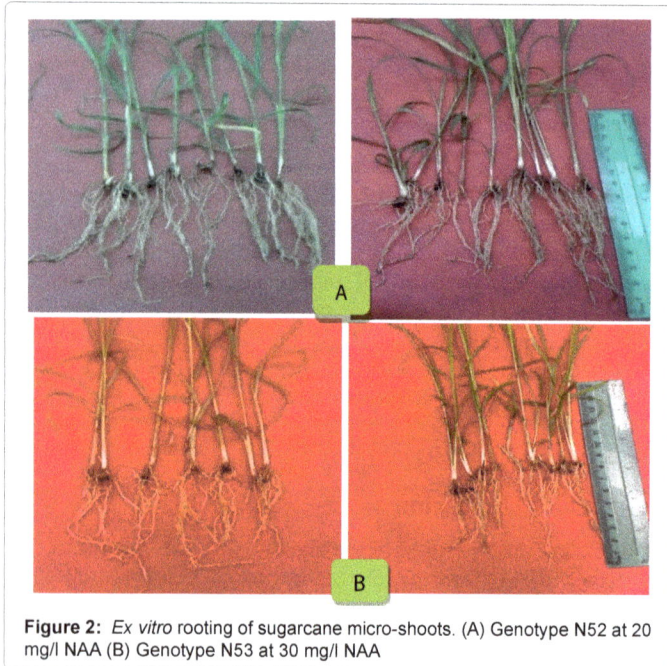

Figure 2: *Ex vitro* rooting of sugarcane micro-shoots. (A) Genotype N52 at 20 mg/l NAA (B) Genotype N53 at 30 mg/l NAA

guineensis Jacq.). These results indicated the significance of treating of microshoots with plant growth regulators during *ex vitro* rooting before culturing in soil medium.

There was a significant response variation in rooting between the two genotypes. Genotype N52 had the highest (76 %) rooting frequency with a maximum (5.88 cm ± 0.04 cm) average root length and 8.06 ± 0.13 average number of roots per shoot on microshoots dipped in 20 mg/l concentration of NAA (Table 2 and Figure 2A). At the same concentration of NAA, N53 had only 60 % rooting frequency with 4.34 cm ± 0.21 cm average root length and 4.08 ± 0.08 average roots number per shoot. On the other hand, genotype N53 showed a maximum rooting frequency (70 %) with 5.42 cm ± 0.08 cm and 4.52 ± 0.19 numbers of roots per shoot on microshoots treated with 30 mg/l concentration of NAA (Table 2 and Figure 2B). At this concentration of NAA, genotype N52 gave almost equal rooting frequency (70 %) with comparable root length (5.04 cm ± 0.05cm) and higher (6.36 ± 0.11) root number per shoot than N53.The result of this experiment revealed that genotype N52 was more responsive than N53 for different NAA concentrations.

The rate of rooting frequency increased from 36 % in control shoots to 76 % when the basal ends of shoots were dipped in a solution of 20 mg/l NAA overnight (Figure 3). Similarly, average root length and average number of roots increased from 4.44 cm ± 0.30cm and 2.14 ± 0.13 to 5.88 ± 0.04cm and 8.06 ± 0.13, respectively, in genotype N52.

However, when NAA concentration was further elevated to higher concentration (40 mg/l), rooting frequency, average root length and average number of roots per shoot, reduced significantly to 56 %, 4.68 ± 0.24 cm and 5.68 ± 0.13, respectively. The same trend was observed in genotype N53, in that rooting frequency increased from 28% in control shoots to 70 % in treated shoots with a solution of NAA at 30 mg/l. The average root length and average roots number also increased from 2.58 cm ± 0.54 cm and 1.74 to 5.42 cm ± 0.08cm and 4.52 ± 0.19 respectively, as the concentration of NAA increased from 0.0 mg/l to 30 mg/l but as the concentration of NAA was increased to 40 mg/l the rooting frequency, root length and root number reduced markedly to 54 %, 4.08 cm ±0.19 cm and 3.88 ± 0.13 respectively (Figures 3-5). This reduction in rooting response could be due to the fact that higher concentrations of NAA promote the biosynthesis and buildup of ethylene at the basal end of the shoot, which have inhibitory effect on the overall rooting response of sugarcane microshoots [21].

The result of the present study on genotype N52 were in agreement with earlier results by Pandey et al. [13], who obtained the highest

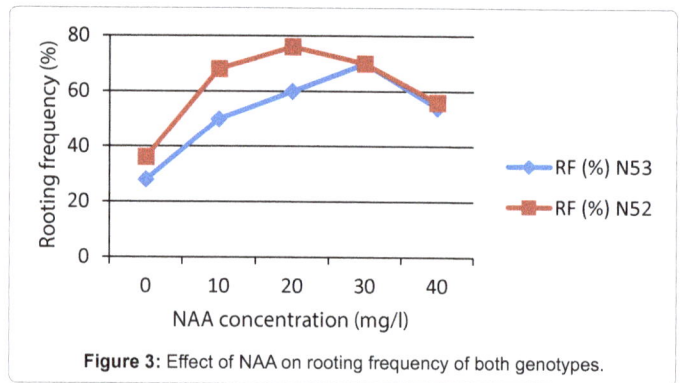

Figure 3: Effect of NAA on rooting frequency of both genotypes.

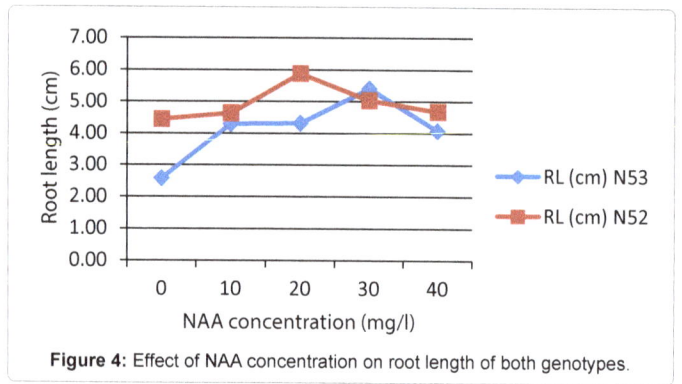

Figure 4: Effect of NAA concentration on root length of both genotypes.

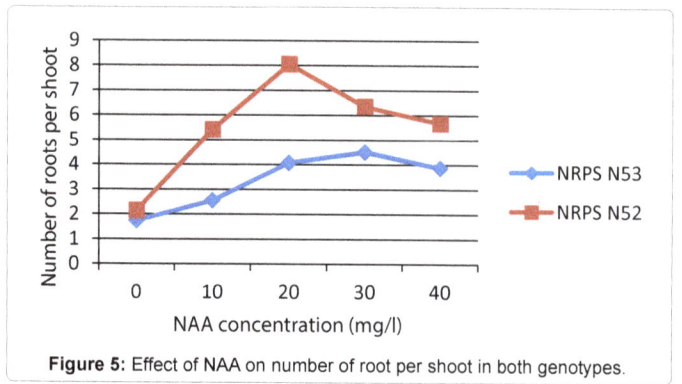

Figure 5: Effect of NAA on number of root per shoot in both genotypes.

rooting frequency, root length and number of roots per shoot from shoots treated in 20 mg/l of NAA concentration in sugarcane genotype CoS96268. Similarly, Martin [16] obtained an average of 5.6 roots per shoot after the microshoots of *Rotula aquatica* Lour were dipped in 0.5 mg/l NAA for 25 days. Biradar et al. [20] found best root formation on shoot treated with 2 mM NAA with 80% rooting frequency in oil palm. However, other authors Martin [22] and Chinnu [23] obtained best result of *ex vitro* rooting by using IBA. In present study, shoots rooted through this method transplanted to small pots containing soil, sand and farmyard manure (1: l: 1) exhibited 100 % survival in both genotypes.

Conclusion

Generally, it appears that rooting of tissue culture-derived rootless sugarcane plantlets can be induced during *ex vitro* acclimatization by dipping in the auxin (NAA) solution overnight. From the five concentrations of NAA tested for *ex vitro* rooting, 20 mg/l NAA was found to be the optimal concentrations for *ex vitro* rooting of genotype N52. It produced the highest rooting frequency (76 %) with an average of 8.06 roots per shoot while in genotype N53, 30 mg/l NAA gave a maximum of 70 % rooting frequency with 4.52 average root numbers per shoot. *Ex vitro* rooting reduces the time of acclimatization and labor cost.

Acknowledgements

We would like to express our thanks to Ethiopian Sugar Corporation (ESC) for financing the study and Holetta National Agricultural Biotechnology Laboratory (HNABL) for provision of Tissue Culture Laboratory with all required consumable chemicals and facilities.

References

1. FAOSTAT (2013) Food and Agriculture Organization of the United Nations Statistics.

2. Ethiopian Sugar Corporation (ESC) (2014) Sweet Newsletter 4:1.

3. Jalaja NC, Neelamathi D, Sreenivasan TV (2008) Micropropagation for Quality Seed Production in Sugarcane in Asia and the Pacific. Food and Agriculture Organization of the United Nations and Asia–Pacific Consortium on Agricultural Biotechnology, Asia Pacific Association of Agricultural Research Institutions.

4. Lorenzo JC, Ojeda E, Espinosa A, Borroto C (2001) Field performance of temporary immersion bioreactor derived sugarcane plant. In Vitro Cellular & Developmental Biology-Plant 37: 803-806.

5. Tilahun M, Mulugeta D, Sharma M (2013) An alternative safer and cost effective surface sterilization method for sugarcane (Saccharum officinarum L.) explants. African J Biotechnol 12: 6282-6286.

6. Belay T, Mulugeta D, Derbew B (2014) Effects of 6-Benzyl aminopurine and Kinetin on in vitro shoot multiplication of sugarcane (Saccharum officinarum L.) varieties. Adv Crop Sci Tech 2: 129.

7. Dereje S, Kassahun B, Tiliye F (2014) Interaction effect of 6-Benzylaminopurine and Kinetin on in vitro shoots multiplication of two sugarcane (Saccharum officinarum L.) genotypes. Adv Crop Sci Tech 2: 143.

8. Gemechu A, Firew M, Adefris T (2014) Effect of genotype on in vitro propagation of elite sugarcane (Saccharum officinarum L.) varieties of Ethiopian sugar estates. International journal of technology enhancements and emerging engineering research 2.

9. Leva A (2011) Innovative protocol for "ex vitro rooting" on olive micropropagation. Central European Journal of Biology 6: 352-358.

10. Hazarika BN (2006) Morpho-physiological disorder in in vitro culture of plants. Sci Hort 108: 105-120.

11. Feyissa T, Welander M, Negash L (2007) Genetic stability, ex vitro rooting and gene expression studies in Hagenia abyssinica. Biologia Plantarum 51: 15-21.

12. Shekafandeh A (2007) Effect of different growth regulators and source of carbohydrates on in and ex vitro rooting of Iranian myrtle. Int J Agricultural Res 2: 152-158.

13. Pandey RN, Rastogi J, Sharma ML, Singh RK (2011) Technologies for cost reduction in sugarcane micropropagation. African J Biotechnol 10: 7814-7819.

14. Bozena B (2001) Morphological and physiological characteristics of micropropagated Strawberry plants rooted in vitro or ex vitro. Sci Hort 89: 195–206.

15. Yan H, Liang C, Yang L, Li Y (2010) In vitro and ex vitro rooting of Siratia grosvenorii, a traditional medicinal plant. Acta physiologiae plantarum 32: 115-120.

16. Martin KP (2003) Rapid in vitro multiplication and ex vitro rooting of Rotula aquatica Lour, a rare rheoephytic woody medicinal plant. Plant Cell Report 21: 415–420.

17. Murashige T, and Skoog F (1962) A revised medium for rapid growth and bio assays with tobacco tissue cultures. Physiologia plantarum 15: 473-497.

18. Pathak S, Lal M, Tiwari AK, Sharma ML (2009) Effect of growth regulators on in vitro multiplication and rooting of shoot culture in sugarcane. Sugar Tech 11: 86-88.

19. Melaku T, Belayneh A, Kassahun B (2016) In vitro shoot multiplication of elite sugarcane (Saccharum Officinarum L.) genotypes using liquid shake culture system. J Biol Agriculture Healthcare 6: 35-40.

20. Sumaryono S, Riyadi I (2011) Ex vitro rooting of oil palm (Elaeis guineensis Jacq.) plantlets derived from tissue culture. Indonesian Journal of Agricultural Science 12: 57-62.

21. Biradar S, Biradar DP, Patil VC, Patil SS, Kambar NS (2009) In vitro plant regeneration using shoot tip culture in commercial cultivar of sugarcane. Karnataka Journal Agricultural Science 22: 21-24.

22. Martin KP (2003) Rapid axillary bud proliferation and ex vitro rooting of Eupatorium triplinerve. Biologia Plantarum 47: 589-591.

23. Chinnu JK, Mokashi AN, Hegde RV, Patil VS, Koti RV (2012) In vitro shoot multiplication and ex vitro rooting of cordyline (Cordyline sp.). Karnataka Journal of Agricultural Science 25: 221-223.

20

Heavy Metals Accumulation in Soil and Agricultural Crops Grown in the Province of Asahi India Glass Ltd., Haridwar (Uttarakhand), India

Vinod Kumar* and A.K. Chopra

Agro-ecology and Pollution Research Laboratory, Department of Zoology and Environmental Science, Gurukula Kangri University, Haridwar-249404 (Uttarakhand), India

Abstract

This investigation was aimed to assess the heavy metals accumulation in soil and agricultural crops *viz.,* Indian mustard (*Brassica juncea* L.), wheat (*Triticum aestivum* L.) and barley (*Hordeum vulgare* L.) irrigated with glass industry effluent. The glass industry effluent was considerably loaded with various plant nutrients and heavy metals. The effluent irrigation significantly ($P<0.05$/$P<0.01$) increased Cd, Cr, Cu, Fe, Mn, Pb, Zn, total bacteria, total fungi, actinomycetes and yeast of the soil used for the cultivation of *B. juncea*, *T. aestivum* and *H. vulgare*. The enrichment of different heavy metals were recorded in the order of Cr>Pb>Cd>Mn>Zn>Cu>Fe for *B. juncea*, Zn>Cd>Cu>Cr>Pb>Fe>Mn for *T. aestivum* and Zn>Cu>Cd>Fe>Mn>Pb>Cr for *H. vulgare* irrigated with glass industry effluent. The translocation of different metals in different parts i.e. root, stem, leaves and fruits were observed in the order of leaves > stem > root > fruit for Cu, Fe, Mn, Pb and Zn; stem > leaves > root > fruit for Cd, root > stem > leaves > fruit for Cr in *B. juncea*, *T. aestivum* and *H. vulgare* irrigated with glass industry effluent. Therefore, glass industry effluent irrigation added heavy metals in the soil and *B. juncea*, *T. aestivum* and *H. vulgare*.

Keywords: Accumulation; Agricultural crops; Enrichment factor; Glass industry effluent; Heavy metals; Soil

Introduction

Contamination of soil and agricultural crops from industrial waste is a result of various types of industrial processes and disposal practices [1-4]. Industries that use large amounts of water for processing have the potential to pollute waterways through the discharge of their waste into streams and rivers, or by run-off and seepage of stored wastes into nearby water sources [5-7]. Other disposal practices which cause water contamination include deep well injection and improper disposal of wastes in surface impoundments. Discharge of industrial effluent without adequate treatment causes severe degradation in the pedosphere, hydrosphere, and atmosphere [8-10]. Effluents from industries contain appreciable amounts of nitrogen (N), phosphorous (P), sodium (Na), potassium (K), calcium (Ca), and magnesium (Mg) along with zinc (Zn), copper (Cu), iron (Fe), manganese (Mn), lead (Pb), nickel (Ni), and cadmium (Cd), and their disposal cause contamination of soil and water environment [2,11-13].

With the rapid enhancement in industrial establishment, there has been substantial increase in the liquid waste which is traditionally discharged either into open land or nearby water courses [14,15]. Industrial effluent is mostly used for the fertigation of agricultural crops, mainly in urban and periurban regions, due to its easy availability, disposal problems and scarcity of irrigation water. In addition, the industrial effluents are the main sources of soil and water contamination [16-18]. Although, effluent irrigation can alleviate water scarcity situations and allows farmers to reduce the purchase of chemical fertilizer and of organic matter that serves as soil conditioner and humus replenishment [10]. However, unregulated irrigation with untreated wastewater poses serious public health risks, as effluent is a major source of heavy metals that cause accumulation in plant parts [13,19,20]. Heavy metals accumulation in agricultural soils is of increasing worldwide concern and particularly in India with the rapid development of industrialization and urbanization [21-23]. The crop plants take up heavy metals and accumulate them in their edible and inedible parts in quantities high enough to cause clinical problems both to animals and human beings consuming these metal-rich plants [4,24,25]. Moreover, a number of serious health problems can develop as

a result of excessive uptake of dietary heavy metals [26,27]. Additionally, the heavy metals such as cadmium (Cd), chromium (Cr), copper (Cu), iron (Fe), nickel (Ni) and zinc (Zn) reaches to the soil through effluent irrigation, and accumulates in plant and vegetable parts, and cause adverse health effects [15,23,28-30]. Keeping in view the present investigation was carried out to the heavy metals accumulation in soil and agricultural crops grown in the Province of Asahi India Glass Ltd., Haridwar (Uttarakhand), India.

Materials and Methods

Study area

Asahi India Glass Ltd. Jhabrera, Roorkee, Haridwar (29°47'50"N 77°48'23"E) was selected for the present study. The glass industry is located about 45 km away from Haridwar at Haridwar Saharanpur via Jhabrera Highway. Asahi India Glass is the largest integrated glass company in India. They manufacture a wide range of international quality automotive safety glass, float glass, architectural processed glass and glass products. The glass industry is using a large quantity of fresh water and discharged huge amount of effluent. The effluent is being used by the farmers in the cultivation of agricultural crops.

Collection of effluent samples and their analysis

The effluent samples were collected from the effluent disposal channel used for the irrigation of agricultural crops in the vicinity of glass industry. Bore well water is considered as control. The samples of

*Corresponding author: Vinod Kumar, Agro-ecology and Pollution Research Laboratory, Department of Zoology and Environmental Science, Gurukula Kangri University, Haridwar-249404, (Uttarakhand), India, E-mail: drvksorwal@gkv.ac.in

glass industry effluent and bore well water were collected in thoroughly cleaned plastic containers of 5 liters capacity separately. The samples were brought to the laboratory and were analyzed for various physico-chemical parameters and microbiological *viz.*, total dissolved solids (TDS), electrical conductivity (EC), pH, biochemical oxygen demand (BOD), chemical oxygen demand (COD), total Kjeldahl nitrogen (TKN), phosphate (PO_4^{3-}), sodium (Na^+), potassium (K^+), calcium (Ca^{2+}), magnesium (Mg^{2+}), cadmium (Cd), chromium (Cr), copper (Cu), iron (Fe), lead (Pb), manganese (Mn), zinc (Zn), total bacteria, total fungi, total coliform and yeast following standard methods [31].

Collection of soil samples and their analysis

The soil samples from the surface (0-20 cm) were collected from the effluent irrigated agricultural fields in the vicinity of Asahi India Glass Ltd. The bore well water irrigated soil was taken as control. The samples were brought to the laboratory and dried in clean plastic trays for 7 days at room temperature and then sieved through a 2 mm sieve. The samples were analyzed for various physico-chemical and microbiological parameters *viz.*, viz., EC, organic carbon (OC), TKN, PO_4^{3-}, Na^+, K^+, Ca^{2+}, Mg^{2+}, Cd, Cr, Cu, Fe, Mn, Pb, Zn, total bacteria, total fungi, actinomycetes and yeast following standard methods [32].

Collection of agricultural crops samples and their analysis

Glass industry effluent irrigated crops *viz.*, Indian mustard (*Brassica juncea*), wheat (*Triticum aestivum*) and barley (*Hordeum vulgare*) grown in the vicinity of Glass Industry were selected for the present study. Bore well water irrigated *B. juncea*, *T. aestivum* and *H. vulgare* were taken as control. The samples of *B. juncea*, *T. aestivum* and *H. vulgare* from effluent and bore well water irrigated agricultural filed were collected in the polythene bags separately. The samples were brought to the laboratory and washed thrice under tap water and followed by distilled water. The crop samples were cut with a sharp knife then air dried and ground and sieved. For the extraction of metals, 10 ml sample

of wastewater, 1.0 g sample of air dried soil/crops was taken in digestion tube. In each sample 3 ml concentrate HNO_3 was added and digested on electrically heated block for 1 hour at 145° C. Then 4 ml of $HClO_4$ was added and heated to 240°C for an additional hour. The aliquot was cooled, filtered through Whatman # 42 filter paper and the volume was made 50 ml with double distilled water and used for analysis. The metals were analyzed by using atomic absorption spectrophotometer (PerkinElmer Analyst 800 AAS) following standard methods. The enrichment factor for metals accumulated in wastewater irrigated soil and vegetables were calculated by following formula.

$$\text{Enrichment factor} = \frac{\text{Mean content of metal in the sample}}{\text{Mean metal content in the control}}$$

Statistical analysis

The values reported here are means of six values. Data were tested at different significant levels using student t-test to measure the variations between the soil parameters before and after irrigation of these crops with wastewater. One way analysis of variance (ANOVA) was used for data analysis to measure the variations of metals in different parts of these crops before and after glass industry effluent irrigation was calculated with the help of MS Excel, 2013.

Results and Discussion

Occurrence of heavy metals in industrial effluents

The physico-chemical and microbiological properties of glass industry effluent are presented in Table 1. The results revealed that the glass industry effluent was considerably rich in various plant nutrients like TKN (384.75 mg L^{-1}), PO_4^{3-} (145.67 mg L^{-1}), Na^+ (185.65 mg L^{-1}), K^+ (220.90 mg L^{-1}), Ca^{2+} (450.87 mg L^{-1}), Mg^{2+} (230.00 mg L^{-1}) and heavy metals Cd (2.75 mg L^{-1}), Cr (1.26 mg L^{-1}), Cu (6.88 mg L^{-1}), Fe (12.89 mg L^{-1}), Mn (1.54 mg L^{-1}), Pb (2.36 mg L^{-1}) and Zn (8.80 mg L^{-1}). The values of TDS (3640 mg L^{-1}), BOD (1450.60 mg L^{-1}), COD

Parameter	Borewell water (control)	Composite effluent	BIS disposal standards
TDS (mg L^{-1})	210.50 ± 2.58	3640 ± 3.64	1900
EC (dS cm^{-1})	0.24 ± 0.04	3.24 ± 0.14	-
pH	7.45 ± 0.05	7.85 ± 0.08	5.5-9.0
BOD_5 (mg L^{-1})	4.59 ± 0.08	1450.60 ± 4.50	100
COD (mg L^{-1})	8.97 ± 0.25	1890.80 ± 5.20	250
TKN (mg L^{-1})	28.90 ± 1.08	384.75 ± 2.96	100
PO_4^{3-} (mg L^{-1})	0.04 ± 0.01	145.67 ± 1.05	-
Na^+ (mg L^{-1})	10.66 ± 0.21	185.65 ± 2.30	-
K^+ (mg L^{-1})	6.80 ± 0.15	220.90 ± 2.40	-
Ca^{2+} (mg L^{-1})	26.90 ± 0.16	450.87 ± 3.74	200
Mg^{2+} (mg L^{-1})	14.66 ± 0.23	230.00 ± 2.66	-
Cd (mg L^{-1})	0.36 ± 0.02	2.75 ± 0.05	2.00
Cr (mg L^{-1})	0.08 ± 0.01	1.26 ± 0.02	2.00
Cu (mg L^{-1})	1.26 ± 0.02	6.88 ± 0.08	3.00
Fe (mg L^{-1})	1.45 ± 0.03	12.89 ± 0.25	1.0
Mn (mg L^{-1})	0.36 ± 0.01	1.54 ± 0.03	-
Zn (mg L^{-1})	2.17 ± 0.54	8.80 ± 0.06	15
Pb (mg L^{-1})	0.24 ± 0.02	2.36 ± 0.04	-
Total bacteria (CFU ml $^{-1}$)	5.64×10^3 ± 4.00	8.57×10^8 ± 14.00	10000
Total fungi (CFU ml $^{-1}$)	3.75×10^2 ± 5.00	5.23×10^5 ± 12.00	-
Total coliform (MPN 100 ml $^{-1}$)	2.23×10^2 ± 3.00	4.26×10^3 ± 6.00	5000
Yeast (CFU ml $^{-1}$)	1.43×10^2 ± 2.00	3.76×10^3 ± 7.00	-

Mean ± SE of six values; BIS- Bureau of Indian Standards

Table 1: Physico-chemical and microbiological characteristics of glass industry effluent.

(1890.80 mg L^{-1}), TKN, Ca^{2+}, Cd, Cu, Fe and Zn were observed beyond the prescribed limit of BIS irrigation standards [33]. Moreover, higher values of microbiological parameter total bacteria (8.57 × 10^8 CFU ml^{-1}), total fungi (5.23 × 10^5 CFU ml^{-1}), total coliform (4.26 × 10^3 MPN 100 ml $^{-1}$) and yeast (3.76 × 10^3 CFU ml^{-1}) were also observed in the glass industry effluent. The higher value of EC (3.24 dS cm^{-1}) is the indicator of more ionic species in the glass industry effluent. The higher values of BOD, COD, TKN and PO$_4$$^{3-}$ in the glass industry effluent are the indicator of higher organic load of the effluent and it is likely due to the mixing of sewage in the glass industry effluent generated from the staff quarters and offices located in the premises of the glass industry. The higher contents of heavy metals the glass industry effluent is might be associated with silicate minerals using as raw materials in the glass manufacturing and various colouring agents to make the coloured glass panels. Kumar and Chopra [13] also reported higher values of BOD (1256.50 mg L^{-1}), COD (2832.50 mg L^{-1}), Cd (2.98 mg L^{-1}), Cr (1.68 mg L^{-1}), Cu (3.56 mg L^{-1}), Fe (15.25 mg L^{-1}), Pb (0.94 mg L^{-1}), Zn (6.42 mg L^{-1}), total bacteria (9.89 × 10^7 SPC mL^{-1}) and coliform bacteria (5.69 × 10^6 MPN 100 mL^{-1}) in the paper mill effluent. Therefore, the glass industry effluent showed higher organic as well as inorganic pollution load.

Occurrence of heavy metals in agricultural soil

Table 2 showed the soil characteristics used in the cultivation of *B. juncea*, *T. aestivum* and *H. vulgare* irrigated with glass industry effluent and bore well water. On perusal of data it is clearly indicated that glass industry effluent significantly (P<0.05/P<0.01/P<0.001) increased EC, OC, TKN, PO$_4$$^{3-}$, Na$^+$, K$^+$, Ca^{2+}, Mg^{2+}, Cd, Cr, Cu, Fe, Mn, Pb, Zn, total bacteria, total fungi, actinomycetes and yeast of the soil used for the cultivation of *B. juncea*, *T. aestivum* and *H. vulgare* compared to bore well water. During the investigation the soil was loamy in texture and no significant change was observed in the soil texture during the study. The pH value is an important parameter, as many nutrients are available only at a particular range of pH for plant uptake. A pH value

of 6.0–8.2 provides predominating bacterial activity and is favorable for maximum yield of crops. The pH of the soil was ranged (7.56-7.68) after irrigation with glass industry effluent. The higher content of OC, TKN and PO$_4$$^{3-}$ were recorded in the soils used for the cultivation of *B. juncea*, *T. aestivum* and *H. vulgare* which is likely due the more organic nature of glass industry effluent. The higher contents of OC (3.45-4.12 mg Kg^{-1}), TKN (268.58-298.50 mg Kg^{-1}) and PO$_4$$^{3-}$ (114.67-173.77 mg Kg^{-1}) were recorded in the soil used for the cultivation of *B. juncea*, *T. aestivum* and *H. vulgare*. The higher values of OC, TKN and PO$_4$$^{3-}$ are the indicator of soil fertility and support the higher population of microbial communities in the soil as earlier reported by Kumar and Chopra [7]. The contents of different heavy metals in the soil used for the cultivation of *B. juncea*, *T. aestivum* and *H. vulgare* were ranged Cd (0.66-0.84 mg Kg^{-1}), Cr (0.24-0.28 mg Kg^{-1}), Cu (4.37-5.84 mg Kg^{-1}), Fe (6.84-7.58 mg Kg^{-1}), Mn (1.38-1.56 mg Kg^{-1}), Pb (0.23-0.29 mg Kg^{-1}), Zn (3.75-4.15 mg Kg^{-1}) after irrigation with glass industry effluent (Table 2). The most contents of heavy metals were recorded in the soil used for the cultivation of *T. aestivum* while the least were found in the soil used for the cultivation of *B. juncea*. Thus it clearly indicated that *B. juncea* was more efficient for the uptake of heavy metals compared to *H. vulgare* and *T. aestivum*. The findings are in line of Chopra et al. [18] who reported higher contents of Cd (4.34 mg Kg^{-1}), Cr (1.65 mg Kg^{-1}), Cu (7.86 mg Kg^{-1}), Fe (10.55 mg Kg^{-1}), Mn (1.87 mg Kg^{-1}) and Zn (8.85 mg Kg^{-1}) in the paper mill effluent irrigated soil. Significant (P<0.05/P<0.01) microbial population in the form of total bacteria (4.85 × 10^7-7.75 × 10^9 CFU g^{-1}), total fungi (3.25 × 10^5-6.35 × 10^6 CFU g^{-1}), actinomycetes (5.88 × 10^3 -9.86 × 10^4 CFU g^{-1}) and yeasts (4.00 × 10^4 -5.65 × 10^5 CFU g^{-1}) were observed in the soil irrigated with glass industry effluent compared to control soil (bore well water irrigated soil) (Table 2). The findings are in accordance with Beligh et al. [16] who reported significantly higher microbial community in the soil irrigated with olive mill wastewater. The organic carbon and pH are principal components that control the accessibility of heavy metals in the soil. Increasing the soil organic carbon can enhance the soil electrical conductivity and effective cation exchange capacity, which is

Parameter	Borewell water irrigated soil (control)	Glass industry effluent irrigated soil		
		B. juncea	*T. aestivum*	*H. vulgare*
EC (dS cm^{-1})	2.18 ± 0.05	3.04* ± 0.06	3.40* ± 0.08	3.13* ± 0.06
pH	7.49 ± 04	7.56 ± 0.03	7.62 ± 0.04	7.68 ± 0.05
OC	0.56 ± 02	3.45** ± 0.10	4.12** ± 0.15	3.87** ± 0.13
TKN (mg Kg^{-1})	33.41 ± 3.94	268.58*** ± 4.27	298.50*** ± 1.84	285.18*** ± 2.84
PO$_4$$^{3-}$ (mg Kg^{-1})	55.54 ± 5.72	114.67*** ± 1.86	173.77*** ± 1.24	121.56*** ± 1.95
Na$^+$ (mg Kg^{-1})	21.32 ± 2.50	28.96* ± 2.17	33.28* ± 1.64	49.84** ± 1.47
K$^+$ (mg Kg^{-1})	169.01 ± 3.06	188.96* ± 1.95	219.97* ± 2.36	210.06* ± 1.06
Ca^{2+} (mg Kg^{-1})	17.73 ± 1.94	118.75** ± 2.36	129.92*** ± 2.05	122.03*** ± 1.16
Mg^{2+} (mg Kg^{-1})	1.72 ± 10	12.67*** ± 1.02	16.75*** ± 1.02	18.94*** ± 1.01
Cd (mg Kg^{-1})	0.24 ± 0.07	0.66** ± 0.03	0.84** ± 0.03	0.75** ± 0.02
Cr (mg Kg^{-1})	0.17 ± 0.06	0.24* ± 16	0.28* ± 16	0.26* ± 0.09
Cu (mg Kg^{-1})	2.37 ± 0.35	4.37* ± 0.91	5.84* ± 0.91	5.26* ± 1.07
Fe (mg Kg^{-1})	3.13 ± 0.44	6.96* ± 0.75	7.58** ± 0.75	6.84* ± 0.0.09
Mn (mg Kg^{-1})	0.11 ± 0.06	1.38** ± 16	1.56** ± 16	1.43** ± 0.03
Zn (mg Kg^{-1})	1.96 ± 17	3.75* ± 0.35	4.15** ± 0.35	4.02** ± 0.08
Pb (mg Kg^{-1})	0.07 ± 0.06	0.23** ± 0.02	0.29** ± 0.02	0.25** ± 0.01
Total bacteria (CFU g^{-1})	2.56×10^6 ± 8.00	7.75×10^9*** ± 6.00	5.69×10^8** ± 9.00	4.85×10^7** ± 7.00
Total fungi (CFU g^{-1})	1.45×10^5 ± 4.00	6.35×10^6* ± 4.00	3.25×10^5* ± 5.00	4.05×10^5* ± 3.00
Actinomycetes (CFU g^{-1})	1.67×10^3 ± 5.00	9.86×10^4* ± 8.00	6.93×10^3* ± 6.00	5.88×10^3* ± 7.00
Yeast (CFU g^{-1})	3.45×10^4 ± 6.00	5.65×10^5* ± 7.00	4.12×10^4* ± 9.00	4.00×10^4* ± 5.00

Mean ± SE of six values; Significant F -***P-0.1%, **P-1% level,*P-5% level of ANOVA to the control.

Table 2: Effects of glass industry effluent irrigation on soil characteristics used for the cultivation of *B. juncea*, *T. aestivum* and *H. vulgare*.

a reason that may affect both soluble and exchangeable metals [2,34]. However, in the present study, the effects of glass on the soil pH seemed to be one of the promising factors, which affects the availability, and accumulation of Cd, Cr, Cu, Fe, Mn, Pb and Zn. Despite the fact that, the contents of Cd, Cr, Cu and Zn were found to be below the maximum levels permitted for Cd (6.0 mg kg^{-1}), Cr (10.0 mg kg^{-1}), Cu (270 mg kg^{-1}) and Zn (600 mg kg^{-1}) for soil in India [33]. However, there was gradual buildup of these metals in the glass industry effluent irrigated soil and this is likely due to the addition of more organic carbon in the soil which provides more binding sites for the attachment of these metals. Therefore, glass industry effluent irrigation significantly altered the contents of heavy metals in the soil used for the cultivation of *B. juncea*, *T. aestivum* and *H. vulgare*.

Occurrence of heavy metals in agricultural crops

During the present investigation, the contents of heavy metals Cd (3.54-5.09 mg Kg^{-1}), Cr (0.71-2.27 mg Kg^{-1}), Cu (9.73-10.99 mg Kg^{-1}), Fe (10.08-12.89 mg Kg^{-1}), Mn (1.55-3.22 mg Kg^{-1}), Pb (1.27-2.42 mg Kg^{-1}) and Zn (8.28-11.60 mg Kg^{-1}) were ranged in *B. juncea*, *T. aestivum* and *H. vulgare* after irrigation with glass industry effluent (Figures 1-3). The maximum contents of Cd, Cr, Cu, Fe, Mn, Pb and Zn were recorded in *B. juncea* while the least contents of these metals were observed in *T. aestivum* after irrigation with glass industry effluent. The enrichment of different heavy metals were recorded in the order of Cr>Pb>Cd>Mn>Zn>Cu>Fe for *B. juncea*, Zn>Cd>Cu>Cr>Pb>Fe>Mn for *T. aestivum* and Zn>Cu>Cd>Fe>Mn>Pb>Cr for *H. vulgare* after irrigation with glass industry effluent (Figure 4). The translocation of metals in different parts i.e. root, stem, leaves and fruits of *B. juncea* was ranged Cd (0.76-1.65 mg Kg^{-1}), Cr (0.42-0.76 mg Kg^{-1}), Cu (2.09-3.74 mg Kg^{-1}), Fe (2.42-4.08 mg Kg^{-1}), Mn (0.64-0.0.96 mg Kg^{-1}), Pb (0.44-0.72 mg Kg^{-1}) and Zn (2.12-3.75 mg Kg^{-1}); in *T. aestivum* Cd (0.45-1.12

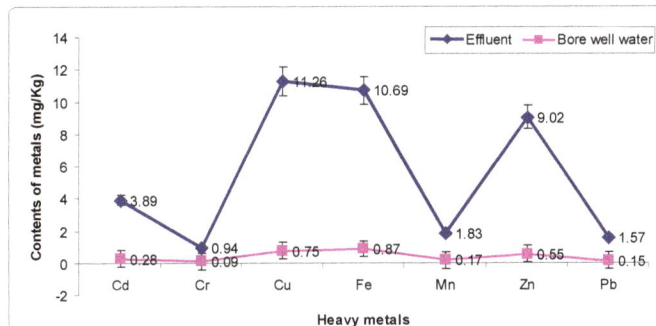

Figure 1: Content of different heavy metals in *B. juncea* irrigated with glass industry effluent. Error bars are the standard error of the mean.

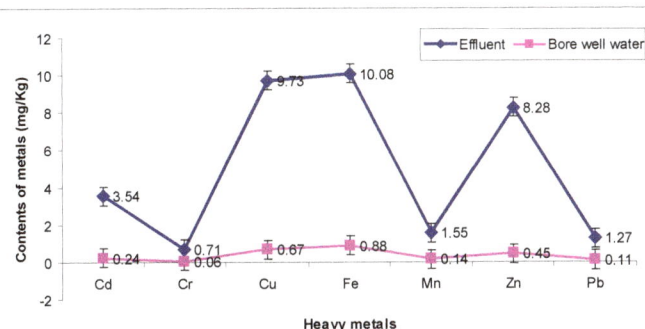

Figure 2: Content of different heavy metals in *T. aestivum* irrigated with glass industry effluent. Error bars are the standard error of the mean.

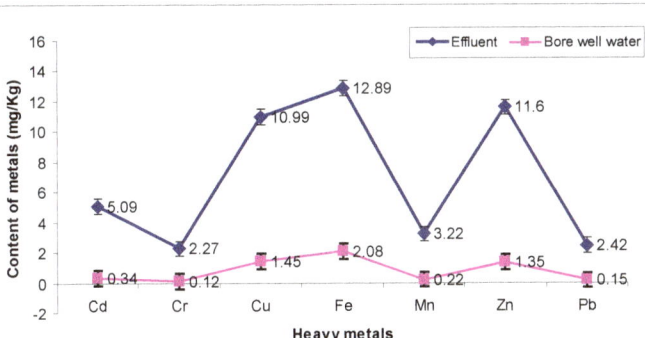

Figure 3: Content of different heavy metals in *H. vulgare* irrigated with glass industry effluent. Error bars are the standard error of the mean.

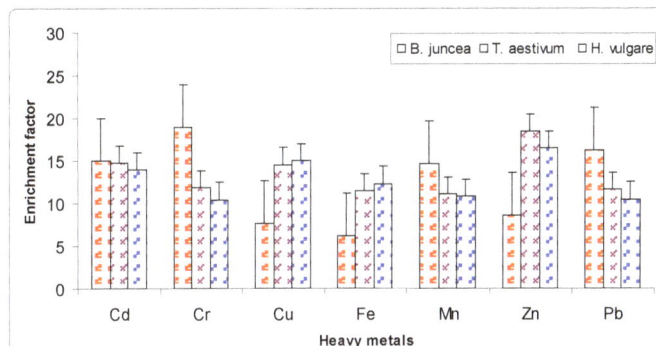

Figure 4: Enrichment factor of different heavy metals in *B. juncea*, *T. aestivum* and *H. vulgare* irrigated with glass industry effluent. Error bars are the standard error of the mean.

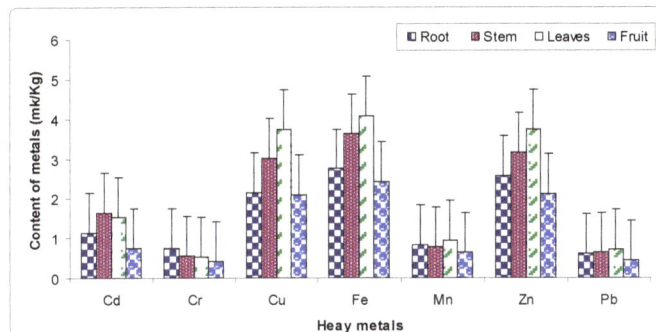

Figure 5: Translocation of heavy metals in different parts of *B. juncea* irrigated with glass industry effluent. Error bars are the standard error of the mean.

mg Kg^{-1}), Cr (0.08-0.28 mg Kg^{-1}), Cu (1.88-3.05 mg Kg^{-1}), Fe (1.65-3.12 mg Kg^{-1}), Mn (0.24-0.52 mg Kg^{-1}), Pb (0.21-0.42 mg Kg^{-1}) and Zn (1.38-2.41 mg Kg^{-1}) and in *H. vulgare* Cd (0.54-1.29 mg Kg^{-1}), Cr (0.11-0.36 mg Kg^{-1}), Cu (1.99-3.08 mg Kg^{-1}), Fe (1.96-3.24 mg Kg^{-1}), Mn (0.28-0.58 mg Kg^{-1}), Pb (0.25-0.52 mg Kg^{-1}) and Zn (1.78-2.56 mg Kg^{-1}) irrigated with glass industry effluent (Figures 5-7). The translocation of different metals in different parts i.e. root, stem, leaves and fruits were observed in the order of leaves > stem > root > fruit for Cu, Fe, Mn, Pb and Zn; stem > leaves > root > fruit for Cd, root > stem > leaves > fruit for Cr in *B. juncea*, *T. aestivum* and *H. vulgare* after irrigation with glass industry effluent (Figures 5-7). The accessibility and bioaccumulation of metals are governed by numerous environmental factors viz. pH, solubility, and chemical speciation of the metal, presence of humic substances, presence of other metals, salinity, soil mineralogy, texture, and amorphous Fe and Al contents [2,35]. However, the observed differences in the metal

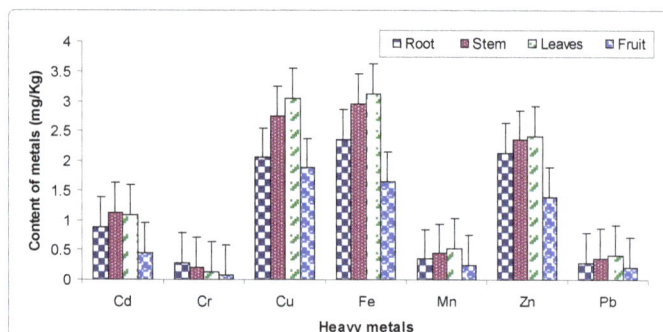

Figure 6: Translocation of heavy metals in different parts of *T. aestivum* irrigated with glass industry effluent. Error bars are the standard error of the mean.

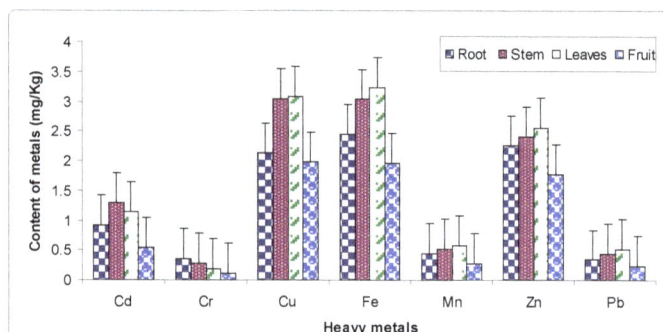

Figure 7: Translocation of heavy metals in different parts of *H. vulgare* irrigated with glass industry effluent. Error bars are the standard error of the mean.

accumulation in the different parts like root, stem, leaves and fruits of the plant suggest different cellular mechanisms of bioaccumulation of heavy metals, which may control their translocation and partitioning in the plant. Cu, Fe, Mn and Zn are associated with the photosynthesis [2,28,36]. The content of Cu, Fe, Mn and Zn was higher in the leaves of *B. juncea*, *T. aestivum* and *H. vulgare* and it is likely due to the higher content of these metals in the glass industry effluent irrigated soil. Cu, Fe and Zn are indispensable for the survival and proliferation of all plants. There is a high demand for Cu, Fe and Zn in the photosynthetic apparatus as a result plants accumulates and translocates Cu, Fe and Zn in their photosynthetic parts [2,36]. Moreover the translocation of Cd, Cr, Cu, Fe, Mn, Pb and Zn are related to the organic matter in the soil, which enhanced their uptake by cultivated plants [2,37].

The contents of Cr were higher in the roots of *B. juncea*, *T. aestivum* and *H. vulgare* than in the aerial parts (i.e., stem, leaves and fruits), indicating that the roots act as barrier for translocation and protect the edible parts from toxic contamination [2]. Deprived translocation of Cr to the shoots and leaves could be due to the sequesterization of most of the Cr in the vacuoles of the root cells to render it non-toxic, which may be a natural protective response of this plant [23]. It must be noted that Cr is a toxic and non-essential element to plants and hence the plants may not possess any specific mechanism to transport the Cr [18,34]. The content of Cd, Cr, Cu, Fe, Mn, Pb and Zn were less in the fruits of *B. juncea*, *T. aestivum* and *H. vulgare* and it is likely due to increases in cation-exchange capacity of the soil, which will decrease the uptake of Cd, Cr, Cu, Fe, Mn, Pb and Zn into *B. juncea*, *T. aestivum* and *H. vulgare* plants. Therefore, among *B. juncea*, *T. aestivum* and *H. vulgare*, *B. juncea* was found most efficient compared to *T. aestivum* and *H. vulgare* to translocate and tolerate the metals contents after irrigation with glass industry effluent. Although the contents of Cr, Cu, Mn and

Zn except Cd in *B. juncea*, *T. aestivum* and *H. vulgare* was noted below the permissible limit of FAO/WHO standards for Cd (0.20 mg Kg⁻¹), Cr (2.30 mg Kg⁻¹), Cu (40.00 mg Kg⁻¹) and Zn (60.00 mg Kg⁻¹) [27]. But long-term application of glass industry effluent in the cultivation of *B. juncea*, *T. aestivum* and *H. vulgare* would carry a risk of progressive accumulation of these heavy metals to toxic levels. The findings are in agreement with Chandra et al. [30] who reported the higher content of heavy metals in wheat (*Triticum aestivum* L.) and Indian mustard (*Brassica campestris* L.) irrigated with distillery and tannery effluents.

Conclusions

This study concluded that glass industry effluent irrigation significantly ($P<0.05/P0.01$) changed the soil properties specially increased the contents of different heavy metals *viz.*, Cd, Cr, Cu, Fe, Mn, Pb and Zn in the soil and further in *B. juncea*, *T. aestivum* and *H. vulgare*. The accumulation and translocation of heavy metals were found to be plant part specific and varied from species to species and it is likely due to the uptake efficiency and tolerance capability of different species. The most accumulation of Cd, Cr, Cu, Fe, Mn, Pb and Zn were recorded in *B. juncea* while the least was in *T. aestivum*. The contents of Cr, Cu, Mn and Zn except Cd in *B. juncea*, *T. aestivum* and *H. vulgare* was noted below the permissible limit of FAO/WHO standards but continuous use of glass industry effluent of these crops would increase these metals in these crop plants. Therefore, care should be taken before the use of glass industry effluent in the cultivation of agricultural crops like *B. juncea*, *T. aestivum* and *H. vulgare* crops. Further studies are required on the use of glass industry effluent in the cultivation of different agricultural crops and their health effects after consumption.

References

1. Qishlaqi A, Moore F, Forghani G (2008) Impact of untreated wastewater irrigation on soils and crops in Shiraz suburban area, SW Iran. Environ Monit Assess 141: 257-273.

2. Kumar V, Chopra AK (2014) Accumulation and translocation of metals in soil and different parts of French bean (Phaseolus vulgaris L.) amended with sewage sludge. Bull Environ Contam Toxicol 92: 103-108.

3. Kumar V, Chopra AK (2013) Response of sweet sorghum after fertigation with sugar mill effluent in two seasons. Sugar Tech 15: 285-299.

4. Chaturvedi M, Kumar V, Singh D, Kumar S (2013) Assessment of microbial load of some common vegetables among two different socioeconomic groups. International Food Research Journal 20: 2927-2931.

5. Bharagava RN, Chandra R, Rai V (2008) Phytoextraction of trace elements and physiological changes in Indian mustard plants (Brassica nigra L) grown in post methanated distillery effluent (PMDE) irrigated soil. Bioresour Technol 99: 8316-8324.

6. Chary NS, Kamala CT, Raj DS (2008) Assessing risk of heavy metals from consuming food grown on sewage irrigated soils and food chain transfer. Ecotoxicol Environ Saf 69: 513-524.

7. Kumar V, Chopra AK (2012) Effect of paper mill effluent irrigation on agronomical characteristics of Vigna radiata (L.) in two different seasons. Communications in Soil Science and Plant Analysis 43: 2142-2166.

8. Sharif A Salah, Suzelle F Barrington (2006) Effect of soil fertility and transpiration rate on young wheat plants (Triticum aestivum) Cd/Zn uptake and yield. Agricultural Water Management 82: 177-192.

9. Clemente R, Dickinson NM, Lepp NW (2008) Mobility of metals and metalloids in a multi-element contaminated soil 20 years after cessation of the pollution source activity. Environ Pollut 155: 254-261.

10. Kumar V, Chopra AK (2011) Impact on physico-chemical characteristics of soil after irrigation with distillery effluent. Archives of Applied Science Research 3: 63-77.

11. Muchuweti M, Birkett JW, Chinyanga E, Zvauya R, Scrimshaw MD, et al. (2006) Heavy metal content of vegetables irrigated with mixture of wastewater and sewage sludge in Zimbabwe: Implications for human health. Agriculture Ecosystem and Environment 112: 41-48.

12. Qadir M, Wichelns D, Raschid-Sally I, McCornik PG, Drechsel P, et al. (2009) The challenges of wastewater irrigation in developing countries. Agriculture Water Management 97: 561-568.

13. Kumar V, Chopra AK (2013) Enrichment and translocation of heavy metals in soil and Vicia faba L. (Faba bean) after fertigation with distillery effluent. International Journal of Agricultural Policy and Research 1: 131-141.

14. Wang JF, Wang GX, Wanyan H (2007) Treated wastewater irrigation effect on soil, crop and environment: wastewater recycling in the loess area of China. J Environ Sci (China) 19: 1093-1099.

15. Pathak C, Chopra AK, Kumar V, Sharma S (2011) Effect of sewage-water irrigation on physico-chemical parameters with special reference to heavy metals in agricultural soil of Haridwar city. Journal of Applied and Natural Science 3: 108-113.

16. Beligh M, Fethi Ben Mariem, Mohamed Baham, Salem Ben Elhadj, Mohamed Hammami (2008) Change in soil properties and the soil microbial community following land spreading of olive mill wastewater affects olive trees key physiological parameters and the abundance of arbuscular mycorrhizal fungi. Soil Biology and Biochemistry 40: 152-161.

17. Tewari PK, Batra VS, Balakrishnan M (2009) Efficient water use in industries: cases from the Indian agro-based pulp and paper mills. J Environ Manage 90: 265-273.

18. Chopra AK, Srivastava S, Kumar V (2011) Comparative study on agro-potentiality of Paper mill effluent and synthetic nutrient (DAP) on Vigna unguiculata L. (Walp) Cowpea. Journal of Chemical and Pharmaceutical Research 3: 151-165.

19. Yadav RK, Goyal B, Sharma RK, Dubey SK, Minhas PS (2002) Post-irrigation impact of domestic sewage effluent on composition of soils, crops and ground water--a case study. Environ Int 28: 481-486.

20. Sharma RK, Agrawal M, Marshall F (2006) Heavy metal contamination in vegetables grown in wastewater irrigated areas of Varanasi, India. Bull Environ Contam Toxicol 77: 312-318.

21. Al-Lahham O, El Assi NM, Fayyad M (2003) Impact of treated wastewater irrigation on quality attributes and contamination of tomato fruit. Agriculture Water Management 61: 51-62.

22. Hati M Kuntal, Anand Swarup, Dwivedi, Misra AK, Bandyopadhyay KK (2007) Changes in soil physical properties and organic carbon status at the topsoil horizon of a vertisol of central India after 28 years of continuous cropping, fertilization and manuring. Agriculture Ecosystem and Environment 119: 127-134.

23. Chopra AK, Srivastava S, Kumar V, Pathak C (2013) Agro-potentiality of distillery effluent on soil and agronomical characteristics of Abelmoschus esculentus L. (okra). Environ Monit Assess 185: 6635-6644.

24. Nastri A, Ramieri NA, Abdayem R, Piccaglia R, Marzadori C, et al. (2006) Olive pulp and its effluents suitability for soil amendment. J Hazard Mater 138: 211-217.

25. Paulose B, Datta SP, Rattan RK, Chhonkar PK (2007) Effect of amendments on the extractability, retention and plant uptake of metals on a sewage-irrigated soil. Environ Pollut 146: 19-24.

26. Fytianos K, Katsianis G, Triantafyllou P, Zachariadis G (2001) Accumulation of heavy metals in vegetables grown in an industrial area in relation to soil. Bull Environ Contam Toxicol 67: 423-430.

27. FAO/WHO (2011) Joint FAO/WHO food standards programme codex committee on contaminants in foods, fifth session 64-89.

28. Barman SC, Sahu RK, Bhargava SK, Chaterjee C (2000) Distribution of heavy metals in wheat, mustard, and weed grown in field irrigated with industrial effluents. Bull Environ Contam Toxicol 64: 489-496.

29. Munir J, Mohammad Rusan, Sami Hinnawi, Laith Rousan (2007) Long term effect of wastewater irrigation of forage crops on soil and plant quality parameters. Desalination 215: 143-152.

30. Chandra R, Bharagava RN, Yadav S, Mohan D (2009) Accumulation and distribution of toxic metals in wheat (Triticum aestivum L.) and Indian mustard (Brassica campestris L.) irrigated with distillery and tannery effluents. J Hazard Mater 162: 1514-1521.

31. APHA (2012) Standard methods for the examination of water and wastewater. American Public Health Association, 21st ed., Washington, DC 2462.

32. Chaturvedi RK, Sankar K (2006) Laboratory manual for the physicochemical analysis of soil, water and plant. Wildlife Institute of India, Dehradun 97.

33. BIS (2010) Bureau of Indian Standards: Indian standards for drinking water-Specification (BIS 10500: 2010).

34. Kim KH, Kim SH (1999) Heavy metal pollution of agricultural soils in central regions of Korea. Water Air and Soil Pollution 111: 109-122.

35. Gu C, Bai Y, Tao T, Chen G, Shan Y (2013) Effect of sewage sludge amendment on heavy metal uptake and yield of ryegrass seedling in a mudflat soil. J Environ Qual 42: 421-428.

36. Porra RJ (2002) The chequered history of the development and use of simultaneous equations for the accurate determination of chlorophylls a and b. Photosynth Res 73: 149-156.

37. Zarcinas BA, Pongsakul P, McLaughlin MJ, Cozens G (2004) Heavy metals in soils and crops in Southeast Asia. 2. Thailand. Environ Geochem Health 26: 359-371.

21

Impact of Long-Term Conventional Cropping Practices on Some Soil Quality Indicators at Ethiopian Wonji Sugarcane Plantation

Alemayehu Dengia[1]* and Egbert Lantinga[2]

[1]*Sugar Corporation, Research and Training, Wonji Research Center, Wonji-Ethiopia, Ethiopia*
[2]*Department of Plant Sciences, Biological Farming Systems Group, Wageningen University, Radix West, 2nd floor, Droevendaalsesteeg, 16708 PB Wageningen, The Netherlands*

Abstract

Over the last 50 years, the sugarcane yield in Wonji plantation has declined by about 40%. Perhaps one of the possible causes for the decline is soil degradation. Thus, the major soil quality indicators were evaluated for the extent of change that might occur due to long-term conventional cropping practices. To that end bio-sequential soil sampling was performed by collecting soil samples from adjacent virgin and cultivated lands of Wonji sugar cane plantation. The samples were analyzed and compared for major soil properties. The result showed that the SOM contents of cultivated land were 53% and 34% lower than the virgin land at 0 cm-30 cm and 30-60 cm depths, respectively. Total N, P Olsen, exchangeable K and soil EC of the cultivated land were also 56%, 84%, 86% and 54% lower than the virgin land at 0 cm-30 cm. The differences were also significant at 30 cm-60 cm. There was no significant change in soil pH at both depths. In general long-term conventional cropping practices depleted the SOM, total N, P Olsen and exchangeable K. However, pH and EC were in the optimum range that soil acidity, salinity and alkalinity were not a problem. As the soil type of the plantation is heavy clay, particularly, the degradation in SOM content might cause the yield decline. In order to fully identify, understand and manage the problems of soil quality deterioration further study is necessary.

Keywords: SOM; Conventional cropping; Yield decline; Soil degradation; Total N, P; Olsen; Exchangeable K; EC; pH 1

Introduction

Sugar cane production is one of the largest and most important agro-industries around the world in general and in developing countries in particular. Nowadays, sugar cane is considered as a *Dollar Earner* for tropical countries due to its immense potential to generate hard currency [1]. Therefore, achieving sustainable cane production is an increasingly important goal in recent years so as to exploit this potential. However, in several sugar cane producing countries around the world, decline in sugar cane yield appears to be the major preoccupation of the agro-industry [2-4]. Likewise, sugar cane yield decline is currently becoming the major area of attention in the Ethiopian sugar cane plantations. Although, Ethiopia is one of the countries with the highest sugar cane yield in the world [5,6] the yield has been declining already for many years. For instance over the last 50 years, the cane yield per ha in Wonji sugar cane plantation has dropped by about 40% (Figure 1). Thus, the future viability of the agro-industry will be doubtful unless the yield decline could be stopped. Therefore, identifying and understanding the cause of the yield decline has paramount importance to design and recommend appropriate management strategies. This is particularly essential in view of the current Ethiopian government ambitious plan to augment the sugar production capacity of the country.

The cause for sugar cane yield decline is a complex issue, as it results from a number of factors. According to De Wit [7] actual crop yield generally depend on growth-defining, growth-limiting and growth-reducing factors. In monoculture sugarcane farming, where long-term conventional cropping practices are widely adopted, these factors may not be sustained at the optimum level for cane growth. Particularly, upon cultivation the deterioration of soil quality is inevitable which may lead to soil degradation and consequently yield decline.

A soil with a high quality is productive and has a stable yield [8] whereas its degradation adversely affects the agronomic productivity [9]. Similarly, in sugar cane plantations, soil degradation was found to be one of the major contributors for yield decline [4,10]. The losses in soil production capacity of a sugar cane field could be mainly ascribed to long-term monoculture, uncontrolled traffic from heavy machineries, excessive tillage before planting [4] pre-harvest cane burning [11], inappropriate irrigation and drainage [12] and excessive utilization of agrochemicals [13]. These conventional cropping practices lead to change in soil biological, physical and chemical properties with concomitant decline in cane yield [13]. Similarly the aforementioned practices were also adopted in Wonji plantation over the last fifty years and May resulted in soil degradation.

Therefore, the objective of this research is to evaluate the major soil quality indicators of Wonji sugar cane plantation so as to understand the possible existence and extent of soil degradation. The approach taken was to collect soil samples from virgin and cultivated lands bio-sequential soil sampling of the sugar cane plantation and to evaluate the differences in the major soil quality indicators soil organic matter, total N, available P, exchangeable K, EC and pH.

Materials and Methods

Site description

The study was conducted between September 2009 and February 2010 in Wonji sugar estate (8°31'N and 39°12'E), which is 110 km southeast of Addis Ababa (Figure 2). The site is located at an altitude of 1540 meter above sea level. The average annual rainfall is 807 mm.

*Corresponding author: Alemayehu Dengia, Sugar Corporation, Research and Training, Wonji Research Center, Wonji-Ethiopia, Ethiopia
E-mail: alexdengia@yahoo.com

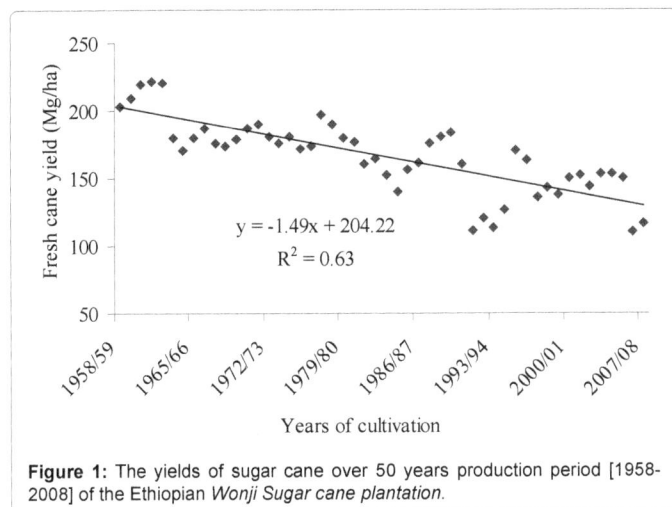

Figure 1: The yields of sugar cane over 50 years production period [1958-2008] of the Ethiopian *Wonji Sugar cane plantation*.

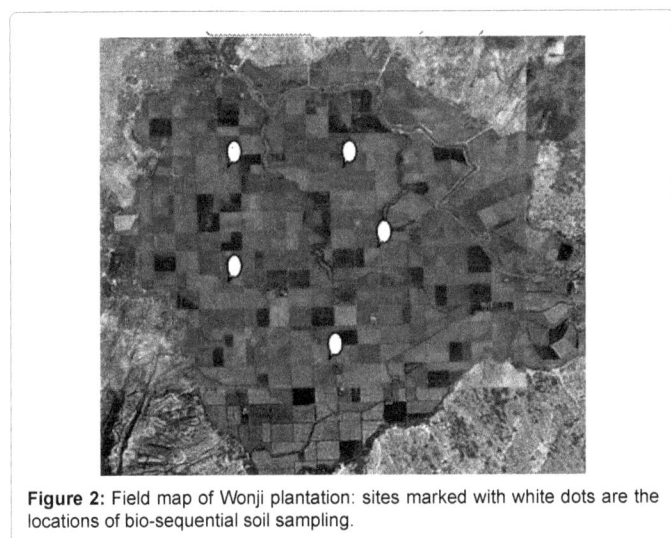

Figure 2: Field map of Wonji plantation: sites marked with white dots are the locations of bio-sequential soil sampling.

while average daily minimum and maximum temperatures are 14.3°C and 27.6°C, respectively. The soils of the study area are predominantly heavy clay. The slope of the land is very gentile and regular. Before the establishment of the estate, the Wonji plain was a sparsely populated area due to flood and malaria hazards. In 1950s the land was given to the Dutch company, HVA, for establishment of a sugar estate and in 1954 sugar production was started [14].

Soil sampling

There are two common methods of studying changes in soil properties of farmlands under long-term cultivation: chronosequential sampling or Type I data and bio-sequential sampling or Type II data [15]. The former is used to monitor soil dynamics over time at the same site while in the latter case soils of cultivated and uncultivated lands are simultaneously sampled and compared. In this study, bio-sequential soil sampling method was used and the underlying assumption was that the soils from adjacent virgin and cultivated land were originally similar and that current differences in the soil physicochemical properties are due to cultivation [15].

In order to collect the required samples, five representative sites were randmly selected from the 7000 ha of Wonji plantation fields (Figure 2). The virgin lands were located near plantation villages of

the sugar estate. Each village has about one ha virgin land which was intended as entertainment place for the villages' residents since the commencement of sugar cane cultivation. Thus, the selected virgin lands were assumed to represent the original soil conditions of the site. Corresponding soil samples were also taken from the cultivated fields, near each of the selected virgin lands.

Method of soil analysis

The soil samples were analyzed in the research directorate of Ethiopian sugar development agency laboratory located in Wonji sugar estate.

Soil pH and EC were measured in a 1:2.5 soil water suspension by a glass electrode pH meter and EC meter, respectively. Total soil N was measured following Kjeldahl procedure which involved digestion of the samples in concentrated H_2SO_4 with a catalyst mixture to raise the boiling temperature and to promote the conversion from organic-N to ammonium-N. Ammonium-N from the digest was obtained by steam distillation, using excess NaOH to raise the pH. The distillate was collected in saturated H_3BO_3 and then titrated with diluted H_2SO_4 to pH 5.0 [16] Organic carbon was determined by Walkley-Black procedure which involves reduction of potassium dichromate by organic carbon compounds and subsequent determination of the unreduced dichromate by oxidation-reduction titration with ferrous ammonium sulphate. Finally, the amount of organic matter was determined according to the approximation: soil organic carbon × 1.72=SOM [17]. Available soil P was determined by sodium bicarbonate method where P was extracted with 0.5M sodium bicarbonate and measured calorimetrically [18]. Exchangeable K was determined by flame photometer after the samples were extracted with Morgan's solution [19].

Data analysis

The data were analyzed by paired samples t-test analyses at 1% probability level using Genstat software statistical packages, 12th edition. Mean comparisons were also performed.

Results and Discussion

SOM, total N, P Olsen, exchangeable K and soil EC of the cultivated land were much lower than the virgin land at both 0 cm-30 cm and 30 cm-60 cm soil depths ($P<0.01$). Nevertheless, no significant difference was observed in soil pH at both depths.

Soil Organic Matter (SOM)

The SOM content of the cultivated land was 53% and 34% lower than the virgin land at 0 cm-30 cm and 30 cm-60 cm depth, respectively (Figure 3). Thus, the SOM of the cultivated land was depleted considerably suggesting that long-term conventional cropping practices have degraded the soil. In agreement with the current result, several data from other long-term cropping systems trials also showed a decline in SOM and deterioration of soil quality under continuous cultivation as compared to native vegetation [20].

The observed differences between SOM contents of the cultivated and virgin land in Wonji sugar cane plantation at 0-30 cm depth was comparable with the differences observed in Philippines and Papua New Guinea. Alaban et al. [21] in Philippines and Hartemink et al. [22] in Papua New Guinea found the differences of 26% and 42% SOM, respectively, over a 20 years period of sugar cane cultivation. In China, Liu et al. reported that during 5, 14 and 50 year cultivation periods soil organic carbon losses were 17%, 28% and 55%, respectively. Contrastingly, Naranjo et al. [23] in Mexico and Bramley et al. [10] in

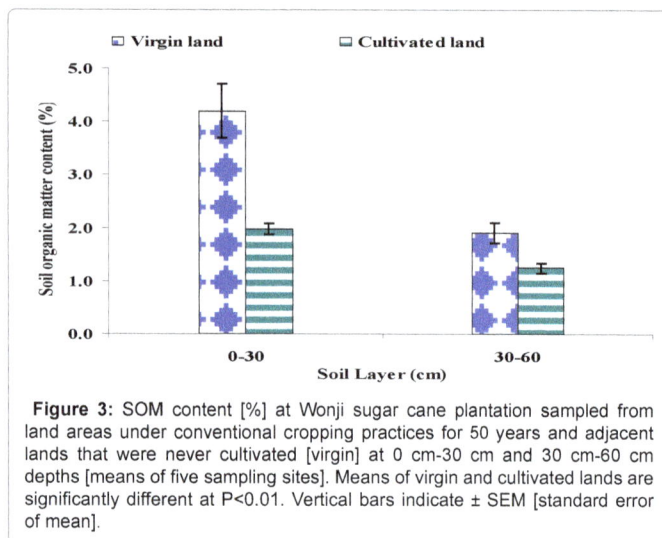

Figure 3: SOM content [%] at Wonji sugar cane plantation sampled from land areas under conventional cropping practices for 50 years and adjacent lands that were never cultivated [virgin] at 0 cm-30 cm and 30 cm-60 cm depths [means of five sampling sites]. Means of virgin and cultivated lands are significantly different at P<0.01. Vertical bars indicate ± SEM [standard error of mean].

northern Australia found non-significant differences in SOM contents after 30 and 20 years of sugar cane cultivation, respectively. This could be explained in the latter case by adoption of recommended practices which have the potential to sustain or increase SOM contents and sugar cane yield [19,23]. These practices mainly include integrated management of nutrients and sugar cane trash retention.

The absence of appropriate practices in the Wonji sugar estate played in all probability a significant role for the substantial differences observed in SOM contents of the tow soil types. For instance due to pre-harvest cane burning, about 14 ton/ha organic matter is lost up on harvesting 120 ton/ha sugar cane [24]. Additionally, excessive tillage results in 17% decline in SOM content within four months. Similar study in Australia also indicated that excessive tillage, insufficient fallowing and burning of crop residues are the major reasons for SOM decline during long-term conventional sugar cane cropping practices [4]. Moreover, the geographical location of Wonji plantation (8°31'N; near equator) might also have contributed to the observed differences. As the area is associated with high temperatures and humidity, the dynamics of SOM could be much higher than in countries like Australia (29°S-46.5°S) and Mexico (23°N) where the temperature is relatively mild. This is because of, as Sanchez and Logan [25] indicated, in tropical countries SOM decomposition rates could be up to five times higher than in temperate regions [26,27].

The main macronutrients

Total N, P Olsen and exchangeable K contents of the cultivated land were 56%, 84% and 81% lower than the virgin land at 0 cm-30 cm depth (Figure 4). The results suggest that there was depletion in the main macronutrients. Unless supplied as organic or inorganic fertilizers, the depletion of these macronutrients can cause substantial yield losses [8].

This observation was in agreement with those of Wood [28] Van Antwerpen and Meyer [2] Hartemink and Wood [15] and Hartemink [27] who reported that uncultivated land had higher levels of total N, available P and exchangeable K than cultivated land. The extent of degradation observed in our case was, however, much higher as compared to other countries. In Mexican fluvisols, after 30 years of sugar cane monoculture the N, P, and K contents of cultivated land was 14%, 43% and-5% lower than uncultivated land, respectively [23].

The significantly lower N, P and K contents of the cultivated soil than the virgin land (Figure 4) could be partly attributed to high nutrient removal by the cane itself and the lack of replenishment [27]. The practice of pre-harvest cane burning might also have played a role as it can cause 70%-95% loss of the dry matter and N from the system [28]. The finding of Alemayehu [24] also indicated that in Wonji plantation up on harvesting 120 ton/ha sugar cane about 66 kg N/ha is lost due to burning. Moreover, the continued decomposition of mineralizeable soil C and N, which is favored by the warm humid climate of the area, excessive tillage and leaching of cations could also be possible causes for the observed differences [29].

At 30 cm-60 cm depth, total N, P Olsen and exchangeable K contents of the cultivated land were 28%, 60% and 81% lower than the virgin land, respectively (Figure 4). The differences were more pronounced at the soil depth of 0 cm-30 cm than 30 cm-60 cm, because sugar cane is shallow rooted and removes little nutrients from deeper soil horizons [30]. Additionally, the extents of the differences between cultivated and virgin lands were much severe in exchangeable K at both depths (81%) and available P at 0 cm-30 cm depth (84%). This might

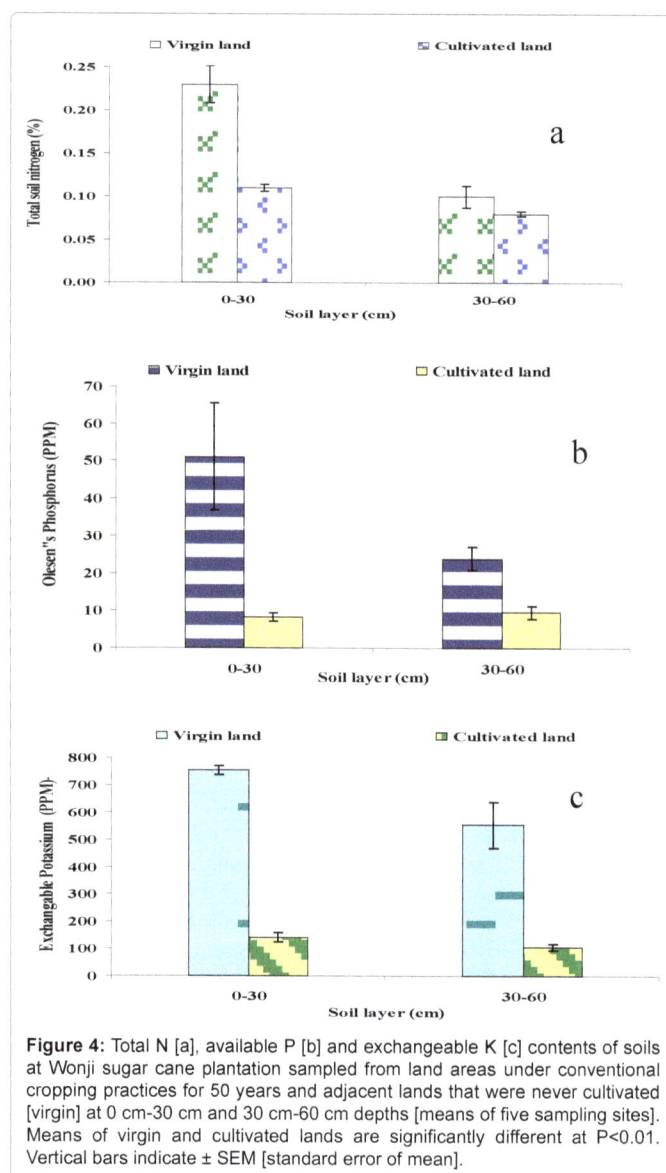

Figure 4: Total N [a], available P [b] and exchangeable K [c] contents of soils at Wonji sugar cane plantation sampled from land areas under conventional cropping practices for 50 years and adjacent lands that were never cultivated [virgin] at 0 cm-30 cm and 30 cm-60 cm depths [means of five sampling sites]. Means of virgin and cultivated lands are significantly different at P<0.01. Vertical bars indicate ± SEM [standard error of mean].

be associated with the fact that no K and P fertilizer applications have been practiced since the inception of Wonji plantation. However, the implication of the depletions of these nutrients for the observed yield decline might be of a minor role. This is because of the fact that the application of N fertilizer in Wonji increases the P and K availability in the soil [24]. Moreover the current levels of both P Olsen (8.22 ppm) and exchangeable K (142 ppm) are not in the range of deficiency. According to Landon [31] sugar cane is among the moderate P demanding crops where deficiency occurs at less than 7 ppm. For K, the critical value ranges from 78 ppm-125 ppm [32]. The values obtained in this study were above these deficiency levels. This is probably because of the cane burning practices that return P and K back to the soil and thus mitigated exhaustion of these nutrients below critical levels [33].

Soil EC

The soil EC of the cultivated land was 59% and 66% lower than the virgin land at 0-30 cm and 30-60 cm soil depths, respectively (Figure 5). This implies that the long-term conventional cropping practices resulted in reduction of the soil salt contents. The results were unexpected as the adopted long-term irrigation practices and accumulation of ashes from cane burning would result in development of salinity [33,34]. The most likely reason for the lower EC of the cultivated land than the virgin land was the good quality of irrigation water used in the plantation [35], which might leach down the salts. The torrential rainfall which used to occur during a summer season might also contribute to the leaching of salts. Thus, the current level of EC (0.51 mScm⁻¹) in Wonji plantation is in the category of optimum range [36].

The soil EC of both the cultivated and virgin land at 30 cm-60 cm depth were much higher than at 0 cm-30 cm depth. This might be attributed to leaching of salts from the top layer of the soil to the lower soil horizon due to the excessive irrigation [37] and rainfall.

Soil pH

The pH values of the virgin and the cultivated lands were not significantly different at both depths (Figure 6). The pH value was expected to decrease, as acid forming fertilizers have been extensively applied in the plantation (96-322 kg urea-N/ha) during the last 50 years. Hartemink and Wood [15] also stated that acid input as ammonium-N fertilizers and alkali removal as uptake of ammonium-N by the plant can result in acidification of the soil. Moreover, considerable removal of bases with the harvested sugar cane and the leaching of cations can also play a role in reduction of soil pH [38,39].

Unlike Wonji plantation, there was a decline in the soil pH in other sugar cane producing countries. For instance in Papua New Guinea, Hartemink et al. [22] reported an 11% decline during 18 years sugar cane production while in Australia, Moody and Aitken [38] reported 18% drop during 15 years cultivation. Masilaca et al. [40] also observed 14.4% drops in soil pH within 5 years production in Fiji. The invariable pH observed in Wonji plantation may be ascribed to the method of irrigation and the type of the soil in the study area. In the plantation, the major method of irrigation is blocked end furrow system. As the soil type of the plantation is heavy clay, this type of irrigation often resulted in severe water logging problem. According to Sun [41] water logging can increase the pH level of a soil through fast depletion of O_2 that leads to anaerobic conditions with concomitant reduction in Eh (redox potential). Reduction reactions use mainly H^+ and thus result in rises in pH. On the contrary, these scenarios might not have occurred in the New Guinea, Australia and Fiji studies, cited before, where the sugar cane is mainly rainfed.

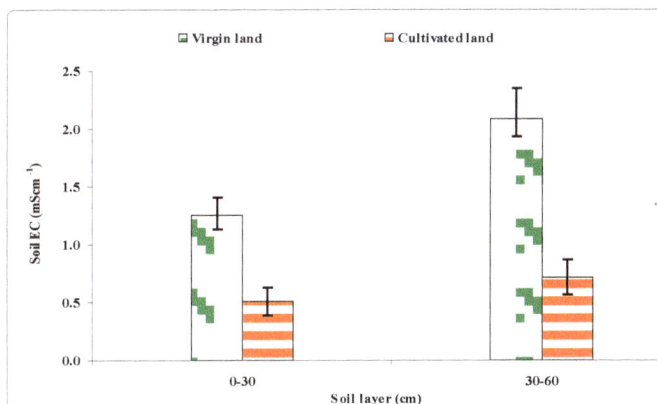

Figure 5: EC [mScm⁻¹] of soils at Wonji sugar cane plantation sampled from land areas under conventional cropping practices for 50 years and adjacent lands that were never cultivated [virgin] at 0 cm-30 cm and 30 cm-60 cm depths [means of five sampling sites]. Means of virgin and cultivated lands are significantly different at P<0.01. Vertical bars indicate ± SEM [standard error of mean].

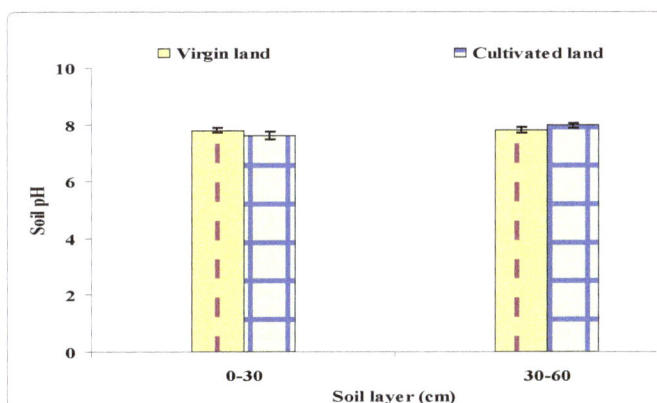

Figure 6: pH of soils at Wonji sugar cane plantation sampled from land areas under conventional cropping practices for 50 years and adjacent lands that were never cultivated [virgin] at 0 cm-30 cm and 30 cm-60 cm depths [means of five sampling sites]. Means of virgin and cultivated lands are not significantly different at P<0.01. Vertical bars indicate ± SEM [standard error of mean].

Conclusion

The long-term conventional copping practices substantially degraded the major soil quality indicators of Wonji plantation i.e., SOM, total N, available P and exchangeable K. Nevertheless, the soil pH and EC were in the optimum range that there are no problems related with soil salinity, alkalinity and acidity. As the soil of Wonji plantation is mainly heavy clay, particularly, the decline in SOM content along with the furrow irrigation system and the excessive traffic of machineries in the sugarcane plantation fields might play a role for the observed yield decline. This could be through exacerbating the detrimental effects of water logging and soil compaction. Thus, improving the SOM content could be essential in partly arresting the declining yield. Moreover an extensive multidisciplinary campaign should be further continued in order to fully identify, understand and manage the problem of soil quality deterioration.

Acknowledgements

I would like to thank Ethiopian Sugar Development Agency for covering all the expenses of this research work. My heartfelt thanks is also forwarded to Netherland fellowship program for grating me a fund to attend my study in the Netherlands. Above all praise be to God for all his assistances.

References

1. Solomon S, Grewal SS, Rui Li Y, Magarey RC, Rao GP (2006) Sugar cane: Production management and agro-industrial imperatives. International Book Distribution Co, India.

2. Van Antwerpen R, Meyer JH (1996) Soil degradation under sugarcane cultivation in northern KwaZulu-Natal. Proceedings of South African Sugar Technologists Association 70: 29-33.

3. Bell MJ, Garside AL, Halpin NV, Berthelsen JE (2001) Yield responses to breaking the sugarcane monoculture. In: Rowe B (eds.) Proceedings of 10th Agronomy Conference, Australian society of Agronomy.

4. Garside AL, Bell MJ, Robotham BG, Magarey RC, Stirling GR (2005) Managing yield decline in sugarcane cropping systems. International Sugar Journal 107: 16-26.

5. Fauconnier R (1993) Sugarcane, The tropical agriculturalist, Macmillan Press Ltd, London.

6. Verma RS (2004) Sugar cane production technology in India. Lucknow, International book distributing co.

7. De Wit CT (1992) Resource use efficiency in agriculture. Agricultural Systems 40: 125-151.

8. Schjønning P, Elmholt S, Christensen BT (2004) Managing soil quality: Challenges in modern agriculture. In: Schjønning P, Elmholt S, Christensen BT (eds.) Soil Quality Management–Concepts and Terms, CABI Publishing, Danish Institute of Agricultural Sciences, Department of Agroecology, Tjele, Denmark, pp: 1-12.

9. Lal R (2009) Soil degradation as a reason for inadequate human nutrition. Food Security 1: 45-57.

10. Bramley RV, Ellis N, Nable RO, Garside AL (1996) Changes in soil chemical properties under long-term sugar cane monoculture and their possible role in sugar yield decline. Australian Journal of Soil Research 34: 967-984.

11. Davies J (1998) The causes and consequences of cane burning in Fiji's sugar belt. The Journal of Pacific Studies, 22: 1-25.

12. Morris DR, Tai PY (2004) Water table effects on sugarcane root and shoot development. American Society of Sugar Cane Technologists 24: 41-59.

13. Pankhurst CE, Magarey RC, Stirling GR, Blair BL, Bell MJ, et al. (2003) Management practices to improve soil health and reduce the effects of detrimental soil biota associated with yield decline of sugarcane in Queensland, Australia. Soil and Tillage Research 72: 125-137.

14. Girma MM, Awulachew SB (2007) Irrigation practices in Ethiopia: Characteristics of selected irrigation schemes. Colombo, Sri Lanka: International Water Management Institute, p: 80.

15. Hartemink AE, Wood AW (1998) Sustainable land management in the tropics: the case of sugarcane plantations. Proceeding of the 16th International Congress of Soil Science, Montpellier.

16. Bremner JM, Mulvaney CS (1982) Nitrogen total. In: Page AL, Miller RH, Keeney DR (eds.) Methods of Soil Analysis. II. Chemical and Microbiological Properties. (2nd edn), pp: 595-624.

17. Walkley A, Black IA (1934) An examination of the Degtjareff method for determining organic carbon in soils: Effect of variations in digestion conditions and of inorganic soil constituents. Soil Sci 63: 251-263.

18. Olsen SR, Cole CV, Watanable FS, Dean LA (1954) Estimation of available phosphorous in soil by extraction with sodium bicarbonate. United States Agricultural Circ., USA.

19. Morgan MF (1941) Chemical diagnosis by the universal soil testing system. Connecticut Agricultural Experimental Station, Bulletin, p: 450.

20. Reeves DW (1997) The role of SOM in maintaining soil quality in continuous cropping systems. Soil Tillage Research 43: 131-167.

21. Alaban RA, Barredo FC, Aguirre AL (1990) An assessment of some indicators and determinants of farm productivity and soil fertility in the VMC district: trends, associations, interactions 1969-1990. Proceedings of the Philippine Sugar Technologists, pp: 64-83.

22. Hartemink AE, Nero J, Ngere O, Kuniata LS (1998) Changes in soil properties at Ramu Sugar Plantation 1979-1996. Papua New Guinea Journal of Agriculture, Forestry and Fisheries 41: 65-78.

23. Naranjo de L, Salgado-Garciá FS, Lagunes-Espinoza LC, Carrillo-Avila E, Palma-López DJ (2006) Changes in the properties of a Mexican Fluvisol following 30 years of sugarcane cultivation. Soil Tillage Research 88: 160-167.

24. Alemayehu D (2010) Sustainable Sugarcane Production in Ethiopia, Exploring Challenges and Opportunities. Biological Farming System Group. MSc Thesis. Wageningen University, The Netherlands.

25. Sanchez PA, Logan TJ (1992) Myths and science about the chemistry and fertility of soils in the tropics. In: Lal R, Sanchez PA (eds.) Myths and science of soil of the tropics. SSSA Spec. Publ. 29. SSSA, Madison, WI, USA, pp: 35-46.

26. Wood AW (1985) Soil degradation and management under intensive sugarcane cultivation in North Queensland. Soil Use and Management 1: 120-124.

27. Hartemink AE (2001) Sustainable land management at Ramu sugar: Assessment and requirements. In: Bourke RM, Allen MG, Salisbury JG (eds.) Food Security in Papua New Guinea, ACIAR Proceedings no. 99, Canberra, pp: 344-364.

28. Thorburn PJR, Van A, Meyer JH, Bezuidenhout CN (2002) The impact of trash management on soil carbon and N: I Modelling long-term experimental results in the South African sugar industry. Proceedings of the South African Sugar Technologists Association 76: 260-268.

29. Brown S, Lugo A (1990) Effects of forest clearing and succession on the carbon and N content of soils in Puerto Rico and US Virgin Islands. Plant Soil 124: 53-64.

30. Smith DM, Inman-Bamber NG, Thorburn PJ (2005) Growth and function of the sugarcane root system. Field Crops Research 92: 169-183.

31. Landon JR (1984) Booker Tropical Soil Manual: A hand book for soil survey and agricultural land evaluation in the tropics. Longman. New York, USA.

32. Orlando FJ (1989) Potassium nutrition of sugarcane. In: Munson RE (eds.) Potassium in Agriculture. American Society of Agronomy, Madison, Wisconsin, USA, pp: 1045-1076.

33. Khan MJ, Qasim M (2008) Integrated use of boiler ash as organic fertilizer and soil conditioner with NPK in calcareous soil. Songklanakarin Journal of Science and Technology 30: 281-289.

34. Smedema LK, Shiati K (2002) Irrigation and salinity: A perspective review of the salinity hazards of irrigation development in the arid zone. Irrigation and Drainage Systems 16: 161-174.

35. Girma A (2006) Evaluation of irrigation water quality in the Ethiopian sugar estates. Ethiopian sugar industries support Centre Co, Research report, Wonji.

36. Yuste MP, Gostincar J (1999) Handbook of agriculture. Marcel Dekker, New York, USA.

37. Habib D, Girma T (2006) Evaluation of irrigation interval and irritation efficiencies in the Ethiopian sugar estates. Research report. Wonji.

38. Moody PW, Aitken RL (1995) Soil acidification and lime use in some agricultural systems in Queensland. In: Date RA (eds.). Plant soil interactions at low pH, Kluwer Academic, Dordrecht, pp: 749-752.

39. Kahlown MA, Ashraf M, Zia-ul-Haq M (2005) Effect of shallow groundwater table on crop water requirements and crop yields. Agricultural Water Management 76: 24-35.

40. Masilaca AS, Prasad RA, Morrison RJ (1986) The impact of sugarcane cultivation on three Oxisols from Vanua Levu, Fiji. Tropical Agriculture 63: 325-330.

41. Sun L, Chen S, Chao L, Sun T (2007) Effects of flooding on changes in Eh, pH and speciation of cadmium and lead in contaminated soil. Bull Environ Contam Toxicol 79: 514-518.

22

Evaluation of *Jatropha curcas L.* Accessions from Shivamogga District, Karnataka for Oil Content and Germination

Rajeshwari N, Kavya Muduvala R*, Bhargavi MK and Ramesh Babu HN

Sahyadri Science College, Kuvempu University, Shivamogga, Karnataka, India

Abstract

Field survey to the *Jatropha* growing taluqs of Shivamogga was undertaken to study the climatic conditions prevailing in those areas followed by collection of seed accessions from surveyed areas. About 200-250 fruits were collected from 15 places representing seven taluqs of Shivamogga. Fruits were processed and about one kilogram of seeds were procured from each place which were labeled as accessions. Fifteen accessions were tested for germination and oil content after storing them in polyethylene bags for two months. Results revealed the hindrance of mycoflora on faster rate of germination and high oil content ranging from 32.9% to 41.9% from the seeds of Esuru having moderate rainfall and temperature with black soil.

Keywords: *Jatropha*; Oil content; Shivamogga; Storage; Germination

Introduction

Self-reliance in energy is a basic requirement for the economic development of any nation. India is not self sufficient in petroleum reserve and has to import about two third of its requirement. Every year India is importing 78% crude oil at a cost amounting Rs.100,000 crore and above. Annual consumption of diesel in the country is around 40 million tons at present. Fossil fuels not only contribute to pollution but it is a major source of Green House Gases. The need to search for alternative sources of energy, which are renewable, safe and nonpolluting is very much necessary. Under such circumstances, options available are to reduce the fossil fuel consumption and increase the energy efficiency, a number of tree-borne oil seeds crops are identified as an alternate to biodiesel. Biodiesel is meant to be used in standard diesel engines and is thus distinct from the vegetable and waste oils used to fuel converted diesel engines [1]. Biodiesel can be used alone or blended with petro-diesel in any proportions. It supports more than 60% jobs across the country. In many rural areas of the country biodiesel plants are the driving force of the local economy. Reducing our dependence on foreign oil. Improving air quality and environment. The growth of biodiesel industry is driving new technologies and feedstock development. Some of the biodiesel plants are neem, flax, hemp, jojoba, olives, Tung tree, peanut, sunflower, safflower, pongamia, oil palm, coconut, cotton, soybean, grand couronne, Jatropha, maize etc., In our country almost neem, pongamia and Jatropha are used. Of these plants, *Jatropha curcas* is one of the prospective biodiesel yielding crops [2]. It doesn't require pesticide due to its pesticidal and fungicidal properties. It lives approximately 40 years and yield starts from 9-12th month after planting. Jatropha may not replace other important food crops like cereals, pulses and oil crops. Since it is meant for flood free waste lands, unutilized fallow lands and less productive lands Simon [3]. It is identified as most suitable oil seed bearing plant due to its various favorable attributes like hardy nature, adaptability in a wide range agro-climatic condition, high oil recovery and quality of oil etc. Planning Commission, government of India has identified two species for mass production of seeds for biodiesel viz., Jatropha and Pongamia. Jatropha is suitable for upland while Pongamia found adaptive for both upland as well as wet land condition. Pongamia seed yield is less comparative to Jatropha.

Materials and Methods

Shivamogga is one of the districts of Karnataka state, India. A major part of Shivamogga district lies in the Malnad region of the Western Ghats. There are seven taluks i.e., Bhadravathi, Hosanagar, Sagar, Sorab, Thirthahalli, Shivamogga and Shikaripur. It is spread over an area of 8407 km^2. Shivamogga lies between the latitudes 13°27' and 14°39' N and between the longitudes 74°38' and 75°45' E at a mean altitude of 640 m above sea level. During our visit information regarding the soil type, habitat of plant and age of the plant were recorded. During our visit temperature and rainfall were recorded. Temperature was recorded by using regular thermometer. Rainfall was recorded by gathering information from meteorological department of Shivamogga. This was done mainly to identify the effect of temperature and rainfall on the economic product that is seed (Table 1). Field visits were made for the collection of seed samples from all the seven taluqs of Shivamogga districts including Shivamogga town. Fruits were collected (Figure 1) from uncultivated wild plants and processed manually for healthy seeds. About 1 kg seed from each location were procured and called it as accession. Further about 15 accessions representing seven taluqs with safe moisture level of 12-13% was stored in laboratory condition in polyethylene bags for two months [4] (Figure 2). Further for testing seed germination, ISTA [5] procedure was followed and for mycoflora studies Standard Blotter Method was followed (Figure 3). 50 seeds at the rate of 10 per plate were plated and five replicates for each sample were made. This method was followed for all accessions. The experimental setup was placed in incubation chamber for 8 days under near ultraviolet radiation for $^{12}/_{12}$ hrs, further the seeds were examined under stereobinocular microscope for the detection of type of fungi. Manual was utilized for further identification. In order to estimate the oil content of the samples from different taluqs, oil estimation was performed [6]. In order to do so, 25 seeds from each sample were taken separately and covered in butter paper, labeled and sent to Main Research Station, Raichur for oil estimation by Near Magnetic Resonance technique.

*Corresponding author: Kavya Muduvala Rudrachar, Sahyadri Science College, Kuvempu University, Shivamogga, Karnataka, India
E-mail: kavyakarthik.2012@gmail.com

S No	Taluq	Village	Temperature (°C)	Rainfall (mm)	Plant count as hedge plant
1.	Bhadravathi	• Bhadravathi	28	63	10
2.	Thirthahalli	• Guddekoppa	28	330	12
		• Guddekeri	27	330	10
3.	Soraba	• Byrekoppa	30	170	10
		• Bhadrapura	30	170	21
4.	Hosanagara	• Arasalu	28	400	08
		• Ripponpete	29	400	36
5.	Shikaripura	• Kattigehalla	27	82	14
		• Esuru	27	82.4	24
		• Shiralakoppa	27	82	04
6.	Shivamogga	• Choradi	26	73.6	14
		• Mandagatta	24	74	26
		• Gajanur	27	70	27
		• Shivamogga city	26	73	16
		• Bioenergy park	26	73	40
7.	Sagar	• Chippali	31	141	30
		• Chikkanallur	33	141	26
		• Settisara	33	141	-

Table 1: Field survey undertaken to different taluqs of Shivamogga.

Figure 1: Fruits collected during field survey.

Figure 3: Mycoflora studies by Standard Blotter Method.

Figure 2: Properly packed seeds for storage.

Results and Discussion

Field Survey was undertaken to a total of 7 taluqs of Shivamogga district with about 18 villages were personally visited to observe the plants, their stage, temperature and rainfall in that particular area. In Shivamogga 5 villages, in Shikaripura and Sagara 3 villages each likewise in Soraba, Hosanagara and Thirthahalli, 2 villages were visited. However we could visit only one village in Bhadravathi. Therefore a total of 18 locations were visited to study the impact of soil, temperature and rainfall (environmental factors) on the yield of the crop. They were also recorded in each visited field. During the study the temperature recorded ranged from 24-33°C where Mandagatta showed a lower temperature of 24°C and Chikkanallur, Settisera showed a temperature of 33°C. However in other fields, it was between the given range. Rainfall ranged from 63 mm-400 mm/yr. except the Bioenergy park, Shivamogga, almost all visited places were with Jatropha as a hedge plant. However in Bioenergy park, the plants are grown as in 2 acres of field. Other than this, the number of plants ranged between 4-40. However in Settisera there was no record of Jatropha plants.

Experiment revealed germination percentage ranged from 60 to 96. Kattigehalla got the least percent i.e., 60 and Guddekoppa, Esuru, Arasalu got the highest percent of 96, 96 and 90 respectively. Since the seeds of Chippali, Chikkanallur were not matured, seeds didn't show germination. Mycoflora ranged between 16% to 49%. Bhadravathi sample showed the least percent mycoflora i.e., 16% and Chordi showed the highest percent of 49. Major fungi found in the samples were *Gonatobotrys simplex corda* with 64%, *Alternaria alternata, Rhizopus stolanifer, Fusarium moniliforme, Aspergillus niger* and *Cladosporium sphaerospermum* ranges from 18-35% [7]. The percentage of each fungi found in all the different places is shown. As per the result, Esuru, Guddekoppa and Arasalu got the highest percent germination of 96 and 90 as we mentioned earlier, mycoflora is below 43%, temperature varies from 27-28°C and the soil type is found to be red gravel [8]. But Bhadravathi accession showed 86% germination, percent mycoflora is 16%, temperature recorded was 28°C with an annual rainfall 63 mm. Soil type is black sandy. Therefore Bhadravathi seed sample is found to be healthy compared to Guddekoppa, Esuru and Arasalu. Whereas the other villages showed below the expected percentage. Based on the healthy seeds and the percent germination, Bhadravathi got the first place and the order can be given to other taluqs is already shown. The mycoflora studied in different taluqs showed varied results. Important mycoflora recorded in almost all samples were found to be *Alternaria alternata, Rhizopus stolanifer* and *Fusarium moniliforme*. However, the dominating species immaterial with the location was *Gonatobotrys simplex corda*. In Sagar taluq, there is no average % mycoflora and germination because the seeds were not matured may be due to the high temperature of above 31°C and annual rainfall 141 mm or the time we have visited to that place.

Oil estimation was performed with the help of Main Research Station, Raichur by Near Magnetic Resonance technique. Literature says that the oil content ranges from 30-40% in Jatropha seeds [9]. However our results revealed the oil content ranged from 32.9-41.9%.

Seeds from Shivamogga area showed lower oil content where as seeds of Esuru showed higher oil content that is about 41.9%. Where it is observed for there was moderate rainfall with moderate temperature which favours the oil content. However the Shivamogga has also showed more or less similar climatic conditions. However the soil of Shivamogga is red soil where as the Esuru with black soil (Table 2).

Conclusion

In viability test, the percentage of germination was calculated by Standard Blotter Method and the same method was followed to study mycoflora. It was studied that the faster rate of germination in all samples. As they were freshly harvested, we could get good rate of germination. The percentage of mycoflora obtained was different in different samples. When we compared the impact of mycoflora on germination, it was interesting to see that if the rate of germination was more, the rate of mycoflora was less (Figure 4). Clearly indicating that the viable seeds avoid mycoflora as there is a continuous physiological process which hinders the growth of mycoflora. As per the literature, the oil content of seeds was 30-40%. However our studies revealed the oil content range from 32.9 to 41.9%. Seeds of Esuru i.e., Shikaripura taluq showed higher percentage of oil content. The environmental factor favoured there may be moderate rainfall with moderate temperature along with black soil. Our results conclude that the seeds from Esuru are good varieties with more oil content and they are very good research material for further study. According to our studies, seeds collected from Esuru showed maximum oil content compared to other accessions. Good germination followed by less growth of mycoflora. It clearly indicates that the seeds of Esuru are the good source of germplasm for further studies. Still there is a need for consistent studies regarding the collection, storage and evaluation of seeds during other seasons.

S No	Taluq	Village	% of mycoflora	% of germination	% of oil content
1.	Bhadravathi	• Bhadravathi	15.8	86	41.8
2.	Thirthahalli	• Guddekoppa	31.5	96	39.8
		• Guddekeri	27	86	40.9
3.	Soraba	• Byrekoppa	31.6	88	41.5
		• Bhadrapura	37.6	76	35.7
4.	Hosanagar	• Arasalu	30.5	90	40.4
		• Ripponpete	34.8	62	37.6
5.	Shikaripura	• Kattigehalla	27.3	60	36.7
		• Esuru	42.8	96	41.9
		• Shiralakoppa	39.6	78	38.9
6.	Shivamogga	• Choradi	49	88	37.3
		• Mandagatta	30	76	40.8
		• Gajanur	41.6	84	41.4
		• Shivamogga city	37.6	68	32.9
		• Bio energy park	23	68	37.9
7.	Sagar	• Chippali	-	-	-
		• Chikkanallur	-	-	-
		• Settisara	-	-	-

Table 2: Table showing the comparison between the surveyed taluq for viability, mycoflora and oil content.

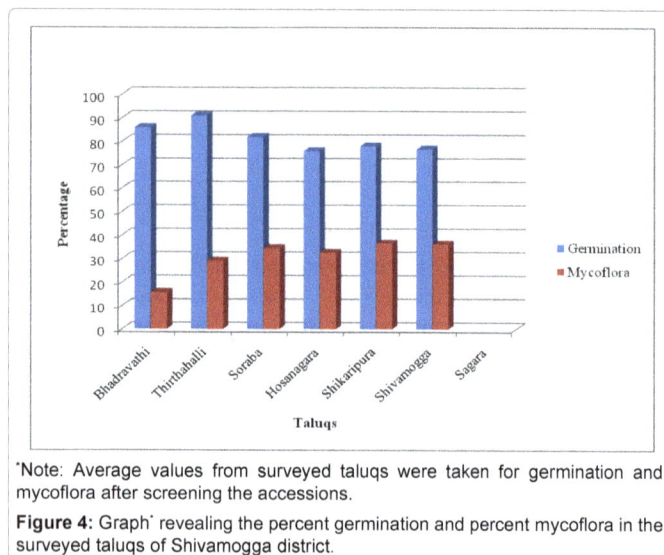

*Note: Average values from surveyed taluqs were taken for germination and mycoflora after screening the accessions.

Figure 4: Graph* revealing the percent germination and percent mycoflora in the surveyed taluqs of Shivamogga district.

Acknowledgements

The authors are grateful for the UGC CPE Project, Sahyadri Science College (A), Shivamogga for funding to carry out this bonafide work.

References

1. Makkar HPS, Becker K (2009) Jatropha curcas a promising crop for the generation of biodiesel and value-added co-products. Eur J Lipid Sci Technol 111: 773-787.

2. Becker K, Makkar HPS (2008) Jatropha curcas: a potential source for tomorrow's oil and biodiesel. Lipid Technology 20: 104-107.

3. Simon G, Reena S, Stephan P, Alok A, Rainer Z (2012) Environmental impacts of Jatropha curcas biodiesel in India. Journal of Biomedicine and Biotechnology, p:10.

4. Kebede W, Yidnekachew H (2014) Effect of seed storage period and condition on viability of Jatropha curcas L. Seed. Research Journal of Forestry 8: 56-63.

5. ISTA (1999) International rules for seed testing. Seed science and Technology 21: 288.

6. Dereje A, Kindie T, Girma M, Birru Y, Wondimu B (2012) The traits oil content and correlation studies of seed and kernel in Jatropha curcas L. African journal of agricultural research 7: 1487-1491.

7. Barnett HL (1960) Illustrated genera of imperfect fungi. Burgess Publishing Company, p: 255.

8. Sanjay K, Satyendra DS, Prakash KG, Shigeru K, Toshinori K (2011) Effect of climate and soil conditions on oil content of Jatropha plants grown in arid areas of India. Journal of Arid Land Studies 21: 51-55.

9. Gairola KC, Nautiyal AR, Dwivedi AK (2011) Effect of temperature and germination media on seed germination of Jatropha curcas L. Adv Biores 2: 66-71.

23

Management of Faba Bean Gall Disease (Kormid) in North Shewa Highlands, Ethiopia

Bitew B* and Tigabie A

Debre Birhan Agricultural Research Center, PO Box 112, Debre Birhan, Ethiopia

Abstract

Production of faba bean is inhibited by several yield limiting factors, among which diseases are the main. In Ethiopia more than 17 disease causing pathogens were reported on faba bean. Major diseases recoded in faba bean includes, chocolate spot (Botrytis fabae), rust (Uromyces viciae-fabae), ascochyta blight (Ascochyta fabae), zonate leaf spot (Cercospora zonatae), and black root rot (Fusarium sp.). A new disease, faba bean gall locally called "Kormid in North Shewa was expanded in the highland faba bean growing areas. Studies showed that seed dressing and foliar fungicides have some effects against faba bean diseases. Field experiment was conducted in North shewa highlands to control faba bean gall disease at farmers' field. The experiment was conducted on farmers' fields in randomized complete block design in six replications. The treatments were arranged with different fungicides (spray and seed dressing), namely Mancozeb, Ridomil, Chlorotalonil, Bayleton wp 25 (Triadimefon 250 g/kg), Thiram, Apron star and control. Fungicides were applied as manufacturers' recommendations. Foliar fungicides were applied three times at seedling, flowering and podding growth stage on local faba bean variety. Disease score and other agronomic data were recorded at different plant growth stage. The highest disease score were recorded in control, Thiram and Apron star in 2013. The highest yield was also recorded in Bayleton and mancozeb sprayed plots respectively. In 2014 the disease prevalence and severity was similar to 2013. Maximum disease score were recorded on Control followed by Thiram and Apron star seed dressing plots. Minimum disease score were also recorded in Baylaton (2.66) and Ridomil gold (2.71) sprayed plots. There was a significant difference between biomass yield and grain yield. The highest grain yield was recorded in Baylaton (3129.8 kg) sprayed plot and followed by Ridomil gold (2708.3 kg) and Mancozeb (2705.7 kg) respectively. There was no significant difference between plots in plant height, pod per plant, seed per pot and seed per plant in both years.

Keywords: Disease; Faba bean; Faba bean gall; Incidence; Prevalence; Severity

Introduction

Ethiopia is the world's second largest producer of faba bean, but its share is only 6.96% of world production and 40.5% of Africa [1]. Faba bean (*Vicia fabae* L.) is the major cool season food legumes produced in Ethiopia next to cereals. It serves as major source of protein and income. The crop also fixes atmospheric nitrogen and improves soil fertility. Because of its wide importance to the nation it is cultivated in large area in the country as well as in Amhara region. Production of faba bean is inhibited by several yield limiting factors, among which diseases are the main [2]. In Ethiopia more than 17 disease causing pathogens are reported on faba bean [3]. Major diseases recoded in faba bean includes, chocolate spot (*Botrytis fabae*), rust (*Uromyces viciae-fabae*), ascochyta blight (*Ascochyta fabae*), zonate leaf spot (*Cercospora zonatae*), and black root rot (*Fusarium sp*).

Currently a new disease faba bean gall *Olpidium viciae* locally called "Kormid in North Shewa is expanded in the highland faba bean growing areas of North. It was first observed in Menze Mama District around Bash kebele in farmers' fields in 2010/2011 main cropping season [4]. Seed dressing and foliar fungicides have some effects against faba bean diseases [5,6]. Report reveals that, chemical control showed better results in controlling gall disease in China and Japan [7]. Hence this study was initiated with the following objectives: to select the right and effective fungicides on gall disease (kormid) on faba bean.

Materials and Methods

The experiment was conducted on farmers' fields in RCB design with six replication (one farmer field was used as one replication in 2013 and 2014 main growing season). The experiment was done with foliar and seed dressing fungicides. The treatments were arranged with different fungicides (spray and seed dressing), namely (a) Mancozeb 80% wp (contact fungicide with preventive activity. It inhibits enzyme activity in fungi by forming a complex with metal-containing enzymes including those involved in production of adenosine triphosphate), (b) Ridomil (Metalaxyl-M 4% +Mancozeb.64%), (c) Chlorotalonil, Bayleton wp 25 (Triadimefon 250 g/kg), (d) Thiram, (e) Apron star (seed treatment fungicide-insecticide mixture and its active ingredient is Thiamethoxam:200 g/kg, Mefenoxam: 200 g/kg , Difenoconazole: 20 g/kg) for controlling seed and soil born disease) & (f) control. The plot size was 3.2 m × 4 m and spacing between rows 0.4 m and 1 m between replications. Fungicides were applied as manufacturers' recommendations. Foliar fungicides were applied three times (at the time of diseases appearance (seedling) and repeated two times before start of flowering and podding stage). Local faba bean variety was used and seed rate was applied as recommendation in row planting.

Data collected

Date of seedling emergence, first date of bean gall disease appearance, faba bean gall score (1-9) scale and; 1 means no or few symptoms and nine means dead plant). Faba bean gall disease recorded

*Corresponding author: Beyene Bitew, Debre Birhan Agricultural Research Center, PO Box 112, Debre Birhan, Ethiopia, E-mail: beyenebitew@yahoo.com

was converted to percent incidence and severity. Plant height, Number of pod per 10 plant, number of seed per pod, thousand seed weight, biomass and seed yield (grain yield) were recorded.

Data analysis

Analyses of variances for the experiment was done and mean comparisons were carried out using Duncan's multiple range test (DMRT) at 5% level of probability. The statistical analysis system (SAS) software [8] was used for all statistical analyses.

Results and Discussion

Foliar fungicides were applied at the beginning of symptom appearance. All foliar fungicides were applied at seedling, flowering and podding growth stage. The disease was very serious in first year (2013). But due to higher rainfall, there was high erosion problem

and hail damage at seedling stage. The symptom of the disease starts from seedling stage and more severe up to flowering growth stage. In severely infected fields, the disease expands to the stem and the whole plant showed shrinked, shortened and died on control plots. The lowest disease score were recorded on Bayleton, chlorotalonil and mancozeb sprayed plots (Table 1). Among fungicides sprayed better grain yield were recorded on bayleton (2124.0 kg), Mancozeb (1702.3 kg), Ridomil gold (1471.7 kg) and chlorothalonil (1470.9 kg) respectively. Apron star and Thiram seed dressing fungicides were not effective against the disease. It was similar with control plots (Table 1).

Values within a column followed by same letter do not differ significantly at 5% level of Duncans multiple range test

In 2014 the disease prevalence and severity was similar to 2013. Disease score and other agronomic data were recorded at different

No.	Treatments	Disease score (1-9)	Ph (cm)	Pod/pl	Sdp (gm)	Hsw (gm)	Gy (kg/ha)
1	Apronstar	4.5ab	37.0ab	8.53b	2.17b	38.86b	811.8bc
2	Baylaton	1.66c	54.8a	19.4a	2.68a	40.13ab	2124.0a
3	Chlorotalonil	2.83c	46.53ab	19.46a	2.55ab	39.83ab	1470.9abc
4	Mancozeb	2.83c	51.93ab	16.66a	2.35ab	42.73a	1702.3ab
5	Ridomil	3.16bc	44.13ab	11.93ab	2.55ab	39.83ab	1471.7abc
6	Thiram	5.16a	34.06b	8.26b	2.50ab	37.66b	1317.4abc
7	Control	5.66a	34.4b	5.73b	2.42ab	39.03b	554.1c
	Mean	3.69	43.26	12.85	2.46	39.72	1351.97
	Cv (%)	23.87	23.64	34.17	8.87	5.04	40.49

Ph=plant hieht, pod/pl=pod per plant, sdp=seed per pod, Hsw=hundred seed weight and Gy=grain yield

Table 1: Disease score and yield of faba bean in North Shewa 2013.

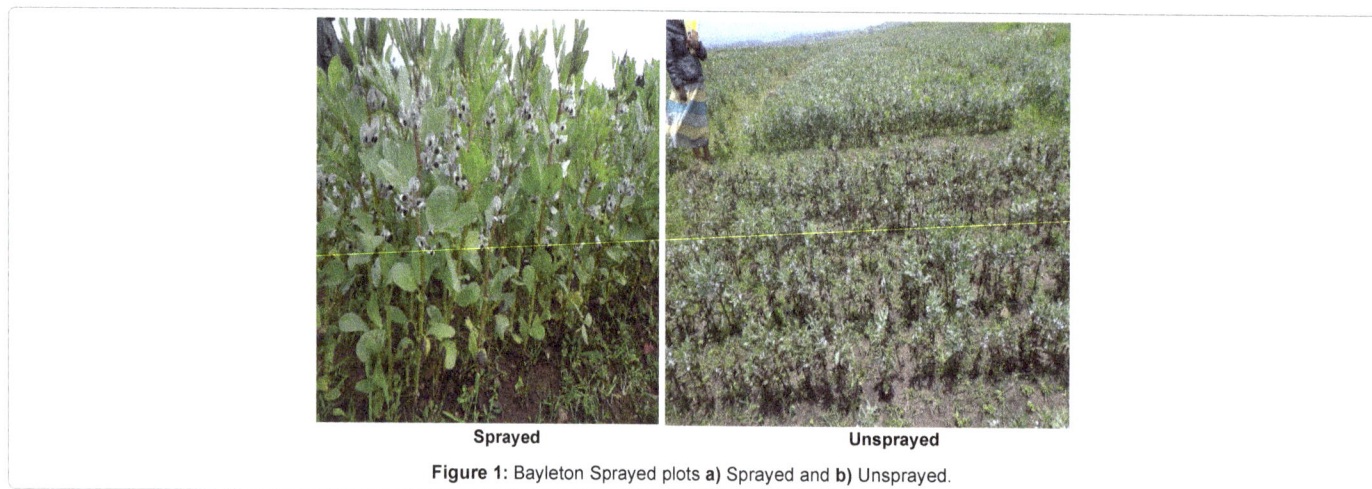

Sprayed Unsprayed

Figure 1: Bayleton Sprayed plots **a)** Sprayed and **b)** Unsprayed.

No.	Treatments	Disease score(1-9)	Ph (cm)	Pod/pl	Seed/pl	Sdp (gm)	Hsw (gm)	Bm (kg/ha)	Gy (kg/ha)
1	Apronstar	4.04b	82.13a	15.56a	41.5a	2.63a	32.28ab	4067.7ab	2247.0abc
2	Baylaton	2.66c	82.86a	14.23a	35.5a	2.48a	32.60a	5446.6a	3129.8a
3	Chlorotalonil	2.91c	79.33ab	15.90a	39.70a	2.42a	29.68bc	5080.7a	2632.9ab
4	Mancozeb	2.91c	74.96ab	14.20a	36.43a	2.68a	31.40abc	5445.7a	2705.7ab
5	Ridomil	2.70c	82.43a	15.13a	38.8a	2.59a	31.61abc	5542.2a	2708.3ab
6	Thiram	6.00a	68.66b	12.80a	32.86a	2.50a	29.46c	2679.1b	1815.16c
7	Control	6.08a	69.70b	12.93a	31.43a	2.39a	31.73abc	2539.5b	1369.2c
	Mean	3.90	77.15	14.39	36.60	2.53	31.39	4400.21	2372.58
	Cv (%)	19.75	13.06	26.83	30.19	9.90	7.27	35.65	37.71

Ph=plant hieht, pod/pl=pod per plant, sdp=seed per pod, seed/pl=seed per plant, Hsw=hundred seed weight, Bmy=biomass yield and Gy=grain yield

Table 2: Disease score and yield of faba bean in North Shewa 2014.

plant growth stage. Maximum disease score were recorded on Control followed by Thiram and Apron star seed dressing plots. Minimum disease score were also recorded in Baylaton (2.66) and Ridomil gold (2.71) sprayed plots. There was a significant difference between biomass yield and grain yield. The highest grain yield was recorded in Baylaton (3129.8 kg) (Figure 1) treated plot and followed by Ridomil gold (2708.3 kg) and Mancozeb (2705.7 kg) respectively (Table 2). There was no significant difference between plots in plant height, pod per plant, seed per pot and seed per plant. Chemical control: treat seeds with fungicides. Studies showed that Thiram, at a dosage of 0.6-1.0 kg/100 kg of seed, 25% Bayleton or 15% Bayleton, at a rate of 0.3% to seed weight, are effective in controlling galls (ICARDA), but thiram was not effective against the disease.

Values within a column followed by same letter do not differ significantly at 5% level of Duncans multiple range test, economic analysis of management options, cost benefit analysis, related terms in cost benefit analysis

Total variable costs: the costs of chemicals, fertilizer and labor.

Gross yield: the total output per hectare of the produce of grain and straw.

Adjusted yield: The difference that gross yield is reduced by 10% from the actual due to risk and uncertainty.

Gross benefit: is the product of output and farm get prices of the produce.

Net benefit: is the difference between the gross benefit and the costs of production that vary.

Rate of return (RR): is the rate net benefit to cost of production or benefit cost ratio (BCR).

Information generated from cost benefit analysis is the most important factor to take into account when making correct decision and lack thereof will inevitably lead to suboptimal allocation of limited resources. The only way of determining how the resources allocated for the production of crops for small holder farmers should be put to use based on evaluating the costs and benefits of all (possible) ventures and selecting treatments having the highest return. Cost benefit analysis can be calculated using partial budget analysis method. Partial budgeting is a planning and decision-making framework used to compare the costs and benefits of alternatives faced by a farm business. Gross yield was adjusted from the output obtained lowered by 10% from the actual yield due to management and other production risks (Table 3).

Partial budget analysis

Partial budget analysis is concerned with evaluating the consequences of changes in treatments that affect only parts than whole. It is budgeting in relation to a partial change to a given farm inputs/budgets. For the evaluation of each treatment only variable costs were included and fixed costs were excluded. The net benefit or farm profit was calculated as net benefit is equal to gross benefit reduced by the costs of inputs and labor. The result was obtained from the farm get price at immediate harvest of grain and straw with 6 and 1.65 Birr per kg, respectively. Labor cost is related to chemical applications and fertilizer cost is the costs incurred for the amount of fertilizer applied based on recommendation. The analysis considers only the treatments that are most candidates having the market prices for full cost information. The result indicated that the treatments that have high net benefit are recommended for future technology packages. From the evaluation bayleton, mancozeb and redomil gold were the most competitive treatment evaluated against the control (Table 4).

Sensitivity analysis

Sensitivity analysis is important to evaluate the impact of such changes on economic parameters on the net returns for each treatment studied. Farm budgets may not expect positive net profits as a result of unexpected changes in yield, market prices or production costs. Those can quickly turn the expected benefit into a loss. Analyzing how changes in key budgeting assumptions/components affect income and cost projections is called sensitivity analysis. One way of trying to handle the problems of applying correct weights to risk is checking the outcomes using sensitivity analysis for future production guarantee. Sensitivity analysis allows the producer to have such information to control probabilities of calculating risks. Such changes are evaluated by creating future assumptions that are more or less dubious "scenarios". Hence, sensitivity analysis was carried out to assess the

Treatments	Treatment cost Birr/ha (ETB)	Gross grin yield/ha	Adjusted yield/ha	Gross biomass yield kg/ha	Adjusted yield kg/ha
Apronstar	980	2247.0	2022.3	4067.7	3660.93
Baylaton	420	3129.8	2816.82	5446.6	4901.94
Chlorotalonil	-	2632.9	2369.61	5080.7	4572.63
Mancozeb	240	2705.7	2435.13	5445.7	4901.13
Ridomil	1575	2708.3	2437.47	5542.2	4987.98
Thiram	-	1815.1	1633.59	2679.1	2411.19
Control	0	1369.2	1232.28	2539.5	2285.55
Mean	643	2372.57	2135.31	4400.21	3960.19

Table 3: Chemical costs and yields of faba bean for treatments.

Treatments	Total Costs that vary Birr/ha (ETB)	Adjusted Yield of grain KG/ha	Adjusted yield of Bio Mass kg/ha	Gross benefit	Net befit
Apronstar	2680	2022.3	3660.93	18235.35	15555.35
Baylaton	2120	2816.82	4901.94	25070.82	22950.82
Chlorotalonil**	1700	2369.61	4572.63	21838.71	20138.71
Mancozeb	1940	2435.13	4901.13	22779.33	20839.33
Ridomil	3275	2437.47	4987.98	22938.12	19663.12
Thiram**	1700	1633.59	2411.19	13820.19	12120.19
Control	1400	1232.28	2285.55	11202.93	9802.93

**The chemicals are not found in the market and the cost of chemical didn't included in the calculation due lack of information.

Table 4: Cost benefits analysis of treatments.

Treatments	Gross benefits and cost of production			Net benefits in different Scenarios		
	1	2	3	1	2	3
Baylaton	22490.21	2332	22490.21	20370.21	22738.82	20158.21
Mancozeb	20427.88	2134	20427.88	18487.88	20645.33	18293.88
Ridomil	20569.49	3602.5	20569.49	17294.49	19335.62	16966.99
Control	10048.35	1540	10048.35	8648.354	9662.93	8508.354

Table 5: Sensitivity analysis result for the best treatments against the changes assumed compared with the check.

Treatments	Net benefits		Cost		Rate of return	
Apronstar	15555.35	13408.9	2680	2948	5.804235	4.548474
Baylaton	22950.82	20158.21	2120	2332	10.82586	8.644172
Mancozeb	20839.33	18293.88	1940	2134	10.74192	8.572577
Ridomil gold	19663.12	16966.99	3275	3602.5	6.004006	4.709782
Control	9802.93	8508.354	1400	1540	7.002093	5.524905

The first columns of net benefit cost and rate of return indicated the values of each component at normal condition while the second column is the value at sensitivity analysis result.

Table 6: Rate of return analysis.

changes in net benefits of treatments based on the scenarios assumed to be changed. The scenarios cover at least a couple of possible outcomes, usually including a "worst case" scenario and "normal" scenario. The normal considers the existing situations in the given time period. This analysis focused on the following scenarios.

1. When yield reduced by 10% but, cost of production remain unchanged.

2. When cost of production increased by 10% while yield is remain unchanged.

3. When both changes happen at a time.

Based on these situations the gross benefit and the net benefit can be changed accordingly. The result indicated that yield reduction of faba bean for all treatments are highly sensitive than cost of production changes. Baylaton and Mancozeb chemicals are more profitable than other management alternatives in all circumstances (Table 5).

Rate of Return (RR) analysis

Rate of return is the rate of change of returns to the rate of change of investment costs. Calculations of returns on investment resulted the value of each crop enterprises tells which production decision allow to make the highest return, taking costs into considerations. As always, it is not wise to overemphasize the usefulness and accuracy of any single measure. The rates of return analysis on total variable cost for those treatments were evaluated. The rate of return analysis result indicated that Bylaton and Mancozeb has high rate of return than other alternative while Redomil has the rate of return value below the control treatment (Table 6).

Conclusion and Recommendation

The primary aim of this experiment was to select the effective fungicide to control faba bean gall. Among treatments the highest disease score were recorded in control, Thiram and Apron star plots in 2013. The highest yield was also recorded in Baylaton and mancozeb sprayed plots respectively. In 2014 the disease prevalence and severity was similar to 2013. Maximum disease score were recorded on Control followed by Thiram and Apron star seed dressing plots. Minimum disease score were also recorded in Baylaton and Ridomil gold sprayed plots. There was a significant difference between biomass yield and grain yield. The highest grain yield was recorded in Baylaton sprayed plot and followed by Ridomil gold and Mancozeb respectively. There was no significant difference between plots in plant height, pod per plant, seed per pot and seed per plant in both years. Seed treatment

fungicides were not effective. Even foliar spray fungicides were also less effective except Bayleton.

Usually no single practice will control faba bean gall disease, when different approaches are combined, losses will be minimized. Consequently sustainable disease management has to be focused on a system approach against this disease by suppressing the pathogens before it reaches to economic threshold level. Such a system approach should incorporate various components like selection of variety, crop rotation, field sanitation and time of planting. Thus by manipulation of different integrated disease management (IDM) approaches, it is possible to minimize the risk of the disease. However the success would depend largely on an effective diseases monitoring system, frequent communication among the various disciplines involved in program and active link between research scientists, extension group and farmers.

Future Directions

The study gives clues about faba bean gall disease management with different fungicides, time of application, and different efficacy of fungicides to control the disease. However, testing this fungicides, current rate and frequency may not be enough to recommend full control package, but this information may lead to start further research direction in the area. Then further disease management studies have to be conducted in the area. Moreover efforts should be focused on applying cultural practice and different disease management activities to minimize the pathogen inoculums level in the field. The epidemiology study, environmentally safe and affordable fungicides should be further studied.

Acknowledgements

The author would like to acknowledge, Debre Birhan Agricultural Research Center and ICARDA ADA for technical and financial support during the execution of the experiments in the field.

References

1. Tilaye A, Demisu B, Getachew T (1994) Genetic and breeding of field pea. First National Cool-season Food Legumes Conference. Addis Ababa, Ethiopia. ICARDA/IAR. ICARDA: Aleppo Syria, pp: 122-137.

2. Beyene B (2015) Survey and identification of new Faba bean Disease (Kormid) in the Highlands of North Shewa Ethiopia. Australian Journal of Industry Research.

3. Gorfu D, Beshir T (1995) First National Cool-season Food Legumes Review Conference. Addis Abeba (Ethiopia), 16-20 Dec. ICARDA, Aleppo, Syria.

4. Dereje G (1999) Survival of Botrytis fabae Sard. Between seasons on crop debris in field soils at Holetta, Ethiopia. Phytopathologya Mediterranean 38: 68-75.

5. ICARDA (1993) International Centre for Agricultural Research in the Dry Areas. Faba Bean in China: State-of-the-art Review. Special study report Aleppo Syria.

6. Samia Z (2006) Production and productivity of pulse crops in Ethiopia. In: Kemal Ali (ed.), Cool–season Food Legumes of Ethiopia. Proceedings of the workshop and forage legumes 2003, Addis Ababa, Ethiopia. ICARDA/EIAR. ICARDA, Aleppo, Syria, pp: 1-5.

7. Sahile S, Fininsa C, Sakhuja PK, Seid A (2008) Effect of mixed cropping and fungicides on chocolate spot (Botrytis fabae) of faba bean (Vicia faba) in Ethiopia. Crop Protection 27: 275-282.

8. SAS Institute (1999) SAS/Stat User's Guide. Version 9.1.3. Cary NC: SAS Institute.

24

Heavy Metals (Cd, Ni and Pb) Contamination of Soils, Plants and Waters in Madina Town of Faisalabad Metropolitan and Preparation of Gis Based Maps

Ghulam Farid[1,3]*, Nadeem Sarwar[2,3], Saifullah[3], Ayaz Ahmad[1], Abdul Ghafoor[3] and Mariam Rehman[4]

[1]Department of Agriculture Extension Gujranwala, Punjab, Pakistan
[2]Nuclear Institute for Agriculture and Biology (NIAB), Faisalabad 38000, Pakistan
[3]Institute of Soil and Environmental Sciences, University of Agriculture, Faisalabad 38040, Pakistan
[4]Lahore College for Women University, Lahore, Pakistan

Abstract

Heavy metal pollution is a great threat to the environment. These metals are enters to the soil-plant environment through anthropogenic sources. A survey study was conducted to assess the heavy metals contamination of soils, plants and waters of Madina town of Faisalabad, Metropolitan area in 2010. Soil, plant and water samples were collected in the vicinity of Faisalabad following 4 × 4 Km grids. Soil samples were taken from 0-15 cm and 15-30 cm depths and prepared for the determination of metals (Cd, Pb and Ni). Plant samples were also taken from the same location and dried and digested in HClO4:HNO3 in the ratio of 1:3. AB-DTPA extract of soil, plants extract and water samples was analyzed on Atomic Absorption Spectrophotometer (Model Thermo S series). The results of the study showed the concentration of metals in Soils ranged from Cd (0.00-0.111 ppm), Pb (0.87-8.97 ppm) and Ni (0.017-1.72 ppm) at 0-15 cm while Cd (0.00-0.88 ppm), Pb (0.43-6.77 ppm) and Ni (0.055-0.852 ppm) at 15-30 cm respectively. Cd, Pb and Ni concentration in the plants ranged from 0.00-2.25 ppm 1.11-5.29 ppm and 1.51-4.96 respectively. Concentration of metal in the ground water ranged from Cd (0.00-0.06 ppm), Pb (0.10-11.10 ppm) and Ni (0.03-0.05). The concentration Pb and Ni was below the permissible limits while concentration of Cd in waters and plants above the permissible limits. Finally it was concluded that soil, plant and water of Madina town were in the safe limits with respect to metals. The use of city effluent is increasing the level of metals into the soils that ultimately contaminate the soils, plants and waters. So, it is suggested that city effluent must be treated for the detoxification of metals before use in irrigation purposes for crops.

Keywords: Heavy metals; Contamination; Soil; Plants and water

Introduction

Pollution of heavy metals directly and indirectly affects the human health. These substances adversely affect the productivity of soils, plants, animals and the entire environment if exceed certain limits [1]. Since quantity of good quality of water for agriculture is decreasing so, peoples are using raw city effluent for the production of different crops especially for vegetables. This raw city effluent contains lot of carcinogenic constituents like heavy metals, organic pollutants, salts and pathogens. Even in low concentration in soil-water system heavy metals persist for longer time in soil from where these enter into food chain through plant uptake. Sources of heavy metals pollution in environment are mainly derived from anthropogenic in nature. Which include vehicle exhaust, tire wearing, weathering street surfaces, power plants, coal combustion, metallurgical industry, auto repair shop, chemicals plant, domestic emission, weathering of building and pavement surface and atmospheric deposits. However, the anthropogenic sources of heavy metals in agricultural soils include mining, smelting, waste disposal, urban effluent, vehicle exhausts, sewage sludge, pesticides and fertilizers application. Among all the heavy metals cadmium (Cd) is a highly toxic for both the plants and animals as well as for human beings. Cadmium enters into soil-plant environment mainly through anthropogenic activities. Compounds of Cd are more soluble than other heavy metals rendering it more available for plant absorption where these could accumulate in edible plant parts.

In Pakistan Cd concentration in soil samples from the Islamabad expressway varied from 5.8 to 6.1 mgkg^{-1} with an average value of 5.95 mgkg^{-1}[2]. The value of Cd in the paddy and straw was ranged from 0.116to 0.370 mgkg^{-1} and 0.315 to0.370 mgkg^{-1} in the areas of Faisalabad

[3]. Nickel (Ni) another heavy metal which is toxic, carcinogenic and dangerous for humans, plants and animals. The Ni released into the soil from copper-nickle smelters, burning of diesel oil containing Ni, city effluent, bio-solid, impurities in fertilizers, mining and smelting [4]. It enters into the soil-environment through anthropogenic activities although small quantities are released during *in-situ* wheathering of parent material. In soil samples from Islamabad expressway, average concentration of Ni was 32 mg kg^{-1} [2]. Chemical analysis of paddy and straw from Sheikhupura contained Ni 0.073-0.093 mg kg^{-1} [5].

Lead (Pb) is a widespread heavy present in soils, plants and waters. It is mostly present in top layer of soil due to the deposition from air containing smoke from vehicles. The Pb is released to the from mining, industrial and agricultural chemicals. In uncontaminated soils, Pb concentrations are generally below 50 mg kg^{-1} [6]. The Pb concentration in vegetation growing on such soils is often less than 10 mg kg^{-1} dry mass. Soil lead ranging from 10 to 293 mgkg^{-1} in agricultural areas and in the areas of pesticide manufacturing companies 57.05 mg kg^{-1} were noted in Rajasthan, area of India [7].

*Corresponding author: Ghulam Farid, Department of Agriculture Extension Gujranwala, Punjab, Pakistan, E-mail: faridghouri@gmail.com

For mapping pollution hit areas different tools are used like geostatistics, multivariate statistical methods and GIS. Geographic information systems (GIS) provide powerful tools for spatial analysis Sweeney [8], Rodda et al [9]. applied GIS based decision support system to predict nitrate leaching to groundwater. Ahn and Chon [10] investigated groundwater contamination and derived spatial relationship between groundwater constituents and pollution sources using GIS. This database will help to locate the pollution of heavy metal in Faisalabad. It may also help the government to develop policies for the contaminated areas in Pakistan.

Faisalabad is one of the third largest city having many textile industries due to which it is called as Manchester of Pakistan. There are 512 large industrial units, out of which 328 are textile, 92 engineering complexes and 92 chemical and food processing units. All these industrial units are releasing huge quantity of untreated city waste water into unlined surface drains. From these drains farmers are using this contaminated water for the production of crops, especially vegetables. Farmers consider this raw city effluent is a good source of water and nutrients, substitute of good quality water and reliable source of irrigation round the clock. Keeping in view the above facts, the studies were designed to investigate the heavy metal contamination in soils, plants and water in the Madina town of Faisalabad, Pakistan followed by preparation of GIS base maps of polluted areas. This efferts will help development agencies to plan out types of sensitive land uses.

Materials and Methods

Study area

This study includes sampling of soil, plant and water. For the collection of soil, plant and water samples Madina Town of Faisalabad, Meteropolitan was selected. The samples was taken in the month of March 2010. Samples were collected from every grid 4-Km apart for soil, plant and water. Samples from each grid was collected and prepared according to the prescribed method and then analysed on Atomic Absorption Spectrophotometer (Model Thermoelectron S-Sreries) for the heavy metals determination. Global Positioning System (GPS) reading of coordinates was taken and then base maps was developed with the help of Geoghrapic Information System (GIS) software Arc GIS v.9.1. Detailed methods for collection, preparation and analysis of soil, plant and water samples are given below.

Collection and preparation of soil samples

Soil samples were collected from different urban and periurban area of Madina Town, Faisalabad after every 4-Km from 0-15 cm and 15-30 cm. soil samples were taken from 3 points at each grid and mixed thoroughly in a plastic bucket. Samples are taken to laboratory air dried, ground with wooden roller and sieved through 2 mm stainless steel sieve. For the determination of heavy metals soil samples were extracted with AB-DTPA (Soltanpur, 1985) and analysed on Atomic Absorption spectrophotometer (Model Thermoelectron S-Sreries). Physiochemical characteristics were also determined ECe, pH_s and SAR [11].

Collection and preparation of plant samples

Plant samples were also collected from the above mentioned places as the soil samples taken. Two Plant samples of vegetables, crops, trees and ornamental plants depending upon the availability of vegetation were taken. Samples were taken to laboratory washed with tap water, diluted HCl water and distilled water to remove the external contamination. Samples were air dried and then placed in Oven at 65°C for drying of samples. After oven drying samples were ground and stored in plastic zipper. A 1 g samples was taken in digestion flask and 12 ml diacid mixture (i.e. Pechloric acid $HClO_4$ and Nitric acid HNO_3 with a ratio of 1:3) were added and kept for overnight stay. Next day samples are digested on hot plate till the plant material digested and color was clear. After digestion sample was cooled and made 25 ml volume with distilled water and stored in air tied bottles for the determination of heavy metals. Samples were analyzed on Atomic Absorption spectrophotometer (Model Thermoelectron S-Sreries). Instrument was calibrated with standard solution of respective metal.

Collection and preparation of water samples

Water samples were collected from the above mentioned sites. For water samples groundwater (tube well, hand pump and motor pumps), surface water (canal) and waste water(sewerage, industies effluent) were taken depending upon the availability in the area but ground water was taken from each site. Water sample was taken to laboratory and filtered with Whatman No.40. The water samples were analyzed for EC, SAR and RSC [11]. After the basic analysis of water samples concentrated HCl was added to the waste water samples and Sodium Haxametaphosphat was added to ground water samples to check the metal precipitation. For the determination of heavy metal Atomic Absorption spectrophotometer (Model Thermoelectron S-Sreries) was used.

Construction of GIS maps

For the construction of GIS (Geographic Information System) maps latitude (X coordinate) and longitude (Y coordinate) reading were taken with the help of Global Positioning System (GPS) model Etrex Germin. From the latitude and longitude reading obtained from the GPS was feed in the GIS software Arc GIS v.9.1 and base maps was drawn.

Results

Metal ions in the soil, plants and water collected from Madina town of Faisalabad

Cadmium (Cd) concentration in Soils

In the soil collected from Madina town of Faisalabad the AB-DTPA Extractable Cd ranged from 0.000 to 0.111 ppm and 0.000 to 0.088 ppm in 0-15 cm and 15-30 cm respectively with mean value of 0.03 and 0.02. The maximum concentration of AB-DTPA Extractable Cd (0.111 ppm) was observed at Crescent sugar mill while minimum concentration of Cd (0.001 ppm) was observed at Chak No. 255 RB, Bogran in 0-15 cm soil depth. The AB-DTPA extractable Cd was high in the upper layer of soil as compare to the lower depth which might be due to the anthropogenic activities. In the meanwhile the maximum concentration of AB-DTPA extractable Cd (0.088 ppm) was also observed at Crescent sugar mill and minimum concentration of Cd (0.000 ppm) was at Kahkashan colony in 15-30 cm soil depth. The high concentration of AB-DTPA extractable Cd may be due to the application of industry water in the lawns of Crescent sugar mill because the soil samples was taken from the industry lawn. Although the AB-DTPA extractable concentration of Cd in all the soil samples collected from the Madina town were below the permissible limits (<0.31 ppm) proposed by Alloway [4]. The results are similar to the findings of Zhou et al. Recent advances in industry and agriculture have led to an increased level of Cd in the agricultural soil environment. Cadmium enters the soil through various anthropogenic sources including application of phosphate fertilizers, waste water, Cd contaminated sewage sludge and manures, and anthropogenic emissions from power stations, metal industries, urban traffic and cement industries The main sources of Cd pollution are enhanced agricultural activities by using phosphatic fertilizers.

Cadmium (Cd) concentration in Plants

Plant samples collected from the Madina town of Faisalabad contained the Cd concentration in the range of 0.000-2.25 with mean concentration of 0.44 ppm in plants. The concentration of Cd was maximum (2.25 ppm) at Nawab town No. 2 in Barseem samples. The high concentration of Cd in the plants might be due to the application of sewage water. Although at Nawab town No. 2 sewage water samples

has the Cd concentration below the permissible limits, but by the continuous application of sewage water build up the concentration of metals into the soil. From the soil metals transfer to the plants and accumulate in the tissues of plants. Cd is more soluble as compare to other metals so, it can accumulate more into the plant tissues. The permissible limits of Cd were (0.10 ppm) in the plant tissue proposed by Macnicol and Beckett. The efficiency of plants to absorb metals can be evaluated by their ability of metal uptake or soil to plant transfer factors. Although soil concentrations may be the source of metals for plants uptake through roots by the process of translocation [12]. Find the Cd concentration in the plant samples above the permissible limits (Map 1).

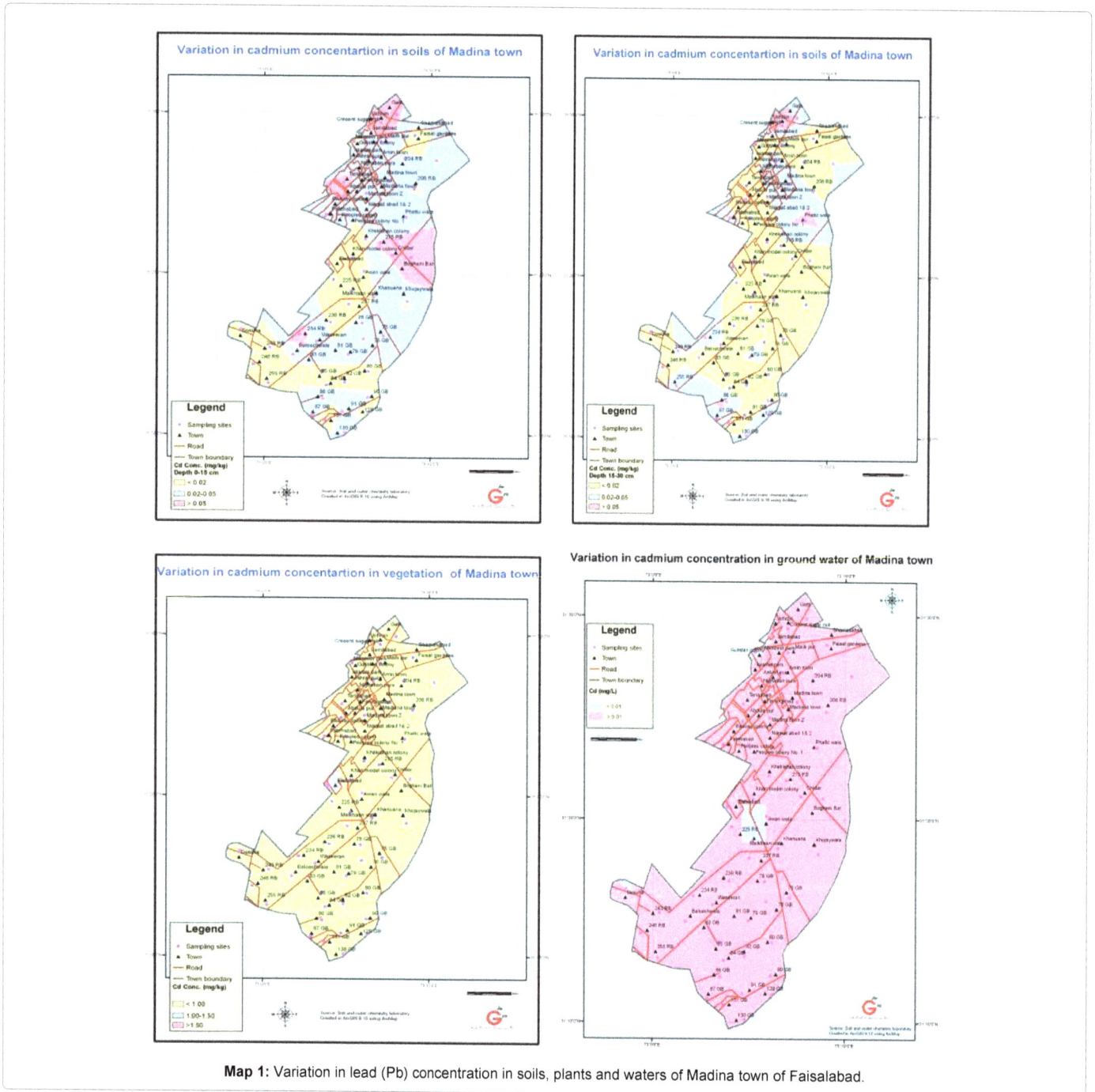

Map 1: Variation in lead (Pb) concentration in soils, plants and waters of Madina town of Faisalabad.

Cadmium (Cd) concentration in waters

In the ground water samples collected from Madina town ranged from 0.00-0.06 ppm with mean concentration of 0.03 ppm. The maximum concentration of Cd 0.06 ppm in ground water was observed at Chak No. 227 JB, Kararwala while Cd concentration was not detected at Gulistan colony and Chak No. 226 RB. The concentration in the drinking water should be less than 0.01 ppm. The concentration of Cd at most of site was found above the critical limits (0.01 ppm). The high concentration of Cd in ground water may be due to the *in situ* weathering of parent material. The results are similar to the findings of Saif, they found Cd concentration in various water samples.

The concentration of Cd in canal water collected from Madina town of Faisalabad ranged from 0.00-0.05 with mean value of 0.02 ppm. The maximum concentration of Cd (0.05 ppm) in canal water was observed at Chak No. 226 RB Malkhanwala while at Chak No. 244 RB and Blochwala the Cd concentration was not detected. The concentration of Cd above the permissible limits might be due to the mixing of city effluent into the canal.

Waste water samples also showed Cd concentration above the permissible limits (0.01 ppm) proposed by Ayers and Westcot. The concentration of Cd in the waste water collected from the Madina town of Faisalabad ranged from 0.02-0.06 with mean value of 0.03 ppm. The maximum concentration of Cd (0.060 ppm) in the waste water was observed at Chak No. 87 GB while minimum concentration of Cd (0.020 ppm) was observed at Chak No. 236 RB Kajlay which is also above the permissible limits. The high concentration in the waste water at Chak No.87 GB might be due to the use of cadmium containing chemical in the industry.

Lead (Pb) concentration in soils

The AB-DTPA extractable Pb in soil samples collected from the Madina town of Faisalabad ranged from 0.87-8.97 ppm and 0.43-6.77 ppm at 0-15 cm and 15-30 cm with mean value 2.64 and 2.07 respectively. The available Pb concentration in the soil samples is much higher in the 0-15 cm depth as compare to the 15-30 cm depth it might be due to the aerial deposition from the vehicle exhaust. The maximum concentration of Pb was observed at Faisal garden (8.97 ppm) while minimum concentration (0.87) was at Chak No. 76 GB at 0-15 cm depth. In all the soils of Madina town the Pb concentration was between the normal range (5-15 ppm) proposed by Alloway [4] reported the high concentration of AB-DTPA extractable Pb at 0-10 and 10-20 cm soil depth around Faisalabad city irrigated with raw city effluent. The high concentration of Pb in the surface layer might be due to the continuous use of city effluent containing metals and also due to the less mobility of Pb within the soils particularly in alkaline soil conditions. The Pb is reported to mobilize more in acidic range of pH because Pb binds strongly with organic matter and oxides of Fe and Mn it is a low mobility metal in the soil. The results of the study are similar to the finding of Hussain and Saif.

Lead (Pb) concentration in plants

Plant samples collected from the Madina town of Faisalabad contained Pb in the range of ppm and 1.11-5.29 ppm with mean value of 2.72 ppm. The maximum concentration of Pb (5.29 ppm) was found at National Textile University in Gardenia shrub while minimum concentration of Pb (1.11 ppm) was minimum at Govt. Islamia high school Ghatti. The normal range (5-10 ppm) for the plant tissue was proposed by Kabata-Pendias and Pendias. The overall concentrations of Pb in the plants of Madina town have Pb concentration between the normal ranges. The Pb concentrations in the samples taken from the urban area of Madina town have high concentration of Pb as compare to the village this might be due to the aerial deposition of Pb from the vehicles exhaust. These results are accordance with Chary et al [12] who reported the concentration of Pb in plants 1.3-3.6 ppm. Sekhar also reported the Pb concentration in the plants ranged from 0.1-5 ppm which are also below the permissible limits.

Lead (Pb) concentration in waters

The concentration of Pb in ground water samples collected from the Madina town of Faisalabad ranged from 0.10-11.10 ppm. The maximum concentration of Pb was high at Chak No. 226 RB Malkhanwala while minimum concentration of Pb (0.10 ppm) was observed at Ahmadabad, Chak No. 130 GB and Chak No. 204 RB. The Pb concentration in the water at Chak No. 86 GB, Maan pur garala and Chak No. 226 RB Malkhanwala was high from the permissible limits (5 ppm) proposed by Ayers and Westcot.

The canal water samples collected from the area of Madina town contained Pd in the range of 0.30-10.70 ppm. The maximum concentration of Pb (10.70 ppm) in canal water was observed at Chak No. 90 GB, Abdullah farm this might be due to the air-borne Pb contamination into the canal water. Minimum concentration of Pb (0.30 ppm) was observed at Chak No. 208 RB. Most of the canal water samples of Madina town have Pb concentration below the permissible limits (5 ppm) proposed by Ayers and Westcot.

The waste water samples collected from the Madina town of Faisalabad have the Pb concentration below the permissible limits (5 ppm) proposed by Ayers and Westcot. The Pb concentration in the waste water of Madina town ranged from 0.10-0.90 ppm. The maximum concentration of Pb (0.90 ppm) was observed at Chak No. 215 RB, Naitheri while the minimum Pb concentration (0.10 ppm) was at Chak No. 86 GB and Chak No. 204 RB. Although in all the samples of waste water have the Pb concentration below the permissible limits. The concentration of Pb in the waste water was between the permissible limits because the most of the samples were from the village waste water (Map 2).

Nickle (Ni) concentration in soils

The AB-DTPA extractable Ni concentration in soil ranged from 0.017-1.072 with mean concentration of 0.33 ppm at 0-15 cm and 0.055-0.852 with mean concentration of Ni 0.27 ppm at 15-30 cm. The AB-DTPA extractable Ni concentration was less at 15-30 cm soil depth as compare to 0-15 cm soil depth. Maximum concentration of Ni was observed in sample collected at Govt. Municipal Degree College Abdullah Pur, Madina town. Nickle concentration at most of the sites was found in the normal range of 0.02-5.2 ppm in the soil [4]. The result showed that soils of Madina town were uncontaminated with nickle. Ni is also essential nutrient for plant growth. Saleem et al. [13] reported the Ni concentration in soil samples 0.92-1.57 ppm the results of study are also similar to the above findings. a study was conducted by Jie-liang et al. and results are in accordance with the present study. Li et al. also reported the Ni contamination in soils of Hong Kong by using GIS-Base approach.

Nickle (Ni) concentration in plants

The result showed that the Ni concentration in the plant samples collected from the Madina town ranged from 1.51-4.96 ppm in plants. The maximum concentration of Ni (4.96) was found at Chak No. 228

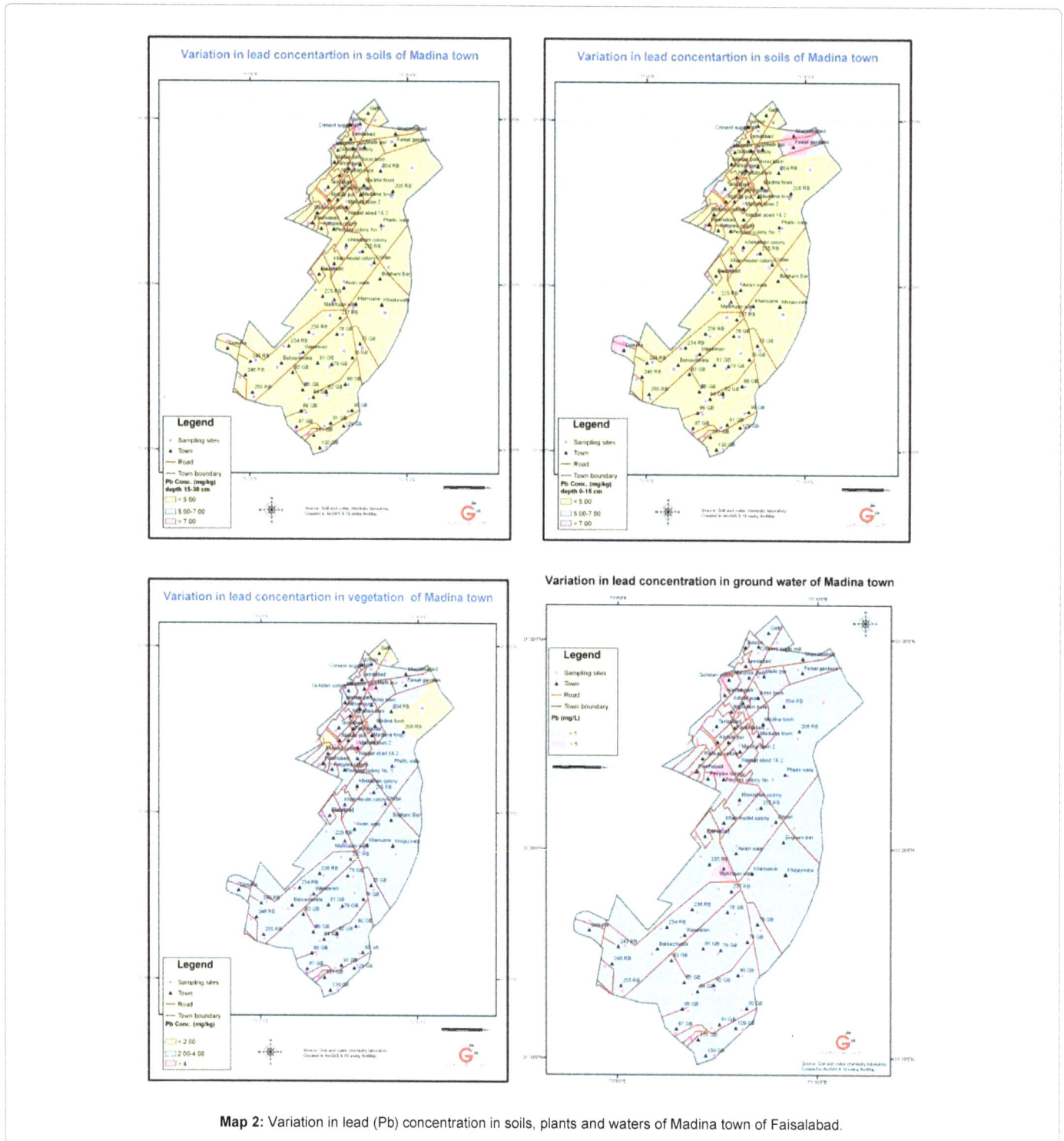

Map 2: Variation in lead (Pb) concentration in soils, plants and waters of Madina town of Faisalabad.

JB while minimum concentration of Ni was observed at Amin town. The plant Ni concentration in the Madina town also remained within the permissible limits (10 ppm) proposed by Macnicol and Beckett. City area has more concentration of Ni in plant as compare to the villages. The concentration of Ni in urban areas might be due to the anthropogenic activities. Hussain et al. reported the concentration of Ni ranged 4.4-20.0 ppm in the vicinity of Faisalabad. The results are similar to the study of Saleem et al [13] who also reported the

concentration of Ni in Korangi area of Karachi. The results are also in line with Sekhar et al.

Nickle (Ni) concentration in waters

The Ni concentration in the ground water samples collected from the Madina town ranged from 0.03-0.05 ppm with mean concentration 0.04. Maximum concentration of Ni (0.05 ppm) was observed at Chak No. 225 RB, Malkhanwala while minimum concentration of Ni (0.030

Heavy Metals (Cd, Ni and Pb) Contamination of Soils, Plants and Waters in Madina Town of Faisalabad Metropolitan...

133

ppm) was at Chack No. 229 JB. The Ni concentration in the ground water samples was below the permissible limits (0.2 ppm) proposed by Ayers and Westcot.

The canal water sample collected from the Madina town showed the Ni concentration ranged from 0.04-0.06 with an average concentration 0.05 ppm. Three canals are passing through the Madina town area viz. Gogera Branch, Jhang Branch and Rakh Branch. The Ni concentration in the canal water was also below the permissible limits (0.2 ppm). The maximum concentration of Ni was observed in canal water also at Chak No. 225 RB, Malkhanwala while the minimum concentration of Ni was

0.04 ppm at Chak No. 225 RB, Dasoha. Sekhar et al. have reported the Ni concentration in the surface water. Saif et al. also reported the Ni concentration in the waters of Korangi area of Karachi 0.02-5.35 ppm.

The results of waste water collected from the Madina town showed that range for the Ni concentration was 0.046-0.060 with mean concentration of 0.05 ppm. The maximum concentration of Ni (0.060 ppm) in waste water was observed at Chak No. 225 RB, Kajlay while the minimum concentration 0.046 was at Chak No. 204 RB. The results of waste water shows that the concentration of Ni below the proposed limits (0.2 ppm) by Ayers and Westcot (Map 3).

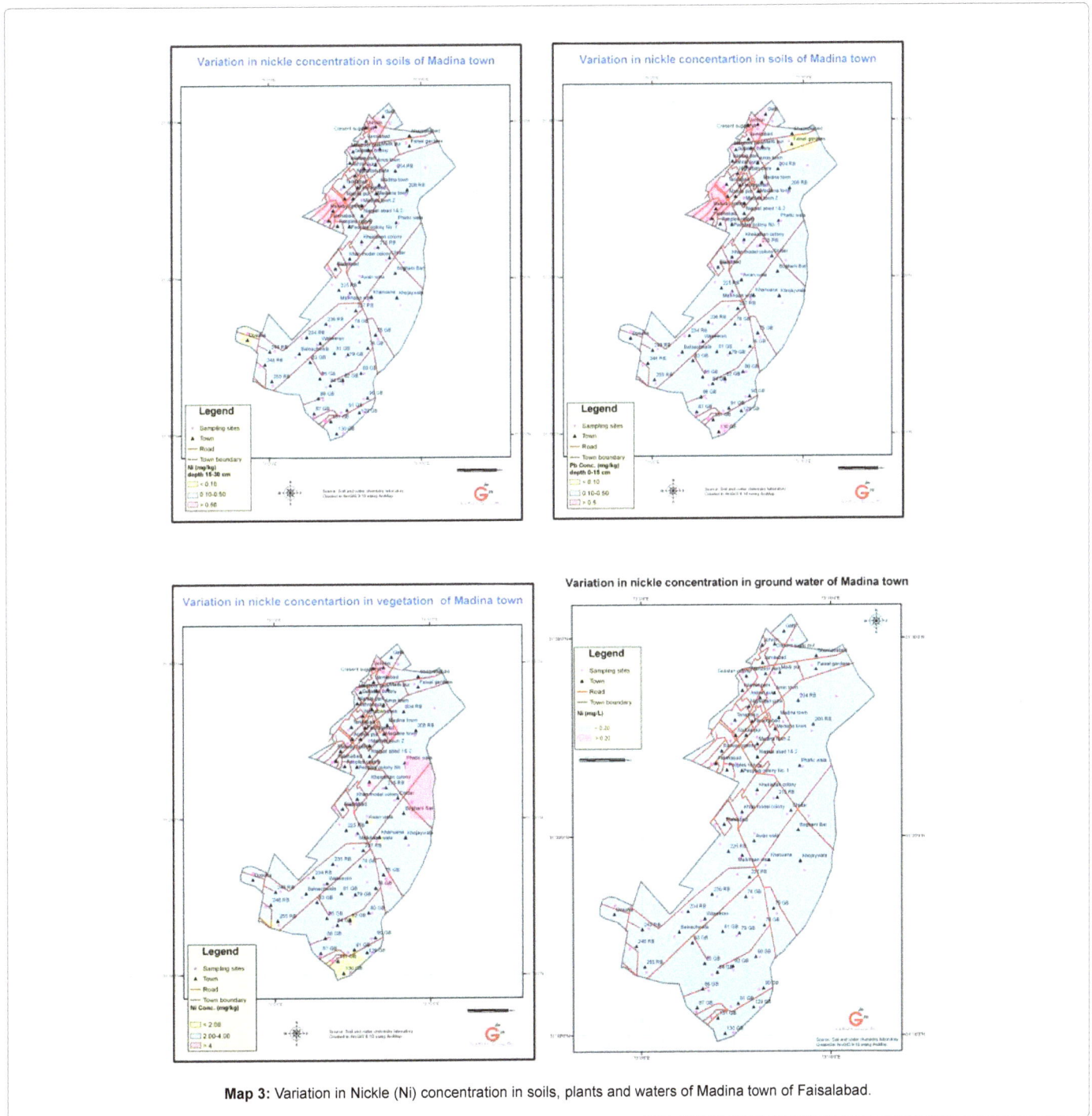

Map 3: Variation in Nickle (Ni) concentration in soils, plants and waters of Madina town of Faisalabad.

Conclusion

Finally it was concluded that in the vicinity of Faisalabad the use of raw city effluent is increasing for the irrigation of crops (due to the shortage of good quality water) which contains many heavy metals. By the continuous application of raw city effluent to the soils the metals are accumulating into the soils and plants as well as surface and ground water is also having these heavy metals concentrations. The production of safe food while using raw city effluent for irrigation requires some management strategies.

References

1. Mapanda F, Mangwayana EN, Nyamangara J, Giller KE (2005) The effect of long term irrigationusing waste water on heavy metal content of soil under vegetables in Harare, Zimbawe. Agric Ecosyst Environ107: 151-165.

2. Faiz Y, Tufail M, Javed MT, Chauhadry M, Siddique N (2009) Road dust pollution of Cd, Cu, Ni, Pb, and Zn along Islamabad expressway, Pakistan. Micro-chem J 92: 186-192.

3. Nawaz A, Khurshid K, Arif MS, Ranjha AM (2006) Accumulation of heavy metals in soil and rice plants (Oryza sativa L) irrigated with industrial effluents. Int J Agri and Biol 391-393.

4. Alloway BJ (1990) Heavy metals in soils. Blackie and Academic Professionals, London, UK.

5. Mehdi SM, Abbas G, Sarfraz M, Abbas ST, Hassan G (2003) Effect of industrial effelunts on mineral nutrition of rice and soil health. Pak J applied sci 3: 462-473.

6. Reimann C, De Caritat P (1998) Chemical elements in the environment-Factsheets for the Geochemist and Environmental Scientist. Springer-Verlag Berlin, Germany.

7. Karishna AK, Govil PK (2005) Heavy metal distribution and contamination in soils of Thane-Belapur industrial development area, Mumbai, Western India. Environ Geol 47: 1054-1061.

8. Sweeney MW (1999) Geographic Information Systems. Water Environ Res 71: 551-556.

9. Rodda HJE, Demuth S, Shankar U (1999) the application of a GIS-based decision support system to predict nitrate leaching to groundwater in southern. Germany. Hydrol Sci 44: 221-35.

10. Ahn HI, Chon HT (1999) Assessment of ground water contamination using geographic information system. Environ Geochem Health 21: 273-289.

11. Allison JE, Bernstein L, Bower CA, Brown JW, Fireman M, et al. (1954) US Salinity Lab Staff, Diagonsis and improvement of saline and alkali soils. USDA Washington DC USA.

12. Chary NS, Kamala CT, Raj DSS (2008) Assessing risk of heavy metals from consuming food grown on sewage irrigated soils and food chain transfer. Ecotoxicol and Environ Safety 69: 513-524.

13. Saleem MS, Haq MU, Memon KS (2005) Heavy metals contamination through industrial effluent to irrigation water and soil in Korangi area of Karachi (Pakistan). Int J Agri & Biol 7: 646-648.

25

Management of Faba Bean Gall Disease through the use of Host Resistance and Fungicide Foliar Spray in Northwestern Ethiopia

Getnet Yitayih Alemu* and Yehizbalem Azmeraw Tadele
[1]Department of Plant Sciences, Debre Tabor University, Ethiopia

Abstract

Most developing countries are hesitated in maximizing tissue culture technology due to the overhead costs involved. Therefore, this research was initiated to evaluate alternative cheap carbon sources and energy in culture media in order to reduce production input costs of sugarcane *in vitro* propagation. The experiment was carried out in completely randomized design (CRD) with 2 × 6 factorial treatment arrangements of genotypes and carbon source in combination. The interaction analysis of genotypes and table sugar concentration significantly influenced *in vitro* sugarcane multiplication. On MS medium with 50 gl^{-1} table sugar, B4906 gave the highest (13.42 ± 0.29) shoots, whereas Pr1013 produced a maximum of 7.78 ± 0.19 shoots at 60 gl^{-1} table sugar. However, 40 gl^{-1} table sugar was optimum to produce usable and separable shoots for further subculture of multiplication. Accordingly results showed that table sugar not only enhanced multiplication but also significantly reduced the production input costs by 94.89% when compared with the analytical grade sucrose.

Keywords: Faba bean gall; Bayleton; Severity; Incidence

Introduction

Faba bean (*Vicia faba* L.) also referred to as broad bean, horse bean and sometimes field bean occupies nearly 3.2 million hectare worldwide [1]. It is believed that the crop was introduced to Ethiopia from the Middle East via Egypt around 5000 B.C., immediately after domestication [2]. In 2003, China leads the world in faba bean production in both area coverage and production followed by Ethiopia, Egypt and Australia [3]. Ethiopia is considered as the secondary center of diversity and also one of the nine major agro-geographical production regions of faba bean [1,2,4,5].

Among pulse crops, faba bean is majorly grown in Ethiopia and leading protein source for the rural people and used to make various traditional dishes. Faba bean is well known by farmers in improving soil fertility by fixing atmospheric nitrogen, and widely use them in rotation with cereals [6]. The crop can be grown for green manure and silage [7]. The majority of the seed produced would be consumed domestically and only a smaller percentage of the crop is delivered to the export market [8]. However, still this small portion of export volume put Ethiopia among the top broad bean exporting countries of the world [9]. Amhara and Oromia are the two major faba bean producing regions. The Amhara region contributes to the highest production (47%) in the country followed by Oromiya region that contributes 39% to national production [10].

Faba bean is an important legume crop and pre-dominantly grown by every individual farmers in Ethiopia. However, there are different faba bean biotic (diseases, insect pests, and weeds) and abiotic production constraints that limit the production and productivity of the crop [6,11]. Among which diseases are the most important biotic factors causing faba bean yield reduction. More than 17 pathogens have been reported so far on faba bean from different parts of the country. Diseases that are economically most important in the major faba bean growing regions including chocolate spot, faba bean rust, and Aschochyta blight [12-14]. In recent years, in additional to the previous common diseases, the crop is threatening by new gall forming disease with typical symptoms of green and sunken on the upper side of the leaf and bulged to the back side of the leaf, and finally develops light brownish color lesion, chlorotic galls, and progressively broaden to become circular or elliptical uneven spots [8,15]. The faba bean gall caused yield loss up to 30-100% [16]. These data showed that faba bean gall is the most destructive disease that causes total yield loss. The disease affects leaves and stems and it affects large areas in the country where faba bean is cultivated and cause considerable losses in quality and quantity of the products [8]. The disease was highly expanding and distributing violently in the country from year to year.

Even though the disease is disseminated at violent rate, there is no management options under took to control the disease. Hence, this needs more attention to find solution for farmers to manage the disease and sustain their life with producing faba bean crops. Therefore, this study was conducted to manage faba bean gall disease through the integrated use of host resistance and fungicide foliar spray.

Material and Methods

Site description

The experiment was conducted at Farta and Tach gaynt in farmers' field during 2014 main cropping season. Farta and Tach gaynt are found an altitude of 2500 and 2880 above sea level in Amhara region, South Gondar Zone, respectively. The average annual rainfall of Farta is about 1750 mm and Tach gaynt is 925 mm. The average annual temperatures of Farta and Tach gaynt are 17°C and 18.5°C, respectively. The major soil type of Farta and Tach gaynt are clay loam and clay loam, loamy sand, respectively.

Experimental design and procedures

The field experiment was conducted within two sites at Farta and Tach gaynt in farmers, fields during 2014 main cropping season.

*Corresponding author: Getnet Yitayih Alemu, Department of Plant Sciences, Faculty of Agriculture and Environmental Science, Debre Tabor University; PO Box-272; Debre Tabor, Ethiopia, E-mail: getnety19@gamil.com

The treatments to be evaluated were 14 faba bean varieties, released (CS20DK, NC58, Kasa, Bulga70, Mesay, Tesfa, Degaga, Adet-Hana., Moti, Gebelecho, Obse, Dosha, Tumsa), and local check. The varieties to be evaluated were collected from Holeta and Adet Agricultural Research Centers. The treatments were arranged in split plot design with three replications. The experiment has 14 main plots and two subplots i.e protected with fungicide and unprotected. The size of main plot was 4 m length × 3.8 m width with having 0.8 m spacing between main plots and the sub-plot size was 4 m length × 1.4 m with having 4 seedling rows (with 2 net rows) and spacing between blocks, was 1 m and spacing between plots, rows and plants was 0.6 m and 0.4 m, 10 cm respectively. The protected plots were sprayed using triadimefon (Bayleton[R] WP 25) systemic fungicide at the rate of 0.125 kg a.i/ha for all varieties starting from 3 weeks after sowing in 15 days interval. 15 g of Bayleton was diluted within 5 liter of water to spray the experimental plots. All necessary agronomic practices were done as required.

Data collection

Crop data: The grain yield of faba bean was measured at maturity, from the central rows of each plot by harvesting the plants manually. Seed yield per plot was measured using a sensitive balance and then adjusted to 9% moisture content, and yield per plot was converted into kg/ ha.

Disease assessment: Faba bean gall incidence and severity assessments were started as soon as the first disease symptoms occurred in every 10 days interval at both locations. Using 10 randomly pre-tagged faba bean plants in the two central rows, severity was rated using standard disease scales of 0-9 [17,18]. The following infection levels on the scale were used: 0, no visible infection on leaves; 1, a few dot-like accounting for less than 5% of total leaf area; 3-4, discrete galls less than 2 mm in diameter, accounting for 6-25% of leaf area; 5, numerous scattered galls with a few linkages, diameter 3-5 mm, on 26-50% of leaf area with a little defoliation; 6, confluent galls formation accounting for 51-75% of leaf/stem area, mild gall formation, half the leaves dead or defoliated; 7, complete destruction of the larger leaves, galls covering more than 76% of leaf area, abundant gall formation; 8, 80% of the defoliated and plants darkened and dead; 9, disease covering more than 80% of the foliar tissue heavy defoliation and plants darkened and dead. Diseases incidence was calculated using the following formula.

$$Disease\ incidence = \frac{Number\ of\ disease\ plants}{Total\ number\ of} \times 100$$

Disease development data were rated using 1-9 rating scale and then converted in to percentage severity index using the formula.

$$PSI = \frac{Sum\ of\ numerical\ ratings \times 100}{Number\ of\ plants\ scored \times maximum\ score\ on\ scale}$$

The area under the disease progress curve (AUDPC) was calculated from percentage severity index using the following formula

$$AUDPC = \sum_{i=1}^{n} \left[0.5 \left(X_i + x_{i+1} \right) \left(t_{i+1} - t_i \right) \right]$$

Where: X_i = the cumulative disease severity expressed as a proportion at the ith observation

t_i=time of the ith assessment , n=the total number of observation

Data analysis

The collected data from the two experimental sites were subjected to ANOVA to determine the treatment effects. AUDPC for each

treatment were evaluated from disease severity values. The severity grades were converted into percentage severity index using the formula stated above. Least significant difference (LSD) value was used to separate the treatment means. All diseased and agronomic data were analyzed by using SAS version 9.1.3 statistical software [19].

Results and Discussion

Incidence of faba bean gall disease

Percentage of disease incidence was showed significant difference (P<0.05) among faba bean varieties at both locations at all dates of assessment. Highly significant difference (P<0.01) was observed on incidence of faba bean gall disease between sprayed and unsprayed plot at both locations at all dates of assessment. Higher disease incidence was recorded from variety Adet-Hana at both initial and final dates of assessment at Farta. Lower disease incidence was scored from variety Tumsa at initial and Degaga and Nc58 at final date of assessment at Farta (Table 1). At Tach gaynt, the disease incidence was higher on variety Bulga70 and Adet-Hana at both initial and final dates of assessment. While lower mean disease incidence was scored from variety Tumsa at both initial and final dates of assessment (Table 1).

The combined use of varieties and Bayleton foliar spray was showed highly significant difference (P<0.01) at Farta and significant difference (P<0.05) at Tach gaynt on disease incidence at the initial and final dates of assessment (Table 2). Higher disease incidence was scored from unsprayed plots of variety Bulga70, Adet-Hana, Dosha, CS20DK and Gebelecho at the final date (66 days after sowing) of disease assessment at Farta. However, lower disease incidence was recorded from unsprayed plots of variety Nc58 and Degaga. At Tach gaynt, the disease incidence was lower on unsprayed plots of variety Tumsa and Moti at final date of assessment. While the disease incidence was higher on unsprayed plots of variety Bulga70, Kasa, Local check, Adet-Hana, Degaga and Mesay at the final assessment. The disease incidence was completely zero in sprayed plots of all faba bean varieties except Adet-Hana, Dosha, Moti and Tesfa at Farta and Adet-Hana, Bulga70, Gebelecho at Tach gaynt at final date of assessment (Table 2). The interaction of variety with Bayleton foliar spray reduced the percentage of disease incidence by 90% and 83.3% on sprayed plot of variety Adet-Hana over unsprayed plot of this variety at Farta and Tach gaynt, respectively, at the final date of assessment. Similarly, Bayleton foliar spray lowers the disease incidence by 80% and 100% on sprayed plot of variety Bulga70 over unsprayed plot of this variety at Farta and Tach gaynt, in that order, at the final date of assessment. The disease incidence was reached 100% on variety Bulga70, Kasa and Local check on unsprayed plots at Tach gaynt at the final date of assessment. This result showed that all the assessed plants of these varieties were infected by the faba bean gall disease. From three surveyed regions by the maximum mean incidence of 43.4% was recorded in South Gondar in Amhara region [8]. The study by Teklay et al. [16] in Tigray showed that the incidence range of the disease varied from 5-100%. In unsprayed plot, mean faba bean gall incidence ranged from 50-93.3% and 56.7-100% at Farta and Tach gaynt, respectively at final date of assessment. At both locations all faba bean varieties were infected by the newly emerged faba bean gall disease. This agreement with the study by Teklay et al. [16] and Hailu et al. [8] who are reported that all improved and local faba bean varieties affected by the diseases indifferently. The prevalence of the new gall disease was in the range of 0 and 100% [8,16].

Severity of faba bean gall disease

Percentage severity index (PSI) calculated from disease severity assessed five times at both locations. Percent severity index was higher

Varieties	Farta				Tach gaynt			
	Incidence (%)DAS		PSI (%) DAS		Incidence (%) DAS		PSI (%) DAS	
	Initial (26)	Final (66)	Initial (26)	Final (66)	Initial (28)	Final (68)	Initial (28)	Final (68)
Adet-Hana	75.0	48.3	17.0	30.7	70.0	51.7	21.5	22.8
Bulga 70	58.3	40.0	14.4	28.0	71.7	55.0	18.2	28.5
CS20Dk	41.7	35.0	6.7	23.3	70.0	46.7	19.8	20.6
Degaga	28.3	25.0	4.1	8.3	53.3	46.7	11.0	17.4
Dosha	33.3	41.7	7.6	19.8	48.3	43.3	12.0	13.7
Gebelecho	43.3	35.0	10.6	15.0	50.0	38.3	15.4	13.7
Kasa	30.0	26.6	1.7	1.3	65.0	50.0	22.6	24.6
Mesay	23.3	28.3	0.6	5.6	63.3	48.3	19.1	22.4
Moti	36.7	35.0	6.9	19.6	43.3	33.3	11.7	13.7
Nc58	26.7	25.0	5.9	13.7	61.7	43.3	20.2	25.2
Obse	30.0	31.7	3.3	6.1	63.3	38.3	14.1	12.4
Tesfa	25.0	28.3	3.7	17.0	41.7	38.3	7.2	13.7
Tumsa	15.0	28.3	1.9	8.3	30.0	28.3	5.9	10.6
Local check	30.0	30.0	1.3	5.2	63.3	50.0	15.9	18.0
CV (%)	26.34	26.36	28.20	29.56	26.90	22.20	18.71	17.61
LSD (5%)	25.59	15.64	12.73	25.44	31.79	12.67	12.83	8.30
Foliar spray								
Sprayed	26.67	1.19	6.03	0.11	44.76	2.14	15.93	0.13
Unsprayed	44.29	64.29	6.19	28.76	68.81	85.24	14.71	36.32
CV (%)	24.39	28.88	29.96	27.84	28.43	22.04	14.14	18.10
LSD (5%)	11.55	5.12	NS	24.08	13.55	18.47	NS	5.92

PSI=percentage of severity index, DAS=days after sowing, means with in a column followed by the same letter were not significantly different, CV=Coefficient of variation, LSD=Least Significant Difference.

Table 1: Effect of faba bean varieties on percentage severity index and incidence of faba bean gall disease at Farta and Tach gaynt under field condition in 2014 cropping season.

in unsprayed plot as compared to sprayed plot at both locations (Table 3). Percentage severity index was showed significant difference (P<0.05) among faba bean varieties at both locations at all dates of assessment. There was also a significant variation (P<0.05) between sprayed and unsprayed plot on percentage severity index of the disease at both locations at all dates of assessment except at initial (26 and 28 days after sowing).

Higher mean severity of faba bean gall was recorded from variety Adet-Hana at both initial and final dates of assessment at Farta. Whereas the mean severity was lower on variety Kasa at both initial and final date of assessment (Table 1). At Tach gaynt, lower mean severity was scored from variety Tumsa at both initial and final dates of assessment. Faba bean gall disease severity was higher on variety Kasa at 28 days after sowing (DAS) and Bulga 70 at 68 DAS disease assessment (Table 1).

The integration of varieties with fungicide spray by Bayleton was showed a significant difference (P<0.05) on percentage severity index (PSI) of faba bean gall all dates of assessment at both sites. The PSI of faba bean gall was reached zero on sprayed plot of variety Degaga, NC58, Obse, Tumsa and Local check at 46 and 48 DAS at Farta and Tach gaynt, respectively. Whereas the mean value of unsprayed plot of these varieties was showed increasing PSI of faba bean gall from the initial to the final date of disease assessment. The percentage severity index of faba bean gall was significantly reduced on sprayed plot than unsprayed at both locations. From unsprayed plot, PSI of faba bean gall was higher on variety Adet -Hana and Bulga70 while lower value of PSI was scored from variety Kasa and Local check at Farta at final date of assessment (Table 3). The use of Bayleton foliar spray reduced PSI of faba bean gall by 60.7%, 55.9% and 46.7% on sprayed Adet Hana, Bulga70 and CS20DK varieties in their order over unsprayed plot of

these varieties. Similarly, Bayleton foliar spray in combination with variety Moti, Tesfa and Gebelecho decreased PSI of faba bean gall by 38.5%, 33.3% and 30.0% in that order over unsprayed plot of these varieties at final date of severity assessment at Farta. At Tach gaynt, Bayleton foliar spray with variety was lower the PSI of faba bean gall on sprayed variety of Bulga70, Nc58, and Kasa by 55.6%, 50.4% and 45.2%, respectively over unsprayed plot of these varieties at final date of assessment. In addition to these, foliar application of Bayleton on Mesay, Adet Hana and CS20DK varieties was reduced the PSI of faba bean gall by 44.8%, 44.8% and 41.1%, in that order over the unsprayed plot of these varieties at the final date of severity record. The maximum mean severity of 57.5% was observed in Awi zone followed by the study area with mean severity of 40.7% [8]. Lower mean value of PSI of faba bean gall was scored from unsprayed plot of Tumsa and Obse at the final date of assessment at Tach gaynt. The disease was more severe in study area as compared to other regions [8]. The plot subjected to foliar spray was reduced the faba bean gall diseases severity in all varieties including local check at all dates of assessment as compare to unsprayed plot at both sites. All faba bean varieties were significantly affected by the disease. This result coincides with Dereje et al., Teklay et al., Hailu et al. [8,15,16]. Generally, Bayleton foliar spray on faba bean varieties has significant effect in reduction of PSI of faba bean gall disease at both locations.

AUDPC (Area under disease progress curve)

The integration of varieties with fungicide spray by Bayleton was showed significant difference (P<0.05) on AUDPC of faba bean gall disease at both sites. The maximum mean AUDPC of the disease was scored on variety Adet-Hana and Bulga 70 on unsprayed plot at Farta and Tach gaynt in their order. Lower value of AUDPC was recorded from variety Local check and Tumsa on unsprayed plot at Farta and

Variety X Foliar spray		Disease incidence at Farta DAS (%)		Disease incidence at Tach gaynt DAS (%)	
		Initial (26)	Final (66)	Initial (28)	Final (68)
Adet-Hana	Sprayed	73.3	3.3	53.3	10.0
	Unsprayed	76.7	93.3	86.7	93.3
Bulga 70	Sprayed	53.3	0	53.3	10.0
	Unsprayed	63.3	80.0	90.0	100.0
CS20Dk	Sprayed	33.3	0	60.0	0
	Unsprayed	50.0	70.0	80.0	93.3
Degaga	Sprayed	26.7	0	40.0	0
	Unsprayed	30.0	50.0	66.7	93.3
Dosha	Sprayed	23.3i	6.7	30.0	0
	Unsprayed	43.3	76.7	66.7	83.3
Gebelecho	Sprayed	26.7	0	43.3	6.7
	Unsprayed	60.0	70.0	56.7	70.0
Kasa	Sprayed	23.3	0	56.7	0
	Unsprayed	36.7	53.3	73.3	100.0
Mesay	Sprayed	10.0	0	50.0	0
	Unsprayed	36.7	56.7	76.7	96.7
Moti	Sprayed	30.0	3.3	50.0	0
	Unsprayed	43.3	66.7	36.7	66.7
Nc58	Sprayed	20.0	0	50.0	0
	Unsprayed	33.3	50.0	73.3	86.7
Obse	Sprayed	16.7	0	50.0	0
	Unsprayed	43.3	63.3	76.7	76.7
Tesfa	Sprayed	13.3	3.3	33.3	0
	Unsprayed	36.7	53.3	50.0	76.7
Tumsa	Sprayed	3.33	0	16.7	0
	Unsprayed	26.7	56.6	43.3	56.7
Local check	Sprayed	20.0	0	40.0	0
	Unsprayed	40.0	60.0	86.7	100.0
CV (%)		24.76	27.85	20.99	23.69
LSD (5%)		31.77	20.27	38.07	16.93

PSI=percentage of severity index, DAS=days after sowing, CV=Coefficient of variation, LSD=Least Significant Difference.

Table 2: Effect of faba bean varieties with fungicide foliar spray on incidence of faba bean gall disease at Farta and Tach gaynt under field condition in 2014 cropping season.

Tach gaynt respectively. Higher mean AUDPC faba bean gall was scored from unsprayed plots than sprayed plots (Table 3). The area under disease progress curve (AUDPC) is a very convenient summary of plant disease epidemics that incorporates into initial intensity, the rate parameter, and the duration of the epidemic which determines final disease intensity [20]. The AUDPC was used to summarize the epidemics of the disease in the different varieties evaluated during this experiment.

Seed yield of faba bean

Faba bean varieties with foliar spray were showed significant variation (P<0.05) on seed yield at both locations. At Farta, higher yield was obtained from sprayed plot of variety Adet-Hana and CS20DK while lower yield was scored from sprayed plot of variety Mesay and Obse. From unsprayed plot, higher yield per hectare was scored from variety CS20DK and lower value was obtained from variety Moti at Farta (Table 4). At Tach gaynt, the highest yield was scored from sprayed plot of variety Kasa (1441.7 kg/ha) and Nc58 (1333.3 kg/ha) where as lower yield per hectare was obtained from sprayed plot of variety of Tesfa (558.3 kg/ha). From unsprayed plot, lower yield was recorded from variety Gebelecho, Tumsa, Moti and Dosha at Tach gaynt. Maximum mean yield was gained from unsprayed plot of variety Local check. The epidemic conditions of the disease have significant implication on the production of faba bean and on the country's Economy [8]. The disease reduced the yield by 1116.7 kg/ha, 700 kg/

ha, 541.6 kg/ha on variety Adet-Hana, Degaga and Gebelecho in their order at Farta. The faba bean gall caused yield loss up to 30-100% [16]. At Tach gaynt, it reduced the yield by 123.4 kg/ha, 291.7 kg/ha, 166.6 kg/ha on the above varieties in their previous order. In this study, the yield of all varieties reduced on unsprayed plots at both locations. Past surveyed study explained that faba bean production in Ethiopia is highly challenged by new faba bean gall forming disease [8].

Conclusion

Faba bean gall disease is the newly emerging and aggressively spread disease in the country and cause 100% yield loss in susceptible faba bean varieties. The combined use of Bayleton foliar spray with variety significantly reduced PSI of faba bean gall by 60.74% and 55.93% on sprayed Adet Hana and Bulga70 varieties in their order over unsprayed plot of these varieties at Farta. At Tach gaynt, the interaction effect was lower the PSI of faba bean gall on sprayed variety of Bulga70, Nc58, and Kasa by 55.56% and 50.37%, respectively over unsprayed plot of these varieties at final date of assessment. The disease reduced the yield by 1116.7 kg/ha and 123.4 kg/ha on unsprayed plot of variety Adet-Hana at Farta and Tach gaynt in their order. It also decreased the yield by 700 kg/ha and 291.7 kg/ha on unsprayed plot of variety Degaga respectively at Farta and Tach gaynt.

The investigation showed that all faba bean varieties were infected by the newly emerged faba bean gall disease at both locations. The

Variety X Foliar spray		Location					
		Farta			Tach gaynt		
		PSI DAS (%)		AUDPC	PSI DAS (%)		AUDPC (% days)
		Initial (26)	Final (66)	(% days)	Initial (28)	Final (68)	
Adet-Hana	Sprayed	19.3	0.4	74.1	21.1	0.4	58.5
	Unsprayed	14.8	61.1	332.6	21.9	45.2	321.5
Bulga 70	Sprayed	17.8	0	74.1	18.9	0.7	54.1
	Unsprayed	11.1	55.9	307.4	17.4	56.3	368.9
CS20Dk	Sprayed	7.8	0	33.3	18.5	0	50.4
	Unsprayed	5.6	46.7	209.6	21.1	41.1	287.4
Degaga	Sprayed	4.1	0	9.6	11.9	0	29.6
	Unsprayed	4.1	16.7	93.3	10.0	34.8	213.3
Dosha	Sprayed	5.2	0.4	32.6	12.6	0.4	32.6
	Unsprayed	10.0	39.3	225.9	11.5	27.0	203.0
Gebelecho	Sprayed	7.4	0	29.6	14.1	0.4	44.4
	Unsprayed	13.7	30.0	205.2	16.7	27.0	210.4
Kasa	Sprayed	3.0	0	14.8	24.4	0	46.6
	Unsprayed	0.4	2.6	15.6	20.7	45.2	345.2
Mesay	Sprayed	0	0	0	21.1	0	44.8
	Unsprayed	1.1	11.1	67.4	17.0	44.8	304.4
Moti	Sprayed	4.4	0.4	19.3	12.6	0	39.3
	Unsprayed	9.3	38.9	196.3	10.7	27.4	214.1
Nc58	Sprayed	5.6	0	14.8	21.5	0.0	54.8
	Unsprayed	6.3	27.4	140.7	18.9	50.4	338.5
Obse	Sprayed	2.6	0	5.9	14.8	0	43.0
	Unsprayed	4.1	12.2	105.2	13.3	24.8	181.5
Tesfa	Sprayed	4.8	0.4	20.0	5.2	0	16.3
	Unsprayed	2.6	33.7	111.6	9.3	27.4	168.9
Tumsa	Sprayed	1.1	0	3.7	7.0	0	24.4
	Unsprayed	2.6	16.7	60.7	4.8	21.1	139.3
Local check	Sprayed	1.1	0	4.4	19.3	0.0	45.2
	Unsprayed	1.5	10.4	50.4	12.6	35.9	260.0
CV (%)		29.07	30.32	28.99	19.90	17.86	11.46
LSD (5%)		14	35.48	199.62	15.01	11.29	100.27

PSI=percentage of severity index, DAS=days after sowing, AUDPC=area under disease progress curve, CV=Coefficient of variation, LSD=Least Significant Difference.

Table 3: Effect of faba bean varieties with fungicide foliar spray on percentage severity index and AUDPC at Farta and Tach gaynt under field condition in 2014 cropping season.

Variety X Foliar spray		Yield (kg/ha) at Farta	Yield (kg/ha) Tach gaynt
Adet-Hana	Sprayed	1900.0	1106.7
	Unsprayed	783.3	983.3
Bulga 70	Sprayed	900	1291.7
	Unsprayed	616.7	1000
CS20Dk	Sprayed	1766.7	1225
	Unsprayed	1166.7	900
Degaga	Sprayed	1466.7	1025
	Unsprayed	766.7	733.3
Dosha	Sprayed	1458.3	658.3
	Unsprayed	841.7	408.3
Gebelecho	Sprayed	1133.3	483.3
	Unsprayed	591.7	316.7
Kasa	Sprayed	733.3	1441.7
	Unsprayed	483.3	1033.3
Mesay	Sprayed	516.7	908.3
	Unsprayed	491.7	725
Moti	Sprayed	583.3	541.7
	Unsprayed	225	400
Nc 58	Sprayed	775	1333.3
	Unsprayed	716.7	1283.3

Obse	Sprayed	550	683.3
	Unsprayed	508.3	550
Tesfa	Sprayed	1008.3	558.3
	Unsprayed	583.3	425
Tumsa	Sprayed	966.7	633.3
	Unsprayed	475.0	391.7
Local check	Sprayed	600.0	1316.7
	Unsprayed	500.0	1308.3
CV (%)		18.35	10.44
LSD (5%)		787.64	970.11

CV=Coefficient of variation, LSD=Least Significant Difference.

Table 4: Effect of faba bean varieties with fungicide foliar spary on yield of faba bean at Farta and Tach gaynt under field condition in 2014 cropping season.

plot subjected to foliar spray was reduced the faba bean gall diseases severity in all varieties including local check at all dates of assessment as compare to unsprayed plot at both sites. In general, Bayleton foliar spray on faba bean varieties has a significant effect in reduction of the severity of faba bean gall disease at both locations. Therefore, foliar spray using Bayleton at rate of 1.5 a.i/ha with variety Adet-Hana, Bulga70 and CS20DK, Nc58 and Kasa provided better yield and the fungicide could be recommended as the best management option for control of faba bean gall disease in south Gondar zone, Northwestern Ethiopia.

References

1. Torres AM, Roman B, Avila CM, Satovic Z, Rubiales D, et al. (2006). Faba bean breeding for resistance against biotic stresses: towards application of marker technology. Euphytica 147: 67-80.

2. Asfaw T, Tesfaye G, Beyene D (1994) Genetics and breeding of faba bean. pp: 122-137.

3. FAOSTAT (2004).

4. Yohannes Degago (2000) Faba Bean (Vicia faba) in Ethiopia. Institute of Biodiversity, Conservation and Research (IBCR). Addis Ababa, Ethiopia, p: 43.

5. Bond DA, Laws DA, Hawtin GC, Saxena MC, Stephens JS (1985) Faba bean (Vicia faba L.). In Summer field, R.J. and Roberts EH (Eds.). Grain Legume Crops. William Collins Sons Co. Ltd., London, UK, pp: 199-265.

6. Sahile S, Ahmed S, Fininsa C, Abang MM, Sakhuja PK (2008) Survey of chocolate spot (Botryty fabae) disease of faba bean (Vicia faba L.) and assessment of factors influencing disease epidemics in northern Ethiopia. Crop Protection 27: 1457-1463.

7. Gemechu K, Mussa J (2002) Comparison of three secondary traits as determinants of grain yield in faba bean on waterlogged Vertisols. J of Genetics and Breeding. 56: 317-326.

8. Hailu E, Getaneh G, Sefera T, Tadesse N, Bitew B, et al. (2014) Faba Bean Gall; a New Threat for Faba Bean (Vicia faba) Production in Ethiopia. Adv Crop Sci Tech 2: 144.

9. Biruk Bereda (2009) Production & Marketing Activity of Broad Bean in Ethiopia. Ethiopia Commodity Exchange Authority study report, Addis Ababa, Ethiopia.

10. CSA (Central Statistics Authority) (2007) Agricultural Sample Survey for 2006 Report Addis Ababa, Ethiopia, p: 349.

11. Mussa J, Dereje G, Gemechu K (2008) Procedures of Faba Bean Improvement through Hybridization. Technical Manual No. 21, Ethiopian Institute of Agricultural Research, p: 48.

12. Berhanu M, Getachew M, Teshome G, Temesgen B (2003) Faba bean and field pea diseases research in Ethiopia. In: Ali K, Kenneni G, Ahmed S, Malhotra R, Beniwal S, et al. (eds.) (2003) Food and forage legumes of Ethiopia: Progress and prospects. Proceedings of the Workshop on Food and Forage Legumes. Addis Ababa, Ethiopia, pp: 278-287.

13. Nigussie T, Seid A, Derje G, Tesfaye B, Chemeda F, et al. (2008). Review of Research on Diseases Food Legumes. In: Abraham Tadesse (Eds). Increasing crop production through improved plant protection 1:85-124.

14. Dereje Gorfu, Tesfaye Beshir (1994) Review of faba bean disease research in Ethiopia. pp: 328-345. In: Proceedings of the First National Review Workshop on Cool season food legumes, IAR, Addis Ababa, Ethiopia.

15. Dereje G, Wendafrash, Gemechu K (2012) Faba Bean Galls: a new disease of faba bean in Ethiopia. pp: 1-6.

16. Teklay A, Tsehaye B, Yemane N, Assefa W (2014) The Prevalence and Importance of Faba Bean Diseases with Special Consideration to the Newly Emerging "Faba Bean Gall" in Tigray, Ethiopia. Discourse J of Agri and Food. 2: 33-38.

17. Bernier CC, Hanounik SB, Hussein MM, Mohamed HA (1984) Rating scale for faba bean diseases in Nile valley. ICARDA Information. Bulletin No 3, p: 37.

18. Guoqing D, Ronghai H, Xuny L, Oifang G, Dazhao Y, et al. (1993) Evaluation and screening of faba bean germoplasm in China. Fabis Newsletter. 32: 8-10.

19. SAS Institute (2003) SAS/STAT guide for personal computers, version 9.1.3 edition, SAS Institute Inc., Cary, North Carolina, USA.

20. Madden LV, Hughes G, Bosch FVD (2008) The Study of Plant Disease Epidemics. American Phytopathological Society. St Paul Minnesota, USA.

26

Fog Water Collection for Agriculture Use (Peanut Irrigation) Under Semi-Arid Region Conditions in North Coast of Egypt

Makled KHM* and Abou El Enin MM

Agronomy, Department of Agriculture, Al-Azhar University, Cairo, Egypt

Abstract

Two field experiments were conducted at Marsa Matrouh Agricultural Research Farm during summer seasons of 2013 and 2014 using drip irrigation system, to evaluate the effect of some fog water harvesting models (f.w.h.m) of, (Double mesh had 220 stitches/cm^2 (model-1), Single layer mesh touching each other had 220 stitches/cm^2 (model-2), Double mesh had 120 stitches/cm^2 (model-3) and Single layer mesh touching each other had 120 stitches /cm^2 model-4) under some farmyard manure rates (20,30,40 m^3)/fad on groundnut productivity. Results indicated that, there were differences between studied factor, (f.w.h) model-1, was significant exposed its superiority on the total water amount harvested it during the two seasons led to give significant greatest values of pods, seeds yield/ Fadden, biological yield/fad., seed and harvest index, and water use efficiency. It is worthy to mention that, also led to enhance yield as compared by the other (f.w.h.m) during the two experimental seasons. Results revealed that, by increasing the amount of the added farmyard manure to improve the most values of the previous peanut traits significantly, during the two seasons. The interaction effect between f.w.h.m and farmyard manure rates showed significant effect, grow in peanut plants under the condition of f.w.h model-1 and fed by 30 or 40 m^3 of that fertilizer gave best significant values for most studied peanut traits compared with other treatments.

Keywords: Fog water; Manure; Peanut; Mesh; Model

Introduction

Fog is an environmental water resource of great importance. It plays an integral role in many diverse ecosystems. A very special part of the fog activities in the world today is focused on fog collection to provide water for managed use. One of the most exciting aspects of this resource is that in many regions the supply of water will be limited only by the number of collectors one chooses to install. In addition, since the source of the fog is normally the movement of marine stratocumulus deck over coastal mountains, the water quality is good and the water can be used for drinking and other domestic and agricultural purposes [1]. The latter experiments indicate that fog has been considered as a water resource in some arid or desert environments but it has never been developed as a serious water supply. Africa has arid and desert conditions in both the extreme north and the extreme south of the continent. Fog-water collection systems may have application at many locations in Africa but of date, there have been few experiments to verify this. One of the most interesting reports was of a tree in the Canary Islands, which as early as 1764, was said to have produced large amounts of fog-water for the islanders [1] The different aspects of the technology and the project results have been documented in the literature and it deserves strong consideration in regions that are arid or seasonally arid. Namibia is the first African country in which the possibility of using fog collection as a water supply for indigenous peoples is being evaluated. Yamen reported that, the potential to collect fog water for fresh water production was investigated in the mountains near Hajja, north capital city of Sana'a and inland from the Red Sea. In 2003, Yamen found that, best sites averaged 4.5 L/m^2/day over the 3-month dry winter period using LFCs fog collectors after successful initiation. The project was given over to the local people and local organization [2]. This application in Egypt depends on finding locations where there are high horizontal fluxes of fog water in regions with an acute water need. Fog has the potential to provide an alternative source of freshwater in semiarid and arid regions if harvested with simple and low-cost collection systems known as fog collectors. This application in Egypt depends on finding locations where there are high horizontal fluxes of fog water in regions with an acute water need. Peanut is planted in arid and semi-arid areas, it is very rich in protein and oil of good quality. Drought is one of the limiting factors to peanut yield in many

countries [3,4] Groundnut (*Arachis hypogaea* L.) has an unique importance in our country either for local use or as foreign exchange earner. The soil texture of Egyptian belts is generally light and well drained. The farmyard manure is one of the very important treatments, which improves the sandy soil properties specially increasing its water hold capacity to save and increase the utilization efficiency of water irrigation. Venkataramana [5] indicated that organic manure has a profound effect on improving soil physical, chemical and biological properties and enhancing productivity of field crops. They also added that, groundnut fed by the application of FYM at 10 to 15 ton/fad increased the pod and haulm yields and improved the yield parameters like shelling percentage, 100 seed weight and sound mature kernel compared to the recommended dose of fertilizers. This study aims to evaluate some fog water harvesting models under different farmyard manure rates on groundnut productivity under the condition of Marsa Matrouh.

Materials and Methods

Two field experiment were conducted during the summer seasons 2013 and 2014 at the farm of Marsa Matrouh Agricultural Research Station, to evaluate the impact of some fog water harvesting models under different farm yard manure rates on yield, yield components and some chemical constituting of peanut (*Arachis hypogaea* L).

Experimental treatments

Fog water harvesting methods

Description of atrapanieblas: The mean structure is called

*Corresponding author: Makled KHM, Agronomy, Department of Agriculture, Al-Azhar University, Cairo, Egypt, E-mail: Magro_modeller@yahoo.com

atrapanieblas (Spanish, meaning trapping fog). It mainly consists of a large meshes made of poly propylene material suspended vertically to the wind direction at 100 m far from the sea water by hang it very taut, between two posts to collect the water droplets out of the fog. As the fog passing through the meshes, the fog with its droplets is pushed through the mesh by the wind. The droplets then collide with the fibers of the mesh and stay attached to them. When the droplets accumulate and grow, they drip down the mesh. Underneath, along the base a drip rail (Figure 1) to collect the fog water, which drips down the mash after it come in contact with the mesh. The dimensions of the mesh are 3 m high 17 m long. Thus, the area of one fog collector is 51 m². The base of the mesh is 2 m above the ground. The collected water in the drip rail is piped through PVC-pipes by gravity to small measured tank for each model. Every day at 7 Am clock the amount of harvested water was estimated, recorded and trans located to special big tank for each model its volume 1000 L (1 m³). That tank was connected by drip irrigation system cover 9 sub plot for each model so, the total amount of harvested water during the growing season started form 20th of April before sowing at 15 days until 15th September the date of stop irrigation can be calculated.

The study covered four models of atrapaniebles as follow

Double mesh had 220 stitches /cm² with shade coefficient of 70%.

- Single layer mesh touching each other had 220 stitches /cm² with shade coefficient for each layer 70%.

- C- Double mesh had 120 stitches /cm² with shade coefficient of 50%.

- D- Single layer mesh touching each other had 120 stitches/cm² with shade coefficient for each layer 50%.

Farmyard manure fertilizer rates: To improve the hold capacity of experimental soil (sandy soil) and to save water irrigation, the study covered three farmyard manure rates as follow:

1-20 m³/fed, 2- 30 m³/fed, 3- 40 m³/fed

The amount of farmyard for each rate was calculated according the area of the sub plot and added during soil preparation.

Soil mechanical and chemical analysis: To be in touch with the soil fertility after applying the three-farm yard manure, soil samples were collected from the experimental site before sowing and after harvesting to the depth of 30 cm and air dried for mechanical and chemical analysis that recorded in Table 1. In both seasons, the treatments were arranged in split plot design in three replications. The main plot was randomly devoted to the fog harvest models. The area of each was 31.5 m² (3.5 m × 9 m) each one consisted of three-sub plot. The sub plot was randomly devoted to the three farmyard manure rates. Each one, area was 10.5 m² (3 m × 3.5 m). It consisted of 5 rows/plot spaced at 60 cm apart and 3.5 m long. Calcium super phosphate (15.5% P_2O_5) at rate of 200 kg/fed, and potassium sulfate (48% K_2O) at rate of 50 kg /fed, Gypsum farm at rate of 500 kg / fed. Were added during land preparation, Sowing

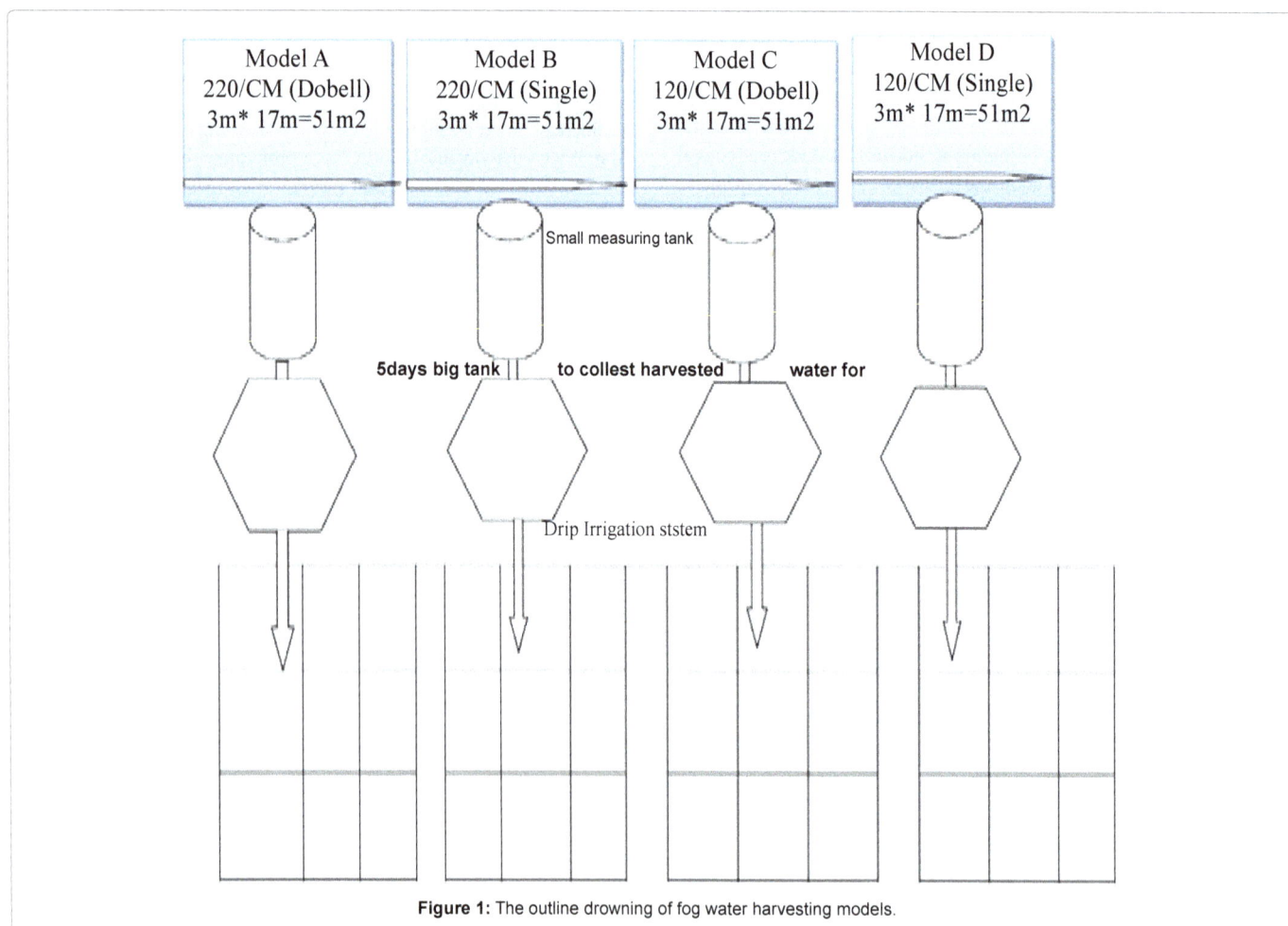

Figure 1: The outline drowning of fog water harvesting models.

Before sowing								
Farmyard manure	Mechanical analysis				Chemical analysis			
	%clay	%silt	%sand	Texture	pH	Ec mm/cm³	%O.M	%O.C
Zero	2	6.6	91.4	sandy	8.2	3.4	0.12	0.07
After harvest								
20 m³	2.3	7.8	89.9	sandy	7.9	3.2	1.23	0.72
30 m³	2.8	8.6	88.6	sandy	8.0	3.0	1.34	0.78
40 m³	3.6	9.8	86.6	sandy	7.9	3.1	1.65	0.96

*Soil, Water and Environ Res Inst. ARC. Giza, Egypt

Table 1: Mechanical and chemical analysis for soil samples of the experimental site before sowing and after harvesting (0-30 cm depth)*.

took place on May 5th each season, the drip irrigation system was used for irrigating the experiment during the two seasons. Each sup-plot contain five GR pipelines of hoses GR-diameter 16 mm at the distance of 20 cm had at rate of 4 L/h apart, so each dripper irrigated two halls. The irrigating was conducted every 5 days after sowing irrigation. It wealthy to mention that, the amount of water harvested form each model during 5 days was used to irrigate that treatment. Peanut seeds were inoculated with *Rhizobium spp* before planting it in hills at 10 cm apart three seeds in each. After germination, the plants in each hall were thinned in two plants.

Characteristics studied

Yield and yield component: All the plants of each plot were harvested and left for air dry then the plants weighted. All the pods of the plants were removed and weight to obtain.

3-Biological yield/Fadden (kg/fed), 4-pods yield Fadden (kg/fed), 5-Seed yield Fadden (kg/fed) 6- Seed index, 7-Harvest index=Economic yield / Biological yield × 100

8-Water use efficiency (kg/m³) Expressed as the weight of air-dried biological yield (kg/fed) or air-dried pods yield (kg/fed) and seed air-dried yield (kg/fed). computed for the different treatment by using the formula of El-Boraie [6], as follow:

WUE=Biological yield or pods yield and Seed yield (kg/fed)/ Evapotranspiration (m³/fed)

Statistical analysis

The analysis of variance was used for this experiment according to Jagdev [7] the least significant differences (L.S.D) test at the 5% level of probability was used to compare the differences between means. Consumptive use (m³/fed), the quantities of added water for the different treatment were recorded.

Results and Discussion

The biological, pods and seeds yield (kg/fed)

Results recorded in Table 2, show the variance between some fog water harvesting models, (F.W.H), farmyard manure rate and the interaction effect between them on biological, pods and seeds

Treatments		Season 2013			Season 2014		
Harvested water (m³/fed)	Manure (m³/fed)	Yield (kg/fed) of			Yield (kg/fed) of		
		Biological	pods	Seeds	Biological	pods	Seeds
*M(1) 1126 m³	20 m³	3289.16	1109.83	730.86	3334.96	1110	715.23
	30 m³	3393.35	1182.33	792.1	3393.76	1178	762.83
	40 m³	3423.33	1165.83	784.2	3444.83	1176.33	749.1
Mean		3368.61	1152.66	769.05	3391.18	1154.8	742.4
*M(2) 1036 m³	20 m³	2942.23	1024.67	658.23	2997.9	1037	668.3
	30 m³	3017.33	1061.67	691.1	3078.26	1068	693.73
	40 m³	3040.67	1088.5	694.2	3081.56	1083	681.9
Mean		3000.08	1058.28	681.18	3052.58	1062.7	681.31
*M(3) 992 m³	20 m³	2578.67	902	553.2	2593.3	904.16	562.63
	30 m³	2627.23	935	585.9	2670.46	923	582.27
	40 m³	2665.52	937	587.46	2701.96	933.66	594.27
Mean		2623.81	924.67	575.52	2655.24	920.28	579.72
*M(4) 880 m³	20 m³	2226.23	741.5	460.27	2286.43	747.33	479.3
	30 m³	2348.33	805	508.53	2380.3	803	527.63
	40 m³	2430.67	804.33	515.43	2463.33	816.33	530.93
Mean		2335.07	783.61	494.74	2376.69	788.89	512.62
GMI		2831.9	979.81	630.12	2868.92	981.65	629.01
Mean of Manure							
20 m³		2759.07	944.5	600.64	2803.15	949.65	606.37
30 m³		2846.56	996	644.41	2880.7	993	641.62
40 m³		2890.05	998.92	645.32	2922.92	1002.3	639.05
LSD at 5%							
Irrigation (I)		42.73	25.8	20.91	66.38	27.28	32
Manure (M)		46.05	23.75	18.9	39.02	20.12	28.45
I × M		61.2	31.56	25.11	51.85	26.72	37.81

*M=Model of atrapanieble (As given in material and methods)

Table 2: Evaluation of some fog water harvesting modes under some farmyard manure rates on the Fadden yield of Biological, pods, and Seeds of peanut in 2013 and 2014 seasons.

yield (kg/fed)during 2013 and 2014 seasons. It was noticed from the results recorded in Table 2 that, biological, pods and seeds yield (kg/fed) were significantly affected by the variance between the total water amount harvested from each F.W.H model during the two experimental seasons. Form studying the results in Table 2, significant positive effect was acquired by growing peanut plants irrigated by the greatest total amount of water harvested by model-1 (1126 and 1144 m³) during 2013 and 2014 seasons respectively which its mesh had the greatest number of stitches /cm² (220) and consisted of double layer of mesh.

That reflect relying on gain the highest values of biological yield/fed (3368.61 and 3391.18 kg/fad), pod yield/fad (1152.70 and 1154.80 kg/fad) and seed yield/fed (769.05 and 742.40 kg/fad) in 2013 and 2014 seasons respectively. It is wealthy to mention that, the previous traits were decreased gradually by decreasing either number of mesh layer or stitches /cm². That companied by decreasing the total amount of harvested water from each during the two experimental seasons. These results are in general, agree with those obtained by Sabino [8], Gohri [9], Aboelill [10].

The remarkable effect was the interaction between F.W.H models and farmyard manure rates, significant effect of that source of variance, was found on the previous studied traits during 2013 and 2014 seasons. Cultivation peanut plants under the condition of irrigate it from the total amount of harvested water of model 1 and fed by 30 or 40 m³ of farmyard manure gave the greatest values of biological (3423.33 kg/fad) pod (1182.33 kg/fad) and seed yield/fad (792.10 kg/fad) compared with the other treatments, in the first growing season. Similar findings had been observed in the second one.

Seed and harvest index

The averages of seed and Harvest index as affected by the different fog water harvesting models (F.W.H.M), some farmyard manure rate and the interaction between them in 2013 and 2014 seasons were recorded in Table 3. Results in Table 3 cleared that, (F.W.H) models significantly varied due to its effect on peanut seed and harvest index; it means the total water amount harvested effect from each one during the two experimental season, F.W.H model 1 and 2 exposed their superiority due to the total water amount harvested by them during 2013 (1126 m³) and 2014 (1144 m³), with its specification which explained before, resulted the greatest pods and seeds yield per plant and Fadden as well as biological yield/fad, shelling % and the lowest number of pod/100 gm. F.W.H. Model 1 increase peanut seed index significantly by (1.17%, 4.85% and 6.54%) as compared by F.W.H models 2, 3 and 4 of water during 2013 season respectively. The results of 2014 season took the same trend. These results may be due to the favor effect of increasing the total water amount added to irrigate peanut plant, led to in favor the vegetative growth, net assimilation rate biological yield /fed and seed yield/fed.

Regarding peanut harvest index as affected by the same factor, results in Table 3 showed significant effect of that factor during 2013 and 2014 seasons F.W.H, model 4. Secured the lowest harvest index (0.336 and 0.332) during 2013 and 2014 seasons compared with the other f.w.h. models that may be due to the harmful effect of the stress condition of drought caused by low total water amount harvested during the two growing seasons compared with model 2 and 3. These results are in the same line with those obtained by Venkataramana [5]. Respecting to the effect of farmyard manure rates, results recorded in ta Table 3 cleared that, during the two experimented seasons, peanut seed index and Harvest index were gradually increased by increasing the farmyard manure amount from 20 m³/fed to 30 m³ /fad. For example,

Treatments		Season 2013		Season 2014	
Harvested water (m³/fed)	Manure (m³/fed)	Seed index	Harvest index	Seed index	Harvest index
*M(1) 1126 m³	20 m³	75.48	0.34	75.81	0.33
	30 m³	76.34	0.35	77.78	0.35
	40 m³	77.63	0.34	78.41	0.34
Mean		76.48	0.34	77.34	0.34
*M(2) 1036 m³	20 m³	74.48	0.35	74.96	0.35
	30 m³	75.66	0.35	76.35	0.35
	40 m³	76.37	0.36	76.18	0.35
Mean		75.59	0.35	75.83	0.35
*M(3) 992 m³	20 m³	71.60	0.35	72.29	0.35
	30 m³	73.48	0.36	72.41	0.35
	40 m³	73.75	0.35	72.79	0.35
Mean		72.94	0.35	72.50	0.35
*M(4) 880 m³	20 m³	70.40	0.33	70.11	0.33
	30 m³	72.12	0.34	71.53	0.34
	40 m³	72.83	0.33	72.26	0.33
Mean		71.78	0.34	71.30	0.33
GMI		74.20	0.35	74.24	0.34
Mean of Manure					
20 m³		72.82	0.34	73.35	0.34
30 m³		74.24	0.35	74.54	0.34
40 m³		75.30	0.35	75.14	0.34
LSD at 5%					
Irrigation (I)		0.80	0.00	0.85	0.00
Manure (M)		0.38	0.00	0.67	0.00
I × M		0.51	0.01	1.33	0.00

Table 3: Evaluation of some fog water harvesting modes under some farmyard manure rates on seed index and Harvest index of peanuts in 2013 and 2014 seasons.

in 2014 season seed index (%) was increased by 1.62% and 2.44% by adding 30 m³ and 40 m³ of farmyard manure as compared by 20 m³ application, the results of Harvest index (%) during the 2013 and 2014 seasons took the same trend, with the exception of 40 m³ of farmyard manure during the two seasons, that rate led to significant reduction on that trait. These results are in agreement with those obtained by Venkataramana [5]. As for the interaction effect between F.W.H models and farmyard manure (F.Y.M) rates showed significant effect on peanut seed and Harvest index during the two experimental seasons. Growing peanut plants irrigated by F.W.H model 1 which supplied peanut plants by 1126 m³ and 1144 m³ of water irrigation during 2013 and 2014 seasons, respectively and fed by 30 m³ or 40 m³ of (F.Y.M) during the two seasons recorded the greatest seed index (%) (76.34 or 77.63) and (77.76 or 78.4) compared by the other treatments. Significant effect was acquired by the interaction effect between F.W.H. models and (F.Y.M) rates on peanut harvest index. Growing peanut plants under the condition of F.W.H. model-2 which supplied the plant by 1036 m³ or 1033 m³ of water irrigation during 2013 and 2014 seasons and/fed by 40 m³ of (F.W.H) gave the greatest harvest index (0.358 and 0.351), respectively compared with the other treatments.

Water use efficiency (kg/m³)

The averages of water use efficiency (W.U.E) depending on Biological, pods and seeds (kg/m³) as affected by the different fog water harvesting models (F.W.H.M), some farmyard manure rates and the interaction effect between them in 2013 and 2014 seasons were tabulated in Table 4. Result in Table 4 cleared that, (F.W.H) models significantly varied due to its effect on W.U.E Biological and seeds (kg/m³) referring to the total water amount harvested from each one during the two

experimental seasons. F.W.H model 1 surpassed the other (F.W.H) models on the total water amount harvested by it during 2013 (1126 m³) and 2014 (1144 m³), it increased peanut W.U.E depending on biological (kg/m³) significantly by 3.1%, 12.83% and 12.83% as compared by F.W.H models 2, 3 and 4 which harvested 1036 m³, 992 m³ and 880 m³ of water during 2013 season respectively. As for peanut W.U.E depending on pod or seed, yield (kg)/fad results in Table 4 observed that model-1 of F.W.H continued its superiority to gave the best W.U.E depending on pod or seed yield (kg)/fad as compared with model 1. These results are previous confirmed by Sabino [8], Gohri [9], Aboelill [10].

Regarding to the effect of farmyard manure rates, results recorded in Table 4 cleared that, during the two experimental seasons, W.U.E depending on biological or pods and seeds yield(kg/fed) were gradually increased by increasing the farmyard manure amount from 20 m³/fed to 30 m³ or 40 m³/fed. For example, in 2014 season, W.U.E depending on biological yield/fad was increased by 2.90% and 4.36% by adding 30 m³ and 40 m³ of farmyard manure as compared by 20 m³ application, the results of W.U.E depending on pods and seeds (kg/m³) during the 2013 and 2014 seasons took the same trend. These results confirmed the aim of adding the farmyard manure under the condition of sandy soil to improve its nutrients content including micronutrients at more appropriate amount and rate to crop, also the slow regular release of nutrient may well better to meet the requirements of peanut crop. Moreover, the utilization efficiency of water irrigation will be increased by increasing sandy soil holding capacity. Similar results had been described by Venkataramana [5]. The interaction between F.W.H models and (F.Y.M) rates showed significant effect on peanut W.U.E biological, pods and seeds (kg/m³) during the two experiment seasons. Growing peanut plants irrigated by F.W.H model-1 which supplied

Treatments		Season 2013			Season 2014		
Harvested water (m³/fed)	Manure (m³/fed)	Water use efficiency (kg/m³)			Water use efficiency (kg/m³)		
		Biological	pods	seeds	Biological	pods	Seeds
*M(1) 1126 m³	20 m³	2.92	0.99	0.65	2.92	0.97	0.63
	30 m³	3.01	1.05	0.70	2.97	1.03	0.67
	40 m³	3.04	1.04	0.70	3.01	1.03	0.66
Mean		2.99	1.03	0.68	2.96	1.01	0.65
*M(2) 1036 m³	20 m³	2.84	0.99	0.64	2.90	1.00	0.65
	30 m³	2.91	1.03	0.67	2.98	1.03	0.67
	40 m³	2.94	1.05	0.67	2.98	1.05	0.66
Mean		2.90	1.02	0.66	2.96	1.03	0.66
*M(3) 992 m³	20 m³	2.60	0.91	0.56	2.62	0.92	0.57
	30 m³	2.65	0.94	0.59	2.70	0.93	0.59
	40 m³	2.69	0.95	0.59	2.73	0.95	0.60
Mean		2.65	0.93	0.68	2.68	0.93	0.59
*M(4) 880 m³	20 m³	2.53	0.84	0.52	2.57	0.84	0.54
	30 m³	2.67	0.92	0.58	2.68	0.90	0.59
	40 m³	2.76	0.91	0.59	2.77	0.92	0.59
Mean		2.65	0.89	0.56	2.67	0.89	0.58
GMI		2.80	0.97	0.62	2.82	0.96	0.62
Mean of Manure							
	20 m³	2.72	0.93	0.59	2.75	0.93	0.60
	30 m³	2.81	0.98	0.64	2.83	0.98	0.63
	40 m³	2.86	0.99	0.64	2.87	0.99	0.63
LSD at 5%							
Irrigation (I)		0.02	0.03	0.02	0.06	0.03	0.03
Manure (M)		0.02	0.02	0.02	0.04	0.02	0.03
I × M		0.03	0.03	0.03	0.05	0.03	0.04

Table 4: Evaluation of some fog water harvesting modes under some manure rates on Water use efficiency (kg/m³) in 2013 and 2014 seasons.

peanut plants by 1126 m³ and 1144 m³ of water irrigation during 2013 and 2014 seasons respectively and fed by 30 m³ or 40 m³ of farmyard manure gave the great W.U.E depending on biological (kg/m³) (3.01 or 3.04) and (2.97 or 3.01) compared by adding 20 m³ of farmyard manure. Peanut W.U.E pods and seeds (kg/m³) took the same trend [11-15].

Acknowledgements

Many thanks, first, are going to our God, the most merciful, the most beneficial and helpful for everyone, and nothing could be achieved without his welling and support.

References

1. Eck HV (1988) Winter wheat response to nitrogen and irrigation. Agron J 80: 902-908.

2. Salama JF, Hanna FR, Ahmed MA (1994) Flower production and yield of groundnut under various concentrations of organic manure and water amounts. Annals of Agric Sci 32:1-19.

3. Awal MW, Ikeda T (2002) Recovery strategy following the imposition of Episodic soil moisture deficit in stands of peanut (Arachis hypogaea L.). J Agron Crop Sci 188: 185-192.

4. Gomez KA, Gomez AA (1984) Statistical procedures for Agricultural Research. 2th edn, Wiley.

5. Venkataramana P, Kiraman NJ (2012) Influence of different levels of organic and inorganic fertilizers on groundnut cultivars. Environment and Ecology 20: 89-91.

6. El- Boraie FM, Abo-El-Ela HK, Gaber AM (2009) Water requirements of Peanut grown in sandy soil under drip irrigation and biofertilization. Australian Journal of Basic and Applied Sciences 3: 55-65.

7. Jagdev S, Singh KP (2000) Effect of Aztobacter FYM and fertility levels on yield nitrogen recovery and use efficiency in spring Sunflower. J Agron 16: 57-60.

8. Sabino A (2007) Fog collection in the natural park of Serra Malagueta. An alternative source of water for the communities. In Proceedings of the 4th International Conference on Fog, Fog Collection and Dew, 22-27 July 2007, La Serena. pp: 425-428.

9. Gohri AA, Amiri E (2011) The effect of nitrogen fertilizer and irrigation management on Peanut (Arachis hypogaea L.) yield in the North of Iran. 21st International Congress on Irrigation and Drainage, 15-23 October 2011, Tehran, Iran.

10. Aboelill AA, Mehanna HM, Kassab OM, Abdallah EF (2012) The response Of Peanut crop to foliar spraying with potassium under water stress Conditions. Applied Sci 6: 626-634.

11. Kumaran S (2001) Response of groundnut to organic manure, fertilizer levels, split application under irrigated conditions. Research on Crops 2: 156-158.

12. Mohamed AG, Usman ARA (2008) Impact of drip irrigation management on peanut cultivated in sandy calcareous soil. Assiut J of Agric Sci 38: 191-206.

13. Schemenauer RS, Osses P, Leibbrand M (2004) Fog collection evaluation and operational projects in the Hajja Governorate, Yemen. In Proceedings of the Third International Conference on Fog, Fog Collection and Dew, 11-15 Oct 2004, Cape Town, South Africa.

14. Singh AL (2004) Mineral nutrient requirement, their disorders and remedies in groundnut. Groundnut Research in India, National Research Centre for Groundnut, Junagadh, India, pp: 137-159.

15. Subrahmaniyan K, Kalaiselvan P, Arulmozhi N (2000) Studies on the effect of nutrient spray and graded level of NPK fertilizers on the growth and yield of groundnut. Intern J Trop Agric 18: 287-290.

27

Evaluation of Insecticides for the Management of Tef Shoot Fly (*Atherigona* spp.) at Sekota, Ethiopia

Anteneh Ademe[1]* and Esmelalem Mehiretu[2]

[1]*Sekota Dry-land Agricultural Research Center, Sekota, Ethiopia*
[2]*Adet Agricultural Research Center, Adet, Ethiopia*

Abstract

The experiment was done for two years on tef shoot fly (*Atherigona spp.*) hot spot areas of Sekota, Ethiopia, with the objective of selecting effective and economically feasible insecticides. Five insecticides were evaluated on DZ-01-99 variety of *Eragrostis tef* in randomized complete block design with three replications. Insecticides were applied twice; seven days after emergence of *E. tef* and 10 days after first application. The result revealed that applications of insecticides were found biologically effective and economically feasible over the unsprayed control. The lowest number of dead heart of 2.82 and 3.09 were recorded on those plots treated with Lambda cyhalothrin and chlorpyrifos-ethyl which gave the highest yield 1267.59 and 1225.77 kg ha[-1], respectively. Therefore, we recommended judicious use of chlorpyrifos-ethyl and Lambda cyhalothrin 5% EC at a rate of 1.5 L and 0.4 L ha[-1], respectively, for the management of tef shoot fly.

Keywords: *Atherigona* spp; *Eragrostis tef*; Tef; Ethiopia

Introduction

Tef (*Eragrostis tef*, (Zucc.) Trotter) is a C_4, self-pollinating, chasmogamous annual cereal crop that belongs to the family Poaceae and is indigenous to Ethiopia. It is a traditional Ethiopian small cereal crop that is adapted to diverse agro-ecological zones including areas with conditions marginal to the production of the other crops [1,2]. Despite the importance of tef in the livelihood of small-scale farmers, its productivity in the country is faced with a number of constraints, as a result of which annual yields are often low and subject to extreme fluctuations. According to the Central Statistical Authority report [3] the average tef yield was 1400 kg/ha and 1170 kg/ha in Amhara Region and Wag-himra zone with 1.09 million and 0.028 million hectares coverage, respectively.

Factors contributing to low tef yields are drought, low soil fertility, soil erosion, poor crop management practices, insect pests and weeds. Insect pests are among the major factors causing low yield. Shoot fly (*Atherigona* spp.), Wello bush cricket (*Decticoides brevipennis* Ragge) and Red tef worm (*Mentaxya ignicollis* (Walker)) are the most important insect pests of tef [4]. Of the above mentioned insect pests, damage by shoot fly infestation is becoming a serious problem in tef production.

Tef shoot fly (*Atherigona* spp.) attacks tef throughout the crops active growing period. However, the seedling and panicle stage is the most critical [5]. Larvae mine in to stems of the central shoots of tef causes "dead heart" symptoms because of internal feeding caused by shoot fly larvae [6]. It caused 42-58% damage to growing panicles in different varieties of tef, which resulted in an estimated loss of 378-522 kg/ha [7]. But, Tef shoot fly (*Atherigona* spp.) is an economically important pest of tef in Wag-Lasta and in some cases can cause up to 100% yield loss.

Nowadays the distribution of rain fall is becoming very erratic due to global climatic change, which is even more dramatic in dry-land areas like Wag-Lasta and the yield loss on tef is too high. Tef shoot fly damage is higher in row-planted tef than broadcast-planted and the pest prefers the tef stalk on row planted because of the vigorous stem growth. At this time, with the expansion of tef row planting technologies, shoot fly infestation is becoming the most important production constraint, especially in dry-land areas. To minimize such serious damage on tef, screening of effective insecticides for management of shoot fly that can be used in combination with other control tactics is urgently needed. Hence, this study was conducted with the objective of selecting effective and economically feasible insecticides for the management of tef shoot fly.

Materials and Methods

This trial was conducted during the main cropping season using a commonly recommended tef variety (DZ-01- 99). Plot size was 12 m² with 1 m spacing between plots and 1.5 m between replicates. Urea was applied twice during the growing season; half at planting and the remaining half at the tillering stage. Weeding was carried out as needed. The experiment was laid out in a randomized complete block design (RCBD) with three replicate. The insecticides used for evaluation were carbaryl (Sevin 85% WP), chlorpyrifos-ethyl (Dursban 48% EC), endosulfan (Thiodan 35% EC), lambda cyhalothrin (Karate 5% EC), dimethote (Ethiothoate 40% E.C) at a rate of 1.5 kg/ha, 1.5 L/ha, 1 L/ha, 0.4 L/ha and 1 L/ha, respectively and they were diluted at 200 to 300 liters of water/ha. The trial was conducted in 2010 and then repeated in 2011. The insecticides were applied twice: the first spray was applied seven days after emergence of tef while the second spray was applied ten days after the first application. Shoot fly damage was recorded based on a 1-5 scale (1=0% damage, 2=25%, 3=50%, 4=75% and 5=more than 75% of total plants damaged in the plot). Data were collected on stand count, number of dead heart plants, number of panicles, damage score, number of tillers and seed yield. Effective insecticides (lambda cyhalothrin and chlorpyrifos-ethyl) selected from the evaluation trial were further verified for one year on sixteen tef

*Corresponding author: Anteneh Ademe, Sekota Dry-land Agricultural Research Center, PO Box 62, Sekota, Ethiopia, E-mail: antish_ad@yahoo.com

shoot fly hot spot farmer fields across Amhara region in 2013 cropping seasons on plot size of 100 square meter using farmers as a replication.

Data analysis

Analysis of variance (ANOVA) was performed using SAS software. Treatment mean differences were tested in Duncan Multiple Range Test (DMRT). Damage score were square root transformed before analysis. The correlation between different yield related parameters and grain yield was calculated using Pearson correlation analysis technique.

Results and Discussion

The analysis of variance of the five foliar applied insecticides tested in 2010 main cropping season showed a significant difference among insecticides. Number of dead heart plants treated with chemicals were significantly lower than ($p<0.05$) that of the untreated plots. The lowest damage was recorded in the lambda cyhalothrin and dimethote treatments (Figure 1). There was significant difference in grain yield between treated and untreated plots and the lowest grain yield of 785 kg/ha was recorded from untreated plots, whereas the maximum grain yield was obtained from the plots treated with chlorpyrifos-ethyl, endosulfan, dimethote and lambda cyhalothrin, yields ranged from 1231.84 kg/ha to 1414.03 kg/ha with no statistical difference among them (Figure 2). The efficacy of some insecticides in both years were not uniform due to the variability in shoot fly infestation which was higher in the second year compared to first year (Figure 1).

In the 2011 cropping season, analysis of variance showed a significant difference ($P<0.05$) among insecticides for both dead heart count and grain yield. The lowest number of dead heart plants was found on plots treated with chlorpyrifos-ethyl and lambda cyhalothrin. The highest grain yield was obtained from plots treated with lambda cyhalothrin. However, the lowest grain yield was recorded from untreated plots (Figure 2).

The combined analysis of variance of insecticides showed a significant difference for all tested parameters. Insecticide sprayed plots had a significantly lower ($p<0.05$) number of dead heart than the untreated plots. Moreover, there were significant variations among the treated plots themselves; those plots that were sprayed with lambda cyhalothrin and chlorpyrifos-ethyl had a significantly lower number of dead heart plants. Likewise, there were significant difference in grain yield between the treated and untreated plots and the lowest yield was recorded from untreated plots. However, the highest grain yield was obtained from plots treated with lambda cyhalothrin and chlorpyrifos-ethyl with no statistical difference between them (Table 1). Similar result was obtained in 2013 from plots sprayed with lambda cyhalothrin and chlorpyrifos-ethyl compared to untreated control. Chemical controls can provide a rapid, effective and dependable means of controlling of insects [8,9].

The result of the correlation analysis indicated that grain yield had negative and significant correlation with number of dead heart plants, damage score and number of tiller and positive and significant correlation with number of panicles and stand count (Table 2). It is well known that in high rainfall areas an increase in the number of tillers can result in increased yields. On the other hand, in an area like Wag-Lasta where there is recurrent moisture stress, tillers are not productive and this results in negative correlation of grain yield with number of tillers. Similarly, the negative correlation of grain yield with dead heart count shows the need for management of tef shoot fly especially in moisture stressed areas. Finally, the partial budget analysis and sensitivity analysis were carried out for insecticides against the control and based on the input and output price the marginal rate of return for chlorpyrifos-ethyl and lambda cyhalothrin were found to be higher than the rest of insecticides (data not shown).

Conclusion and Recommendation

This study revealed that the application of insecticides was found to be biologically and economically advantageous over the untreated check in the management of tef shoot fly. In an area like Wag-Lasta and similar agro-ecological zones with recurrent tef shoot fly infestation,

Treatments	Stand count 100 cm²	No. dead heart plants 100 cm²	No. panicles 100 cm²	Damage score	No. tillers	Grain yield (kgha⁻¹)
Carbaryl	37.55[CD]	3.91[C]	47.52[A]	2.67(1.60[B])	14.54[CD]	983.77[B]
Chlorpyrifos-ethyl	56.98[A]	3.09[DE]	45.30[B]	0.83(1.07[D])	11.59[E]	1225.77[A]
Endosulfan	35.99[D]	4.38[B]	47.95[A]	2.50(1.53[BC])	16.99[B]	1035.70[B]
Lambda- cyhalothrin	36.51[D]	2.82[E]	43.00[C]	1.17(1.18[CD])	15.58[C]	1267.59[A]
Dimethote	54.99[A]	4.01[BC]	45.51[B]	1.50(1.26[BCD])	18.44[A]	1079.67[B]
Control	39.92[C]	5.93[A]	42.40[C]	3.83(2.07[A])	15.20[C]	655.97[C]
Mean	44.07	3.91	45.71	1.47	15.08	982.71
DMRT(0.05)	**	**	**	**	**	**
CV (%)	5.13	8.23	2.87	20.14	7.57	10.17

Means followed by a common letter within a column are not significantly different with DMRT at 5%; Values in brackets are square root transformed

Table 1: The effect of different insecticide treatments on stand count, number of dead heart plants, number of panicles, damage level, number of tillers and grain yield of tef combined over two years.

	Stand count	No. dead heart plants	No. panicles	Damage score	No. tillers	Grain yield (kgha⁻¹)
Stand count	1	-0.296	0.316*	-0.452*	-0.263	0.33*
No. dead heart plants		1	-0.766*	0.873*	0.698*	-0.671*
No. panicles			1	-0.66*	-0.55*	0.475*
Damage score				1	0.67*	-0.764*
No. tillers					1	-0.309*
Grain yield (kgha⁻¹)						1

*Significant at p=0.05

Table 2: Correlation coefficients of grain yield, pest damage and other agronomic parameters of tef combined over two years.

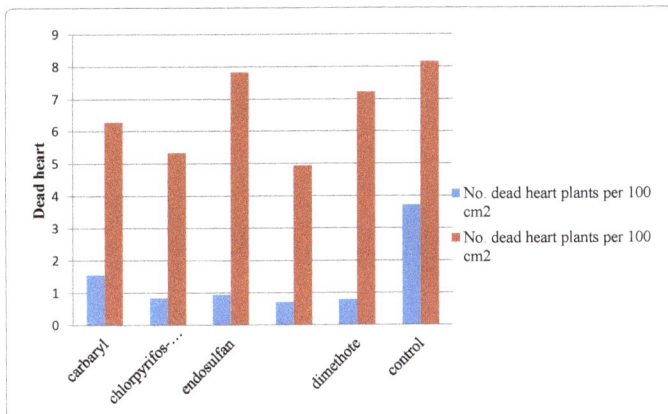

Figure 1: Effect of insecticides on number of dead heart of tef per 100 cm^2; insecticides have significant difference at p=0.05.

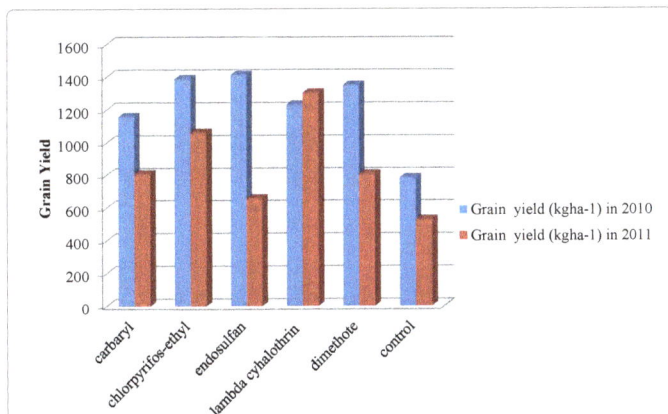

Figure 2: Effect of insecticides on the grain yield of tef in 2010 and 2011; insecticides have significant difference at p=0.05.

insecticide application seven days after tef emergence and ten days after this first application can help to achieve higher productivity through minimizing crop loss due to the tef shoot fly. Among the insecticides tested, chlorpyrifos-ethyl (1.5 L ha^{-1}) and lambda cyhalothrin (0.4 L ha^{-1}) were more profitable than the rest with the yield advantage of 81.87% and 111.63%, respectively. Therefore, we recommended judicious use of chlorpyrifos-ethyl (1.5 L ha^{-1}) and lambda cyhalothrin (0.4 L ha^{-1}) for the management of tef shoot fly.

References

1. Ketema S (1997) Tef Eragrostis tef (Zucc.) Trotter. Promoting the conservation and use of underutilized and neglected crops 12. Institute of Plant Genetics and Crop Plant Research, Gatersleben/International Plant Genetic Resources Institute, Rome, Italy, pp: 36-46.

2. Hailu T, Seyfu K (2001) Production and importance of tef in Ethiopia agriculture. Proceedings of the International workshop on Tef Genetics and Improvement, pp: 16-19.

3. Central Statistical Authority (CSA) (2013) Agriculture Sample Survey. Report on Area and Production for Major Crops, Statistical Bulletin No 532 CSA, Addis Ababa, Ethiopia.

4. Hein B (1989) Insect pests of cereals in Ethiopia identification and control methods. Crop Protection and Regulatory Department, Ministry of Agriculture, Addis Ababa, Ethiopia.

5. Bahyeh M, Biruk W, Gezahegne G, Belay E (2009) The significance of tef shoot flies on tef and their control in Western and South western Zones of Shoa, Central Ethiopia. Annual plant protection society of Ethiopia, Addis Ababa, Ethiopia.

6. Davidson A (1969) Effect of some systemic insecticides on an infestation of the barley fly, Delia arambourgi in Ethiopia. E Afr Agr Forestry J 34: 422-425.

7. Gudeta S (1997) Biology of the Tef Shootfly, Atherigona hyalinipennis Van Emden in Eastern Ethiopia. Int J Trop Insect Sci 17: 349-355.

8. Ascher KRS (1993) Noncoventional insecticidal effects of pesticides available from the neem tree, Azadirachata indica. Arch Insect Biochem Physiol 22: 433-449.

9. Dzemo WD, Niba AS, Asiwe JAN (2010) Effects of insecticide spray application on insect infestation and yield of cowpea (Vigna unguiculata L. Walp) in the Transkei, South Africa. Afr J Biotechnol 11: 1673-1679.

28

Participatory Variety Evaluation of Red Common Bean (*Phaseolus vulgaris L.*) in Borecha District of Southern Ethiopia

Abraham Mulu[1]*, Andergachew Gedebo[2] and Hussien Mohammed[2]

[1]*Department of Dryland Crop Science, College of Dryland Agriculture, Jigjiga University, Ethiopia*
[2]*Hawassa University, Awassa, Ethiopia*

Abstract

Red colour seed common beans are more demanded over the white beans at Borecha district. However seed producers established in the district were not started red common bean seed production due to lack of awareness in improved red common bean varieties. Therefore, farmer's participatory varieties evaluation and selection methods were applied to identify preferred common bean variety in Borecha district. Eight red common bean genotypes were planted in Randomized Complete Block Design as mother trial replicated on four villages of Borecha district to identify farmer's selection criteria; to popularize the variety and to enhance seed produces capacity in managing varieties portfolios. Farmer's preference related data was collected for eight red common bean traits. Common bean pre harvest traits such as earliness, seed per pod (>5 seed per pod), pod load, up right growth habit and height of basal pod from the soil are indentified as the best descriptors to accept and reject the varieties. Uniformity of red seed colour was identified as major decisive criteria for accepting and rejecting common bean variety after harvest. Hawassa dume and Dimtu were variety scored the best ranks for all criteria. Farmers selected and discarded the varieties at various stage of common bean growth with fairly high degree of precision.

Keywords: Common bean; Genotypes; Seed producers; Pre-harvest traits; Uniformity

Introduction

Over the millennia, farmers grew complex mixtures of bean types against drought, disease, and pest attacks. During this process, farmers have been using limitless genetic array of bean seed with a wide variety of colours, shape, and sizes to meet the growing conditions and taste preference of different growing regions. Seed color and size are important characters of consumers' preference. In Tanzania, Malawi, Kenya and Ethiopia consumers prefer large brownish/purple or reddish colour seeded beans Beebe et al. [1]. Red seed beans are preferred because of the red colour imparts to the food after cooking.

For instance, farmers in central rift valley prefer white bean while farmers of Southern Ethiopia prefer red bean Ferris and Kaganzi [2]. A local variety 'Red woliata' a red seeded bean is popular and almost the only improved variety grown by 91.4% farmers. The main reasons for its popularity were due to its earliness in maturity and its dual importance. First, it provides an excellent nutritional complement to maize and enset which is the main local dishes, and secondly it is major cash crop of the Borecha district LSB [3]. Furthermore, the demand for red beans in northern Kenya, associated with drought in these areas, has encouraged production of red beans in this region [2]. The local variety 'red woliata' passed many generations of natural and human selection and found to be low yielding and susceptible to insects. Beside to it is demand on the local market and for consumption, seed producer cooperative farmers were established in the district with a mandate of producing quality and demand driven common bean seed. However, improved common bean varieties developed at research institute has not evaluated for major common bean quality traits at seed producers farm condition of Borecha district. The seed producers has not informed about improved red beans. Therefore, this research is aimed to popularize improved variety of common bean to seed producers and their customers. It is also initiated to identify and select the most preferred common bean cultivars.

Materials and Methods

Descriptions of study site

The experiment was conducted in 2010 under rain fed condition in Boricha district with Kayyo seed producer cooperative (SPC). The site is located in Sidama Zone of Southern Ethiopia, which is selected as one of innovation site for Local Seed Business (LSB) Project. It is located 65 km away from Hawassa and 324 km from the capital city Addis Abeba in the South West direction. According to the Bureau of Agriculture and Rural Development of the Borecha district, 37,500 hectares of the land were cultivated in the area. Out of the land for cultivation, the share for maize and common bean was 38% and 37%, respectively. Common bean is grown as relay and intercropping. This practice makes the area covered by common bean very similar to that of maize. Followed to maize and common bean, enset accounts 21%, demonstrating the importance of this crop as the areas staple food Tesfaye et al. [4].

Experimental materials

Before selecting the experimental materials local needs were assessed using semi-structured interview. Based on their preference, early maturing varieties of red and red mottled common bean (Tables 1 and 2) were obtained from Pulse Crops Improvement Sections of Awassa and Melkasa Agricultural Research Centres.

*Corresponding author: Abraham Mulu Oljira, Department of Dryland Crop Science, College of Dryland Agriculture, Jigjiga University, PO Box 1020, Ethiopia
E-mail: abrahammulu09@gmail.com

Data collection and analysis

Farmer's preferences were identified using Focus group discussions (FGD) held with 30 households. The households were randomly selected from seed producer and their customers. The introduced cultivars were visually evaluated at vegetative stage, maturity and after harvest. Farmers discussed and agreed on criteria they thought to be important for selecting a given variety at a particular development stage. Field days were arranged at different growth stages to collect the data using agreed criteria by research participant. To select varieties, the farmers were asked to give a score for major common bean traits on a scale of 1 to 6 i.e., (1=Excellent; 2=Best; 3=Very good; 4=Good and 5=Poor, 6=Bad). Then descriptive statistics were used for rating and means of score were used for comparison among genotypes. The picture also used to demonstrate the red beans varieties.

Results and Discussion

Farmer's preference

Farmers attended the participatory variety evaluation and selection had diversified selection criteria to accept and reject bean variety (Table 2). This diversity during selection is an indication of the complexity of users' preference. Asrat [5] reported that, when there is more diversity in selection criteria, there is better chance of maintaining on farm diversity since positive traits are seldom found on single variety. However, the result from FGD revealed that uniformity in red seed color after maturity (Figure 1) and high biological yield were the major decisive criteria in retaining and rejecting the variety. Asrat [5] reported similar findings where pure red and red mottled seed color and high yielding variety were reported as the major decisive criteria to accept or reject common bean in Southern Ethiopia.

The research participant farmers categorized common bean traits such as earliness, seed per pod (>5 seed per pod), pod load, up right growth habit and height of basal pod from the soil as descriptors of good common bean varieties. Asrat and Teshale et al. [5,6] reported similar findings for common bean variety evaluated by farmers. The score ranking indicated that Hawassa dume and Dimtu were 1st and 2nd respectively for various traits evaluated by seed producers and their customers (Table 2).

Varieties selected for seed production

Hawassa dumme and Dimtu fulfilled criteria of pure red seed colour after harvesting. This criterion is identified as major decisive criteria for accepting and rejecting common bean variety after harvest (Figure 1).

No	Variety	Year of release	Description	Source	Days to Maturity
1	Red woliata	1974	Local check	Farmer	90-100
2	Hawassudume	2008	Tested variety	AwARC	85-90
3	Omo-95	2003	Tested variety	AwARC	104
4	Ibado	2003	Tested variety	AwARC	-
5	Nazir	2003	Tested variety	MARC	86-88
6	Dimtu	2003	Tested variety	MARC	86-88
7	Melkadima	2006	Tested variety	MARC	80-90
8	Dinkinash	2006	Tested variety	MARC	95-100

Source: Awassa Agricultural Research Centre (AwARC) and Melkasa Agricultural Research Centre (MARC), 2010
Table 1: Details of common bean varieties used in this study.

Trait	Melkadima	Dinkinash	Omo-95	Dimtu	Naser	Hawassadume	Ibado	Redwoliata(check)
Pods load	4	6	5	2	5	1	3	6
Earliness	4	6	5	2	1	3	5	4
Yield	5	6	5	1	3	1	3	4
Seed/pod	3	6	4	2	3	3	2	1
Seed size	2	6	5	2	5	2	1	4
Red color	3	2	5	3	3	2	6	1
Redness after moisture deficit	5	2	6	1	3	1	6	2
Salk strength	1	5	6	2	6	2	1	6
Market value	4	3	6	1	4	1	4	2
Seed shape	4	6	6	1	5	1	4	5
Seed plumpness	3	6	5	1	5	1	5	3
Synchrony of maturity	3	4	6	1	2	1	3	4
Disease Resistance	6	5	2	1	4	1	1	2
Insect Resistance	5	6	4	1	3	1	1	4
Pod appearance	1	5	6	2	3	1	1	5
Green leaf	6	1	5	1	3	1	4	3

Varieties local names

Shattering	1	5	4	1	6	1	2	3
Snow rainfall resistance	6	3	2	1	1	1	5	3
Lodging	1	5	6	1	5	1	1	6
Days emergency	3	2	3	2	1	2	3	5
Days to pod	1	6	4	3	3	2	3	4
Height of basal pod from soil	1	3	4	2	6	2	1	5
Leaf shading	1	6	5	1	1	1	4	5
Whole plant vigor	1	6	3	3	5	2	1	5
Leafiness	4	5	1	2	3	2	1	3
Suitability for Intercropping	6	2	1	4	5	4	5	2
Mean preference rating	**3.23**	**4.53**	**4.38**	**1.69**	**3.61**	**1**	**2.80**	**3.73**
Rank	**4**	**8**	**7**	**2**	**5**	**1**	**3**	**6**

Table 2: Common bean traits used for variety evaluation and selection.

Figure 1: 3 Pictures of varieties selected and rejected for seed production.

Conclusion and Recommendations

Common bean seed production can be enhanced through selection of bean varieties suitable both for seed producers and customers as the users have been diversified preference, which may result in varietal diversity for seed producer cooperatives. Several PVS experiments on different crops have shown the importance of variety evaluation and its selection with farmers. Thus doing participatory varietal evaluation and selection with seed producer's and their customer was very important to enhance seed producer research capacity, product line and strength the link between their customers since preferred trait was included during selection process.

Thus, introducing new varieties through PVS help the farmers to choose the variety that possesses the character preferred by SPC and customer on market. Besides, the training given during PVS process enhances capacity of the SPC for managing varietal diversity. It was observed that as the area is highly market-oriented, high-yielding varieties were still top on their selection list with uniform pure red bean seed. However, most of the customers were looking for red bean which is suitable for intercropping. In view of this, PVS was conducted at Borecha district common bean seed producer cooperative and their customers in 2010 to evaluate eight common bean varieties. Seven varieties and one local check were used for mother trial four replication were used. Seed producer established in the district were producing white haricot bean seed from Awash one and Awash melka. Even though the demand for red bean is high the production is not started.

Additionally, there is one local red bean variety in Borecha district. The varieties that have "passed" the evaluation of farmers during participatory evaluation were spread in the community by commercial system. In addition the varieties will be part of seed producers' product line and help them as additional varieties which increase diversity of their product range. Markowitz [7] recommend that, "seeding a large percentage of land by a variety is not recommended; planting several varieties minimizes the risk of damage from adverse weather and disease and pest epidemics and increases the chance for quality seed with maximum yields which resulted in positive economic benefits for seed producer". Lanier et al. [8] also states that, application of portfolio theory to variety selection is new, but it helps seed producer's potential to increase yield and decrease yield variability simultaneously. Three varieties: Hawassa dume, Dimtu and Ibado were selected by majority of the scored criteria. Four varieties namely Red woliata (local check), Omo-95, Dinkinash and Nasir were discarded since they did not fulfil characteristics that the SPC and customers preferred [9]. Thus, introducing new varieties through PVS help the farmers to choose the varieties those posses the character preferred both by SPC and their customer. Further research is recommended to identify potential red bean varieties since most of the customers were looking for red bean which is suitable for intercropping.

Acknowledgements

The authors acknowledge financial support of Local seed Business Project of Hawassa University to conduct this study. Authors are also grateful to Ministry of Education and farmers who participated on the study.

References

1. Beebe S, Rao I, Blair M, Butare L (2009) Breeding for abiotic stress tolerance in Common bean: Present and Future challenges. Proceedings of the 14th Australia Plant Breeding & 11th SABRAO Conference, 10 to 14 August, Brisbane, Australia. pp: 10-72.

2. Ferris S, Kaganzi E (2008) Evaluating marketing opportunities for haricot beans in Ethiopia. IPMS, ILRI Working Paper 7, Nairobi, Kenya, p: 68.

3. Local Seed Business Newsletter (2009) Seed multipliers" cooperative with a potential to become a local seed business. LSB Newsletter, pp: 1-9.

4. Tesfaye T, Tefera Z, Mata G, Abebe T, Solomon B (2008) Wamole farmers "Cooperative for maize seed production in Boricha woreda, SNNPR". In: Farmers, seeds and varieties: supporting informal seed supply in Ethiopia. Wageningen International, pp: 329-301.

5. Asrat A (2008) Participatory varietal evaluation and breeding of the common bean in the Southern region of Ethiopia. In: Farmers, seeds and varieties: supporting informal seed supply in Ethiopia. Wageningen International, p: 348.

6. Teshale A, Girma A, Chemeda F, Bulti T, Abdel Rahman M, et al. (2005) Participatory Bean Breeding with Women and Small Holder Farmers in Eastern Ethiopia. World Journal of Agricultural Sciences 1: 28-35.

7. Markowitz H (1959) Portfolio Selection: Efficient Diversification of Investments. NewYork, USA, p: 15.

8. Lanier LN, Andrew B, Brad W, Jeffery H (2009) Enhancing Farm Profitability through Portfolio Analysis: The Case of Spatial Rice Variety Selection. Journal of Agricultural and Applied Economics 41: 641-652.

9. Katungi E, Farrow A, Mutuoki T, Gebeyehu S, Karanja D, et al. (2010) Improving common bean productivity: An Analysis of socioeconomic factors in Ethiopia and Eastern Kenya. Baseline Report Tropical legumes II. Centro International de Agricultura Tropical (CIAT), Cali, Colombia.

29

Food Seed Health of Chick Pea (*Cicer arietinum L.*) at Panchgaon, Gurgaon, India

Narendra Kumar*

Amity Institute of Biotechnology, Amity University Haryana, Manesar, Gurgaon, Haryana, India

Abstract

The chickpea (*Cicer arietinum*) is a legume of the family Fabaceae, subfamily Faboideae. It is also known as gram, or Bengal gram, garbanzo or garbanzo bean and sometimes known as Egyptian pea, ceci, cece or chana, or Kabuli chana (particularly in northern India). The 20 food seed samples of chick pea (*Cicer arietinum* L) were collected from farmer markets of Panchgaon (Gurgaon) Haryana. The mycobiota analysis through Std. blotter paper, Agar plate, Seed washates of food seed revealed the presence of 16 fungal species viz., *Aspergillus flavus, A. fumigatus, A. niger, A. sydowi, A. ochraceous, A. terreus, A. nidulans, Cladosporium macrocarpum, F. oxysporum, F. semitectum, Macrophomina phaseolina, Penicillium notatum, Sclerotium rolfsii, Rhizoctonia solani, R. batatiocola, Rhizopus arrhizus*. These species showed diversity in terms of Percent (%) frequency of fungi. In these species *Aspergillus flavus, A. niger, A. ochraceous, A. terreus* showed dominance in terms of Percent (%) frequency of fungi. The insect analysis revealed presence of one species of insect *Callosobruchus chinensis*. These species showed reduction in terms of weight loss, germination, protein content and carbohydrate content.

Keywords: Mycobiota; *Callosobruchus chinensis*; Chick pea (*Cicer arietinum*)

Introduction

Pulse seeds are reported to carry many moulds both in fields and during storage and association of this adversely affects quality and health of the seeds and under storage bring about several undesirable changes making them unfit for consumption and sowing. The chick pea (*Cicer arietinum*) is a legume of the family *Fabaceae*, subfamily *Faboideae*. It is also known as gram, or Bengal gram, garbanzo or garbanzo bean, and sometimes known as Egyptian pea, ceci, cece or chana, or Kabuli chana (particularly in northern India). Its seeds are high in protein chickpea seeds are eaten fresh as green vegetables, parched, fried, roasted, and boiled; as snack food, sweet and condiments; seeds are ground and the flour can be used as soup, dhal, and to make bread; prepared with pepper, salt and lemon it is served as a side dish and grown over 6.66 m ha of land in India.

Chickpea seed has 58.9% carbohydrate, 3% fiber, 5.2% oil, 3% ash, 0.2% calcium, and 0.3% phosphorus. It furnishes an important food for lower classes and the flour is quite nutritious. Many fungal species viz., *Alternaria porri, A. alternata, Aspergillus amstelodami, A. flavus, A. fumigatus, A. nidulans, A. niger, A. sydowi, A. wentii, Botrytis cinerea, Cladosporium macrocarpum, Curvularia lunata, Fusarium equiseti, F. moniliforme, F. oxysporum, F. semitectum, Macrophomina phaseolina, Myrothecium roridum, Penicillium notatum, Rhizoctonia sp.,* and *Rhizopus arrhizus* been reported from chickpea Ahmad et al. [1].

Therefore present study was undertaken in order to study seed mycobiota in terms of frequency of the fungus and insect responsible for chickpea damage during storage.

Materials and Methods

The 20 seed samples of chickpea were collected from farmer stores of Panchgaon (Gurgaon).

Moisture content estimation

The weight (100-seeds) of *Cicer arietinum* were recorded on a randomly using an electronic balance. The seed moisture content was estimated following oven dry method using two replications each of 20 g ISTA [2]. After estimating the initial moisture content of seeds, about 200 g seed sample in each accession was kept in muslin cloth bags, to permit free flow of air, and placed in a seed drying room maintaining a constant temperature of 15°C and 15% RH. Seed samples were drawn at an interval of seven days to estimate the moisture content. Observations on mean 100-seed weight, moisture content are presented in Table 1.

It is evident from Table 1 that the 100-seed weight of the seed of *Cicer arietinum* ranged from 9.0 to 39.0 g with a mean of 18.5 g which indicates the seed size diversity. Differences in seed moisture contents of *Cicer arietinum* under constant drying environment were significant among the three seed sizes. After seven days of incubation, small seeds showed moisture content of 9.01 to 6.71%, medium seeds from 9.27 to 7.32% and large seeds from 9.71 to 6.81%.

Detection of seed mycobiota

The mycobiota of small, medium and large sized seed was analysed following Standard blotter paper method, Agar plate method and Seed washates method.

Blotter method: The collected small, medium and large seed samples of chick pea (*Cicer arietinum*) were analyzed for the presence of major seed borne fungal pathogens by blotter method following the International rules for Seed Testing ISTA [3]. Two hundred mixed

Seed size	100-seed Weight (g)	Seed moisture content (%)	
		0 days	7 days
Small Mean	9.0	9.01	6.71
Medium Mean	17.5	9.27	7.32
Large Mean	39.0	9.71	6.94
Overall Mean	18.5	9.33	6.94

Table 1: Mean 100-seed weight, changes in moisture content and viability of small-, medium- and large-seeded chickpea under constant seed drying environment.

*Corresponding author: Narendra Kumar, Amity Institute of Biotechnology, Amity University Haryana, Manesar-122 413, Gurgaon, Haryana, India
E-mail: narendra.microbiology@rediffmail.com

seeds were placed on three layers of moist blotting paper (Whatman No. 1) in each glass petridish containing 5 seeds. The petridishes were incubated at 25 ± 1°C under 12/12 hrs light and darkness cycle for 7 days. Each seed was observed under stereo microscope in order to record the presence of fungal colony. In doubtful cases temporary slides were prepared from the fungal colony and observed under compound microscope. For isolation of internal fungi seeds were treated with 1% sodium hypochlorite solution followed by 2-3 washings with sterile water. The developing fungal colonies were examined. The colour of the colony on the reverse side is also examined as it is characteristic for some fungi which are identified on this basis.

Agar plate method: In the agar plate method, two hundred mixed (small, medium and large) seeds of chick pea (*Cicer arietinum*) were tested for each maintaining four replications. For internal seed fungi Surface disinfected seeds (1% sodium hypochlorite solution) were plated on the Czapek Dox Agar medium and the plated seeds were usually incubated for 5-7 days at 28 ± 2°C under 12 h altering cycles of light and darkness. Sucrose 30.0; Sodium nitrate 2.0; Dipotassium phosphate 1.0; Magnesium sulphate 0.5; Potassium chloride 0.500; Ferrous sulphate 0.010; Agar 15.0 Gms / Litre; Final pH (at 25°C) 7.3 ± 0.2. These constituents were dissolved in 1000 ml distilled water. Heated to boiling to dissolve the medium completely, sterilized by autoclaving at 15 lbs pressure (121°C) for 15 minutes and mixed well and poured into sterile Petri plates. At the end of the incubation period, fungi growing out from the seeds on the agar medium were examined and identified. Identification was done based on colony characters and morphology of sporulation structures under a compound microscope. In the agar plate method more than one type of fungal colonies were produced. In this case, identification was done on the most frequently occurring colony present in all the petridishes and then the second most frequent, the third most frequent and so on. Thereafter, the identification of the different colonies were done visually and then under a stereomicroscope and followed by an examination of the fruiting structures under a compound microscope.

Seed washates method: 100 mixed seeds (small, medium and large) of chick pea (*Cicer arietinum*) were taken in flask with sterile distilled water for their soaking. The flasks were subjected to mechanical shaker for 5-10 minutes. 1 mL of seed washing, thus obtained was placed on PDA medium for growth of individual spore of fungus. The seed washing contains spores of the fungi. The plates were incubated at room temperature for development of colonies and observations were made. Fungi developed within 3 days. These colonies were immediately transferred to Czapek Dox Agar slants for further study. The various moulds appeared on seeds in blotter tast, agar plates and seed washates were isolated and maintained on Czapek Dox Agar. For identification of fungi temporary slides were prepared from the fungal colony and observed under compound microscope at 100X and 400X and identified with the help of keys suggested by Ref. [4-11]. The fungi from the incubated seeds were also transferred to PDA when needed.

The percent frequency was calculated by using following formulae.

$$\text{Frequency (\%)} = \frac{\text{No. of plates in which individual fungal species occurred}}{\text{Total no. of plates studied}} \times 100$$

Histopathological techniques

To find the location of fungi in the seed, component plating was done. The seeds were soaked in water and thus obtained soft seeds were sectioned. The sections were stained and examined.

Culture of insects: The cultures of *Callosobruchus chinensis* (L.) were established from infested stored chick pea seeds collected from twenty farmer places and identified by literatures following Drees and Jackman [12] and Beck and Blumer [13]. The cultures of insects were maintained subsequently on insecticide free newly harvested chick pea seeds at laboratory (25 ± 2°C temperature) in darkness to obtain same aged insects.

Effect of storage pest on chick pea seeds

The deterioration caused by dominant fungal species viz., *Aspergillus flavus*, *A. niger*, *A. ochraceous*, *A. terreus* with respect to weight loss and seed germination was evaluated. For this purpose freshly harvested sterilized chick pea seeds were taken in presterilized polyethylene bags (200 g seeds/bag) and inoculated by two disc (5 mm diam) of different fungal species separately. Likewise 5 insect – *Callosobruchus chinensis* were inoculated separately in presterilized polyethylene bags. The inoculated matar seed samples were stored for 20 days under laboratory conditions at room temperature. Experiments were revised and contained five replicates. The protein content was studied following Lowry [14] using bovine serum albumin as standard. The optical density of each specimen was measured at 650 nm. Carbohydrate estimation was done following anthrone method of Thimmaih [15].

Results and Discussion

It is evident from Table 1 that the 100-seed weight of the seed of *Cicer arietinum* ranged from 9.0 to 39.0 g with a mean of 18.5 g which indicates the seed size diversity. Differences in seed moisture contents of *Cicer arietinum* under constant drying environment were significant among the three seed sizes. After seven days of incubation, small seeds showed moisture content of 9.01 to 6.71%, medium seeds from 9.27 to 7.32% and large seeds from 9.71 to 6.81%.

A total of 16 fungal species viz., *Aspergillus flavus*, *Aspergillus fumigatus*, *Aspergillus niger*, *Aspergillus sydowi*, *Aspergillus ochraceous*, *Aspergillus terrus*, *Aspergillus nidulans*, *Cladosporium macrocarpum*, *Fusarium oxysporum*, *Fusarium semitectum*, *Macrophomina phaseolina*, *Penicillium notatum*, *Rhizoctonia solani*, *Rhizoctonia batatiocola*, *Rhizopus arrhizus* were isolated (Table 2). In these species *Aspergillus flavus*, *Aspergillus niger*, *Aspergillus ochraceous* and *Aspergillus terreus* showed higher per cent frequency. All fungi developed on agar and blotter paper, except for *P. notatum* which developed on agar plate only. Blotter method showed greater incidence of fungi on different parts of seeds followed by agar plate method and seed washates method.

The fungal species were reduced in surface sterilized seeds, which indicate that most of fungi were located on seed coat. Blotter method showed greater incidence of fungi on different parts of seeds followed by agar plate and deep-freezing method (Table 3). Component plating of chickpea seeds showed that seed coat and cotyledons were infected by greater number of fungi (16) followed by axis (radicle+plumule) (12). *Aspergillus flavus*, *Aspergillus niger*, *M. phaseolina* and *R. solani* were also isolated from seed coat, cotyledons and axis of seed.

As evident from Table 4, *Aspergillus flavus*, *A. niger* and insect – *Callosobruchus chinensis* played important role in seed weight loss and seed germination. The *Aspergillus flavus* inoculated seeds showed 13%, *A. niger* 14% while insect inoculated showed 18% protein content respectively. *Aspergillus flavus* inoculated seeds showed 16.10, *A. niger* 16.37 g while insect inoculated 17.10 g/100 g respectively. On account of wide occurrence and their pathogenicity these were selected as test organisms.

S No	Name of Fungi	Percent (%) frequency of fungi		
		Std. blotter Paper Mean ± SE	Agar Plate Mean ± SE	Seed washates Mean ± SE
1	*Aspergillus flavus* Link	50.7 ± 0.23	45.1 ± 0.21	35.6 ± 0.23
2	*Aspergillus fumigatus* Fresen	20.1 ± 0.00	14.1 ± 0.11	10.1 ± 0.13
3	*Aspergillus niger* van Tieghem	49.2 ± 0.12	44.3 ± 0.13	40.1 ± 0.14
4	*Aspergillus sydowi*(Bainier and Sartory)	5.1 ± 0.25	2.3 ± 0.11	2.1 ± 0.10
5	*Aspergillus ochraceous* Wilhelm	45.1 ± 0.01	40.3 ± 0.12	37.1 ± 0.13
6	*Aspergillus terreus* Thom	41.0 ± 0.13	38.1 ± 0.25	38.3 ± 0.26
7	*Aspergillus nidulans* (Eidam) G.	10.0 ± 0.25	7.3 ± 0.23	3.3 ± 0.21
8	*Cladosporium macrocarpum* Preuss	5.0 ± 0.00	3.0 ± 0.00	1.0 ± 0.00
9	*Fusarium oxysporum* von Schlechtendal	8.1 ± 0.21	2.0 ± 0.22	1.1 ± 0.21
10	*Fusarium semitectum* Berk. and Ravenel,	2.2 ± 0.00	1.1 ± 0.10	1.0 ± 0.00
11	*Macrophomina phaseolina* (Tassi) Goid	1.0 ± 0.00	1.2 ± 0.01	0.8 ± 0.00
12	*Penicillium notatum* Thom	-	1.1 ± 0.00	1.0 ± 0.00
13	*CephalosporumAcromonium Corda*	2.0 ± 0.00	1.5 ± 0.01	0.1 ± 0.00
14	*Rhizoctonia solani* J.G. Kühn	1.6 ± 0.10	0.5 ± 0.00	
15	*Rhizoctoniabatatiocola(Taubenh.) E.J.Butler*	1.1 ± 0.00	1.0 ± 0.00	
16	*Rhizopus arrhizus* A. Fisch	1.2 ± 0.01	1.1 ± 0.02	0.3 ± 0.01

Table 2: Percent frequency of mycobiota on unsterilized seeds of chickpea.

S No	Name of Fungi	Percent (%) frequency of fungi		
		Std. blotter paper Mean ± SE	Agar plate Mean ± SE	Seed washates Mean ± SE
1	*Aspergillus flavus*	30.1 ± 0.31	34.1 ± 0.21	25.1 ± 0.23
2	*Aspergillus fumigatus*	10.1 ± 0.01	10.1 ± 0.02	5.1 ± 0.03
3	*Aspergillus niger*	39.2 ± 0.21	24.3 ± 0.23	10.1 ± 0.11
4	*Aspergillus sydowi*	5.1 ± 0.12	2.3 ± 0.02	2.1 ± 0.05
5	*Aspergillus ochraceous*	35.1 ± 0. 07	30.3 ± 0.19	17.1 ± 0.17
6	*Aspergillus terreus*	31.0 ± 0.75	20.1 ± 0.37	16.3 ± 0.21
7	*Aspergillus nidulans*	7.0 ± 0.31	5.3 ± 0.21	1.3 ± 0.23
8	*Cladosporium macrocarpum*	3.0 ± 0.31	2.0 ± 0.21	0.1 ± 0.34
9	*Fusarium oxysporum*	3.1 ± 0.31	1.0 ± 0.32	0.1 ± 0.33
10	*Fusarium semitectum*	1.2 ± 0.31	0.1 ± 0.22	
11	*Macrophomina phaseolina*	1.0 ± 0.21	0.3 ± 0.01	
12	*Penicillium notatum*	-	0.4 ± 0.02	
13	*Cephalosporum acromonium*	1.8 ± 0.00	0.1 ± 0.00	
14	*Rhizoctonia solani*	1.7 ± 0.00	0.7 ± 0.00	
15	*Rhizoctonia batatiocola*	1.7 ± 0.03	1.0 ± 0.00	
16	*Rhizopus arrhizus*	1.5 ± 0.04	0.5 ± 0.00	

Table 3: Percent frequency of mycobiota on sterilized seeds of chickpea.

Fungal species/insect	Weight loss (in/g)		Germination%		Protein %		Carbohydate per 100 g	
	C	T	(C Mean ± SE)	(T Mean ± SE)	(C Mean ± SE)	(T Mean ± SE)	(C Mean ± SE)	(T Mean ± SE)
Aspergillus flavus	nil	0.186	84.43 ± 0.01	45.17 ± 0.21	25.3 ± 0.11	13 ± 0.12	27.40 ± 0.13	16.10 ± 0.12
A. niger	-	0.179	86.13 ± 0.13	49.30 ± 0.12	25.6 ± 0.12	14 ± 0.00	26.42 ± 0.03	16.37 ± 0.04
A. ochraceous	-	0.130	81.0 ± 0.00	74.31 ± 0.03	25.8 ± 0.50	20 ± 0.00	26.41 ± 0.04	23.40 ± 0.03
A. terreus	-	0.05	85.00 ± 0.00	75.23 ± 0.32	26.6 ± 0.23	20 ± 0.07	26.41 ± 0.07	22.10 ± 0.13
Insect-*Callosobruchus chinensis*	-	0.184	86.00 ± 0.00	45.23 ± 0.13	28.0 ± 0.11	18 ± 0.00	26.46 ± 0.17	17.10 ± 0.12

C: Control; T: Treatment

Table 4: Fungal species /insect species *vis-à-vis* weight loss, germination and protein content of chickpea seeds after 20 days storage.

Patil [16] observed that fungi associated with seeds of Chickpea were *Alternaria alternata, Aspergillus flavus, Aspergillus. niger, Aspergillus carboniferus, Cladosporium herbarum, Chaetomium globosum, Curvularia lunata, Fusarium moniliforme, Fusarium. oxysporum, Fusarium. semitectum, Fusarium roseum, Mucor* sp. *Penicillium citrinium, Phytophthora* sp., *Pythiurn* sp., *Rhizoctonia solani, Rhizopus stolonifer, Trichoderma viride* while Singh [17] found nine fungal species, namely, *Alternaria alternata, Aspergillus flavus, Aspergillus niger, Curvularia lunata (Cochliobolus lunatus), Fusarium moniliforme (Gibberella moniliformis), Helminthosporium sativum (Cochliobolus sativum) Mucor* sp, *Penicillium notatum* and *Rhizopus nigricans (R. stolonifer)*, were observed on seven seed samples of chickpea. All fungi developed on agar and blotter paper, except for *P. notatum* which developed on agar plate only.

But Ghangoaker and Kshirsagar [18] reported many fungal species viz. *Alternaria alternata, Aspergillus terrus, A. flavus, A. fumigatus, A. niger, Botrytis* sp, *Cladosporium, Curvularia lunata, Fusarium solani, F. moniliforme, F. oxysporum, Macrophomina phaseolina, Penicillium notatum, Rhizoctonia* sp. and *Rhizopus nigricans* from *Cicer arietinum* Razia [19] reported twenty one fungal species viz., *M. sphaerosporus, R. arrhizus, C. cucurbitarum, A. niger, A. flavus, A. terreus, Afumigatus, P. vermiculatum, A. alternata, A. sonchi, A. clamydospora, C. cladosporioides, C. herbarum, C. clavata, D. australiensis, D. hawaiiensis, D. halodes, H. fuscoatra, F. equiseti, F. oxysporum* and *Fusariella* spp. from the external seed surface of damaged seeds of gram. Highest frequency value (12.50) and relative abundance (9.50) were recorded for *A. niger* and lowest frequency (1.00) and relative abundance (2.00) were recorded for *M. sphaerosporus, A. fumigatus, C. clavata,* and *F. equiseti.* During a seed borne mycoflora of five cultivars of Cicer under blotter paper method all varieties were found more susceptible to *Fusarium, Aspergillus niger, Aspergillus flavus, Botrys cinerea, Sclerotium rolfisii* Margeret et al. [20]. In a study Zaidi [19] isolated thirty fungal species were among these were *Alternaria alternata, Chaetomium spp., Penicillium citrinum, Aspergillus niger, A. flavus, Rhizopus nigricans, Fusarium oxysporum.* The no of fungal species were reduced in surface sterilized seeds which indicate of that many of the fungi were located on seed coat. Blotter method showed greater incidence of fungi on different parts of seeds followed by agar plate method. The seed mycoflora devalue the seed quality, reduce its nutritional value and cause a germination failure of the seedlings and of the crop raised from such infected seeds. Narayan [18] reported *Alternaria alternata, Chaetomium spp., Penicillium citrinum, Aspergillus niger, A. fumigatus, A. flavus, Rhizopus nigricans, Fusarium oxysporum, F. moniliform, F. solani, Chaetomium* sp, *Curvularia lunata, Macrophomin* sp, *Monilia* sp., *Penicillium* sp., *Rhizoctonia* sp, *Trichoderma* etc. from gram seeds. Sontakke and Hedawoo [21] found thirteen different fungi like *Actinomucor repens, Alternaria alternata, Aspergillus flavus, A. fumigatus, A. niger, A. ochraceus, Cladosporium* sp., *Fusarium oxysporum, Fusarium* sp., *Mucor varians, Penicillium notatum, Phoma herbarum, Rhizopus stolonifer* which were isolated in variable frequencies. Frequency of the individual species ranges between 1.11-8.19%. Of which, *Fusarium oxysporum* (8.19%), *Rhizopus stolonifer* (7.63%), *Phoma herbarum* (5.69%) and *Aspergillus flavus* (5.44%) were found to be predominant. Blotter paper method was found to be more effective than agar plate method. The percent germination of the Chickpea seeds was evaluated by the standard rolled paper towel method. Higher incidence of fungi on the seeds of chickpea adversely affected its germination. Many fungal species viz., *Alternaria porri, A. alternata, Aspergillus amstelodami, A. flavus, A. fumigatus, A. nidulans, A. niger, A. sydowi, A. wentii, Botrytis cinerea, Cladosporium macrocarpum, Curvularia lunata, Fusarium equiseti, F. moniliforme, F. oxysporum, F. semitectum, Macrophomina phaseolina, Myrothecium roridum, Penicillium notatum, Rhizoctonia* sp., and *Rhizopus arrhizus* have been reported from chickpea. These fungal diseases can kill chickpea crops and is difficult to remove once it sets in Ahmad [1] and Singh [22] also isolated seven fungal species such as *Alternaria alternata, Aspergillus flavus, A. niger, A. fumigatus, Curvularia lunata, Fusarium monoliforme* and *Rhizocton ia solani.*

Component plating of chickpea seeds showed that seed coat and cotyledons were infected by greater number of fungi followed by axis (radicle+plumule). *Aspergillus flavus, Aspergillus niger, M. phaseolina* and *R. solani* were also isolated from seed coat, cotyledons and axis of seed. Similarly Shahnaz et al. [23] reported that seed coat and cotyledons were infected by greater number of fungi followed by axis (radicle+plumule). *M. phaseolina* and *R. solani* were also isolated from seed coat, cotyledons and axis of seed. The fungal species were reduced in surface sterilized seeds, which indicate that most of fungi were located on seed coat.

Seed germination reduced significantly in artifically inoculated seeds with *Fusarium oxysporum* f. sp. ciceri as compared to uninoculated seeds Lily and Trivedi [24]. Similar, trend of results could be due to different response of different varieties due to their susceptible and resistant reactions and thus, indicating more seed borne infection in susceptible variety Singh et al. [17]. Rathod [25] studied standard blotter paper, agar plate and seed washates methods for seed mycoflora study. Among the three methods, the agar paper method was found to be suitable as in less incubation; there was higher percent incidence of seed mycoflora. The variation in fungal species may be due to different climatic conditions, isolation periods and different storage containers.

Use of Sodium hypochloride helped in minimizing the incidence of superficial and fast growing as well as common seed borne fungi like *Aspergillus* spp., *Chaetomium* spp., *Cladosporium* spp., *Rhizopus* spp., *Cephalosporium* spp. Similar results were obtained by Dawar and Ghaffar [26] on sunflower seeds. Surface disinfection of seed with 1% $Na(OCl)_2$ reduced the incidence of *Aspergillus* spp. A number of fungi isolated in the present study are known to produce mycotoxins which are harmful for human health. Mycotoxins can cause severe damage to liver, kidney and nervous system of man even in low dosages Rodricks [27].

Aspergillus flavus produces aflatoxin B_1, B_2, G_1, G_2 which are carcinogenic and produce liver cancer Purchase, Diener and Davis (Pestka and Bondy) [28-30]. *Fusarium solani* cause corneal ulcer while *F. oxysporum* produce Zeralenone α and β causing haemorrhage and necrosis in bone marrow. *F. proliferatum* and *F. verticillioides* cause epidemiologically human esophageal cancer Desjardins et al. [31]. Significant decrease in protein content due to attack of seed-borne fungi like *Aspergillus flavus* and *Fusarium semitectum* has been observed in seeds of Black gram and Green gram Bilgrami et al. [32]. Prasad and Pathak [33] reported loss in protein content of cereals like Wheat, Maize and Barley seeds affected by *Fusarium oxysporum* and *Fusarium semitectum* under different storage condition. Similarly in present investigation *Aspergillus flavus, A. niger* showed a decrease in protein and carbohydrate content.

Conclusion

There is need for proper storage of chickpea seed to minimize the fungal infestation and mycotoxin production during storage and provide disease free food seeds for human consumption.

Acknowledgements

Author is thankful to Director, Amity Institute of Biotechnology, Amity University Haryana for providing Library and Laboratory facilities.

References

1. Ahmad I, Iftikhar S, Bhutta AR (1993) Seed borne microorganism in Pakistan. A checklist 1991. Pakistan Agricultural Research Council, Islamabad, Pakistan, p: 32.

2. ISTA (1993) International rules for seed testing. Seed Science and Technology 21: 1-288.

3. ISTA (1966) International Rules of Seed Testing. Proc Int Seed Test Assoc 32: 565-589.

4. Malone GP, Muskette AE (1964) Seed Borne Fungi: Description of 77 Fungal Species. Proc Int Seed Test Assoc 29: 180-183.

5. Booth C (1971) The genus Fusarium. Common Wealth Mycological Institute, Kew Survey, England, p: 237.

6. Raper KB, Thom C (1949) A Manual of the Penicillia. Boulliere, Tindall and Cox, London, UK, p: 875.

7. Gillman JC (1967) A manual of soil fungi. Oxford and JBH Publishing Co, India.

8. Raper KB, Fennell DI (1965) The genus Aspergillus. The Williams and Wilkins Company, Baltimore, USA, p: 686.

9. Ellis MB (1971) Dematiaceous hyphomycetes. Commonwealth Mycological Institute, Kew, Surrey, England, UK, p: 608.

10. Ellis MB (1976) More dematiaceous hyphomycetes. Common Wealth Mycological Institute, Kew Surrey, England.

11. Chidambaram PS, Mathur SB (1975) Deterioration of Grains by Fungi. Ann Rev Phytopathol 3: 69-89.

12. Drees BM, Jackman J (1999) Field Guide to Texas Insects. Gulf Publishing Company, Houston, Texas, USA.

13. Beck CW, Blumer LS (2007) A hand book of bean beetles, Callosobruchus maculatus. Bean Beetles.

14. Lowry OH, Rosebrough NJ, Farr AL, Randall RJ (1951) Protein measurement with the Folin phenol reagent. J Biol Chem 193: 265-275.

15. Thimmaiah SK (1999) Standard methods of Biochemical analysis-Anthrone method. Kalyani Publishers, Ludhiana, India, pp: 54-55.

16. Patil DP, Pawar PV, Muley SM (2012) Mycoflora associated with Pigeon pea and Chickpea. International Multidisciplinary Research Journal 2: 10-12.

17. Singh K, Singh AK, Singh RP (2005) Detection of seed mycoflora of chick pea (Cicer arietinum L.). Annals of Plant Protection Sciences 13: 167-171.

18. Narayan M, Ghangaoker A, Kshirsagar D (2013) Study of Seed Borne Fungi of Different Legumes. Trends in Life Sciences 2: 32-35.

19. Razia K, Zaidi, Pathak N (2013) Evaluation of seed infection of fungi in Chickpea. e-Journal of Science & Technology, p: 8.

20. Margaret, Neeraja PV, Rajeswari B (2013) Screening of Seed Borne Mycoflora of Cicer arietinum. Int J Curr Microbiol App Sci 2: 124-130.

21. Sontakke N, Hedawoo (2014) Mycoflora associated with seeds of chickpea. Journal of Life Sciences 2: 27-30.

22. Singh VK (2014) Detection of mycoflora associated with Cicer arietinum seeds by agar plate method with PDA. Weekly Science Research Journal 1: 1-4.

23. Shahnaz D, Farzana S, Ghaffar A (2007) Seed borne fungi associated with chickpea. Pak J Bot Pakistan 39: 637-643.

24. Lily T, Rathi YPS (2015) Detection of seed mycoflora from chickpea wilt complex seedborne Fusarium oxysporum f.sp. ciceri diseased seeds. World Journal of Pharmacy and Pharmaceutical sciences 4: 1242-1249.

25. Rathod LR, Jadhav MD, Mane SK, Muley SM, Deshmukh PS (2012) Seed Borne Mycoflora of Legume Seeds. International Journal of Advanced Biotechnology and Research 3: 5.

26. Dawar S, Ghaffar A (1991) Detection of Seed borne mycoflora of sunflower. Pak J Bot 23: 173-178.

27. Rodricks JV (1976) Mycotoxins and other fungus related food problems. Advance in Chemistry, Series 149. American Chemicals Society, Washington, USA, p: 239.

28. Purchase IRH (1974) Mycotoxin. Elsevier Scientific Publ. Amsterdam, p: 443.

29. Diener UL, Davis ND (1969) Relation of environment to aflatoxin production from Aspergillus flavus. pp: 15-34.

30. Pestka JJ, Bondy GS (1990) Alteration of immune function following dietary mycotoxin exposure. Can J Physiol Pharmacol 68: 1009-1016.

31. Desjardins AE, Busman M, Proctor R, Stessman R (2006) Wheat kernel black point and fumonisin contamination by Fusarium proliferatum. National Fusarium Head Blight Forum Proceedings, p: 115.

32. Bilgrami KS, Jamaluddin, Rizvi MA (1979) Fungi of India. Today and Tomorrow's printers and Publishers, New Delhi, India.

33. Prasad T, Pathak SS (1987) Impact of various storage systems on biodeterioration of cereals. Indian Phytopath 40: 39-46.

30

Evaluation of Some Botanicals and *Trichoderma harzianum* for the Management of Tomato Root-knot Nematode (*Meloidogyne incognita* (Kofoid and White) Chit Wood)

Belay Feyisa[1], Alemu Lencho[1], Thangavel Selvaraj*[1] and Gezehegne Getaneh[2]

[1]*Department of Plant Sciences, College of Agriculture and Veterinary Sciences, Ambo University, Ambo, Post Box No: 19, Ethiopia, East Africa*
[2]*Addis Ababa University, Salale Campus, Addis Ababa, P. B. No: 2003, Ethiopia*

Abstract

Root-knot nematode disease caused by *Meloidogyne incognita* (Kofoid and White) Chit wood) is one of the major constraints for successful cultivation of tomato (*Lycopersicon esculentum* Mill.) in Ethiopia. Hence, the present study was conducted to evaluate the effect of leaf and seed extracts of four botanicals viz., Rape seed (*Brassica napus* L.), Lantana (*Lantana camara* L.), African marigold *(Tagetes erecta* L.) and Neem *(Azadirachta indica* L.) at 5% and 10% concentrations and *T. harzianum* at 5% plus control were tested on root-knot nematode under *in vitro* and also to evaluate their against root-knot nematode development and their role on plant growth parameters of tomato under *in vivo* condition. Plant extracts were more effective and significantly inhibited egg hatching and immobilizing the J_2 larval mortality of *M. incognita* than *T. harzianum*. Aqueous extract of all the tested plants inhibited egg hatching of nematode and resulted 84.67-100% mortality of the J_2 juveniles of *M. incognita in vitro* at the 10% concentration after 72 h of exposure time. There were no significant differences among the treatments of rape seed leaf (84.7%) at 10% concentration and *Lantana camara* (87%), African marigold (86.3%) and Neem leaf (85%) at 5% concentration after 72 h. Aqueous seed extracts of *A. indica* more significantly inhibited egg hatching and larva mortality of the J_2 of *M. incognita in vitro* at the 10% concentration and immobilized by 89, 93 and 100% after 24, 48 and 72 h of exposures, respectively, while at similar concentration of *T. erecta, B. napus* and *L. camara* leaf extracts exhibited 92, 89 and 93.2% inhibition of egg hatching and 75, 62.1 and 73% larval mortality, respectively. The effect of different botanicals and *T. harzianum* singly and in combination were studied for the management of tomato root-knot nematode under greenhouse condition. There was a significant difference in the reduction of root-knot nematode incidence, root-knot nematode population, nematode reproduction rate (NRR), number of galls and egg masses per plant were recorded. In pot culture condition, the application of leaf extract of individual plant in the presence of the nematode significantly enhanced the growth of tomato seedlings in comparison to the control. A significant increase in plant height, shoot weight and root weight of the seedlings were observed at the 10% concentration of leaf extracts in comparison to control. There was a significant difference in the reduction of root-knot nematode population, nematode reduction rate, number of galls and egg masses per plant of *L. camara* combined with *T. harzianum*. The mean fruit weight and total yield were observed highest in the combination treatment of *L. camara* combined with *T. harzianum*. This study results revealed that the test plants are readily available to farmers at no cost and able to reduce nematode population below economic threshold.

Keywords: Tomato seedlings; Botanical leaf; Seed extracts; *Trichoderma harzianum*; Root-knot nematode; Egg hatching; Larval mortality; Growth

Introduction

Tomato (*Lycopersicon esculentum* Mill.) is one of the most widely grown vegetable in the world and regarded as one of the top priority vegetable which is widely cultivated in tropical, sub-tropical and temperate climates [1]. Tomato fruits contributes to healthy, because it is rich in minerals (potassium, magnesium, calcium, iron and zinc), proteins (essential amino acids), citric acid, sugars, dietary fibers (pectin) and high levels of vitamin C, lycopene, and beta-carotene which are anti-oxidants against oxygen radicals that probably cause cancer, aging and arteriosclerosis [2,3]. In Ethiopia, tomato is also among the most important vegetable crops and its production has shown a marked increase since it became the most profitable crop providing a higher income to small scale farmers compared to other vegetable crops [4]. The total area under production reaches 51,698 hectares and annual production is estimated to be more than 230,000 tons in Ethiopia [5]. However, the national average yield of tomato in the country is very low which is around 7 tons/ha [6] and less than 50% of the current world average yield of about, 27 tons/ha [7]. Its production is hampered by poor soil fertility, unreliable rainfall

patterns, poor marketing structures, post-harvest handling problems and most important pests and diseases.

Tomato crops are more susceptible to diseases as compared to other vegetable and cereal crops. Bacterial, fungal and nematode diseases of vegetables are common problem of all agro climatic zones and it is worldwide problem [8]. The root-knot nematode disease *(Meloidogyne incognita)* is the most destructive and widespread diseases of solanaceous vegetables in Ethiopia [9]. Furthermore, the percent incidence of root-knot nematode disease is as high as 65% on tomato and it was recorded in major tomato producing areas of Ethiopia. This

***Corresponding author:** Thangavel Selvaraj, Department of Plant Sciences, College of Agriculture and Veterinary Sciences, Ambo University, Ambo, Post Box No: 19, Ethiopia, East Africa, E-mail: tselvaraj_1956@yahoo.com*

report gives good indication of the losses due to the disease can cause in major tomato producing areas of Ethiopia particularly in Ambo and Toke Kutaye district of West Showa, Ethiopia. *Meloidogyne* spp. is one of the most harmful nematode pests in both tropical and sub-tropical crop production regions and cause extensive economic damage worldwide [10,11]. Many workers have attempted to assess crop losses caused by plant parasitic nematode species in Ethiopia [12,13]. Tomato root-knot nematode species viz., *Meloidogyne incognita*, *M. javanica* and *M. ethiopica* have been reported to occur in Ethiopia [14,15]. Tadele and Mengistu [15] reported that the occurrence of *M. incognita* on tomato in the Eastern part of the country, particularly in Eastern Hararghe, where many vegetable crops were attacked by root-knot nematodes. Apart from the Eastern parts of the country, root-knot nematode, *M. incognita* is the major problem in tomato cultivation in the Central and Western parts of the country [16].

Several methods are known to manage the root-knot nematode which includes the use of nematicide, organic amendments, resistant cultivars, soil solarization and biological control and these have been used with different levels of success on tomatoes [17,18]. However, detrimental environmental effects associated with chemical control and the recent losses of methyl bromide as a multipurpose soil fumigant have spurred research into nematode control alternatives [19]. In view of the uneconomical and hazardous effects of chemical nematicide, researchers have focused their attention to adopt biological control of *Meloidogyne* spp. [20]. The persistent pressure on farmers to adopt strategies that do not pollute the environment has increased urgency in the search for alternative sustainable methods [21,22]. Bio-control appears to offer an environmentally safe and ecologically feasible option for plant protection with great potential for promoting sustainable agriculture. The bio control efficiency depends on the nematode species, plant host and their root exudates, and other crops in rotation [23].The beneficial effects of certain types of plants derived materials and microorganisms in soil have been attributed to a decrease in the population densities of plant-parasitic nematodes [24].

Several fungi have been identified and classified according to their nematophagous properties. Fungi that have toxic effects on nematodes include *Aspergillus* spp. and *Trichoderma* spp. *Trichoderma viride* which were reduced egg-hatching [25] and trade formulations have also proven to be efficacious in tropical greenhouse conditions [26]. Some species of *Trichoderma* have been used widely as bio-control agents against soil-borne plant diseases [27] and also they have activity towards root-knot nematode [28,29]. Al Kader [30] reported a high nematicidal effect of the fungus *Paecilomyces lilacinus* culture filtrate on J2 of *M. incognita*, with 99% of J2 immobilized after 2 days of treatment. *Trichoderma* spp. has been reported to produce chitinase into the culture [31], which might help in the inhibition of egg hatching.

Botanicals, plant-based pesticides are favored as alternatives to chemical pesticides in recent times. When French marigold was planted immediately after the termination of a *Meloidogyne* susceptible host, bitter melon (*Momordica charantia* L.) and marigold suppressed approximately 50% of *M. incognita* compared to the bare ground treatment [32]. Several higher plants and their constituents have been successful in plant disease control and have proved to be harmless and non-phytotoxic, unlike chemical fungicides [33]. The fresh leaf extracts of *Azadirachta indica*, *Allium sativum* (Garlic) and *Tagetes erecta* (African marigold) were examined against *M. incognita* on tomato *in vitro* and *in vivo* conditions. All treatments immobilized juveniles (J$_2$), the highest effect caused by neem leaves extract after 24 and 48 h of exposure. In soil, all treatments significantly reduced the root galling,

nematode population, and enhanced the plant growth and yield [34].

So far, little efforts have been made to exploit locally available botanicals and antagonistic fungal organisms for the control of root-knot nematode on crops in Ethiopia. Even if few works were done by botanicals in Ethiopia, their combination with biological and their synergistic effect with antagonistic fungi are not studied. The management of plant parasitic nematodes using plant products and their derivatives are gaining importance in the light of increased awareness of environmental and human health hazards associated with nematicidal chemicals, biodegradability, and selective toxicity to target pests, safety to non target organisms. Therefore, the present study was conducted to find out the effect of different botanicals leaf and seed extracts and *T. harzianum* on egg hatching and juvenile mortality of root-knot nematode under *in vitro* condition and also to compare the effect of different botanicals and *T. harzianum* on individual and in combination for the management of tomato root-knot nematode development and their role on plant growth under greenhouse condition.

Materials and Methods

Description of the study area

Both *in vitro* and *in vivo* experiments were conducted at Ambo Plant Protection Research Center (APPRC), Ambo, Ethiopia. This center is found in Ambo District, West Shewa Zone of Oromia Regional State, Ethiopia, which is away from Addis Ababa 115 km, with an altitudes of 2100 m, latitude 8° 57' 58"N and longitude 37°5'33"E. The minimum and maximum temperature was about 10°C and 28°C, respectively. The average annual rainfall was about 1260 mm with the relative humidity of 70.

Botanicals, antagonistic fungus and tomato cultivar used: Rapeseed, *Lantana* and marigold were collected from Ambo university campus, Ambo. Neem seeds and leaves and the seeds of tomato cv. *Marglobe* were obtained from Melkassa Research Centre, Melkassa, Ethiopia. *Trichoderma harzianum* (Jimma isolate) was obtained from Department of Mycology, APPRC, Ambo, Ethiopia.

Sample collection and estimation: Diseased root samples of tomato were collected from pure culture pots grown at APPRC green house. For confirmation roots of tomato infested with root-knot nematode were thoroughly washed, cut into small pieces and stained with Acid Fuchsin in lacto phenol [35]. After cooling to normal temperature, they were keeping in lacto phenol overnight for partial de-staining. Root pieces were dissected under stereomicroscope and adult females were taken out and placed in lacto phenol. The perineal region of females were cut with a sharp razor blade and adhering tissue clear off with a fine pick and the perineal sections were examined under microscope. The ten female patterns of root knot nematode were examined and estimated [36,37].

Multiplication, extraction and counting number of juveniles

Egg masses of *M. incognita* were picked up from pure culture pots of infected roots and placed into the sterilized plastic plates with sterile water and kept on the laboratory benches at room temperature (20-23°C) and allowed to hatch for 3-6 days. Two weeks old transplanted seedlings of tomato cv. Marglobe, raised in autoclaved soil in the wire house were inoculated with the juveniles that emerged out of the egg-mass. Inoculation was done by removing top soil (1-2 cm) around the seedlings two sides, make the hole and nematodes in the water poured on partly exposed root-system with pipette. The removed soils were again placed on sides of the seedlings. To get regular supply and sufficient culture of root-knot nematode for subsequent experiments, *M. incognita* was sub-cultured by inoculating juveniles to freshly transplanted tomato seedlings raised in sterilized soil in pots.

The pure cultures of *M. incognita* were raised from single egg mass and maintain on tomato roots in wire house. Infected plants then uprooted from soil and the entire root system was dipped in water and washed gently to remove adhering soil particles. Egg masses of nematodes were picked up and kept it in small sieves. Then the sieves were placed in sterilized plastic plates and pour the water up to neck of the sieves and kept in the laboratory at room temperature. After 2 to 7 days, eggs were hatched and active juveniles cross the sieve and settle down in plastic plates. The J_2 juveniles were collected and then counted by using eelworm nematode counting dish for experimental study. Population densities of J_2 were determined from one ml aliquot of an inoculum suspension. 100 J_2 and 10 egg masses of root knot nematode were used for each treatment in *in vitro* experimental study and 2000 J_2 was used for *in vivo* culture study for each treatment.

Raising and maintenance of tomato plants and inoculation with nematode: The seeds of tomato cv. *Marglobe* were axenized by NaOCl method. About 100 seeds were placed in sterilized beaker containing a mixture of 95% ethanol and 5.25% NaOCl in the ratio of 1:1. The mixture was stirred gently and the seeds were allowed to soak for about 10 minutes. The mixture was drained off and the seeds were rinsed thrice with distilled water. Seeds of cultivar *Marglobe* were sown on sterilized soil in plastic pots under greenhouse. Three leaf stage/ one month-old seedlings were transplanted to plastic pots (15 cm dia.) containing 3 kg of sterilized soil with 1:2:3 proportions of sand, compost and clay, respectively. Each pot was planted only one tomato seedling. Fresh roots of tomato were taken from pure culture developed in the wire house and brought to Plant Pathology Laboratory. Egg masses were picked up by using sterile forceps and dissecting needle and placed to Petri dish having sterile water then kept on laboratory benches at room temperature (20-23˚C) till hatching was completed. Appropriate suspension of nematode was prepared in a beaker and 3 ml was taken from the total suspension and placed on counting dish, then the number of juveniles of the suspension was determined under stereomicroscope at the magnification of 50 ×. The population of nematode per ml was calculated from one ml aliquot of an inoculum suspension for *in vitro* and *in vivo* experiments. Finally, seedlings of tomato were inoculated with the 2 ml suspension of M. *incognita* at 2000 juveniles/pot after one week of transplanting. For inoculation, 1-2 cm of top soil was separated out and nematode suspension was poured around the plant. Each treatment was replicated three times and the pots were arranged in a randomized complete design. Un-inoculated set of plants were served as control.

Preparation of botanical test plants extracts

The test plants leaves and seeds (Table 1) were shade dried and separately powdered using an electric grinder and 20 g powder of each plant powder was soaked separately in 100 ml of distilled water for 24 h in 500 ml Erlenmeyer flask. After 24 h of soaking, they were filtered through Whatmann No.1 filter papers and then the filtrate was centrifuged at 2000 rpm for 10 min for *in vitro* experiments. Each extract was considered as a standard solution of "S" (100% concentration) and then kept in the refrigerator until use for further studies. Suspensions of the concentrations of 0, 5, and 10% were prepared with distilled water [38]. 5 ml and 10 ml of plant extracts were incorporated in to each pot with different treatments.

Common name of the botanicals	Botanical name	Parts used
Rape seed	*Brassica napus L.*	Leaf
Lantana	*Lantana camara L.*	Leaf
African marigold	*Tagetes erecta L.*	Leaf
Neem	*Azadirachta indica L.*	Leaf and seed

Table 1: List of botanicals used.

Mass multiplication of Trichoderma harzianum: Mass multiplication of T. *harzianum* was performed by the method described by Tiwari and Mukhopadhyay [39]. Pure culture of T. *harzianum* was cultured on Potato Dextrose Agar (PDA) media. 5 mm blocks of the 10 day old pure cultures of T. *harzianum* were placed upside down at the center of each plate and the Petri dishes were kept in the growth chamber at 22ºC temperature. After 10 days, an aliquot of 10 ml of sterile water was added to each plate and the mycelium was scraped with a spatula until the culture surface was free from mycelia and the suspension was collected in a 100 ml conical flask. Spores/conidial suspension were separated from mycelia by sieving through cheese cloth and the spore/conidial suspensions were then adjusted to the desired concentration (10^6spores/ml) after counting spore density using a haemocytometer [40]. The pure cultures of T. *harzianum* was attained by inoculating in one liter jar containing sterile sorghum seeds, sand and water with spore suspensions. Spore suspensions were obtained by adding 20 ml sterile distilled water to three- week old cultures and scraping gently with spatula. The spore suspension of T. *harzianum* was inoculated in to sterilized one litter jar containing sorghum seeds and transferred or inoculated to water medium and preserved at 20˚C for 3 days.

In vitro experimental study

Egg bioassay: Test tube bio-assay was carried out to determine the effect of different concentrations of botanical extracts and T. *harzianum on* hatching of *M. incognita* egg masses under *in vitro* [41] condition. Root-knot nematode infected tomato plants from the pure culture pots were up-rooted and washed gently under running tap water. Egg masses of *M. incognita* were picked up from the root using dissecting needle and forceps. Ten uniformly sized egg masses of *M. incognita* were transferred to 5 ml and 10 ml of each concentration of plant extracts and 5 ml of T. *harzianum* alone and combined in sterilized test tubes. Egg-masses in distilled water were only served as control. The experiment was laid out in completely randomized design with three replications. All the test tubes containing the suspensions and the egg masses were kept at room temperature on laboratory bench for seven days to allow eggs hatching.

Juveniles (J2) bioassay: 2 ml suspensions (100 J_2 juveniles) were placed in each test tube containing 5 ml and10 ml of each botanical and 5 ml of T. *harzianum* alone and in combinations. Each treatment was replicated three times. The number of dead J2s were recorded every 24 hours for three days. After 24, 48 and 72 hours, active and inactive J2s were counted in each test tube and sterilized distilled water was served as control [38]. Juveniles were considered dead if they were not move when probed with fine needle and body become straight [42]. Percent J_2 mortality in the test tube was calculated as:

$$\text{Percent J2 mortality} = \frac{\text{No. of inactive (dead) J2s}}{\text{Total J2s in a tube}} \times 100$$

***In vivo* experimental study:** 20 cm wide plastic pots were filled with 3 kg/pot of sterilized mixed soil (sandy clay loam, sand and compost as 2:1:1 (v/v). Seeds of susceptible tomato cultivar were sown at germination pot and after 21 days, seedlings were transplanted to the green house pots. One seedling per pot was maintained at the center. The experiments were laid out in Complete Randomized Design (CRD) with three replications. Tomato potted plant soils were inoculated with 2 ml suspension of 2000 freshly hatched second stage juveniles (J_2) of *M. incognita* and also infested with 10 ml of each botanical and 20 ml of *T. harzianum* suspension [43]. Then proper watering was provided and the pots were kept at 20°C ± 2°C. Applications of botanicals and *T. harzianum* were also repeated after once in 20 days [38].

In vivo experiment consisted of the following thirteen treatments:

T1- Application of Rape seed leaf extract alone,

T2- Application of *Lantana* leaf extract alone,

T3- Application of African marigold leaf extract alone,

T4- Application of neem leaf extract alone,

T5- Application of neem seed extract alone,

T6- *T. harzianum* alone,

T7- Rapeseed+*T. harzianum*,

T8- Lantana+*T. harzianum*,

T9- Marigold+*T. harzianum*,

T10- Neem leaf+*T. harzianum*,

T11-Neemseed+*T.harzianum*

T12- Un inoculated control and

T13- Nematode only inoculated control.

After 90 days of the growth, the plants were uprooted, thoroughly washed and then the plant height, fresh and dry weight of shoot and roots, root-knot nematode population, nematode reproduction rate (NRR), number of galls/plant and egg masses per plant were recorded. The number of fruits per pot was counted. The galling index and the number of egg masses (gall) per plant in each pot were determined using a scale following the rating scale described by Taylor and Sasser [44] and Colyer et al. [45]. Scale 0=0, 1=1-2; 2=3-10; 3=11-30; 4=31-100; 5=>100. Galling index: 0=no galls, 1=slight infection, 2=moderate infection, 3=moderately severe, 4=severe, 5=very severe. The numbers of egg masses per plant on infected roots were counted after staining with Phloxin B [46]. The nematode population was recorded in soils of each treatment separately, after 90 days. The final population density of nematode was determined based on Cobb's sieving and decanting method [47]. The number of nematodes per pot was counted using counting dish.

The reproduction factor (R_F) was calculated by the formula [48]: $R_F = P_F/P_I$

Where P_f is the final population and P_i is the initial population.

Data analysis: Data on plant height, fresh shoot weight, fresh root weight and dry shoot weight, number of galls, egg mass/root, and final nematode population / pot were statistically analyzed as described by Gomez and Gomez [49]. The data were subjected to an Analysis of Variance (ANOVA) procedures using Statistical Analysis system [50] (version.9.1.3, SAS Institute Inc, Cary, NC, USA). All data were subjected to analysis of variance and Duncan's New Multiple Range Test used to separate means at 5% level of probability.

Results and Discussion

In vitro effect of botanicals and *Trichoderma harzianum* against juvenile mortality

The results of the treatments with plant extracts and *T. harzianum* individually and in combination immobilized with *M. incognita* J_2 after 24, 48 and 72 h of exposures are given in (Table 2). The percentage mortality of the second stage juveniles of *M. incognita* under *in vitro* tests as affected by aqueous plant extracts and *T. harzianum* at 24 h showed that there was a significant difference between the treatments. Plant extracts were more effective in immobilizing J_2 than *T. harzianum*. The neem seed extracts were effective in causing J_2 mortality with 10% concentration being more efficacious. Neem seed extract at 10% concentration caused significant mortality of *M. incognita* J_2 24 h after treatment application when compared to all the other treatments. Neem seed extracts which applied at 10% concentration immobilized J2 by 89, 93 and 100% after 24, 48 and 72 h of exposure, respectively. Similar results were reported by Agbenin, [51] after 24 h of exposure of all the treatment levels of dry leaf neem extract caused 100% mortality of larvae except in the control where 85% of larvae remained alive. Parmar, [52] also reported that aqueous extracts of leaf, flower, fruit, bark, root and gum of neem were reported to be highly toxic to nematodes with fruit extract showing the most lethal activity followed by leaf extract. In the present study, at 5% concentration of botanicals, the highest juvenile mortality within 24 h was shown in neem seed and followed by *L. camara*, African marigold and neem leaf respectively. After 48 h of application both at 5 and 10% concentrations, the highest mortality was shown in neem seed and the lowest mortality was shown in rape seed+*T. harzianum*, respectively. After 72 h treatment application the highest and the lowest percent mortality was found in neem seed and rape seed+*T. harzianum*, respectively. On the other hand, all botanicals which combined with *T. harzianum* and applied at both 5% and 10%

Percent mortality of J2 of *M. incogita*				
Treatments	Con. %	24h	48h	72h
Rape seed leaf extract alone	5	72.33e	76.33e	78.67d
	10	77.67d	81.33d	84.67c
Lantana leaf extract alone	5	82.67bc	86.33bc	87.33c
	10	82.60bc	85.33bcd	96.00b
African marigold leaf extract alone	5	80.33cd	83.33cd	86.33c
	10	84.00b	88.00b	95.00b
Neem leaf extract alone	5	79.00d	82.00d	85.33c
	10	83.33bc	86.67bc	94.67b
Neem seed extract alone	5	84.33b	88.67b	94.00b
	10	89.00a	93.00a	100.00a
T. harzianum suspension alone	5	64.33f	70.00f	80.67d
Rape seed + *T.harzianum*	5	50.33i	53.00e	57.33g
Lantana + *T. harzianum*	5	58.67g	61.67g	65.67f
African marigold + *T. harzianum*	5	55.667gh	58.67gh	64.00f
Neem Leaf + *T. harzianum*	5	54.00h	56.00hi	59.00g
Neem Seed + *T. harzianum*	5	62.30f	67.00f	73.67e
Distilled Water (Control)	5	0.00j	0.00j	0.00h
LSD		3	3.69	3.54
CV (%)		2.68	3	2.78

Note: Means in each column followed by the same letter were not significantly different at (P<0.0001), according to Duncan's Multiple Range Test (DMRT)

Table 2: Percentage mortality of the J2 of *M. incogita* under *in vitro* test using botanicals and *T. harzianum*.

concentrations showed less mortality of juveniles than individually applied within 24, 48 and 72 h. There were no significant differences among treatments of rape seed leaf at 10% concentration, *L. camara*, African marigold and neem leaf at 5% concentration within 72 h. Effects of all treatments and *T. harzianum* on J2 mobility continued as exposure time increased, although the differences were not significant as such after 24 h (Table 2). Generally, the mortality rates of juveniles increased with an increase in exposure time. A similar result was reported by Elbadri et al. [53].

There were significant differences between treatments in number of infective juveniles/egg mass of *M. incognita* (Table 3). Different botanicals applied at different concentrations and *T. harzianum* individually and in combination adversely inhibit egg/juvenile hatching. Among botanicals applied, neem seed at concentration of 10% can inhibit egg mass hatching to juveniles, because this concentration had least number of infective juveniles per 10 egg masses, in comparison to 301 juveniles in control. There were no significance difference between the treatment of Rape seed, *L. camara*, African marigold because there were 30-36 number of juveniles per ten egg masses at the concentration of 10%, respectively, but *L. camara* were more effective next to neem seed and neem leaf than other treatment at 10% concentrations. On the other hand there were no significance difference statistically between *T. harzianum* which applied at 5% concentration individually and combination with other botanicals but *T. harzianum* with combination of neem seed at both 5% concentration each were inhibit juvenile hatching. Rape seed leaf applied at both concentrations was less effective because 56 juveniles were hatched per 10 egg masses when compared to control. In general botanicals applied at concentration of 10% was more effective than botanicals applied at 5% concentration on egg mass hatching than *T. harzianum* applied at 5% concentration. Neem seed, neem leaf and *L. camara* at both concentrations, African marigold at 10% concentrations reduce the hatching maximum (>90%) over the control. Both at 10 and 5% concentrations, the greatest percentage of hatching inhibition (96%) and (92%) was achieved by neem seed,

neem leaf followed by *L. camara* and African marigold (90%). Among the botanicals the least egg mass inhibition was obtained by rape seed leaf at both concentrations individually (88%) and combination with *T. harzianum*. Susan and Noweer, Susan AH and Noweer EMA [54] reported that the plant extracts of basil, marigold, pyrethrum, neem and china berry proved to be effective against *M. incognita*. Also, the inhibitory effect of the extracts might be due to the chemicals present in the extracts that possess ovicidal and larvicidal properties [55]. These chemicals either affected the embryonic development or killed the eggs or even dissolved the egg masses. Similar results were reported that the extracts contained alkaloids, flavonoids, saponins, amides including benzamide and ketones that singly and in combination inhibit egg mass hatching [56,57]. Also, Salawu EO [58] reported that the neem seed extracts to inhibit egg hatch, and juvenile activity. In the present study, the neem seed was acted as the highest in juvenile mortality and egg mass hatch inhibition by *in vitro*. Meira et al. [59] reported that the soluble plant extracts were very effective in inhibiting egg-hatch and larval motility of nematodes. The active principles of neem viz. nimbidin and thionimone were reported to be highly active against nematodes. Fatema and Ahmad [60] have been reported that the extracts of neem leaf and garlic bulb completely inhibited hatching of egg masses of *M. incognita* and were lethal to larvae. In this study, the neem leaf extracts can inhibit 90% of egg hatching. The inability of the egg mass to hatch is as a result of ingress/entrance of plant extracts into the egg mass. Larvae in the egg mass were exposed to the toxic effect of the extract resulting first in reduced mobility and finally death or moribund state. Once this state is reached the larva cannot pierce through the wall with its stylet hence hatching ceases. The egg mass which is a part of the perineal region of the female in root-knot is permeable to the active ingredient in the extracts [61]. These compounds act by various mechanisms like blocking molting of larvae, disrupting mating and sexual communication of nematodes, reducing the motility of gut and by inhibiting the formation of chitin [62]. Sharon et al. [63] showed that eggs adhered with *Trichoderma* conidia became non-viable, thus decreasing the eclosion rate. In this study, the botanicals used only they were effective but when they were used in combination they show less effective so it is evident that as extract was diluted, toxicity was decreased resulting in correspondent decrease in inhibition and any inhibition was observed in distilled water.

In vivo effect of botanicals and *Trichoderma harzianum* against *M. incognita*

Plant height: The treatments did not showed any negative effects on plant growth. There were significant differences in the height of tomato plants treated with aqueous plant extracts and *T. harzianum* over inoculated control plants (Table 4). The highest plant height was observed in pots treated with combination of *L. camara* and *T. harzianum* followed by neem seed and neem leaf with *T. harzianum* over inoculated control. The lowest height of plants was recorded in pots treated with rape seed leaf. The highest plant height was 160% increased in pots treated with combination of *L. camara* and *T. harzianum* over inoculated control. Pots treated with combination of botanicals and *T. harzianum* showed more height than botanicals applied alone or without fungus. The addition of botanicals to soil leads to a better environment for the growth of the roots. This enhances the utilization of soil nutrients, as a consequence of which the nematode damage might have been markedly reduced [64]. These botanicals may be act as substrate for the growth and multiplication of *T. harzianum*. Some *Trichoderma* isolates were reported to do both enhanced plant growth and reduced root-knot nematode damage [65]. It has been reported that *Trichoderma* has not only been proved to parasitize

Treatments	Con.	No. eggs hatched to J₂ after 7days	Z**
Rape Seed Leaf extract alone	5	36.00cdef	88
	10	33.60cdef	89
Lantana leaf extract alone	5	30.67cdef	90
	10	32.60cdef	91
African marigold leaf extract alone	5	33.00cdef	89
	10	27.00ef	91
Neem Leaf extract alone	5	33.33cdef	89
	10	29.00def	90
Neem Seed extract alone	5	25.00f	92
	10	10.67g	96
T.harzianum suspension alone	5	38.67cde	87
Rape seed + *T.harzianum*	5,5	56.33b	81
Lantana + *T. harzianum*	5,5	40.30cd	87
African marigold + *T. harzianum*	5,5	43.00c	86
Neem Leaf + *T. harzianum*	5,5	54.67b	82
Neem Seed + *T. harzianum*	5,5	39.33cde	87
Distilled Water (Control)	5	301.33a	
LSD		11	
(CV)%		13	

Note: means in column with the same letter are not significantly different (P<0.0001) by DMRT.
Z** Hatching inhibition over the control in percent

Table 3: Egg mass hatching and hatching inhibition of *Meloidogyne incogita* by *in vitro* test using botanical aqueous plant extracts and *Trichoderma harzianum*.

nematodes and inactive pathogen enzymes but also help in tolerance to stress condition by enhanced root development. It participates in solubilization of inorganic nutrients [66]. The shortness' of the plant height might be due to the stunting action of *M. incognita*. Jinfa et al. [67] also reported that this kind of height reduction caused by root-knot nematode. In inoculated control, the lowest growth performances by the control plants could be as result of the combined effect of nematodes and availability of nutrients [68]. The galls on the root system might disturb important root functions like uptake and transport of water and nutrients [13].

Fresh and dry shoot weight: The higher fresh shoot weight was significantly obtained in seedlings treated with aqueous plant extracts and *T. harzianum* over inoculated control (Table 4). The highest and the lowest shoot fresh biomass was observed in plants treated with the combination of *T. harzianum* with *L. camara* and rape seed leaf, 146 and 80 g, respectively, when compared with inoculated control. The results of the present experimental study was not agree with Agbenin et al. [51], Neem seed powder increased root and shoot weights and heights and decreased root galling index and presence of mycelium on root. Generally, *T. harzianum* individually and combination with botanicals showed more effective on plant fresh shoot weight than botanicals. Dry shoot weight of plants after 90 days were significantly lower in inoculated control plants than inoculated treated plants (Table 4). There were no significance difference between pots treated with all botanicals applied individually, rape seed and neem leaf with combination of *T. harzianum* and un inoculated control but they were significant difference when compared with inoculated control. The maximum total plant shoot dry weights were recorded in pots treated with combination of *L. camara* and *T. harzianum* followed by *T. harzianum* with combination of neem seed and neem leaf over inoculated control. The lowest dry shoot weight was observed in pants treated with rape seed than other treatments (Table 4).

Fresh root weight: There were highly significance differences among recorded fresh root weight between pots treated with aqueous plant extracts and *T. harzianum* when compared with inoculated

control (Table 4). The highest fresh root weight was recorded by plants grown on pots with inoculated control followed by neem leaf with *T. harzianum* and African marigold jointly with *T.harzianum*. The lowest weight was recorded in pots with un inoculated control or negative control when compared with inoculated control. *Trichoderma* spp. found in close association with roots contributes as plant growth stimulators [69]. In the present study, the root weight of inoculated control was greater than that of un inoculated weight. Wong and Mai [70] reported that differences in root weight may be explained by gall development, gall mass being heavier than an equivalent linear length of similar non-galled roots. Perry et al. [71] reported similar results that root weight increased in untreated infected plants compared with those amended with herbal powder due to the formation of galls and giant cells.

Number of galls per root system: The number of galls per root system was observed significantly reduced between the pots treated with aqueous plant extracts and *T. harzianum* over inoculated control (Table 5). Maximum inhibition of gall formation was observed in pots treated with combination of lantana and *T. harzianum* and followed by neem seed with *T. harzianum* .The highest and lowest reduction of number of galls per root system was observed in pots treated with combination of *L. camara* with *T. harzianum* and rape seed leaf because they reduced number of galls by 88 and 37% over inoculated control, respectively. Combination of *T. harzianum* with neem seed, neem leaf and the fungus only also showed gall reduction next to combination of neem seed with *T. harzianum* that shows gall reduction 83, 79 and 75%, respectively. Generally the highest reductions in number of galls per root were observed in pots treated with combination of botanicals and *T. harzianum* than botanicals treated individually. The lowest growth rate, high galling due to nematode activity at root zone resulting in giant cell formation, high population of nematodes because the nematodes larvae were able to penetrate roots freely and reproduce without any inhibition. A reduction in root knot development could be attributed to poor penetration of the second stage juveniles and later retardation in their activities, for example feeding and /or reproduction

Treatment	Cons.	Plant height (cm)	Z**	Fresh shoot weight (g)	z**	Dry shoot weight(g)	Z**	Fresh root weight(g)
Rape seed leaf extract alone	10	68.00d	50	80.00d	46	23.17e	71	41.00cd
Lantana leaf extract alone	10	80.0bcd	66	109.00c	88	26.00e	86	29.33efg
African marigold leaf extract alone	10	80.bcd	66	108.00c	86	25.50e	82	27.00fg
Neem leaf extract alone	10	73.33cd	52	86.00d	48	25.00e	78	24.00gh
Neem seed leaf extract alone	10	72.33cd	41	85.00d	38	24.00e	64	20.00h
T.harzianm **suspension only**	20	83.0bcd	73	110.00c	89	27.00e	93	30.00efg
Rape seed + *T.harzianum*	10+10	88.0bcd	83	120.00bc	106	30.00de	114	35.00de
Lantana + *T. harzianum*	10+10	125.00a	160	146.47a	151	62.00a	342	10.00i
African marigold + *T. harzianum*	10+10	89.00bc	85	121.00bc	108	38.00cd	171	45.00bc
Neem Leaf + *T. harzianum*	10+10	91.67bc	90	128.00b	120	40.00bc	185	48.00b
Neem Seed + *T. harzianum*	10+10	95.00b	98	130.00b	124	48.00b	242	32.00ef
UC	-	84.67bcd		111.00c		28.00e		19.00h
IC CV (%)	- 12.82	48.00e 8.2	16	58.00e	12.7	14.00f		55.00a
LSD	17.84	14.76	8.5		6.8			

Note: means in column with the same letter are not significantly different (P<0.0001) DMRT.
Z** increase over inoculated the control in percent.
Values are averages of three replicates.
Significance is given compared to positive controls (inoculated control).
Control positive = control in conjunction with inoculation of *M. incognita* juveniles.
Control negative = control without inoculation of *M. incognita* juveniles

Table 4: Effect of aqueous plant extracts and Trichoderma harzianum Table 4 Effect of aqueous plant extracts and Trichoderma harzianum on growth of tomato plants against root-knot nematode infested soil under green house condition on growth of tomato plants against root-knot nematode infested soil under green house condition.

Treatment	Con$_{st}$	Gall/root	X**	Eggmass/root	X**	Final Nematodepopulation./pot	X**	Reproduction factor(R=PF/P)
Rape Seed leaf extract alone	10	203.00b	37	175.00b	39	650.00b	79	0.325b
Lantana leaf extract alone	10	115.00ef	64	91.00de	68	513.00c	83	0.26de
A. marigold leaf extract alone	10	138.00d	57	128.00c	55	548.00c	82	0.27c
Neem leaf extract alone	10	158.00c	51	129.00c	55	560.00c	81	0.275c
Neem seed extract alone	10	120.00de	63	96.00d	66	521.00c	83	0.261cde
T.harzianum suspension alone	20	79.00hi	75	62.00g	78	532.00c	82	0.266f
Rape seed+*T.harzianum*	10+10	99.00fg	69	83.00ef	71	509.00c	83	0.25de
Lantana+*T.harzianum*	10+10	39.00k	88	35.00i	87	303.00d	90	0.15g
African margold+*T.harzianum*	10+10	90.00gh	72	78.00f	73	490.00c	84	0.245ef
Neem leaf+*T.harzianum*	10+10	68.00ij	79	53.00h	81	463.33c	85	0.23cd
Neem seed+*T.harzianum*	10+10	55.00jk	83	42.00i	85	341.00d	89	0.17h
UC	-	0.00l		0.00j		0.00e		0.00e
IC	-	325.00a		290.00a		3080.00a		1.54a
CV (%)		9.81		5.44		8.2		
LSD		18.85		8.86		89.8		

Note: means in column with the same letter are not significantly different (<0.0001) PDMRT
Each pot contains 3000 cc sterilized soil.
X**: Reduction over inoculated control in percent

Table 5: Effect of aqueous plant extracts and *Trichoderma harzianum* on nematode population, gall and egg mass on tomato plants (cv. Marglobe) in root-knot nematode infested soil under green house.

as suggested by Abdi M [72].

Number of egg masses per root system: There were significant differences between treatments on egg mass reduction over inoculated control (Table 5). Similarly all the treatments were found to be highly effective in their ability to reduce egg mass per root system when compared with inoculated control/untreated plants. The highest and the lowest egg mass reduction was observed in pots treated with *L. camara* combined with *T. harzianum* and botanical rape seed leaf over inoculated control. The highest percentage of egg mass reduction was observed with pots treated with combination of *L. camara* and *T. harzianum* (87%) and followed by combination of neem seed with *T. harzianum* (85%) and *T. harzianum* alone (80%). Concerning the effect of rape seed on nematodes it is true with results reported by Johnson et al. [73].

Final nematode population and reproduction factor: The suppressive effect of aqueous plant extracts and *T. harzianum* was recorded as the nematode population in the soil at the end of the experiment 90 days after nematode inoculation. Significantly, the less number of parasitic nematodes was observed in the soil samples obtained from pots treated with *L. camara* with *T. harzianum* as compared to the control. Among treatments, pots treated with *L. camara* and *T. harzianum* showed more effective in reducing the final nematode population over inoculated control. For this reason it is suggested that the use of plants residue too would be more efficient against nematodes when used in combination with other management practices that are currently available. Except rape seed there were no significant difference between pots treated with botanicals each other. The maximum and minimum final nematode population was recorded from combination of rape seed leaves and *T. harzianum* and *Lantana* leaves with *T. harzianum,* respectively, (Table 5). The highest percentage (90%) of final nematode population reduction was shown in pots treated with combination of *L. camara* combination with *T. harzianum* followed by combination of neem seed with *T. harzianum* (89%) over inoculated control. The lowest nematode population reduction was observed in pots treated with rape seed leaf when compared with other treatments (Table 5). A reduction in root-knot development could be attributed to poor penetration of the second stage juveniles and later retardation in their activities, for example feeding and/or reproduction

as suggested by Abdi [72]. Nematotoxic compounds especially the Azadirachtin released through gradual decomposition of the neem seeds [29] and suppress nematode populations throughout the whole period of the nursery stage. Nematode population in nematode + fungus treatment was 532 but *L. camara* has decreased population to 303. Except rape seed leaf extract, there were no significant differences between the pots treated with all botanicals including *T. harzianum* which applied individually. In this study, African marigold and neem seed treated pots were reduced nematode population in the soil 82 and 83%, respectively. Similarly, Hasabo and Noweer [74] found that the soil treatment with aqueous extracts of marigold leaves and neem seeds significantly reduced *M. incognita* J$_2$ in soil and roots of egg plants. The J2 population in roots was reduced by 90% and 75% respectively, 4 months after treatments, applied at 50 ml/plant as soil drench. Begum et al. [75] and Qamar et al. [76] observed that isolated chemical constituents such as lantanoside, lantanone, camaric acid and oleanolic acid from aerial parts of *L. camara*, possessing nematicidal activity against *M. incognita*. Ahmad et al. [77] also noted that various concentrations of leaf extract of *L. camara* were deleterious to *M. incognita*.

Reproduction factor: The reproduction rate of *M. incognita* was significantly suppressed by all the treatments as compared to untreated inoculated plants (Table 6). Reproduction rate of *M. incognita* was 0.23 in nematode+ fungus but its decrease to 0.15 by *L. camara*. Nematode reproduction factor was reduced in pots treated with combination of lantana and *T. harzianum* followed by pots treated with neem seed and *T .harzianum* when compared with inoculated control than other treatments. The highest nematode reproduction factor was observed in pots treated with rape seed leaf aqueous extracts than other treatments. This is because we suggest that *L. camara* act as substrate for the growth and multiplication of *T. harzianum*. Decomposed leaves have been found to support greater sporulation and multiplication of *T. harzianum* and *P. chlamydosporia* [78]. Several authors have been shown the potential of using plant extracts in the control of plant parasitic nematodes [79,80,81]. The reduction in population of *M. incognita* in this investigation may be due to the accumulation of nematicidal components and/or to increase host resistance. This significant reduction on the final nematode population density in the soil could be due to the chemicals present in the extracts that possess ovicidal or larvicidal properties resulting in inhibition of its multiplication. *T. harzianum* and *L. camara* not only could decrease nematode population but also increase growth parameters of tomato.

Fruit number per pot: Highest number of tomato fruit was found in pots treated with lantana camara + *T. harzianum* followed by neem seed+ *T. harzianum* over inoculated control (Table 7). Among all treatment the lowest number of fruit was recorded from pots treated with rape seed leaves. There were no significance difference between pots treated with only botanicals but they were significant difference from inoculated control. The inability of the control plants to flower and fruit is probably due to the combined action of the nematode and inadequate availability of nutrients.

Yield of tomato per hectare: Application of aqueous plant extract and *T. harzianum* on root knot nematodes; *Meloidogyne incognita* infested pot shows significant difference ($P < 0.0001$) on yield of tomato over control (Table 7). The highest yield of tomato was observed in pots treated with combination of *Lantana camara* with *T. harzianum* and followed by neem seed with *T. harzianum* over inoculated control, respectively. The lowest yield was observed in pots treated with aqueous rape seed leaf extracts. The presence of nematode on tomato plants significantly affected their yield, un inoculated plants had 82% yield higher ($P < 0.0001$) than inoculated plants. Root colonization by *Trichoderma* spp. frequently enhances root growth and development, crop productivity, resistance to a biotic stresses and uptake and use of nutrients [82,83]. Cuevas VC [84] showed that the presence of the fungus in the soil in sufficient population resulted in the uptake of more mineral nutrients especially P and Zn available for plant use that increased crop growth and yield in the screen house and farmers' field.

Conclusions

For its management, different plant species (botanicals) and an antagonistic fungus, *T. harzianum* were being tried in different forms as an alternative to nematicide. Water extract of all tested plants significantly inhibited egg hatching of nematode and resulted in 100% mortality of the second juveniles of *M. incognita in vitro* after 72 h of exposure. Results on mortality of infective juveniles (J_2) up to 100%

Treatment	Con$_s$	Total no. of fruits/pot	Kg/Pot	t/he
Rape Seed leaf extract alone	10	7.50h	0.75ef	14.00f
Lantana leaf extract alone	10	10.00fgh	0.60fg	18.00def
A. marigold leaf extract alone	10	9.00gh	0.90def	16.00ef
Neem leaf extract alone	10	8.00h	0.80ef	15.00ef
Neem seed extract alone	10	11.00e-h	1.10cde	22.00cdef
T.harzianum suspension alone	20	14.00edf	1.15cde	23.00cdef
Rape seed+*T.harzianum*	10+10	15.00cde	1.20cde	24.00cde
Lantana+*T.harzianum*	10+10	34.00a	2.2a	44.00a
African margold+*T.harzianum*	10+10	17.33cd	1.30bcd	26.00bcd
Neem leaf+*T.harzianum*	10+10	19.00bc	1. 5bc	30.00bc
Neem seed+*T.harzianum*	10+10	22.00b	1.70b	34.00b
UC	-	13.00d-g	1.12cde	22.40cdef
IC	-	3.00i	0.20g	4.00g
CV (%)		17.5	22.6	22.1
LSD		4.13	0.4	8.35

Note: means in column with the same letter are not significantly different (P<0.05) by DMRT

Table 6: Effect of aqueous plant extracts and *Trichoderma harzianum* on yield of tomato plants (cv. Marglobe) in root-knot nematode infested soil under green house.

	HT	FSW	DSW	FRW	G.root	E.Mas	N.po	N.Fruit	Kg/pot	Ton/ha
HT	1	0.83	0.81	-0.48	-0.69	-0.70	-0.60	0.85	0.77	0.75
FSW		1	0.80	-0.35	-0.79	-0.79	-0.65	0.81	0.79	0.78
DSW			1	-0.31	-0.62	-0.61	-0.46	-0.90	-0.83	-0.81
FRW				1	0.54	0.55	0.59	-0.38	-0.42	-0.43
G.root					1	0.99	0.86	-0.68	-0.75	-0.73
E.Mas						1	0.88	-0.67	-0.74	-0.73
N.pop							1	-0.47	-0.57	-0.57
N.Frui								1	0.86	0.87
Kg/pot									1	0.97
Tons/ha										1

** Correlation is significant at the 0.0001 level
* Correlation is significant at the 0.05 level.
Where
HT=Height; FSW=Fresh dry weight; DSW=Dry shoot weight; FRW=Fresh root weight; G. root=Number of gall per root system; E.mas=number of egg mass per root system,; N.pop=Final nematode population; N. fruit=Number of fruit per plant; Kg/pot=kilogram per pot; Ton/ha=ton per hectare

Table 7: Correlation of plant height, fresh shoot weight, dry shoot weight, fresh root weight, yield/tons/hectare, egg masses, final population, and number of gall Meloidogyne incognita in tomato plant under green house.

in72 h and egg hatch inhibition up to 96% in seven day duration were observed in test tubes treated with neem seed and *L. camara* individually in laboratory experiment. Egg inhibition and larval mortality decreased with increase in dilution of all the extracts. Juvenile mortality increased corresponding to an increased time of exposure. Similarly a prominent reduction in final nematode population density, egg mass, galls per root system and a significant increase yield per plant and total yields of tomato were observed from plants treated with the combination of *T. harzianum* with *L. camara* and neem seed extracts compared to any other treatments. These results suggest that in laboratory experiment application of aqueous neem seed and *L. camara* in green house combination of *T. harzianum* jointly with *Lantana* leaf and neem seed would be a good alternative to manage root-knot nematode populations in tomato production. These combinations not only reduce nematode infestation and population buildup on tomato but also increase soil fertility. Therefore, bio-control is suggested to be a safer solution. Botanicals are more effective jointly with fungus than applied individually in green house because some botanicals act as a substrate for the growth and multiplication of *T. harzianum*. Non chemicals and eco-friendly management such as bio-control management system by using *T. harzianum* and botanicals mentioned above were gaining importance and also greater attention which are easily available, less cost effective with no pollution hazards.

References

1. Food and Agricultural Organization (2006) FAO production year book, Basic Data Unit, Statistics division, FAO, Rome, Italy 55: 125-127.

2. Naika S, Jeude JL, Goffau M, Hilmi M, Dam B (2005) Cultivation of Tomato: Production, Processing and Marketing. Agromisa Foundation and CTA,4th edition, Wageningen, Netherlands.

3. FAOSTAT (2011) Statistical database of the food and agriculture of the United Nations. FAO, Rome, Italy.

4. Lemma D, Yayeh Z, Herath M (1992) Agronomic studies in tomato. Horticulture Research and Development in Ethiopia: Proceedings of the Second National Horticultural Workshop of Ethiopia. Addis Ababa, Ethiopia.

5. Central Statistics Authority (2009) Sample Survey 2008/2007. Report on area and production of crops (Private peasant holdings, main season). Stat. Bull, Addis Ababa, Ethiopia, 01-446.

6. Central Statistics Agency (2011) Agriculture in Figures: Key Findings of the 2008/09-2010/11 Agricultural Sample Surveys for All Sectors and Seasons, Country Summary.

7. FAOSTAT (2007) FAOSTAT on line. Rome: United Nations Food and Agriculture Organization.

8. Lanny G (2001) Fruit vegetables. In: Raemaekers RH, (eds).Crop production in Tropical Africa Brussels, Belgium. 1540.

9. Tesfaye T, Habtu A (1986) a Review of Vegetable Diseases Research in Ethiopia. In: Tsedeke, a (ed.). A Review of Crop Protection Research in Ethiopia. Proceedings of the First Ethiopian Crop Protection Symposium, 4-7 February 1985. Addis Ababa, Ethiopia 495-518.

10. Howard RJ, Allan Garland J, Llyod seaman W (1994) Diseases and Pests of Greenhouse Crops. In: Garland JA and Seaman WL. (Eds.). Diseases and pests of vegetable crops in Canada. The Canadian Phytopathological and Entomological Society of Canada 303-305.

11. Ministry of Agriculture and research Development (2009) Rural Capacity Building project. Course for Training of trainers on improved horticultural crop technologies. 5-19.

12. Mutitu EW, Muiru WM, Mukunya DM (2008) Evaluation of Antibiotic Metabolites from Actinomycete Isolates for Control of Late Blight of Tomatoes under Greenhouse Conditions. Asian Journal of Plant Sciences 7: 284-290.

13. Sikora RA, Fernandez E, Bridge J, Luc M (2005) Nematode parasites of vegetables. In: Plant Parasitic Nematodes in subtropical and Tropical Agriculture, ed, CABI Publishing, Wallingford, UK 319-392.

14. Trifonova Z, Karadjova J, Georgieva T (2009) Fungal parasites of the root-knot nematode, Meloidogyne spp. in Southern Bulgaria, Estonia. Journal of Ecology 58: 47-52.

15. Tadele T, Mengistu H (2000) Distribution of Meloidogyne incognita (root-knot nematode) in some vegetable fields in eastern Ethiopia. Pest Management Journal of Ethiopia 4: 77-84.

16. Wondirad M, Tesfamariam M (2002) Root-knot nematodes on vegetatble crops in central and Western Ethiopia. Pest Management Journal of Ethiopia 6: 37-44.

17. Ahmad F, Rather MA, Siddiqui MA (2010) Nematicidal Activity of Leaf Extracts from Lantana camara L. against Meloidogyne incognita (Kofoid and White) Chitwood and its use to Manage Roots Infection of Solanum melongena L. Brazil Arch Biology and Technology 53: 543-548.

18. Zia-Ul-Haq M, Nisar M, Shah MR, Akhter M, Qayum M, et al. (2011) Toxicological screening of some selected legumes seed extracts. Legume Research 34: 242-250.

19. Wondirad M, Kifle D (2000) Morphological variations of root-knot nematode population from Ethiopia. Pest Management Journal of Ethiopia 4: 19-28.

20. Randhawa N, Sakhuja PK, Singh I (2001) Management of root-knot nematode Meloidogyne incognita tomato with organic amendments. Plant Disease Research 16: 274-276.

21. Sakhuja PK, Jain RK (2001) Nematode diseases of vegetable crops and their management. In: Diseases of fruits and vegetables and their management (ed.) TS. Thind, Kalyani Pub, Ludhiana (India).

22. Nico AI, Jimenez-Diaz RM, Castillo P (2004) Control of root-knot nematode by composted agro-industrial wastes in potting mixtures. Crop Protection 23: 581-587.

23. Singh S, Mathur N (2010) In vitro studies of antagonistic fungi against root-knot nematode, Meloidogyne incognita, Bio-control Sciences and Technology 20: 275-285.

24. Pinkerton JN, Ivors KL, Miller ML, Moore LW (2000) Effect of solarization and cover crops on population of selected soil borne plant pathogens in Western Oregon. Plant Disease 84: 952- 960.

25. Hallman J, Davies KG, Sikora R (2009) Biological control using microbial pathogens, endophytes and antagonists. In: Root-knot Nematodes. Perry RN, Moens M, Starr JL (eds.). Wallingford, UK, CAB International: 380-411.

26. Akhtar M (2000) Approaches to Biological Control of Nematode Pests by Natural Products and Enemies. Journal of Crop Production 3: 367-395.

27. Goswami BK, Mittal A (2004) Management of root-knot nematode infecting tomato by Trichoderma viride and Paecilomyces lilacinus. Indian Phytopathol

57: 235-236.

28. Meyer SLF, Roberts DP, Chitwood DJ, Carta LK, Lumsden RD (2001) Application of Burkholderia cepacia and Trichoderma virens, alone and in combinations, against Meloidogyne incognita on bell pepper. Nematropica 31: 75-86.

29. Sharon E, Bar-Eyal M, Chet I, Herrera-Estrella A, Kleifeld O, et al. (2001) Biological Control of the Root-Knot Nematode Meloidogyne javanica by Trichoderma harzianum. Phytopathology 91: 687-693.

30. Al Kader MAA (2008) in vitro studies on nematode interaction with their antagonistic fungi in the rhizosphere of various plants. Ph.D. Thesis 58. Albert-Ludwigs-Universität, Germany.

31. Abo-Elyousr, Kamal A, Zakaullah Khan, Magd El-Morsi Award, M F Abedel-Moneim (2010) Evaluation of plant extracts and Pseudomonas spp. for control of root-knot nematode, Meloidogyne incognita on tomato. Nematropica 40: 289-299.

32. Whipps JM1 (2001) Microbial interactions and biocontrol in the rhizosphere. J Exp Bot 52: 487-511.

33. Cuadra R, Ortega J, Morfi OL, Soto L, Zayas MDIA, et al. (2008) Effect of the biological controls Trifesol and Nemacid on root-knot nematodes in sheltered vegetable production. Rev. Protección Veg 23: 59-62.

34. Alam S, Akhter N, Begun F, Banu MS, Islam MR, et al. (2002) Antifungal activities (in vitro) of some plant extracts and smoke on four fungal pathogens of different hosts. Pakistan Journal of Biological Sciences 5: 307-309.

35. Barker KR, Carter CC, Sasser JN (1985) an advanced treatise on Meloidogyne: Volume II. North Carolina State University Graphics. 223.

36. Seinhorst JW (1998) the common relation between population density and plant weight in pot and microplot experiments with various nematode plant combinations. Fundamental and Applied Nematology 21: 459-468.

37. Orisajo SB, Okeniyi MO, Fademi OA, Dongo LN (2007) Nematicidal effects of water extracts of Acalypha ciliate, Jatropha gosssypifolia, Azadiractha indica and Allium ascalonicum on Meloidogyne incognita infection on cacao seedlings. Journal of Research Biosciences 3: 49-53.

38. Taye W, Sakhuja PK, Tefera T (2012) Evaluation of plant extracts on infestation of root- knot nematode on tomato (Lycopersicon esculentum Mill), Journal of Agricultural Research and Development 2: 086-091.

39. Tiwari AK, Mukhopadhyay AN (2001) Testing of different formulations of Gliocladium virens against chickpea wilt complex. Indian Phytopathology 54: 67-71.

40. Niranjana SR, Lalitha S, Hariprasad P (2009) Mass multiplication and formulations of biocontrol agents for use against Fusarium wilt of pigeon pea through seed treatment. International Journal of Pest Management 55: 317–324.

41. Nitao JK, Meyer SL, Chitwood DJ (1999) In-vitro Assays of Meloidogyne incognita and Heterodera glycines for Detection of Nematode-antagonistic Fungal Compounds. J Nematol 31: 172-183.

42. Siddiqui IA, Shaukat SS (2004) Trichoderma harzianum enhances the production of nematicidal compounds in vitro and improves biocontrol of Meloidogyne javanica by Pseudomonas fluorescens in tomato. Lett Appl Microbiol 38: 169-175.

43. Elbadri GAA, Lee DW, Park JC, Choo HY (2009) Nematicidal efficacy of herbal powders on Meloidogyne incognita (Tylenchida: Meloidogynidae) on potted watermelon. Journal of Asian –Pacific Entomology 12: 37-39.

44. Taylor AL, Sasser JN (1978) Biology, identification and control of root-knot nematodes (Meloidogyle spp.). Coop. Publ. Dept. Plant Pathology, North Carolina State University, Graphics, Raleigh, NC. 111.

45. Colyer PD, Kirkpatrick TL, Caldwell WD, Vernon PR (2008) Root-knot nematode reproduction and root galling severity on related conventional and transgenic cotton cultivars. Journal of Cotton Science 4: 232-236.

46. Charchar JM, Santo GS, O'bannon JH (1984) Life cycle of Meloidogyne chitwoodi and M. hapla on potato in controlled temperature tanks and microplots. In: International Congress of Nematology, I, Guelph, Canada, 18.

47. Southey JF (1986) Laboratory Methods for Work with Plant and Soil Nematodes. Ministry of Agriculture, Fisheries and Food, HMSO, London.

48. Zhang F, Schmitt DP (1994) Host Status of 32 Plant Species to Meloidogyne

konaensis. J Nematol 26: 744-748.

49. Gomez KA, Gomez AA (1984) Statistical Procedures for Agricultural Research, 2nd Edn, John Willey and Sons, New York, 680.

50. SAS Institute (2002) The SAS system for windows, version 9.1. SAS, Institute, Cary, NC.

51. Agbenin NO, Emechebe AM, Marley PS (2004) Evaluations of neem seed powder for Fusarium wilt and Meloidogyne control on tomato. Arch. Phytopathol. Plant Protection 37: 319-326.

52. Parmar BS (1987) An overview of neem research and use in India during the years 1983-1986; in: Natural Pesticides from the Neem Tree and other Tropical Plants; Proceedings of the 3rd International Neem Conference, Nairobi (1996), (eds) Schmutterer H, Ascher KRS.;; GTZ Eschborn, Germany 55–80.

53. Elbadri GAAD, Lee W, Park JC, Yu HB, Choo HY (2008) Evaluation of various plant extracts for their nematicidal efficacies against juveniles of Meloidogyne incognita. Journal of Asia-Pacific Entomology 11: 99–102.

54. Susan AH, Noweer EMA (2005) Management of root-knot nematode, Meloidogyne incognita on eggplant with some plant extracts. Egyption Journal of Phytopathology 33: 65-72.

55. Adegbite AA (2003) Comparative effects of carbofuran and water extracts of Chromoleana odorata on growth, yield and food components of root knot nematode infested soybean (Glycine max (L), Merill). Ph.D Dissertation, University of Ibadan, Nigeria.

56. Goswami BK, Vijayalakshmi V (1986) Nematicidal properties of some indigenous plant materials against root-knot nematode Meloidogyne incognita on tomato. Indian Journal of Nematology 16: 65-68.

57. Mousa EM, Mahdy ME, Younis, Dalia M (2011) Evaluation of some plant extracts to control root-knot nematode Meloidogyne spp. on tomato plants. Egypt Journal of AgroNematology 10: 1-14.

58. Salawu EO (1992) Effect of neem leaf extract and ethoprop singly and in combination on Meloidogyne incognita and growth of sugarcane. Pakistan Journal of Nematology 10: 51-56.

59. Meira B, Edna S, Yitzhak S (2006) Nematicidal activity of Chrysanthemum coronarium. European Journal of Plant Pathology 114: 427-433.

60. Fatema S, Ahmad MU (2005) Comparative efficacy of some organic amendments and a nematicide (Furadan-3G) against root-knot on two local varieties of groundnut. Plant Pathology Journal 4: 54-57.

61. Hirschmann, H (1985) The classification of the family Meloidogynidae; in: An Advanced treatise on Meloidogyne: Vol. 11, Methodology, (eds) Sasser JN and Carter, CC. North Carolina State University Department of Plant Pathology and the United States of Agency for International Development, Raleigh.

62. Ramasamy I (2008) High quality biopesticides for cost effective pest management Agricultural Info Technology, Tamil Nadu, India.

63. Sharon E, Chet I, Viterbo A, Meira BE, Harel N, et al. (2007) Parasitism of Trichoderma on Meloidogyne javanica and role of the gelatinous matrix. European Journal of Plant Pathology 118: 247–258.

64. Abubakar U, Adamu T, Manga SB (2004) Control of Meloidogyne incognita (Koffoid and white) Chitwood (root-knot nematode) of Lycopersicon esculentum (tomato) using cowdung and urine. African Journal of Biotecnology 3: 379-381.

65. Meyer SLF, Huettel RN, Liu XZ, Humber RA, Juba J, et al. (2004) Activity of fungal culture filtrates against Soybean cyst nematode and root-knot nematode egg hatch and juvenile motility. Nematology 6: 23-32.

66. Sharma P, Pandey R (2009) Biological control of root-knot nematode, Meloidogyne incognita in the medicinal plant, Withania somnifera and the effect of bio-control agents on plant growth. African Journal of Agricultural Research 4: 564-567.

67. Jinfa Z, Waddell C, Sengupta GC, Potenza C, Cantrell RG (2006) Relationships between root-knot nematode resistance and plant growth in upland cotton galling index as a criterion. Crop sci 46: 1581-86.

68. Netcher C, Sikora RA (1990) Nematodes parasites of vegetables, in: Plant Parasitic Nematodes in Subtropical and Tropical Agriculture, (eds.) Luc M, Sikora RA, Bridge J. C.A.B. International.

69. Ousley MA, Lynch JM, Whipps JM (1994) Potential of Trichoderma spp. as consistent plant grow stimulators. Biology and Fertility of Soils 17: 85-90.

70. Wong TK, Mai WF (1973) Pathogenicity of Meloidogyne hapla to Lettuce as Affected by Inoculum Level, Plant Age at Inoculation and Temperature. J Nematol 5: 126-129.

71. Perry RN, Maule AG (2009) Physiological and biochemical basis of behavior. In: Gaugler, R and Bilgrami, A.L. (eds.) Nematode Behaviour. CAB International, Wallingford, UK, 197–238.

72. Abdi M (1996) Studies on the control of root-knot nematode (Meloidogyne incognita) with botanical toxicant. Ph.D Thesis. University of Karachi, Karachi-75270, Pakistan 375.

73. Johnson AW, Golden AM, Auld DL, Sumner DR (1992) Effects of Rapeseed and Vetch as Green Manure Crops and Fallow on Nematodes and Soil-borne Pathogens. J Nematol 24: 117-126.

74. Akhtar M, Mahmood I (1994) Potentiality of phytochemicals in nematode control: a review. Bioresource Technology 48: 189-201.

75. Hasabo AH, Noweer EMA (2005) Management of root-knot nematode Meloidogyne incognita on eggplant with some plant extracts. Egyptian Journal of Phytopathology 33: 65-72.

76. Qamar F1, Begum S, Raza SM, Wahab A, Siddiqui BS (2005) Nematicidal natural products from the aerial parts of Lantana camara Linn. Nat Prod Res 19: 609-613.

77. Ahmad S, Akhter M, Zia-Ul-Haq M, Mehjabeen AS (2010) Antifungal and nematicidal activity of selected legumes of Pakistan. Pakistan Journal of Botany 42: 1327-1331.

78. Khan MR, Khan N, Khan SM (2001) Evaluation of agricultural materials as substrate for mass culture of fungal biocontrol agents of fusarial wilt and root-knot nematode diseases. Annals of Applied Biology 22: 50-51.

79. Opareke AM, Dike MC, Amatobi CI (2005) Field evaluation of extracts of five Nigerian species for control of post flowering insect pest of cowpea, Vigna unguiculata (L.) Walp. Plant Protection Science 41: 14-20.

80. Abbasi WM, Ahmed N, Zaki JM, Shaukat SS (2008) Effect of Barleria acanthoides Vahl. On root-knot nematode infection and growth of infected okra and brinjal plants. Pakistan Journal of Botany 40: 2193-2198.

81. Okeniyi MO, Fademi OA, Orisajo SB, Adio SO, Otunoye AH, et al. (2010) Effect of botanical extract on root-knot nematode (Meloidogyne incognita) infection and growth of cocoa seedlings. Journal of Applied Biosciences 36: 2346-2352.

82. Yedidia I, Srivastva AK, Kapulnik Y (2001) Effect of Trichoderma harzianum on microelement concentrations and increased growth of cucumber plants. Plant and Soil 235: 235-242.

83. Harman GE1, Howell CR, Viterbo A, Chet I, Lorito M (2004) Trichoderma species--opportunistic, avirulent plant symbionts. Nat Rev Microbiol 2: 43-56.

84. Cuevas VC (2006) Soil Inoculation with Trichoderma pseudokoningii Rifai enhances yield of rice. Philippines Journal of Sciences 135: 31-37.

31

Influence of Different Growth Media on the Morphometric Characters of *Sansevieria liberica*

Okunlola AI[1]* and Ogungbite OC[2]

[1]Department of Crop, Soil and Pest Management, Federal University of Technology, Nigeria
[2]Centre for Continuing Education, Federal University of Technology, Nigeria

Abstract

The performance of *Sansevieria liberica* was determined on different growth media in the nursery. The growth media used include topsoil (TS), sandy soil (SS), rice husk (RH), topsoil plus rice husk (TS+RH), sandy soil plus rice husk (SS+RH), topsoil plus sandy soil (TS+SS) and top soil plus rice husk plus sandy soil (TS+SS+RH). The media were prepared in ratio 1:1 and their effect was observed on the height, stem girth, root length, root number and leaf number after six weeks of planting. The proximate and anti-nutritional analysis of the plant as well as the pH and mineral composition of the growth media were determined. The (TS+SS) medium showed the highest performance as regard plant height, stem girth, root length, root number and leaf number as it was significantly ($p<0.05$) different from other media. (TS+SS) growth medium had the highest pH value of 5.88 and the *S. liberica* grown on it recorded higher value of the mineral contents than those grown on other media. There was strong correlation between the performance of the plant and the pH value of the growth media as reflected by the linear regression analysis. Plant grown on RH recorded highest value of the anti-nutritional component tested. Thus, the rate at which *S. liberica* from different growth media contain the anti-nutritional components can be arranged as RH>SS+RH>TS+RH>TS+SS. Base on the result obtained, the growth of *S. liberica* can be enhanced using topsoil with sandy soil as growth media.

Keywords: Growth media; Soil pH; Rice husk; Mineral composition; Sandy soil; Topsoil

Introduction

Botanicals play vital and integral role in the wellbeing of heterotrophs as their social, cultural, economic and environmental importance cannot be over emphasized. In both urban and rural settings, botanicals ranging from horticultural to agricultural to timber species have shown significant impact on the survival of humans and their livestock because of some benefits such as source of living, control of erosion, landscape enhancement, provision of recreational and cultural facilities, watershed protection, supply of fruits and seeds and fuel-woods derivable from them [1]. In addition, before the discovery of many nowadays synthetic drugs and insecticides in the early 1930s, the extracts of the botanicals have been the major means of healing and major weapon in farmer's armory [2-4].

In recent years, the use of plants and plant products are gaining more attention because of the perils associated with many synthetic drugs and chemicals. Many diseases, fungi, bacteria and even insects and other pests have developed resistances to many popular synthetic drugs and pesticides [5,6]. For example, many malaria drugs are no longer effective as before Basco and Ringwald [5]. Also, the residue effects of these synthetic chemicals and drugs on both human and environmental health have become major factors encumbering their widespread use. Therefore, because of the public awareness of the downsides of these synthetic drugs, fungicides, bactericides and insecticides, researches have been shifted toward the use of herbal cure and plant base pesticides as a new boulevard of disease, infection and pest control to outwit these associated cons [1,7,8]. Moreso, that the parts of different plant species are believed to contain myriads of secondary metabolites that could be useful as drug sources, natural fungicides, bactericides, insecticides, natural food flavourings and colouring agents and natural fragrances [2,3,9,10]. Thus these have increased the demand for plants and plant products.

Hitherto, despite the importance of the botanicals to human existence, this weighty natural endowment has been facing a lot of challenges thwarting their growth and large scale production. Deforestation due to urbanization, climate change and insect infestation as well as low attention from government and individual towards the production of this imperative natural gift are the major factors exacerbating their large scale production especially in the developing countries where there is high rate of deforestation than afforestation [2,11,12]. However, despite the low production of botanicals in many parts of the world including Nigeria, the demand for their use has increased incessantly over the years probably because of the public knowledge of their importance. Hence, this has led to competition between different companies and individual that depends on plants as their major source of raw materials as well as their source of living. Therefore, to increase the production of these valuable resources become an important subject.

To increase the yield of this natural endowment, different strategies are being employed. Manipulation of growth media is one of the strategies introduced to increase yield of ornamental plants and botanicals in general because the quality and quantity of growth media is directly proportional to the performance of the plant. James and Michael [13] as well as Bhardwaj [14] opined that growth media have direct effects on the functional rooting system and that for a plant to perform well, the growth media used must be able to reduce water content and yet retain sufficient water to reduce watering frequency, must be able to sufficiently anchor or support the plant and

*Corresponding author: Okunlola AI, Department of Crop, Soil and Pest Management, Federal University of Technology, PMB 704, Akure, Ondo State, Nigeria, E-mail: okunlolaa1.hort@gmail.com

must be able to serve as reservoir for nutrients necessary for growth [15]. *Sansevieria liberica* is an ornamental plant with high medicinal values. It is used for the treatment of colic, cold and fever, diarrhea, rheumatism, microbial infections, snake bite, gonorrhea, convulsion, eczema, menorrhagia, sexual weakness, sedative abdominal pains, hypertension, conjunctivitis, asthma and hemorrhoids [16-19]. Considering the importance associated with this pertinent ornamental plant, this research investigated the growth of *Sansevieria liberica* on different growth media in order to recommend best growth media that could enhance large production of the plant.

Materials and Methods

Study location

The experiment was conducted at the green house of the Department of Crop, Soil and Pest Management, Federal University of Technology, Akure, Ondo State (Lat. 5°N and 15°E). The location is characterized by two peaks of rainfall that occur in the month of June and September/October with annual mean temperature of 27°C. The dry season is usually witnessed of Akure between November and March, while the rainy season ranged from April to October.

Collection of plant

Sanseviera liberica root was obtained from a healthy root stock of the plant in an open field in the Royal Garden, Akure, Ondo State, Nigeria. The root was uprooted early in the morning (6-7 AM) and was carefully packed in a polythene bag before being transferred to the study location. The roots were planted on different growth media on that same morning (7:30-8:30 am) and watering of the plant was done ones in three days Planting was immediately carried out that same morning, after planting watering was continuously carried out till the end of the experiment.

Preparation of growth media and experimental procedure

The growth media used in this study include topsoil (TS), sandy soil (SS), rice husk (RH), topsoil plus rice husk (TS+RH), sandy soil plus rice husk (SS+RH), topsoil plus sandy soil (TS+SS) and top soil plus rice husk plus sandy soil (TS+SS+RH). These media were prepared in ratio 1:1. The topsoil used was collected from Reliable Horticultural Garden, Akure while the sandy soil used was collected from Wisdom Garden, Igem, FUTA South gate, Akure. The rice husk used was obtained from a milling company in Ogbese, Ondo State. The media were thoroughly mixed on a dry concrete surface and were filled into polythene pots of diameter 11.4 cm and length 20 cm. The root of *S. liberica* of about 2 cm was planted horizontally on each medium. The experiment was arranged in a complete randomized design and each treatment was replicated four times. The plant height, stem girth, number of leaves, number of roots and root length was observed six weeks after plant.

Proximate and mineral content analysis

The moisture content, ash content, fat content, crude protein content and crude fibre content of plant from different growth media was carried out using the method described by AOAC [20]. The minerals analyzed in different plant from different growth media include K, Na, Ca, Mg and phosphorus. These minerals were analyzed as described by AOAC [20].

Determination of growth media pH

Five grams of sieved air-dried soil was weighed into a 250 ml

beaker, 35 ml of the extracting solution was added to the soil, shaken and allowed to react under 30 min and then filtered. 10 ml of the filtered was pipetted into a 50 ml standard flask, 16 ml of Murphy and Riley solution was added and then made up to level with distilled water. Standard solutions of different concentration of phosphorus were prepared from $KHPO_4$ solution and their respective absorbent readings were obtained from the photometer.

Determination of phytochemical component of plant from different growth media

The phytochemicals present in each of the plant of *S. liberica* was analyzed using the method of Sofowora [21] as described by Ileke [22]. The phytochemicals analyzed include Alkaloids, Cardiac glycosides, Phenol, Phytate, Flavonoid and Saponins.

Statistical analysis

All data were subjected to one-way analysis of variance and means were separated using New Duncan's Multiple Range Test. Also, linear regression analysis was carried out to check the correlations between growth media pH and the morphometric characters of the plant. SPSS version 17 was used for the analysis.

Results

Effect of different growth media on morphometric characters of *S. liberica*

Height, stem girth, root length, root number and leaf number of *S. liberica* grown on different media were presented in Table 1. These morphometric characters varied with the type of growth media used. Growth was observed in plant grown on all the media except those planted TS+RH, SS+RH and TS+SS+RH. *S. liberica* planted on TS+SS recorded the highest height, stem girth, root length, root number and leaf number of 5.07 cm, 0.74 cm, 6.40 cm, 37.45 and 6.62 respectively. The effect of TS+SS as a growth medium for *S. liberica* was significantly ($p<0.05$) different from other growth media.

pH and mineral composition of the different growth media used for the growth of *S. liberica*

Table 2 presented the pH and mineral composition of the different growth media used for the growth of *S. liberica*. Variation existed in the pH and mineral component of the growth media. The pH of all the media was on the acidic region of the pH scale. However, growth medium TS+SS recorded the highest pH of 5.88 and its effect was significantly ($p<0.05$) different from other media except TS and SS which recorded 5.42 and 5.12 respectively. The lowest pH value of 2.26 was recorded in growth medium TS+SS+RH. Regardless of the growth media, potassium recorded the highest proportion of the mineral composition of the growth media. However, TS+SS recorded the highest value of 475.00, 187.00, 3.70, 34.80 and 6.09 mmol/kg of potassium, sodium, calcium, magnesium and phosphorus respectively. The amount of mineral compositions of TS+SS was significantly ($p<0.05$) different from all other growth media. The order at which the growth media varied in their pH and mineral composition can be arranged thus TS+SS>TS>SS>RH>TS+RH>SS+RH>TS+SS+RH.

Correlation between growth media pH and morphometric characters of *S. liberica*

The correlation between the growth pH and morphometric characters of *S. liberica* was presented in Table 3. There is great correlation between the pH and the morphometric characters of *S.*

Growth media	Measurement in cm			Root number	Number of leaf
	Plant height	Stem girth	Root length		
TS	2.86 ± 0.14[c]	0.37 ± 0.01[c]	5.67 ± 0.22[cd]	32.00 ± 0.00[d]	5.41 ± 0.01[b]
SS	4.32 ± 0.18[d]	0.53 ± 0.01[d]	4.12 ± 0.28[bc]	27.24 ± 0.22[c]	5.37 ± 0.11[b]
RH	1.04 ± 0.12[b]	0.16 ± 0.02[b]	3.41 ± 0.24[b]	15.28 ± 0.14[b]	3.21 ± 0.12[b]
TS+RH	0.00 ± 0.00[a]	0.00 ± 0.00[a]	0.00 ± 0.00[a]	0.00 ± 0.00[a]	0.00 ± 0.00[a]
SS+RH	0.00 ± 0.00[a]	0.00 ± 0.00[a]	0.00 ± 0.00[a]	0.00 ± 0.00[a]	0.00 ± 0.00[a]
TS+SS	5.07 ± 0.16[d]	0.74 ± 0.02[e]	6.40 ± 0.22[d]	37.45 ± 0.2[e]	6.62 ± 0.02[c]
TS+SS+RH	0.00 ± 0.00[a]	0.00 ± 0.00[a]	0.00 ± 0.00[a]	0.00 ± 0.00[a]	0.00 ± 0.00[a]

Each value is mean ± standard error of four replicates. Values followed by the same letters are not significantly (p>0.05) different from each other using New Duncans Multiple Range Test.

Table 1: Effect of different growth media on morphometric character of S. liberica.

Growth media	pH	Values in mmol/kg				
		Potassium	Sodium	Calcium	Magnesium	Phosphorus
TS	5.42 ± 0.01[c]	398.00 ± 0.11[b]	124.00 ± 0.13[c]	2.94 ± 0.23[b]	22.00 ± 0.11[c]	3.15 ± 0.15[b]
SS	5.12 ± 0.01[c]	346.00 ± 0.12[c]	119.00 ± 0.11[bc]	2.80 ± 0.14[b]	24.50 ± 0.20[c]	3.08 ± 0.11[b]
RH	3.60 ± 0.02[ab]	286.00 ± 0.09[b]	108.00 ± 0.11[b]	1.80 ± 0.14[a]	12.00 ± 0.16[b]	1.86 ± 0.10[b]
TS+RH	2.44 ± 0.01[a]	145.00 ± 0.11[a]	66.00 ± 0.11[a]	1.12 ± 0.16[a]	5.90 ± 0.10[a]	0.27 ± 0.12[a]
SS+RH	2.63 ± 0.01[a]	138.00 ± 0.07[a]	63.00 ± 0.08[a]	1.10 ± 0.07[a]	7.70 ± 0.14[a]	0.13 ± 0.08[a]
TS+SS	5.88 ± 0.02[c]	475.00 ± 0.10[d]	187.00 ± 0.10[d]	3.70 ± 0.09[c]	34.80 ± 0.23[d]	6.09 ± 0.08[c]
TS+SS+RH	2.26 ± 0.01[a]	136.00 ± 0.12[a]	57.00 ± 0.12[a]	0.89 ± 0.12[a]	5.30 ± 0.12[a]	0.16 ± 0.10[a]

Each value is mean ± standard error of four replicates. Values followed by the same letters are not significantly (p>0.05) different from each other using New Duncan's Multiple Range Test.

Table 2: pH and mineral composition of the different growth media used for the growth of S. liberica.

liberica as reflected by their R value which is tending towards 1. The R^2 value showed that only 78.3, 76.7, 79.7, 98.1 and 98.6% of the plant height, stem girth, root length, root number and leaf number can be explained by the pH value respectively. The R^2 reflected high correlation as the values are large. However, the correlation between pH and plant height, pH and stem girth as well as pH and root length was not significant at 0 p<0.05. Moreover, the correlation between pH and root length as well as correlation between pH and leaf number is significant at p<0.01 and p<0.05 respectively.

Proximate and anti-nutritional component of S. liberica grown from different growth media

Figures 1 and 2 presented the proximate composition and the anti-nutritional composition of the plant respectively. There were no proximate and anti-nutritional components recorded for plant grown on TS+RH, SS+RH and TS+SS+RH. S. liberica on TS+SS recorded the highest value of 53.21, 2.15, 15.24, 0.28, 23.23 and 14.28% of moisture content, ash, crude fibre, fat, protein and carbohydrate respectively. The order at which the S. liberica from different growth media contain the proximate and anti-nutritional composition can be arranged as TS+SS>TS>SS> RH>TS+RH=SS+RH=TS+SS+RH. The rate at which the anti-nutritional components present in S. liberica varied with the type of growth media used. Plant grown on TS+SS recorded the lowest proportion of alkaloid, cardiac glycoside, phytate, flavonoid and saponins. Thus, the rate at which S. liberica from different growth media contain the anti-nutritional components can be arranged as RH>SS+RH>TS+RH>TS+SS.

Discussion

Botanicals have been the closest companions of human as more than 90% of human's life depend on them. In fact, abundance of different species of botanicals is directly proportional to the wellbeing of humans and animals. However, the high demand for this weighty gift of nature from different quarters has been the major obstacle to its abundance especially in developing countries where less attention is given to afforestation. Also, the advancement in technology has contributed immensely to the climate change which has direct effect on soil composition [23,24]. Since soil is the major growth medium for plants, there is need for investigating the growth medium that will enhance the performance of different species of plant as this could increase their abundance.

The result obtained showed that the performance of S. liberica varied with the type of growth media used. However, no growth was observed on S. liberica planted on TS+RH, SS+RH and TS+SS+RH. The highest height, stem girth, root length, root number and leaf number was observed in the plant planted on TS+SS growth medium. The inability of the plant planted on TS+RH, SS+RH and TS+SS+RH could be due to the RH admixture with the soil because it has been noted that rice husk when added to the soil have some negative effect on growth of plants as suggested by Moyin-Jesu [25]. Also, rice husk has been noted to contain high amount of ash and this could result in the shift of bacteria that helps in decomposition of materials in the soil [26]. However, the works of Jeon et al. [27] and Milla et al. [28] as well as Badar and Qureshi [29] revealed that rice husk could be a very good soil substitute when used in a carbonized form. In addition, the low or

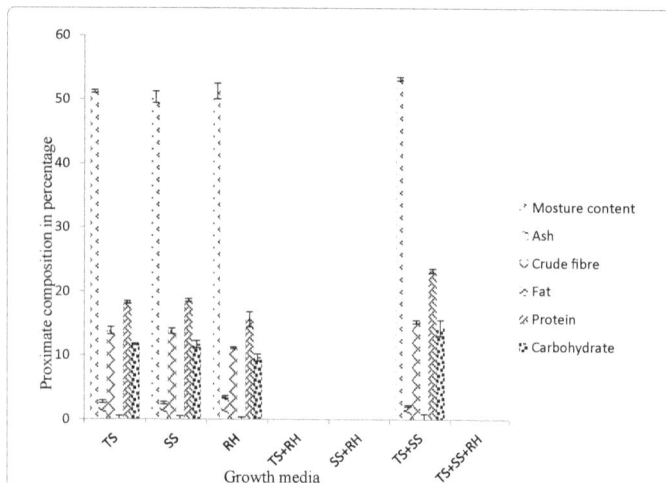

Figure 1: Proximate composition of *S. liberica* grown from different growth media.

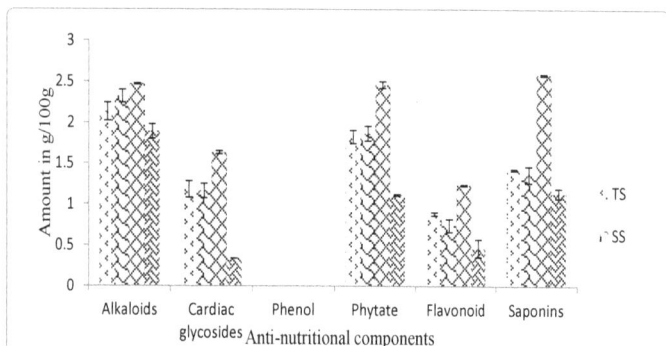

Figure 2: Anti-nutritional components present in *S. liberica* grown from different growth media.

Parameters	R	R²	K ± S.E	R$_c$ ± S.E	R$_E$	Sig.
Plant height	0.885	0.783	-4.65 ± 3.01	1.59 ± 0.59	-4.65+1.59(pH)	0.115
Stem girth	0.876	0.767	-0.64 ± 0.43	0.22 ± 0.09	-0.64+0.22(pH)	0.124
Root length	0.893	0.797	-1.32 ± 2.25	1.24 ± 0.44	-1.32+1.24(pH)	0.107
Root number	0.990	0.981	-19.43 ± 4.75	9.47 ± 0.94	-19.43+9.47(pH)	0.01
Leaf number	0.986	0.972	-1.94 ± 0.86	1.42 ± 0.17	-1.94+1.42(pH)	0.014

K=Constant; R$_c$=Regression coefficient; R$_E$=Regression equation.
Table 3: Correlation between growth media pH and morphometric characters of *S. liberica*.

no growth of *S. liberica* on RH, SS+RH and TS+RH could be due to the low nitrogen content of the rice husk as reported by Kumar et al. [30] which estimated the nitrogen content of rice husk to be less than 0.24% compared to its ash content (about 29%).

Furthermore, the result obtained showed that TS+SS recorded the highest pH value while the TS+SS+RH recorded the lowest pH value. The high pH value noted in TS+RH, SS+RH and TS+SS+RH may due to the RH used as supplement. PH is an important factor in determining the availability of mineral elements in the soil [31]. Moyin-Jesu and Adekayode [32] opined that soil pH can either positively or negatively affect plant growth. The result obtained showed that the growth media with low pH recorded low growth of *S. liberica* compared to those that have higher values of ph The low phosphorus and other macronutrients in TS+SS+RH, TS+RH and SS+RH as well as RH could be due to the low pH present in them (Webb, Loneragan and Moyi-Jesu) and

(Londo et al.; Zeng et al.) [25,31,33,34] reported that macronutrients are affected by the increase or decrease in pH of soil. These authors reported that at low pH plants take up little amount of nitrogen because the microbial conversion of NH_4^+ to nitrate (nitrification) will be slow. In the same vein, the amount of phosphorus in the soil is pH dependent because at high pH more phosphorus is available to the plant in the soil [32]. Therefore, the low root length and root number observed on *S. liberica* grown on RH could be due to low pH in the medium. The result obtained on root performance of *S. liberica* acquiesced with the findings of Haller and Sutton [35] as well as Conde et al. [36]. The low number of RH, TS, SS compared to TS+SS could be due to the effect of the pH on the macronutrient of the media which in turn affected the height of the plant. This agreed with the work of Moyin-Jesu and Ayodele [32]. However, the regression analysis of the result of this work revealed that there is strong correlation between the pH of the growth media and root number as well as the leaf number. Therefore, the low height, root length and stem girth recorded on *S. liberica* grown on RH could be attributed to the low number of the leaf (source) and the root (sink). This agreed with the work of Moyin-Jesu and Adekayode [32]. Also, the result of this work agreed with the findings of Valipour [37] in which regression analysis was used to compare mass transfer-based models to determine the best model under different weather conditions. Igbal et al. [38] reported that the nutritional and anti-nutritional components of plants are affected by the pH of their media. They opined that low pH reduces both nutritional and anti-nutritional components of plant and vise verse. Therefore, the low pH recorded in RH medium could be responsible for the low anti-nutritional component present in the *S. liberica* grown on it.

Conclusion

The result of the work showed that the use of RH as supplement for the growth of *S. liberica* on TS and SS in the nursery may not yield good result. Considering the necessity for the rapid production of this pertinent ornamental plant, the mixture of TS and SS could be the best growth medium for *S. liberica* in the nursery and could be recommended for farmers.

References

1. Fuwape JA, Onyekwelu JC (2011) Urban forest development in West Africa: benefits and challenges. Journal of Biodiversity and Ecological Sciences 1: 77-94.

2. Kirakosyan A, Cseke LJ, Kaufman PB (2009) The use of plant cell biotechnology for the production of phytochemicals. Recent Advances in Plant Biotechnology, pp: 15-32.

3. Zibaee A (2011) Botanical insecticides and their effects on insect biochemistry and immunity. Pesticides in the world - Pests control and pesticides exposure and toxicity assement, pp: 55-68.

4. Ogungbite OC, Ileke KD, Akinneye JO (2014) Bio-pesticide Treated Jute Bags: Potential Alternative Method of Application of Botanical Insecticides against *Rhyzopertha dominica* (Fabricius) Infesting Stored Wheat. Molecular Entomology 5: 30-36.

5. Basco L, Ringwald P (2000) Drug-resistant malaria: problems with its definition and technical approaches. Sante 10: 47-50.

6. Ashamo MO, Odeyemi OO, Ogungbite OC (2013) Protection of cowpea, *Vigna unguiculata* L. (Walp.) with Newbouldia laevis (Seem.) extracts against infestation by Callosobruchus maculatus (Fabricius). Archives of Phytopathology and Plant Protection 46: 1295-1306.

7. Forim MR, Da-silva MFGF, Fernandes JB (2012) Secondary metabolism as a measurement of efficacy of botanical extracts: The use of Azadirachta indica (Neem) as a model. In: Perveen F (ed.), Insecticides-Advances in Integrated Pest Management, pp: 367-390.

8. Ileke KD, Ogungbite OC (2014) Entomocidal activity of powders and extracts

of four medicinal plants against *Sitophilus oryzae* (L), *Oryzaephilus mercator* (Faur) and *Ryzopertha dominica* (Fabr.). Jordan Journal of Biological Sciences 7: 57-62.

9. Oni MO, Ogungbite OC, Akindele AK (2015) The effect of different drying methods on some common Nigerian edible botanicals. International Journal of Advanced Research in Botany 1: 1-8.

10. Ogungbite OC (2015). Entomopoison efficacy of fume of different parts of *Newbouldia laevis* against *Callosobruchus maculatus* in storage. International Journal of Research Studies in Microbiology and Biotechnology 1: 6-14.

11. Ogungbite OC (2008) Growth of *Khaya ivorensis* under different light intensities. BSc Dissertation submitted to the Department of Plant Science and Biotechnology, Adekunle Ajasin University Akungba-Akoko, Ondo State, p: 62.

12. Gbadamosi AE, Okere AU, Ogungbite OC (2009) Growth of Khaya ivorensis (A. Chev) as Influenced by Different Light Regimes. Science Research Annals, AAUA 1: 6-12.

13. James AR, Michael RE (2009) Growing media for container production in green house or nursery. Agriculture and Natural Resources.

14. Bhardwaj RL (2014) Effect of growing media on seed germination and seedling growth of papaya cv. Red lady. African Journal of Plant Science 8: 178-184.

15. Abad M, Noguera P, Puchades R, Maquieira A, Noguera V (2002) Physico-chemical and chemical properties of some coconut dusts for use as a peat substitute for containerized ornamental plants. Bioresources and Technology 82: 241-245.

16. Gill LS (1992) Ethnomedical uses of plants in Nigeria. UniBen Press, Benin City, Nigeria, p: 509.

17. Amida MB, Yemitan OK, Adeyemi OO (2007) Toxicological assessment of the aqueous root extract of *Sanseviera liberica* Gerome and Labroy (Agavaceae). Journal of Ethnopharmacology 113: 171-175.

18. Ikewuchi CC, Ikewuchi JC (2009) Amino acid, mineral, and vitamin composition of *Sansevieria liberica*. Gérôme and Labroy. Journal of Science Technology 10: 477-482.

19. Adeyemi OO, Akindele AJ, Ogunleye EA (2009) Evaluation of the antidiarrhoeal effect of *Sanseviera liberica* Gerome and Labroy (Agavaceae) root extract. Journal of Ethnopharmacology 123: 459-463.

20. AOAC (1970) Official methods of Analysis. 12th edn. AOAC, Arlington, VA. Association of Official Analytical Chemists, Washington DC, USA.

21. Sofowora A (1993) Phytochemical screening of medicinal plants and traditional medicine in Africa. 2nd edn. Spectrum Books Limited, Nigeria, pp: 150-156.

22. Ileke KD (2014) Anti-nutritional factors in cowpea cultivars and their effects on susceptibility to *Callosobruchus maculatus* (Fab.) [Coleoptera: Bruchidae] infestation. Bioscience Methods 5: 1-8.

23. Brinkman R, Brammer H (1990) The influence of a changing climate in soil properties. In: Trans. 14th ISSS Congress, Kyoto, pp: 283-288.

24. Varallyay G (2010) The impact of climate change on soils and on their water management. Agronomy Research 8: 385-396.

25. Moyi-Jesu EI (2004) Comparative Evaluation of Different Plant Residues on the Soil and Leaf Chemical Composition, Growth, and Seed Yield of Castor Bean (Ricinus communis). Pertanika Journal of Tropical Agricultural Studies 27: 21-29.

26. Bougnom BP, Insam H (2009) Ash additives to compost affect soil microbial communities and apple seedling growth.

27. Jeon WT, Seong K, Lee J, Oh I, Lee J, et al. (2010) Effects of green manure and carbonized rice husk on soil properties and rice growth. Korean Journal of Soil Science and Fertilizer 43: 484-489.

28. Milla OV, Rivera EB, Huang WJ, Chien CC, Wang YM (2013) Agronomic properties and characterization of rice husk and wood biochars and their effect on the growth of water spinach in a field test. Journal of Soil Science and Plant Nutrition 13: 251-266.

29. Badar R, Qureshi SA (2014) Composted Rice Husk Improves the Growth and Biochemical Parameters of Sunflower Plants. Journal of Botany 2014: 1-6.

30. Kumar A, Mohanta K, Kumar D, Parkash O (2012) Properties and Industrial Applications of Rice husk: A review. International Journal of Emerging Technology and Advanced Engineering 2: 86-90.

31. Londo AJ, Kushla JD, Carter RC (2006) Soil pH and tree species suitability in the South. Southern Regional Extension Forestry 2: 1-5.

32. Moyin-Jesu EI, Adekayode FO (2010) Comparative Evaluation of Different Organic Fertilizers on Soil Fertility Improvement, Leaf Mineral Composition and Growth Performance of African Cherry Nut (*Chrysophyllum Albidium* L) Seedlings. Journal of American Science 6: 217-223.

33. Webb MJ, Loneragan JF (1985) Importance of environmental pH during root development on phosphate absorption. Plant Physiology 79: 143-148.

34. Zeng W, Zeng M, Zhou H, Li H, Xu Q, et al. (2014) The effects of soil pH on tobacco growth. Journal of Chemical and Pharmaceutical Research 6: 452-457.

35. Haller WT, Sutton DL (1972) Effect of pH and high phosphorus concentrations on growth of water hyacinth. University of Florida, Agricultural Research Center, Fori Louderdale, Florida, USA, pp: 59-61.

36. Conde LD, Chen Z, Chen H, Liao H (2014) Effects of phosphorus availability on plant growth and soil nutrient status in the rice/soybean rotation system on newly cultivated acidic soils. American Journal of Agriculture and Forestry 2: 309-316.

37. Valipour M (2015) Calibration of mass transfer-based models to predict reference crop evapotranspiration. Applied Water Science, p: 1-11.

38. Igbal M, Hussain I, Habib A, Ashraf MA, Rasheed R (2015) Effect of semiarid environment on some nutritional and antinutritional attributes of calendula (*Calendula officinalis*). Journal of Chemistry 2015: 1-8.

32

Exploiting the Insecticidal Potential of the Invasive Siam Weed, *Chromolaena odorata* L. (Asteraceae) in the Management of the Major Pests of Cabbage and their Natural Enemies in Southern Ghana

Ezena GN[1], Akotsen-Mensah C[1,2]* and Fening KO[1,3]

[1]*African Regional Postgraduate Programme in Insect Science, PMB L59, University of Ghana, Legon, Ghana*
[2]*Forest and Horticultural Crops Research Centre, University of Ghana, Kade, Ghana*
[3]*Soil and Irrigation Research Centre, University of Ghana, Kpong, Ghana*

Abstract

Cabbage is an important leafy vegetable widely cultivated and consumed in Ghana. It offers a good source of vitamins and minerals to the human body. Despite its importance cabbage production is constrained by insect pests attack. Chemical control has being the main strategy, but without much success, thus the need for alternative options for pest management. Field experiments were conducted in the major and minor rainy seasons of 2014 to evaluate the insecticidal potential of Siam weed, *Chromolaena odorata* L. at three concentrations (10, 20 and 30 g/L w/v) in the management of the key pests of cabbage and its effect on their natural enemies. Neem seed extract (50 g/L) and Lambda cyhalothrin (Sunhalothrin® 2.5 ml/L) were used as reference insecticides with tap water as a control. The key pests recorded during the major season were the cabbage aphid, *Brevicoryne brassicae*, and cabbage webworm, *Hellula undalis*, whiles Diamondback moths (DBM), *Plutella xylostella* and *B. brassicae* were recorded in the minor season. Generally, the three concentrations of *C. odorata* were efficacious in controlling aphids and DBM than the tap water and conventional insecticide, Sunhalothrin® in both the minor and major seasons. However, the 10 and 20 g/L *C. odorata* recorded the highest buildup of natural enemy populations and also had higher yield as compared to 30 g/L *C. odorata*, Sunhalothrin®, and tap water. The cabbage plots sprayed with neem obtained the highest yield and was the most economical to adopt, followed by the *Chromolaena*, and the least being the synthetic insecticide. Crude extracts from neem and *C. odorata* could become an integral part in the Integrated Pest Management of vegetables, especially in smallholder farms.

Keywords: *Plutella xylostella*; *Hellula undalis*; *Brevicoryne brassicae*; Natural enemies; *Chromolaena odorata*; Neem seed extract; Cost benefit analysis

Introduction

Cabbage (*Brassica oleracea var* capitata) is an important crop for many smallholder farmers in Africa and Asia, due to its nutritional and financial benefits Asare-Bediako et al. [1]. It is one of the popular vegetables consumed and cultivated by both the urban and rural dwellers in Ghana Timbilla and Nyarko [2]. The crop has replaced many indigenous green vegetables and its cultivation has provided a good source of employment in Ghana, Timbilla and Fening [3,4]. The cabbage plant has medicinal properties such as its anti- inflammatory property; and can also prevent cancer and a good remedy for ulcer Norman and Shealy [5]. Although there seems to be numerous studies conducted on cabbage pest management, cabbage production in Ghana, is still faced with numerous constraints especially insect pest attack De Lannoy and Mochiah et al. [6,7]. The diamondback moth (DBM), *Plutella xylostella* L. (Lepidoptera: Plutellidae), the cabbage webworm, *Hellula undalis* (Fab) (Lepidoptera: Crambidae) and the cabbage aphid, *Brevicoryne brassicae* L. (Hemiptera: Aphididae) are the most important pests of cabbage in Ghana Ninsin, (Obeng- Ofori and Ankrah) and (Timbilla and Nyarko) [2,8,9]. The DBM, for instance, has been one of the greatest threats to crucifer production in the tropics causing over 90% crop losses Charleston [10], if not managed. Farmers have limited control options and have therefore resorted to the use of synthetic insecticides which have led to insect pest resistance problems, toxic residues in cabbages posing high risks to humans, animals and the environment Ninsin, Amoako and Fening et al. [8,11,12].

Botanicals, which are derived from plant products are reported to be effective against many insect pests and are considerably cheap and biodegradable when applied Amoabeng et al. [13]. Studies in some African countries suggest that, extracts of locally available plants can be effective as crop protectants Isman [14]. Extracts from marigold have been used effectively against the bruchid beetle from cowpeas in storage in Uganda Kawuki et al. [15]. In Benin, the bushmint, *Hyptis suaveolens* (L.) Poit (Lamiales:Lamiaceae) extract has been used for the control of pink stalk borer, *Sesamia calamistis* H. (Lepidoptera: Noctuidae) on maize in the field. In Ghana, several botanicals such as neem seed extracts, *Azadirachta indica*; hot chilli pepper, *Capsicum frutescens*; garlic extract, *Allium sativum*; tobacco, *Nicotiana tabacum*; cassia plant, *Cassia sophera*; physic nut, *Jatropha curcas*; castor oil plant, *Ricimus communis*; basil, *Ocimum gratissimum*; goat weed, *Ageratum conyzoides*, Siam weed, *Chromolaena odorata* and Cinderella weed, *Synedrella nodiflora*, etc., have been used to protect foodstuffs such as vegetables in the field and stored grains against insect pest infestations Obeng- Ofori, Ankrah, Fening et al. and Amoabeng et al. [9,12,13].

Most plant species with potential insecticidal properties are in the families of Meliaceae, Rutaceae, Asteraceae, Piperaceae, Compositae,

*Corresponding author: Godfred Nwosu Ezena, Ghana Ministry of Food and Agriculture, PO Box 43, Bawku, Bawku-Zebilla Road, Ghana
E-mail: goddie191@yahoo.com (or) cakotsenmensah@ug.edu

Lamiaceae, Euphorbiaceae, Combretaceae and Annonaceae. They are common weed, shrub and tree species on and around farms (Devanand and Rani) and Amoabeng et al. [13,16]. Pests control using these local materials offer the farmer the opportunity to reduce production cost, as the plants often grows in the wild and around farms and so can be obtained with little effort and zero or minimal cost Amoabeng et al. [13].

However, some botanicals may adversely affect beneficial insects Buss & Park- Brown, Dubey et al. [17,18]. Amoabeng et al. [13] confirmed this in a field experiment using 30 g/L C. odorata to control DBM larvae among other botanicals. They showed that although the 30 g/L C. odorata extract was effective in reducing pest populations, this concentration was however detrimental to natural enemies and non-targeted organisms in the field. Fening et al. [19] have also cautioned that higher concentrations of botanicals could possibly have some detrimental effects on the natural enemy populations because their numbers reduce with increasing concentrations. Hence, the current research was conducted to optimize the concentrations of crude extract from Siam weed, C. odorata for effective control of the key pests of cabbage with minimal or no adverse effect on the natural enemies of these pests and other non- target organisms.

Materials and Methods

Study site

The field experiments were carried out at the University of Ghana's Forest and Horticultural Crops Research Center (FOHCREC), Kade, in the Eastern Region, Ghana. The Centre is located (06º 09' 26N, 000º 55' 00W) in the forest transition ecological zone. It is characterized by humid climate associated with a bimodal rainfall ranging between 1200-1500 mm. This provides the appropriate environmental conditions for many plantain cultivars to do well in this region (wet seasons and a dry spell with average elevation of 150 m). The mean annual temperature range at this location is 24-38°C. Relative humidity is around 70-80% in most part of the year.

Preparation of extracts/treatments

Fresh leaves of Siam weed, C. odorata were obtained from around FOHCREC and the leaves were washed with tap water to remove sand, dust and other possible chemical contaminants. The leaves were then dried under shade on concrete platform. A mortar and a pestle were used to ground the leaves into coarse powder which was weighed into the different concentrations (10, 20 and 30 g/L w/v) of C. odorata and then soaked in 30 litre plastic buckets. The solutions were then stirred continuously for a few minutes and left to stand overnight. Prior to applications, the extracts were filtered using muslin cloth and three drops of natural vegetable oil (0.15 ml per litre of sunflower oil) and local soap (alata samina- is a mixture of roasted cocoa pods, red palm oil, coconut oil, sea salt and shea butter with fragrances added to make it more appealing) were added to enhance the stickiness on the leaf surface of the cabbage.

Neem seed extract (50 g/L w/v) was also prepared and used as a standard botanical treatment. Fresh neem seeds were collected along Winneba to Apam road and dried in the sun for a week. The kernel were removed from the shells and pounded in a mortar to obtain a coarse powder. The recommended concentration was soaked in a plastic container. The solution was left overnight and filtered through a fine linen material before application. A reference conventional insecticide, Sunhalothrin˙ 2.5 ml/L (a.i. lambda cyhalothrin) and tap water, as control were also prepared for application.

Nursery establishment, land preparation and transplanting of seedlings

Cabbage seeds (cv. oxylus) were obtained from AGRIMAT Ltd. in Accra, Ghana. The seeds were nursed in rows in seed trays raised 1 m above the ground. The soil used in the nursery was made of a mixture of sandy loam with carbonated rice husks in the ratio 2:1. Watering was done manually with a watering can, twice daily. The experimental fields were manually cleared and later sprayed with herbicide (Gramoxone) before cabbage seedlings were transplanted on raised beds. The entire field measuring 19 × 11.5 m was pegged, labeled with 3 replications. Cabbage seedlings were transplanted at 4-5 true leaf stage (about 30 days after sowing) in a spacing of 0.5 × 0.5 m. The individual plots measured 1.5 × 2.5 m resulting in 24 plants per plot and a 2 meter wide unplanted alley was left between each plot to avoid spray drift. Standard cultural and agronomic practices such as weed control, watering and earthing-up of soil to improve aeration were employed during the growing periods.

Experimental design, treatment application and data collection

A randomized complete block design was used with three replications in each of the growing seasons. There were six treatments including: three concentrations of C. odorata (10, 20 and 30 g/L), neem seed extract (50 g/L of water), Sunhalothrin˙ (2.5 ml/ L) and tap water. A 15 L capacity Knapsack sprayer (JA- 15 model/ type obtained from AGRIMAT Ltd., Madina- Accra) was used to apply each of the treatments at recommended concentrations. Application of treatments commenced 2 weeks after transplanting of seedlings. Data collections started 2 weeks after transplanting and were done on weekly basis for all the growing seasons. Data on insect population such as DBM, aphid and other pests and natural enemies were counted and recorded from the middle row excluding the border plants. In all, an average of 8 plants per plot was used for yield and insect damage assessment. Data collection was done between 06:00 and 08:00 GMT.

Cost and benefit assessments were also carried out for both seasons, with the production costs and yield (sold at prevailing prices on the local market) were computed into formulae used by Amoabeng et al. [13]. Income was converted to a per hectare basis by extrapolating the plant population of each plot based on plant spacing 0.5 × 0.5 m. A total plant population of 64,000 per ha were assessed, following the procedures used by Amoabeng et al. [13].

Statistical analysis

All data were subjected to analysis of variance (ANOVA) using JMP statistical package (JMP version 10, SAS 2010). The number of insect counts and the percentage yield data were used to determine the effectiveness of the treatments. All the data were tested to ensure that they meet the assumptions of ANOVA. Where the assumptions of ANOVA were found to have been violated the data were transformed using $\sqrt{x+0.5}$ for insect counts and arcsine $\sqrt{x+0.5}$ for percentage yield. Treatment means were separated using Tukey-Kramer Honestly Significant difference (HSD) test. Significant difference was set at $P \leq 0.5$.

Results

Insect fauna encountered on cabbage

Some insect species were recorded during the entire seasons of the cabbage production. These include diamondback moth, P. xylostella, cabbage aphid, B. brassicae, cabbage webworm, H. undalis, whitefly, *Bemisia tabaci* (Genn.) (Homoptera: Aleyrodidae), variegated grasshopper, *Zonocerus variegatus* (Orthoptera: Pyrgomorphidae), cabbage flea beetle, *Phyllotreta* spp. (Coleoptera: Chrysomelidae) and cabbage looper, *Trichoplusiani* (Hübner) (Lepidoptera: Noctuidae). Other pests such as snails (class Gastropoda) and millipedes (class Diplopoda) were also observed attacking the cabbage. Some natural enemies of cabbage pests such as *Diaeretiella rapae* (Stary) (Hymenoptera: Braconidae) *Cotesia plutellae* (Kurdjumov) (Hymenoptera: Braconidae), hoverflies (Diptera: Syrphidae), ladybird beetles, *Cheilomenes* spp. (Coleoptera: Coccinellidae) and other beneficial arthropod such as spiders (order Araneae) was identified and observed attacking some insect pests.

Effects of treatments on insect pest population

Infestations of aphid and cabbage webworm were high compared with other pest in the major season (Table 1). The results showed that, there were significant differences among treatments in the aphid infestation (Table 1). The three concentrations of *C. odorata* extracts and neem seed extract had significantly lower number of aphids (0.67 ± 0.22, 1.58 ± 0.29, 1.58 ± 0.22, 1.08 ± 0.28) than the conventional insecticide, Sunhalothrin* (3.33 ± 0.43) (Table 1). Also, 10 g/L *C. odorata* extract (0.67 ± 0.22) had less number of aphids than the plots sprayed with tap water (2.58 ± 0.58) (Table 1). The effect of all the botanicals and the conventional insecticide, Sunhalothrin* were effective in managing the cabbage webworm than tap water treated plots (Table 1). The results also indicated that, the *C. odorata* extracts applied at 10 and 20 g/L (w/v) and sunhalothrin (2.5 ml/ L) were better than tap water plots in managing the infestation of other pests (Table 1). Means within a column for each treatment under each concentration followed by different letters differ significantly from each other ($P \leq 0.05$). Other pests: cabbage flea beetle (*Phyllotreta* spp.), cabbage white butterfly (*Pieris rapae*), cabbage looper (*Trichoplusia ni*), and Cotton leafworm (*Spodoptera litorallis*).

The two major pests of cabbage identified and recorded with high infestations during the minor season were the diamondback moth and the cabbage aphid (Table 2). There were significant differences in the infestation of aphids in the minor season (Table 2). The result showed that the botanicals had significantly lower number of aphids than the conventional insecticide, Sunhalothrin* and the tap water plots in the minor growing season (Table 2). However, there were no significant differences among all treatments in the management of *P. xylostella* and the other pests (Table 2).

Effects of treatments on natural enemies

The result showed that 20 g/L (w/v) of *C. odorata* had the highest number of *Diaretiella rapae* as compared to 30 g/L (w/v) of *C. odorata*, neem seed extract and sunhalothrin* (Table 3). The result again exhibited that, 20 g/L *C. odorata*, neem and the tap water (control) had the highest number of other natural enemies as compared to the plots treated with the conventional insecticide, sunhalothrin* (Table 3). Meanwhile, the results revealed that there were no significant differences among all treatments in the number of spiders and hoverflies sampled (Table 3).

The results for the minor rainy season showed that there were no significant differences in the numbers of *D. rapae*, *C. plutellae* and hoverflies when the various treatments were applied (Table 4). However, 20 g/ L of *C. odorata* and neem had significantly high number of spiders compared to the other treatments (Table 4). Also, for other natural enemies sampled, 20 g/L *C. odorata* treated plots had significantly high populations than that of sunhalothrin* (Table 4).

	Mean ± SE insect pests/ plant		
Treatments	B. brassicae	H. undalis	Other pests[1]
10 g/L C. odorata	0.67 ± 0.22 c	1.38 ± 0.27 c	1.83 ± 0.30 b< 0.00*
20 g/L C. odorata	1.58 ± 0.29 bc	1.38 ± 0.28 b	2.46 ± 0.31 b
30 g/L C. odorata	1.58 ± 0.22 bc	1.79 ± 0.28 b	2.63 ± 0.39 ab
50 g/ L Neem	1.08 ± 0.28 c	1.13 ± 0.21 b	3.04 ± 0.43 ab
Sunhalothrin®	3.33 ± 0.43 a	1.13 ± 0.23 b	1.83 ± 0.34 b
Tap water	2.58 ± 0.58 ab	4.0 ± 0.56 a	4.08 ± 0.35 a
F	7.44	11.13	5.56
P	< 0.00*	< 0.00*	< 0.00*

Means within a column for each treatment under each concentration followed by different letters differ significantly from each other ($P \leq 0.05$). [1]Other pests: cabbage flea beetle (*Phyllotreta* spp.), cabbage white butterfly (*Pieris rapae*), cabbage looper (*Trichoplusia ni*), and Cotton leafworm (*Spodoptera litorallis*).

Table 1: Mean (± *SE*) populations of insect pests on cabbage sprayed with botanical and conventional insecticides and tap water as control, during 2014 major rainy season.

	Mean ± SE insect pests/ plant		
Treatments	B. brassicae	P. xylostella	Other pests[1]
10 g/L C. odorata	1.83 ± 0.22 b	4.50 ± 0.73 a	1.67 ± 0.35 a
20 g/L C. odorata	1.21 ± 0.25 b	3.13 ± 0.69 a	1.13 ± 0.25 a
30 g/L C. odorata	1.58 ± 0.32b	3.79 ± 0.64 a	1.13 ± 0.25 a
50 g/ L Neem	1.46 ± 0.26 b	3.08 ± 0.54 a	1.13 ± 0.31 a
Sunhalothrin®	3.96 ± 0.64 a	4.13 ± 0.09 a	1.46 ± 0.30 a
Tap water	3.75 ± 0.62 a	5.08 ± 1.01 a	1.96 ± 0.32 a
F	7.61	1.02	1.67
P	< 0.00*	0.41	0.15

Means within a column for each treatment under each concentration followed by different letters differ significantly from each other ($P \leq 0.05$). [1]Other pests: cabbage flea beetle (*Phyllotreta* spp.), cabbage white butterfly (*Pieris rapae*), cabbage looper (*Trichoplusia ni*), cabbage webworm (*Hellula undalis*) and Cotton leafworm (*Spodoptera litorallis*).

Table 2: Mean (± *SE*) populations of insect pests on cabbage sprayed with botanical and conventional insecticides and tap water as control during 2014 minor rainy season.

	Mean ± SE natural enemies/ plant			
Treatments	D. rapae	Spider	Hoverfly	Other N.E[1]
10 g/L C. odorata	1.88 ± 0.57 ab	2.13 ± 0.30a	0.63 ± 0.25a	2.29 ± 0.42 ab
20 g/L C. odorata	3.75 ± 0.63 a	2.00 ± 0.26a	0.04 ± 0.05a	3.50 ± 0.56 a
30 g/L C. odorata	0.83 ± 0.21 b	1.63 ± 0.38a	0.08 ± 0.09a	2.83 ± 0.42 ab
50 g/ L Neem	1.25 ± 0.38 b	2.08 ± 0.28a	0.17 ± 0.09a	3.50 ± 0.51 a
Sunhalothrin®	2.33 ± 0.75 b	1.17 ± 1.22a	0.46 ± 0.24a	1.46 ± 0.26 b
Tap water	2.29 ± 0.57 ab	2.00 ± 0.32a	0.54 ± 0.29a	3.38 ± 0.31 a
F	3.51	1.55	1.68	3.65
P	0.01*	0.18	0.14	0.00*

Means within a column for each treatment under each concentration followed by different letters differ significantly from each other ($P \leq 0.05$). [1]Other natural enemies: ladybird beetle (*Cheilomenes lunata*), earwigs, red ants (*Oecophylla* sp.) and mason wasp.

Table 3: Mean (± *SE*) numbers of natural enemies of pests of cabbage sprayed with botanical and conventional insecticides during 2014 major rainy season.

Effects of treatments on cabbage head damage

Head damage was mainly caused by insects such as DBM, cabbage webworms, the variegated grasshoppers and the millipedes in the growing seasons (Table 5). There were no significant differences in the head damage among the treatments in the major rainy season (Table 5). On the other hand, there was significant difference in the

	Mean ± SE natural enemies/ plant				
Treatments	*D. rapae*	*C. plutellae*	Spider	Hoverfly	Other N.E1
10 g/L *C. odorata*	1.67 ± 0.64a	1.58 ± 0.62a	2.33 ± 0.41 ab	1.0 ± 0.35	1.25 ± 0.8 ab
20 g/L *C. odorata*	2.38 ± 0.78a	1.83 ± 0.50a	2.58 ± 0.43 a	1.08 ± 0.22	1.50 ± 0.26 a
30 g/L *C. odorata*	1.0 ± 0.43a	0.83 ± 0.39a	1.92 ± 0.40 ab	0.58 ± 0.25	0.42 ± 0.20 ab
50 g/ L neem	1.13 ± 0.45a	1.25 ± 0.34a	2.67 ± 0.41 a	1.17 ± 0.35	1.13 ± 0.24 abc
Sunhalothrin®	2.29 ± 0.97a	1.88 ± 0.53a	0.83 ± 0.21 b	0.42 ± 0.15	0.29 ± 0.11 c
Tap water	2.5 ± 0.86a	2.21 ± 0.58a	1.96 ± 0.31ab	1.25 ± 0.38	0.67 ± 0.20 abc
F	0.84	0.95	3.22	1.27	4.82
P	0.52	0.45	0.01*	0.28	0.00*

Means within a column for each treatment under each concentration followed by different letters differ significantly from each other (*P* ≤ 0.05).[1]Other natural enemies: ladybird beetle (*Cheilomenes lunata*), earwigs, red ants (*Oecophylla* sp.) and mason wasp.

Table 4: Mean (± *SE*) numbers of natural enemies of pests of cabbage sprayed with botanical and conventional insecticides during 2014 minor rainy season.

	Mean ± SE head damage/ plant	
Treatment	Major rainy season	Minor rainy season
10 g/L *C. odorata*	3.17 ± 0.33 a	2.5 ± 0.28 ab
20 g/L *C. odorata*	3.21 ± 0.29 a	2.96 ± 0.30 ab
30 g/L *C. odorata*	3.13 ± 0.37 a	2.79 ± 0.32 ab
50 g/ L Neem	1.96 ± 0.25 a	1.83 ± 0.27 b
Sunhalothrin®	3.04 ± 0.35 a	2.75 ± 0.27 ab
Tap water	3.04 ± 0.31 a	3.38 ± 0.31 a
F	2.20	3.03
P	0.06	0.01*

Means within a column for each treatment under each concentration followed by different letter differ significantly from each other.

Table 5: Mean (± *SE*) scores for head damage during 2014 major and minor seasons.

mean head damage among treatments in the minor rainy season (Table 5). Cabbage heads from tap water treated plots had the highest head damage as compared to neem seed extract which had the lowest damage (Table 5). Means within a column for each treatment under each concentration followed by different letter differ significantly from each other.

Effects of treatments on multiple head formation, weight and yield of cabbage

The results on the percentage number of cabbage with multiple heads for the major season showed that the 30 g/L (w/v) of *C. odorata* and tap water treated plots had the highest number, followed 10, 20 g/L *C. odorata* and Sunhalothrin, with the neem seed extract having the lowest number of multiple heads formed (Table 6). The result also showed that, there were no significant differences among treatments on the weight per cabbage head and yield, with the exception of the neem seed extract plots which had the highest head weight and yield (Table 6).

In the minor season, the result depicted that there were no significant differences in multiple head formation among all treatment plots. The head weight of cabbage had no significant differences among all treatments with the exception of neem seed extracts which had the highest head weight. The yield (t/ha) had significant differences among treatments, with neem seed extract having the highest yield, followed

by 10 and 20 g/L (w/v) *C. odorata* and then Sunhalothrin and 30 g/L (w/v) *C. odorata*. The tap water treated plots had the lowest yield (Table 7).

Cost: benefit ratio

The results indicated that, the 10 and 20 g/L (w/v) *C. odorata* had higher cost benefit ratios as compared to 30 g/L *C. odorata* for both seasons (Tables 8 and 9). The neem seed extract had the highest cost: benefit ratio as compared to the conventional insecticide, sunhalothrin' (Tables 8 and 9).

Discussion

Our goal was to evaluate the efficacy of the various Siam weed as a potential botanical insecticide in providing acceptable control of pest of cabbage in southern Ghana. In both seasons, all materials tested resulted in the production of marketable cabbage heads with considerably lower pest pressure and crop damage ratings compared with untreated control plots which never yielded marketable produce. These results indicate that all five botanicals at the different concentrations were effective in controlling the pests, mainly the lepidopterans. The presence and abundance of insect presence differed during both seasons, and this could be attributed to seasonal and climatic differences. Surprisingly, diamondback moth, *P. xylostella* was not found on the treatment plots during the major season. This may be attributed to the occurrences of natural elements such as rainfall, which washes off the eggs, larvae, pupae and the adults from the plant to the soil where they are destroyed and may have led to the disruption of the lifecycle of the pest during the period.

Populations of the aphid, *B. brassicae*, were dominant throughout both seasons and were reduced as a result of the application of the

Treatments	No. of cabbage with multiple heads (%)	Weight per cabbage head (kg)	Total yield of cabbage (t/ ha)
10g/L *C. odorata*	3.13 ± 1.14 ab	0.21 ± 0.05b	8.37 ± 1.83 b
20g/L *C. odorata*	2.08 ± 0.95 ab	0.26 ± 0.05b	10.50 ± 2.19 b
30g/L *C. odorata*	6.25 ± 1.30 a	0.24 ± 0.05b	9.67 ± 1.95 b
50g/L Neem	1.56 ± 0.87 b	0.52 ± 0.06a	20.75 ± 2.39 a
Sunhalothrin®	2.60 ± 1.07 ab	0.24 ± 0.05b	9.43 ± 1.88 b
Tap water	6.25 ± 1.29 a	0.20 ± 0.05b	3.30 ± 0.58 b
F	3.44	4.93	8.96
P	0.01*	0.00*	< 0.00*

Means within a column for each treatment under each concentration followed by different letters differ significantly from each other (*P* ≤ 0.05).

Table 6: Mean (± *S.E*) number of multiple heads, weight of cabbage heads and yield of cabbage under different spray treatments during 2014 major rainy season.

	Mean ± SE		
Treatments	No. of cabbage with multiple heads (%)	Weight per cabbage head (kg)	Total yield of cabbage (ton/ ha)
10g/L *C. odorata*	3.65 ± 1.18 a	0.25 ± 0.03 b	10.04 ± 1.08 b
20g/L *C. odorata*	2.60 ± 1.05 a	0.25 ± 0.04 b	10.13 ± 1.60 b
30g/L *C. odorata*	5.73 ± 1.29 a	0.24 ± 0.03 b	9.56 ± 1.31 bc
50g/ L Neem	3.13 ± 1.14 a	0.47 ± 0.06 a	18.80 ± 2.45 a
Sunhalothrin®	4.69 ± 1.25 a	0.22 ± 0.03 b	8.99 ± 1.17 bc
Tap water	5.73 ±1.30 a	0.17 ± 0.03 b	3.49 ± 0.62 c
F	1.20	7.14	10.82
P	0.31	< 0.00*	< 0.00*

Means within a column for each treatment under each concentration followed by different letters differ significantly from each other (*P* ≤ 0.05).

Table 7: Mean (± *S.E*) number of multiple heads, weight of cabbage heads and yield of cabbage under different spray treatments during 2014 minor rainy season.

Treatment	Mean total yield per plant (ton/ha)	Marketable Head Yield (ton/ha)	Unmarketable head yield (ton/ha)	Cost of Treatment (US$/ha)	Income from Mktable head (US$/ha)	Income from Unmktable head (US$/ha)	Total Income (US$/ ha)	Net benefit (US$/ha)	Benefit Over Ctrl.	Cost: Benefit Ratio
10 g/L C. odorata	10.04 ± 1.08	6.01 ± 1.04	4.02 ± 0.42	100.00	3,756	1,005	4,761	4,661	3,159	1:32
20 g/L C. odorata	10.13 ± 1.60	8.04 ± 1.02	2.10 ± 0.44	100.00	5,025	525	5,550	5,450	3,948	1:41
30 g/L C. odorata	9.56 ± 1.31	6.07 ± 1.12	3.45 ± 0.45	100.00	3,794	864	4,656	4,556	3,053	1:31
50 g/L Neem	18.80 ± 2.45	15.06 ± 1.95	3.20 ± 0.58	128	9,413	800	10,213	10,085	8,583	1:67
Sunhalothrin	8.99 ± 1.17	4.54 ± 1.07	4.02 ± 0.55	113	2,838	1,005	3,843	3,730	2,228	1:20
Tap water	3.49 ± 0.62	1.74 ± 0.42	1.66 ± 0.48	0.0	1,088	415	1,503	1,503	0	-

Table 8: Evaluation of cost and benefit in managing cabbage pests with crude water extracts of Siam weed, neem and conventional insecticide, Sunhalothrin® during 2014 minor rainy season.

Treatment	Mean total yield per plant (ton/ha)	Marketable Head Yield (ton/ha)	Unmarketable head yield (ton/ha)	Cost of Treatment (US$/ha)	Income from Mktable head (US$/ha)	Income from Unmktable head (US$/ha)	Total Income (US$/ ha)	Net benefit (US$/ha)	Benefit Over Ctrl.	Cost: Benefit Ratio
10 g/L C. odorata	8.37 ± 1.83	7.05 ± 1.02	1.32 ± 0.60	88	2,644	165	2,809	2,721	1,946	1:22
20 g/L C. odorata	10.50 ± 2.19	8.32 ± 1.61	2.15 ± 0.65	88	3,120	269	3,389	3,301	2,526	1:29
30 g/L C. odorata	9.67 ± 1.95	6.31 ± 1.19	3.22 ± 0.81	88	2,366	403	2,769	2,681	1,906	1:22
50 g/L Neem	20.75 ± 2.29	16.45 ± 1.25	4.33 ± 1.05	115	6,169	541	6,710	6,595	5,820	1:51
Sunhalothrin	9.43 ± 1.88	7.17 ± 1.05	2.55 ± 0.95	113	2,689	319	3,008	2,895	2,120	1:19
Tap water	3.30 ± 0.58	1.65 ± 0.45	1.25 ± 0.55	0.0	619	156	775	775	0	-

Table 9: Evaluation of cost and benefit of managing cabbage pests with crude water extracts of Siam weed, neem and conventional insecticide, Sunhalothrin® during 2014 major rainy season.

botanical treatment. This was especially observed in the 10 g/L C. odorata and neem seed extracts plots in both the minor and major seasons. This reduction was due to the toxic effect of the plants. Botanicals and plant - based insecticides have been noted for their larvicidal effects Sanda et al., Ogendo et al. and Agboka et al. [20-22]. The effectiveness of botanicals in the study was generally better as compared to the conventional insecticide in the control of aphid. This may be due the fact that, B. brassicae may gradually be developing resistance to the conventional insecticides Fening et al. [12]. The performance of the plant extracts in reducing the population of B. brassicae indicates their usefulness in controlling insect pests when incorporated into Integrated Pest Management (IPM). The efficacy of the botanical treatments against B. brassicae is supported by the findings of previous studies on this insect pest. For instance, the extracts of Azadirachta indica and Melia azedarach have been successfully used to control infestations of cabbage aphids Rando et al. and Kibrom et al. [23,24].

The number of natural enemies of the insect pests observed on the cabbage treated plot was somewhat not different from that of the control. In the minor season, D. rapae and C. plutellae showed no significant difference due to less toxicity of the treatments on the beneficial insects. For instance, neem products have been proved to be harmless to beneficial insects and environmentally friendly Borror et al. [25]. However, the lower concentrations of C. odorata (10 and 20 g/L w/v) were more benign to the natural enemies, ensuring the survival of D. rapae as compared to the 30 g/L C. odorata treated plots. This may be attributed to the low toxicity of the treatment at low concentrations on the beneficial insects, hence making them available and thereby promoting high level parasitism and predation of pest Charleston [10].

In the study, it was observed that the botanicals and the conventional insecticides, sunhalothrin˙ were most active against insect pests causing damage to cabbage heads as compared with control plot in the both

seasons. Fening et al. [19] however, revealed that the use of plant extracts in IPM provides added advantage over the use of synthetic insecticide; as, they are not persistent in the environment, readily available, affordable and easily made. The findings on high number of multiple head formed on untreated plots as compared to treated plots indicated that, cabbage cannot be cultivated without making an attempt to control insect pests. This is because, like other crucifers, they contain mustard oil and glucosides Gupta and Thorsteinson [26] which make them palatable and more susceptible to insect pest attack.

High yield of cabbage head was recorded on treated plots against that of tap water treated plot, with Neem seed extract treated plots recording the highest mean weight of cabbage heads due its insecticidal ability. This result confirms the findings of Landis et al. [27] that aqueous neem seed extracts (ANSE) applied at 50-70 g/L provided good protection against collard insect pests and increased dry matter content significantly.

The cost: benefit ratio, the total income and the benefit obtained from each treatment is greatly influenced by the price of the commodity. The price of cabbage heads were about 50% higher in the minor season harvest than the major season yield. The study has shown that the 10 g/L and 20 g/L of C. odorata extracts were effective in the management of insect pests, less detrimental to natural enemies and also economical to use than the higher rate, 30 g/L C. odorata extracts. This finding has therefore offered a remedy to Amoabeng et al. [13] who applied C. odorata extract at 30 g/L and was effective in the control of insect pests but was more detrimental to the survival of natural enemies. Thus, C. odorata applied at 10-20 g/L (w/v) and also neem at 50 g/L are recommended for use by small scale cabbage farmers (especially those in organic farming systems) to ensure food and environmental safety Fening [4].

Acknowledgements

We are grateful to the Technical staff, of the Entomology Unit of the Forestry and Horticultural Crops and Research Centre (FOHCREC), University of Ghana, Ghana.

References

1. Asare-Bediako E, Addo-Quaye AA, Mohammed A (2010) Control of Diamond back moth (Plutella xylostella) on cabbage (Brassica oleracea var capitata) using intercropping with Non-Host Crops. American Journal of Food Technology 5: 269-274.

2. Timbilla JA, Nyarko KO (2004) A survey of cabbage production and constraints in Ghana. Ghana Journal of Agricultural Science 37: 93-101.

3. Timbilla JA (1997) Integrated pest management of DBM. P xylostella (Lepidoptera: Plutellidae) in Ghana. Masters of Philosophy Thesis, KNUST - Kumasi, Ghana, p: 125.

4. Fening KO (2013) Improving Maternal and Child Health: The Role of Food Safety in the Developing World. In: Maternal-Child Health-Interdisciplinary Aspects within the Perspective of Global Health. Uwe Groß and Kerstin Wydra (eds.), Göttingen International Health Network (GIHN), Universitätsverlag Göttingen, Germany, pp: 53-61.

5. Norman C, Shealy MD (2007) Illustrated Encyclopedia of Healing Remedies. (online) PHD, Elements Book Inc. 160 North, Washington Street. Boston MA 02114. Available from: http://www.hcun.com/healthandhealing/kcabbage/indexhtm.

6. De Lannoy G (2001) Leafy Vegetables in Crop Production in Tropical Africa. Raemaekers RH (ed), Directorate General for International Co-operation, Brussels, Belgium, pp: 403-511.

7. Mochiah MB, Baidoo PK, Owusu-Akyaw M (2011) Influence of different nutrient applications on insects populations and damage to cabbage. Journal of Applied Bioscience 38: 2564- 2572.

8. Ninsin DK (1997) Insecticides use pattern and residue levels in cabbage cultivated within the Accra-Tema Metropolitan areas of Ghana. Master of Philosophy Thesis, Insect Science Programme. University of Ghana, Legon, p: 85.

9. Obeng-Ofori D, Ankrah DA (2002) Effectiveness of aqueous neem extracts for the control of insect pest of cabbage (Brassica oleracea var capitata L) in the Accra plains of Ghana. Agriculture and Food Science Journal Ghana 1: 83-94.

10. Charleston D (2004) Botanical Pesticides and their impact on info chemicals. IK International Publishing House Pvt. Ltd., p: 546.

11. Amoako PK (2010) Assessment of Pesticides used to control insect pests and their effects on storage of cabbage (Brassica oleracea var capitata). A Case Study in Ejisu- Juaben Municipal Area. Masters' Thesis submitted to School of Graduate Studies, KNUST.

12. Fening KO, Amoabeng BW, Adama I, Mochiah MB, Braimah H, et al. (2013) Sustainable management of two key pests of cabbage, Brassica oleracea var. capitata L. (Brassicaceae) using home-made extracts of garlic and pepper. Organic Agriculture 3: 163- 173.

13. Amoabeng BW, Gurr GM, Gitau CW, Nicol HI, Munyakazi L, et al. (2013) Tri-Trophic Insecticidal Effects of African Plants against Cabbage Pests. PLoS ONE 8: e78651.

14. Isman BM (2008) Perspective Botanical insecticides: for richer, for poorer. Pest Management Science 64: 8-11.

15. Kawuki RS, Agona A, Nampala P, Adipala E (2005) A comparison of effectiveness of plant-based and synthetic insecticides in the field management of pod and storage pests of cowpea. Crop Protection 24: 473-478.

16. Devanand P, Rani PU (2008) Biological potency of certain plant extracts in management of two lepidopteran pests of Ricinus communis L. Journal of Biopesticides 1: 170-176.

17. Buss EA, Park-Brown SG (2002) Natural products for insect pest management. UF/IFAS Publication ENY-350.

18. Dubey N, Shukla R, Kumar A, Singh P, Prakash B (2011) Global scenario on the application of natural products in integrated pest management programmes. In: NK Dubey. Natural Products in Plant Pest Management, Wallingford: CABI Publishing, pp: 1-20.

19. Fening KO, Owusu-Akyaw M, Mochiah MB, Amoabeng B, Narveh E (2011). Sustainable management of insect pests of green cabbage, Brassica oleraceae var. capitata L. (Brassicaceae), using homemade extracts from garlic and hot pepper. Third Scientific Conference of the International Society of Organic Agriculture Research (ISOFAR), 17th IFOAM Organic World Congress: Namyangju, Korea: Organic crop production, pp: 567-570.

20. Sanda K, Koba K, Poutouli W, Iddrissou N, Agbossou AB (2006) Pesticidal properties of Cymbopogon schoenatus against the Diamondback moth P. xylostella L, (Lepidoptera: Hyponomeutidae). Discovery and Innovation 18: 220- 225.

21. Ogendo JO, Kostyukovsky M, Ravid U, Matasyoh JC, Deng AL, et al. (2008). Bioactivity of Ocimum gratissimum L. oil and two of its constituents against five insects attacking stored food products. Journal of Stored Products Research 44: 328-334.

22. Agboka K, Agbodzavu KM, Tamo M, Vidal S (2009) Effects of plant extracts and oil emulsions on the maize cob borer Mussidia nigrivenella (Lepidoptera: Pyralidae) in laboratory and field experiments. International Journal of Tropical Insect Science 29: 185- 194.

23. Rando JSS, De Lima CB, De Almeida Batista N, Feldhaus DC, De Lourenço CC (2011) Extratos vegetais no controle dos afídeos Brevicoryne brassicae (L.) Myzus persicae (Sulzer). Semina: Ciências Agrárias 32: 503-512.

24. Kibrom G, Kebede K, Weldehaweria G, Dejen G, Mekonen S (2012) Field evaluation of aqueous extract of Melia azedarach Linn. seeds against cabbage aphid, Brevicoryne brassicae Linn. (Homoptera:Aphididae), and its predator Coccinella septempunctata Linn.(Coleoptera: Coccinellidae). Archives of Phytopathology, Plant Protection 45: 1273-1279.

25. Borror DJ, Triplehorn CA, Johansson NF (1992) An introduction to the study of insects. 6th edn. Harcourt Brace College publishers, USA, p: 879.

26. Gupta PD, Thorsteinson AJ (1960) Food plant relationship of diamondback moth, Plutella maculipennis (Curt.) II Sensory relationship of oviposition of adult of the female. Entomology of Experiments and Application 3: 305-314.

27. Landis DA, Wratten SD, Gurr GM (2000) Habitat Management to conserve natural enemies of arthropod pests in Agriculture. Annual Review of Entomology 45: 175-201.

33

Phenotypic Characterization of Selected Kenyan Purple and Yellow Passion Fruit Genotypes Based on Morpho-Agronomic Descriptors

Matheri F[1*], Mwangi M[2], Runo S[1], Ngugi M[1], Kirubi DT[2], Fred Teya[1], Mawia AM[1], Kioko FW[1] and Kamau DN[3]

[1]Department of Biochemistry and Biotechnology, School of Pure and Applied Sciences, Kenyatta University, Nairobi, Kenya
[2]Department of Agricultural Science and Technology, School of Agriculture and Enterprise Development, Kenyatta University, Nairobi, Kenya
[3]Department of Microbiology, School of Pure and Applied Sciences, Kenyatta University, Nairobi, Kenya

Abstract

Phenotypic characterization is crucial in determination of variability of hybrid varieties and their parents. The objective of this study was to determine phenotypic variation among known genotypes of both parent and KPF hybrids, as well as genotypes collected mainly from Embu County which is one of the growing areas of hybrid varieties developed by KALRO. Analysis was done using Minitab 17.0 software. Six out of seven morpho-agronomic descriptors evaluated, showed significant differences among the genotypes under study. A dendrogram based on the 7 morpho-agronomic descriptors discriminated the genotypes into two main clusters with one main cluster (II) carrying only 2 genotypes. Principal component analysis corroborated the findings of the dendrogram, distantly placing the two genotypes further from the other genotypes.

Keywords: Phenotypic; Morpho-agronomic descriptors

Introduction

The passion fruit is considered a high value crop in Kenya, ranking third (8%) after avocado (62%) and mango (26%) in terms of foreign exchange earnings for the country [1]. Kenya is considered as the market leader of fruit juice exports in East Africa and is also listed among the large producers of passion fruit globally with its major regional market being Uganda [2,3].

If production is carried out efficiently, passion fruit enterprises have good returns, with a gross margin of Ksh. 629,850 per hectare (approximately 6298 USD) [4]. After orchard establishment, production is expected to increase subsequently from the first to the third year and can therefore be used productively during this period [5]. The relatively high gross margin makes passion fruit a high value crop with potential for poverty alleviation since it is mainly grown by farmers owning 0.5 - 2 acres of land [6,7]. Passion fruit farming is also preferred due to the fast maturity period of 9 months (flowering period) and the minimal labor and land space requirements [8].

The passion fruit is native to the Southern Brazil, Paraguay and North Argentina, thus this region is considered as the main center of genetic diversity of the *Passiflora species* [9,10]. The plant was introduced to Kenya by the white settlers in the early 20th century after which cultivation was limited to plantations owned by the European settlers [11,12]. The passion fruit only gained significant economic importance as an income generating crop in the 1990s when Kenya started bulk export of fruits and vegetables to the international markets [13].

Passion fruit production in Kenya had been increasing gradually from the beginning of the 21st century until 2007 when it started to decline. There was notable increase in production between 2005 and 2007 when production doubled after which it declined in 2008 with fluctuations in production through the subsequent years. This decline is attributed to perennial challenges that lead to the sector operating below potential and as such, lagging behind other global competing producers like Australia and South Africa [14,15].

Insufficient knowledge on good agricultural practices as well as pest and disease management as well as the inaccessibility of pathogen-free planting materials are some of the major challenges that face passion fruit production [16,17]. Changing climate patterns are also contributing to the decline in passion fruit production by favoring population densities and emergence of new species of pests [16].

There is need to identify and document existing and new passion fruit varieties especially those perceived to have superior traits such as tolerance to *Fusarium*. A description of variety should help in resolution of identification conflicts that may arise during registration and protection of cultivars [18]. Morphological and agronomic characterization of germplasm as well as new varieties using descriptors is a key consideration in breeding programs. The term descriptor is used to refer to a character or attribute that is used to discriminate between varieties, with redundant descriptors being seen during evaluation of many traits and thus many descriptors are judged accordingly as unnecessary due to their low contribution to variability [18-20]. Elimination of redundant descriptors is an important strategy in that it ensures reduction of the work required to collect data without causing significant losses in genotype discrimination [20,21].

Some of the techniques used to determine the descriptors with high information content include regression [22], discriminant analysis [23] as well as principle components [24]. The distribution of variation is associated with the nature and number of characters that are used in the analysis and is concentrated in the first components only when few agronomically important traits are evaluated [25].

The current study aimed at characterizing hybrid cultivars that were recently developed by KALRO as well as their parents, using quantitative morpho-agronomic descriptors.

*Corresponding author: Felix Matheri, Department of Biochemistry and Biotechnology, School of Pure and Applied Sciences, Kenyatta University, PO Box 43844-00100, Nairobi, Kenya, E-mail: felmat06@yahoo.com

Materials and Methods

Collection of plant material

Fully ripe fruits and fully expanded leaves were collected from healthy vigorously growing vines of KPF 4, KPF 11, KPF 12, Brazil, and purple passion fruit genotypes in KALRO- Kandara. All samples were assigned to populations based on the variety. Samples were also collected from different geographic locations in Embu County, Kenya and Kenyatta University School of agriculture farms; all orchard ranging between 2-3 years since establishment. Replication was done five times per plant for each trait under study with three biological replicates covering the two main seasons in Kenya.

Morpho-agronomic descriptors

Seven quantitative morpho-agronomic traits developed by IPGRI (now Bioversity International) were evaluated in this study. These traits included; leaf length, leaf width, fruit length, fruit diameter and seed length whose data was recorded in centimeters. Fruit mass was recorded in grams.

Data management and analysis

The data for all the three biological replicates for each genotype were combined and analyzed statistically using Minitab 17.0 software. The differences in means of the 7 traits were separated through ANOVA followed by Tukey's post hoc. Cluster analysis was used to develop a dendrogram and scatter plot for examination of the phenotypic relatedness among the genotypes under study.

Results

The measurement of the leaf, fruit and seed traits for the 54 passion fruit genotypes and their mean values are shown in Appendix 1. From the tabulated results, KR4-1 which belongs to the KPF-4 variety, had the highest value of leaf length (14.82 cm) while KRC-3 which was a coastal yellow passion fruit variety had the lowest value (8.60 cm). In relations to the leaf width, there was no significant difference among all the genotypes although the data range indicated wide variation. KR4-1 had the highest mean value of leaf width (13.36 cm) while genotype JSE-N3 had the lowest mean value (9.62 cm) (Appendix 1).

On the other hand, longitudinal fruit length also had wide variations

ranging from 5.10 cm (PKS-TD1) to 10.58 cm (KR12-3). Equatorial fruit diameter ranged from 3.74 cm (KMD-TD1) to 10.18 cm (KR12-1). There was no significant difference among the mean fruit diameter for most genotypes. Fruit Rind thickness also had wide variation and ranged from, 0.52 (SGD-TD1) to 0.86 cm (KR12-1). In relations to seed length, there was variation among the genotypes, ranging from 0.36 cm (KR12-4) to 0.78 cm (PKS-TD1). There was also wide variation in fruit mass among the populations. The mean fruit mass ranged from 28.62 g (BMM-TD2) to 121.82 g (KR11-1) (Appendix 1).

Cluster analysis

The genotypes were discriminated into two major clusters; I and II with cluster I comprising two sub-clusters as shown in Figure 1. Each of the sub-clusters in cluster I was further been divided into several sub-clusters carrying various genotypes. From the dendrogram, most of the known genotypes were clustered together. For example, KRP-1, KRP-2, KRP-3 and KRP-4, all purple genotypes were clustered together. KR11-1, KR11-2, KR11-3 and KR11-4 were also clustered together at a similarity value close to 100%. However some of the known genotypes lacked homogeneity in clustering. For example, despite Brazil genotypes; KRC-2, KRC-3 and KRC-4 being clustered close together in the tree, KRC-1 which belonged to the same variety was clustered distantly. Other known varieties with non-homogeneity in clustering were KR12 (KPF-12) and KR4 (KPF-4). Genotypes belonging to the undetermined population were clustered together with those of known populations. For example, PKS-N3 which was an undetermined genotype, clustered with the coastal genotypes (KRC-2, KRC-3 and KRC-4). SGE-N1, SNV-N1, JSE-N3, MMN-N1 clustered together with KR4-4 and KR4-1 at a similarity value close to 100%, an indication that they may be related.

Principal component analysis (PCA)

The first three Eigen values were 3.5880, 1.6130 and 1.0062 respectively. The first principal component (PC1) accounted for 51.3% of total variance, while the second principal component (PC2) accounted for 23.0%. The third principal component (PC3) accounted for 14.4% variability for the 7 morpho-agronomic traits evaluated (Table 1). There was positive correlation between the Eigen values for PC1 while those of PC2 was negatively correlated to leaf length and leaf width and positively correlated to the remaining traits. The Eigen

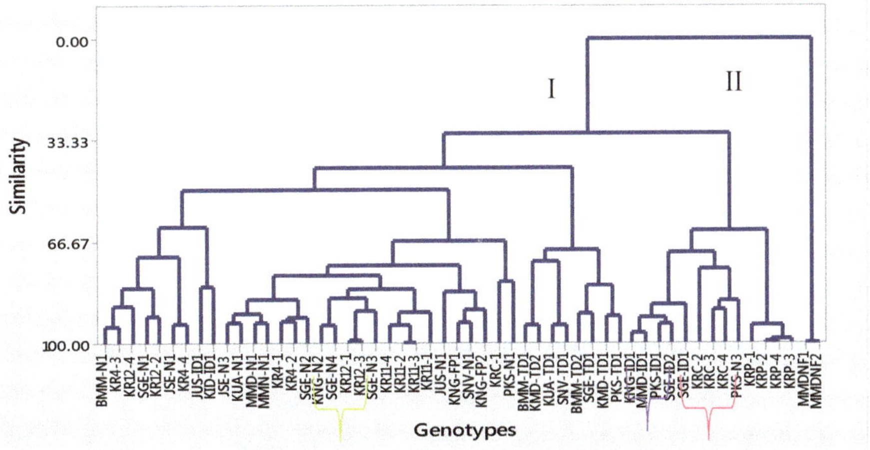

Figure 1: Eucledian distance based dendrogram developed from mean values of the 7 morpho-agronomic traits.

	PC1	PC2	PC3
Eigenvalue	3.5880	1.6130	1.0062
% Total Variance	51.3	23.0	14.4
%Cumulative	**51.3**	**74.3**	88.7
Traits	PC1	PC2	PC3
Leaf Length (cm)	0.119	**-0.691**	**-0.015**
Leaf Width (cm)	0.070	**-0.597**	**-0.524**
Fruit Length (cm)	0.511	0.036	0.095
Fruit Diameter (cm)	0.512	0.055	0.110
Rind Thickness (cm)	0.082	0.391	**-0.824**
Fruit Mass (g)	0.497	0.005	0.110
Seed Length (cm)	0.451	0.096	**-0.118**

Table 1: Eigen vectors, Eigen values, total variance and cumulative variance for 54 genotypes based on 7 morpho-agronomic traits.

value for PC3 was negatively correlated to leaf length, leaf width rind thickness as well as seed length. This value was however, positively correlated to the remaining morpho agronomic traits. The traits with negative correlation were retained in analysis since the all Eigen values for PC1 which accounted for much of the variation were positively correlated to the traits (Table 1).

Scatter plot

A scatter plot of the genotypes under study complemented the findings of the dendrogram that some of the genotypes had similarity value close or equal to 100%. For example, MMD-NF1 and MMD-NF2 which were clustered together on the dendrogram were also placed graphically on the scatter plot (Figure 2). The lack of homogeneity of clusters was also seen on the scatter plot with some of the known genotypes being on a far graphic location from the other genotypes of the same population.

Discussion

Measurement of genetic variability of passion fruit species by accessing markers such as morphological descriptors is a fundamental activity for both plant breeding and conservation programs of many species [26].

Such descriptors include fruit size, which can be described through fruit length and equatorial diameter. The size of fruit is important in the physical quality of the fruits destined for markets and industry [27].

From the tabulated results, the difference in leaf lengths of KR4-1 which belongs to the KPF-4 variety (14.82 cm) and KRC-3 which was a coastal yellow passion fruit variety (8.600 cm) can be explained by genotypic variation that is known to exist between hybrids and parent genotypes. Hybrid genotypes have been reported to have higher values for leaf length [28]. The lack of significant difference in leaf width of all the 54 genotypes indicates lack of agronomic and environmental influence on this trait.

The wide variation of seed length with an overall mean of 0.67 indicates a variation in seed fitness where the size of the seed affects fitness of the plant growing from it. This variation in seed length can be explained by difference in position on the inflorescence or the fruit [29,30]. The mean value for seed length was slightly higher than that obtained in related studies [28].

The mean value for rind thickness (0.48 cm) was lower than that obtained by Santos et al. [28]. The value was also slightly higher than that obtained in studies [30]. Conversely, the value was lower than that obtained by Silva et al. [31] and Cavalcante et al. [32]. Breeding programs seek to select genotypes with reduced rind thickness, which may be used to indicate greater amount of pulp which is regarded as a relevant factor in fruit ranking [28,33]. Therefore, based on these criteria, genotypes SGE-ID1 and SGE ID2 were favorable in terms of pulp and juice yield and can be adopted for crosses targeting higher juice and pulp yield. On the other hand, the wide variation in fruit diameter can be attributed to environmental and agronomic influence. The mean fruit diameter was equal to that obtained by Santos et al. [28] and close to that obtained by Silva et al. [31].

The wide variation in fruit length has also been reported in other studies [28,34]. The mean fruit length (8.06 cm) of the 54 genotypes was close to that obtained for *Passiflora edulis* (8.15 cm) [28] and slightly lower than the values obtained in function of the genotypes and fruit weight in passion fruit [31].

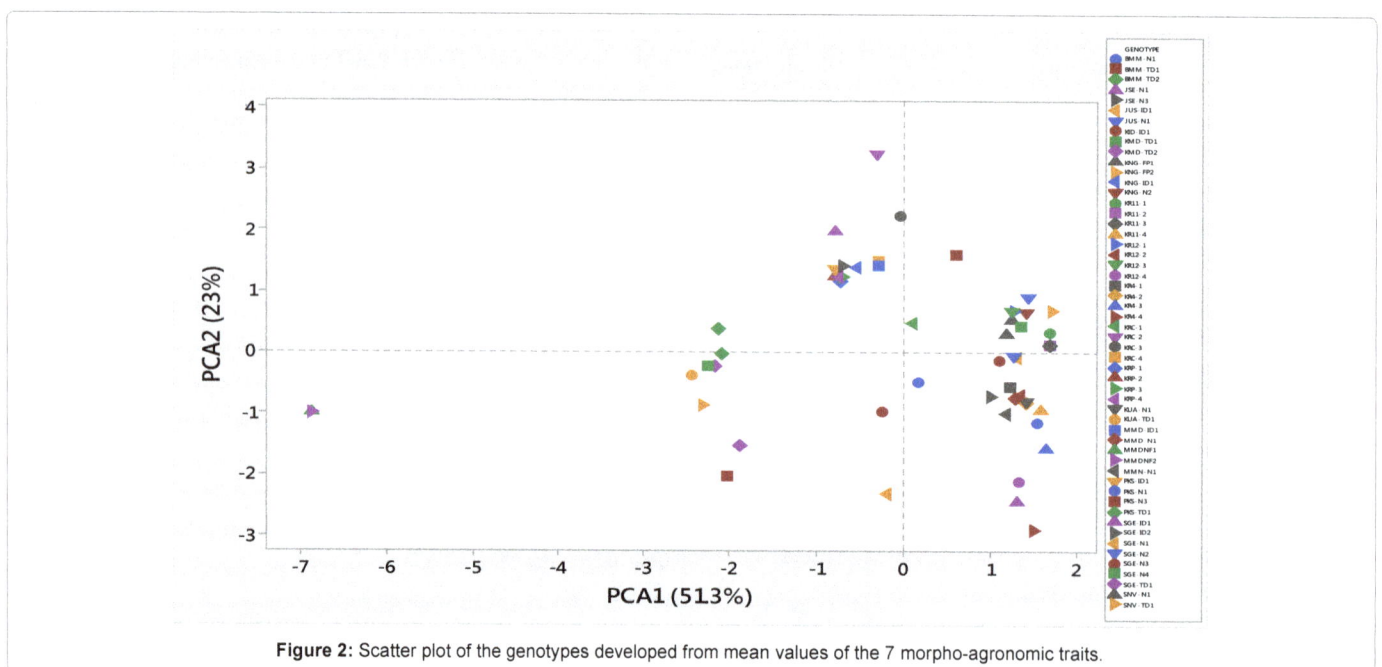

Figure 2: Scatter plot of the genotypes developed from mean values of the 7 morpho-agronomic traits.

Fruit length and width are important attributes of passion fruit where higher length than in width, is the preferred fruit form for the consumer market [9]. The fruit form index is an important aspect that is useful in classification and standardization of passion fruit in the fruit market where it influences the acceptance and judgment of the product in some markets [35,36] and external aspect such as size and shape [9,37].

The results of principal component analysis indicated the contribution of each principal component to overall variation. The principal component technique is useful in phenotypic variability studies in that it allows the evaluation of importance of each trait/character of the accessions being studied over total variation, hence allowing elimination of less discriminating characters. The first principal component was responsible for much of the overall variation, having accounted for more than half (51.3%) and as such was reliable in discrimination of the genotypes based on the seven traits. The high cumulative variance obtained with only the first three principal components may be explained by the fact distribution of variation is associated with the nature and number of characters used in the study. This variation is concentrated in the first principal component especially when evaluating few agronomically important traits or certain groups such as flowers and fruits [25]. Quantitative descriptors should be discarded when they have high correlation with principal components of the lowest variance. However, none of the traits had a high correlation with the lowest variance (PC3) and as such not necessary to discard.

From the dendrogram in Figure 1; clustering of a majority of genotypes in cluster I is an indication of the great divergence among genotypes in the cluster. Discrimination of only two genotypes into main cluster II can be explained by their lack of fruit based traits, thus bringing wide variation between them and the rest of the genotypes. Their clustering at 100% similarity can be interpreted to mean that the two genotypes had common ancestry with a probability of being full siblings. The lack of homogeneity in clustering of the known genotypes can be attributed to underlying genetic basis, since the genotypes shared the same agronomic and environmental influence. Clustering of the undetermined genotype, PKS-N3, at a similarity of 100% could also be interpreted to mean that it belonged to the coastal variety.

The distribution of the genotypes on the scatter plot confirms the results of the dendrogram. The graphical position of genotypes, MMD-NF1 and MMD-NF2 confirms their wide variation compared to other genotypes. The graphical closeness of the two genotypes on the principal axis is an indication of their biological relatedness, which could be interpreted to indicate that they are biological replicates. Moreover, the graphical representation of the genotypes confirms the lack of homogeneity of the some of the known populations where some genotypes are clustering far from others.

Conclusion

Despite experiencing possible variation in environmental and agronomic conditions, this study was able to show existing morpho-agronomic variation existing between the genotypes under study. This is confirmed by the fact that the discrimination did not stratify the genotypes to the respective orchard and even plants in the same orchard could be separated. Moreover, that lack of significant difference in leaf width as well as the significant difference seen in the studied genotypes confirms, the little influence of environment and agronomic practice on the evaluated traits.

Acknowledgements

The authors wish to acknowledge the Kenya Agricultural Productivity and Agribusiness Project (KAPAP) for funding this research work. We also thank the, Kenya Agricultural and Livestock Research Organization (KALRO) and MAKI farmers' group who provided the plant material.

References

1. HCDA (2011) 2010 Horticulture Validated Report. Nairobi: HCDA.

2. KHCP (2011) Kenya's Intra Africa Horticulture Trade. Nairobi: USAID.

3. Sebstad J, Snodgrass D (2008) Impacts of the KBDS and KHDP projects on the tree fruit value chain of Kenya. Micro Report no129. Nairobi: USAID-Kenya.

4. Kibet N (2011) The Role of Incentives and Drivers in Crop Trade-Off: Evidence from Passion Fruit Uptake in Uasin-Gishu District, Kenya.

5. Fintrac (2009) USAID-KHDP Kenya Horticultural Development Program October 2003–2009. Nairobi: USAID, Kenya.

6. Otipa M, Amata R, Waiganjo M, Ateka E, Mamati G, et al. (2008) Incidences and severity of viruses in passion fruit production systems in Kenya: 1st African Biotechnology Congress, Nairobi, Kenya.

7. Kibet N, Lagat J, Obare G (2011) Identifying Efficient and Profitable Farm Enterprises in Uasin-Gishu County in Kenya. Asian Journal of Agricultural Sciences 3: 378-384.

8. Kahinga N, Muthoka NK, Chege K, Mbugua W (2006) Training Manual for Passion Fruits. Thika: KARI/ HDP.

9. Oliveira EJ, Santos VS, Lima DS, Machado MD, Lucena RS, et al. (2008) Seleção em progênies de maracujazeiro amarelo com base em índices multivariados. Pesquisa Agropecuária Brasileira 43: 1543-1549.

10. Kilalo DC, Olubayo FM, Ateka EM, Hutchinson JC, Kimenju JW (2013) Monitoring of Aphid Fauna in Passionfruit Orchards in Kenya. International Journal of Horticultural and Crop Science Research 3: 1-18.

11. Morton J (1987) Passion fruit. In: Fruits of warm climates. Miami, FL, pp: 320-328.

12. Wangungu C, Maina M, Gathu K, Mbaka J, Kori N (2010) In: Mwangi M. Contributions of agricultural sciences towards achieving the Millenium Development Goals. FaCT Publishing, Nairobi, Kenya, pp: 58-64.

13. Muigai S (2002) Passion fruit breeding: Annual report. Thika-Kenya: KARI Thika.

14. HDC (2005) Horticultural Update. Nairobi: USAID, Kenya.

15. Mbaka J, Waiganjo M, Chege B, Ndungu B, Njuguna J, et al. (2006) A Survey of the Major Passion Fruit Diseases in Kenya. 10th KARI biennial scientific conference.

16. EU (European Union) (2009) Working Paper: Impacts of climate change on human and plant health.

17. Kleemann G, Chege P, Krain E (2010) Growing Value: Achievements in Value Chain Promotion for Passion Fruit Value Chain in Kenya. Nairobi: PSDA.

18. Castro JA, Neves CG, Jesus ON, Oliveira EJ (2012) Definition of Morpho-agronomic Descriptors for the Characterization of Yellow Passion Fruit. Scientia Horticulturae 145: 17-22.

19. Daher RF, Moraes CF, Cruz CD, Pereira AV, Xavier DF (1997) Selecao de caracteres morfologicos discriminantes em capim-elefante. Rev Bras Zoo 26: 247-254.

20. Oliveira EJ, Dias NLP, Dantas JLL (2012) Selection of morpho-agronomic descriptors for characterization of papaya cultivars. Euphytica 185: 253-265.

21. Oliveira MSP, Ferreira DF, Santos JB (2006) Seleção de descritores para caracterização de germoplasma de açaizeiro para produção de frutos. Pesq agropec bras Brasília 41: 1133-1140.

22. Beale EM, Kendall MG, Mann DW (1967) The discarding of variables in multivariate analysis. Biometrika 54: 357-366.

23. Mardia RF, Kent JT, Bibby JM (1979) Multivariate analysis. Academy Press, London.

24. Cruz CD, Regazzi JA, Carneiro PCS (2004) Divergencia genetica. In Cruz CD, Regazzi JA, Carneiro PCS (Editors) Modelos Biometricos aplicados ao Melhoramento Genetico. Universidade Federal de Viços 371-413.

25. Pereira AV, Vencovsky R, Cruz CD (1992) Selection of Botanical and Agronomical descriptors for the characterization of Cassava (Mannihot esculenta crantz) germplasm. Rev Revista Brasileira de Genetica 15: 115-124.

26. Paiva CL, Viana AP, Santos EA, Freitas JCO, Silva RNO, et al. (2014) Genetic variability assessment in the genus Passiflora by SSR markers. Chilean J Agric Res 74: 355-360.

27. Freire JLO, Cavalcante LF, Rebequi AM, Dias TJ, Souto AGL (2011) Necessidade hídrica do maracujazeiro amarelo cultivado sob estresse salino, biofertilização e cobertura do solo. Revista Caatinga 24: 82-91.

28. Santos EA, Viana AP, Freitas JC, Souza MM, Paiva CL, et al. (2014) Phenotyping of Passiflora edulis, P. setacea, and their hybrids by a multivariate approach. Genet Mol Res 13: 9828-9845.

29. Giles BE (1990) The effects of variation in seed size and reproduction in the wild Barley, Hordeum vulgare ssp spontaneum. Journal of Heredity 64: 239-250.

30. Santos C, Bruckner CH, Cruz CD, Siqueira DLDE, Pimentel LD (2009) Passion fruit physical traits in function of the genotypes and fruit weight. Revista Brasileira de Farmacognosia 31: 1102-1119.

31. Silva MAP, Plácido GR, Caliari M, Carvalho BS, Silva RM, et al. (2015). Physical and chemical characteristics and instrumental color parameters of passion fruit (Passiflora edulis Sims). Afr J Agric Res 10: 1119-1126.

32. Cavalcante LF, Santos GD, Olivieira FA, Cavalcante IHL, Gondim FC, et al. (2007) Growth and production of yellow passion fruit in a soil of low fertility treated with liquid biofertilizers. Revista Brasileira cienca agricultura 2: 15-19.

33. Negreiros JRS, Araújo-Neto SE, Álvares VS, Lima VA (2008) Caracterização de frutos de progênies de meiosirmãos de maracujazeiro-amarelo em Rio Branco - Acre. Revista Brasileira de Genetica 30: 431-437.

34. Cerqueira-Silva CBM, Cardoso-Silva CB, Nonato JVA, Corrêa RX (2009) Genetic dissimilarity of 'yellow' and 'sleep' passion fruit accessions based on the fruits physical-chemical characteristics. Crop Breeding Applied Biotechnology 9: 210-218.

35. Purquerio LFV, Cecilio- Filho AB (2005) Concentracao de nitrogenio na solucao nutritive e numero de frutos sobre a qualidade de frutos de melao. Hortic Bras 23: 831-836.

36. Meletti LMM, Soares-Scott MD, Bernacci LC, Azevedo FJA (2002) Desempenho das cultivares IAC-273 e IAC-277 de maracujazeiro amarelo (Passiflora edulis f. flavicarpa Deg.) em pomares comerciais. In: Reunião Técnica de Pesquisa em Maracujazeiro Amarelo, Viçosa. AnaisViçosa 3: 166-167.

37. Hafle OM, Ramos JD, Lima LCO, Ferreira EA, Melo PC (2009) Produtividade e qualidade de frutos do maracujazeiro amarelo submetido à poda de ramos produtivos. Revista Brasileira de Fruticultura 31: 763-770.

34

Response of Potato (*Solanum tuberosum* L.) Varieties to Nitrogen and Potassium Fertilizer Rates in Central Highlands of Ethiopia

Egata Shunka*, Abebe Chindi, Gebremedhin W/giorgis, Ebrahim Seid and Lema Tessema

Ethiopian Institute of Agricultural Research, Holetta Agricultural Research Center, Ethiopia

Abstract

Field experiment was conducted at Holetta and Jeldu Agricultural Research Station in the central highlands of Ethiopia to determine the rates of Nitrogen (N) and Potassium (K) fertilizers on growth, yield and yield components of potato. 4 × 3^2 factorial treatment was arranged in completely randomized block design with three replications on plot size of 3 m × 3 m during 2014-2015 cropping season. Nitrogen (87 kg, 110 kg and 133 kg/ha), Potassium (0, 34.5 kg, 69 kg and 103.5 kg/ha) and potato varieties (Betete, Gudenie and Jalenie) were used. Data were analyzed by using SAS software Version 9.2. The interaction effect of potassium and nitrogen fertilizers did affect marketable tuber number and plant height significantly. Gudenie produced the highest marketable yield (30.53 ton/ha) in 2015 with application of 69 kg/ha potassium and 110 kg/ha nitrogen rates while lowest marketable yield (16.67 ton/ha) was obtained from Belete variety at 0 kg/ha potassium rate and 87 kg/ha nitrogen rate. From these results, it can be concluded that interaction of nitrogen and potassium rates affected significantly plant height and marketable tuber numbers. Therefore, it is better to apply 69 kg/ha potassium and 110 kg/ha nitrogen for potato production to obtain reasonable economic yield at sites similar to experimental locations.

Keywords: Potato; Potassium and nitrogen rates; Marketable; Unmarketable yield; Ethiopia

Introduction

Potato (*Solanum tuberosum* L.) is one of the most important food crops worldwide. It ranks third after rice and wheat in terms of human consumption [1]. Among root and tuber crops, potato ranks first in volume of production and consumption, followed by cassava, sweet potato and yam. Annual world production of potato is about 330 million metric tons from 18,651,838 ha area coverage and in Africa total production is about 17,625,680 tons from total area coverage of 1,765,617 ha [2]. In Ethiopia, total area coverage of potato is nearly 0.18 million hectare from which 1.62 million ton is harvested [3]. According to Yilma [4], about 70% of cultivated agricultural land of Ethiopia is suitable for potato production. Despite high potential production environments and marked growth, the national average potato yield in farmers field in Ethiopia is only 11.1 t ha^{-1}, which is lower than the experimental yields of over 38 t ha^{-1}, which is very low compared to the world average of 17.6 t ha^{-1} [1,2,5]. The main contributing factors for under production and utilization of potatoes are lack of high yielding and diseases tolerant varieties, unavailability of quality seed and poor agronomic practices such as optimum nutrition and irrigation etc.

Low soil fertility in general and deficiency of Nitrogen (N) and Phosphorus (P) in most Ethiopian soils in particular is the most important constraint limiting potato production in Ethiopia [2]. The authors reported that, the soil fertility decline is attributed to continuous cropping, abandoning of fallowing, reduced crop rotation, removal of nutrients together with the harvested crops, reduced use of animal manure and crop residues due to their use as a fuel, which should be added to the soil and erosion coupled with low inherent fertility. The situation is exacerbated by the inherently high soil acidity with pH values of 4.02 to 4.6 being common. Most of the potato growing areas in Ethiopia have a soil pH of less than 5.5 [6,7]. A pH of less than 5.5 severely limits availability of potassium, nitrogen, phosphorus, sulphur, calcium and magnesium, while availing excessive levels of aluminum, manganese, boron, iron, copper and Zinc [8]. It is possible that this problem of low soil pH has led to nutrient imbalance hence reducing potato yields even further.

In Ethiopia, some farmers use inorganic fertilizers for increasing potato yields. However, they use only nitrogen (as Urea) and phosphorus (as DAP) since these are the only fertilizers commercially available in the local market. In addition, application of these fertilizers to potato crop is based on blanket recommendations that were formulated for potato grown on soils of certain sites in the country decades ago, that is, 165 kg Urea/ha (111 kg N/ha) and 195 kg DAP/ha (40 kg P/ha). These recommendations wholly disregard the specific physico-chemical characteristics of the varied soils on which the crop is grown as well as the dynamic nature of soil nutrient status. But application of 138 kg N and 20 kg P/ha is found to be the appropriate rate for optimum productivity of Gorebella variety on the vertisols of Debere Berhan in the central highlands of Ethiopia under rain fed conditions, which can be an insight to conduct trials for other varieties to develop optimum rate enhancing economic return [9]. When excessive nitrogen is applied, crop yield is reduced; cost of production increased and environment is polluted, especially soil and ground water is acidified [10].

Early reports by Murphy described that, favorable potassium supply, except in a few acutely deficient soils, have led researchers and farmers to ignore the need for potassium in many parts of East Africa. Consequently, potassium fertilizer is not entirely applied to crops by farmers in Ethiopia. Without application of phosphate and potassium, the yield response to increasing levels of nitrogen was smaller than when adequate amounts of P and K were applied. Therefore, all the essential nutrients should be available to the crop to realize maximum

---

*Corresponding author: Egata Shunka, Ethiopian Institute of Agricultural Research, Holetta Agricultural Research Center, PO Box 2003, Addis Ababa, Ethiopia, E-mail: egata.shunka2007@yahoo.com

Potassium levels (kg ha⁻¹)	Total Tuber Yield (ton/ha)	Marketable yield (ton/ha)
0	18.72b*	15.67b*
34.5	19.83ab*	16.17b*
69	20.25ab*	16.44b*
103.5	21.72a*	18.58a*
CV%(0.05)	17.80	20.24

Means followed by same letter(s) are not significantly different from each other at p ≤ 0.05. *- indicate means which are significantly different at 5% level of probability. CV% - Coefficient of Variance

Table 1: Effect of potassium rates on total tuber and marketable yields.

Year	Unmarketable Tuber Number/plot	Unmarketable Tuber weight (ton/ha)	Total Tuber Yield (ton/ha)	Plant height (cm)
2014	46.79b**	1.81b**	25.93b**	56.19b**
2015	63.16a**	3.61a**	35.24a**	63.34a**
CV% (0.05)	30.49	23.21	20.5	14.95

Means followed by same letter(s) are not significantly different from each other at p ≤ 0.05**- indicate means which are significantly different at 1% level of probability. CV% - Coefficient of Variance

Table 2: Effect of growing year on performance of potato varieties.

Year	Total Tuber Number/plot	Marketable Tuber Weight (ton/ha)	Total Tuber Yield (ton/ha)	Unmarketable Tuber Number/plot	Unmarketable Tuber Weight (ton/ha)	Plant Height (cm)
Holetta	175a**	27.9a**	37.55a**	68a**	4.31a**	77.02a**
Jeldu	159b**	18.15b**	23.62b**	42b**	1.1b**	42.50b**
CV%(0.05)	20.9	24.14	20.5	30.49	23.21	14.95

Means followed by same letter(s) are not significantly different from each other at p ≤ 0.05. **- indicate means which are significantly different at 1% level of probability. CV% - Coefficient of Variance

Table 3: Effect of growing location on performance of potato varieties.

yields. In addition to this, the information about Potassium fertilizer and its rates on potato product is also scarce in Ethiopia. Furthermore, fertilizer recommendations do not cater for potassium yet some studies have indicated response of potatoes to potassium addition on some highland parts of the country [6]. Even though the crop requirement of potassium is higher than N and P rates, the cultivation is done without Potassium fertilizer application in major potato growing areas [11]. Significantly increase in leaf potassium (K) content was indicated with applied K and showed positive correlation with tuber yield and negative correlation with frost score [12]. On the other hand potassium deficient potato crop is found less resistance to diseases and other pests, frost damage, low yielder and poor quality even though varying with variety [12]. Improved potato varieties that have been recently released in Ethiopia may differ in nutrient efficiency, and could have different optima of balanced macro-nutrient requirements for maximum yield of good quality seed tubers. However, there is limited information on the optimum requirements of balanced NK nutrition of improved potato varieties in the country. Hence, this study was initiated to investigate the main effects of nitrogen and potassium fertilizers on yield and yield component of potato varieties.

Materials and Methods

An experiment was conducted in 2014 and 2015 main cropping season at Holetta Research and Jeldu sub-center containing three factors (Jalenie, Gudanie and Belete potato varieties; 87, 110, 133 kg/ha nitrogen and; 0, 34.5, 69, 103.5 kg/ha potassium levels) which were managed in completely randomized block design with three replications. The fertilizers source used were urea (CO ([NH$_2$]$_2$) (46% N) and 195 kg /ha of DAP (46% P$_2$O$_5$) and Potassium nitrate (KNO$_3$=13% N and 46% K$_2$O). The Average PH H$_2$O (1:2.5), Exch. Acidity (cmol(+)/kg, Buck Density (g/cm³). Total available nitrogen b and organic matter in %. Available P (PPm) were 4.26, 0.38, 1.18, 0.15, 1.50 and 6.92 respectively for Holetta growing location. Planting was carried out using sprouted tubers having uniform size for the three varieties with 10 cm depth and 75 cm distance between rows and 30 cm between plants on 3 m × 3 m

plot size. The nitrogen fertilizer was applied in two split: half at planting and half at 45 days after planting as side dress at 5 cm around the root zones as reported in Teriessa [13,14]. Whole Phosphorus (as DAP) at rate of 195 kg /ha and whole Potassium fertilizer was applied during planting in the first year while whole potassium was applied at 45-55 days after planting in second growing season. Redoml at 2 kg/ha was applied to control late blight following incidence of 24-36 hours. Others cultural practices were done in the same practice as Holetta Agricultural research center recommended practice for potato production. Tuber harvesting was done once at proper physiological maturity (70% leaves withering). Data collected from those middle rows were plant height (cm), tuber fresh weight (ton) and dry weight (gm), marketable tuber number, marketable tuber yields (ton), unmarketable tuber number and weight (ton) and total tuber number. Data was subjected to analysis of variance using proc GLM (general linear model) procedure of SAS 9.2 software [15]. The means were compared with Duncan's Multiple Rage Test at 5% significance level.

Results

Based on 2014 data result, the potassium rates were affected both marketable and tuber yield ton/ha significantly (P<0.05) as indicated in Table 1. The highest marketable tuber yield and total tuber yield was obtained from 103.5 kg/ha even though the later was not statistically different from result of 34.5 and 69 kg/ha of potassium rates while lowest yield in ton/ha was produced from control (Table 1). The growing years also affected unmarketable tuber number and weight, total yield ton/ha and plant height highly significantly (P<0.01) (Table 2). In all parameters, the 2015 cropping season exceeded the 2014 production year (Table 2). This is probably due to lime application in Holetta growing location as whole and time of application of potassium rates in addition to other climatic variation between the two consecutive years. Likewise the growing location also influenced significantly (P<0.05) the total tuber number/plot, marketable tuber weight ton/ha, total yield ton/ha, unmarketable tuber number/plot and weight ton/ha, and plant height cm (Table 3). Regarding the location, Holetta research station

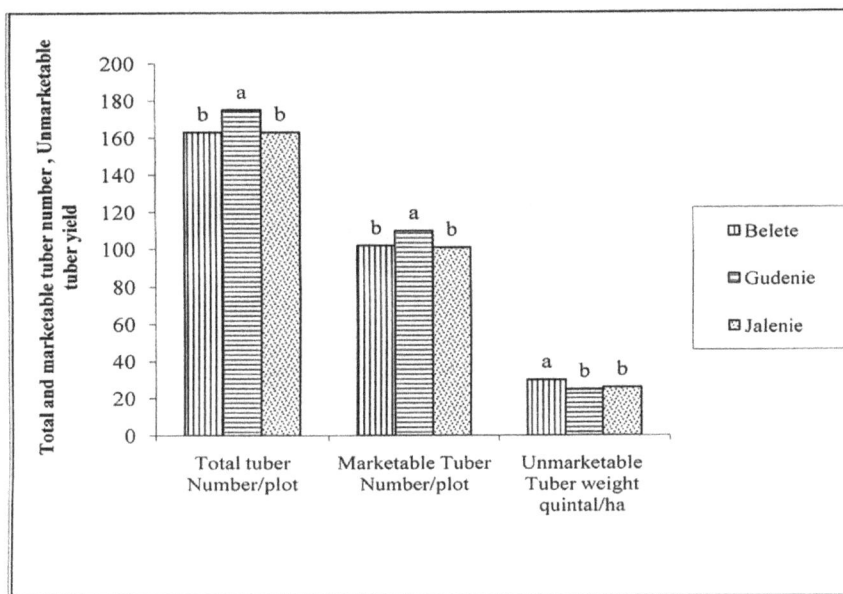

Figure 1: Varieties total and marketable tuber number, unmarketable tuber weight in quintal/ha.

Potassium levels (kg ha⁻¹)	Nitrogen levels (kg ha⁻¹)	Marketable Tuber Number/plot	Plant Height (cm)
0	87	100bc*	59.27bc**
	110	99bc*	59.3bc**
	133	110ab*	57.23c**
34.5	87	115a*	57.45bc**
	110	105abc*	57.57bc**
	133	101bc*	63.89a**
69	87	107ab*	59.93abc**
	110	99bc*	62.29ab**
	133	106abc*	58.77bc**
103.5	87	94c*	60.11abc**
	110	108ab*	62.66ab**
	133	109ab*	58.69bc**
CV% (0.05)		26.04	14.95

Means followed by same letter(s) are not significantly different from each other at p ≤ 0.05. * and ** -indicate means which are significantly different at 5 and 1% level of probability, respectively. CV % - Coefficient of Variance in percent.

Table 4: Interaction effects of potassium and nitrogen rates.

provided higher value in all above mentioned parameters than Jeldu research site (Table 3).

As indicated in Figure 1, the total and marketable tuber numbers affected highly significantly ($P<0.01$) by varieties. The higher total and marketable tuber number was obtained from variety Gudanie while the lower was obtained from both Belete and Jalenie varieties as the two produced statistically the same, numerically they are different.

In addition, the interaction of potassium and nitrogen was affected marketable tuber number significantly ($P<0.05$) and plant height highly significantly ($P<0.01$) (Table 4). The highest marketable tuber number was obtained from 34.5 kg potassium and 87 kg/ha nitrogen while lowest yield at 0 kg/ha and 87, 110 kg/ha nitrogen as well as 103.5 kg/ha potassium and 87 kg/ha nitrogen. The highest plant height was recorded at 34.5 kg/ha potassium and 133 kg/ha nitrogen; 69 kg/ha with 87 kg/ha and 110 kg/ha nitrogen; 103.5 kg/ha with 87 kg/ha and 110 kg/ha while lowest was at 0 kg/ha potassium with all rates of nitrogen.

The interaction of potassium and variety was highly significant

($P<0.01$) as indicated by Figure 2. Potato variety, Gudanie produced the highest total and marketable tuber number at 0 kg/ha while Belete at 34.5 kg/ha and Jalenie at 69 kg/ha potassium rates. The lowest total and marketable tuber number was obtained at 34.5 kg/ha from Gudanie, 69 kg/ha from Belete and 103.5 kg/ha potassium from Jalenie varieties, respectively.

As indicated in Table 5 above, the interaction of nitrogen rates and varieties affected highly significantly ($P<0.01$) the total and marketable tuber number. At 87 kg/ha nitrogen Belete variety yielded lower total and marketable tuber number than both Gudanie and Jalenie varieties, respectively. The latter varieties were produced statistically not different total and marketable tuber number. On the other hand, at 110 kg/ha nitrogen Belete and Gudanie produced highest total and marketable tuber number than Jalenie variety. But at 133 kg/ha nitrogen all varieties were not produced significantly different total and marketable tuber number.

The growing year, nitrogen and variety affected marketable tuber number significantly ($P<0.05$). Highest marketable tuber number was

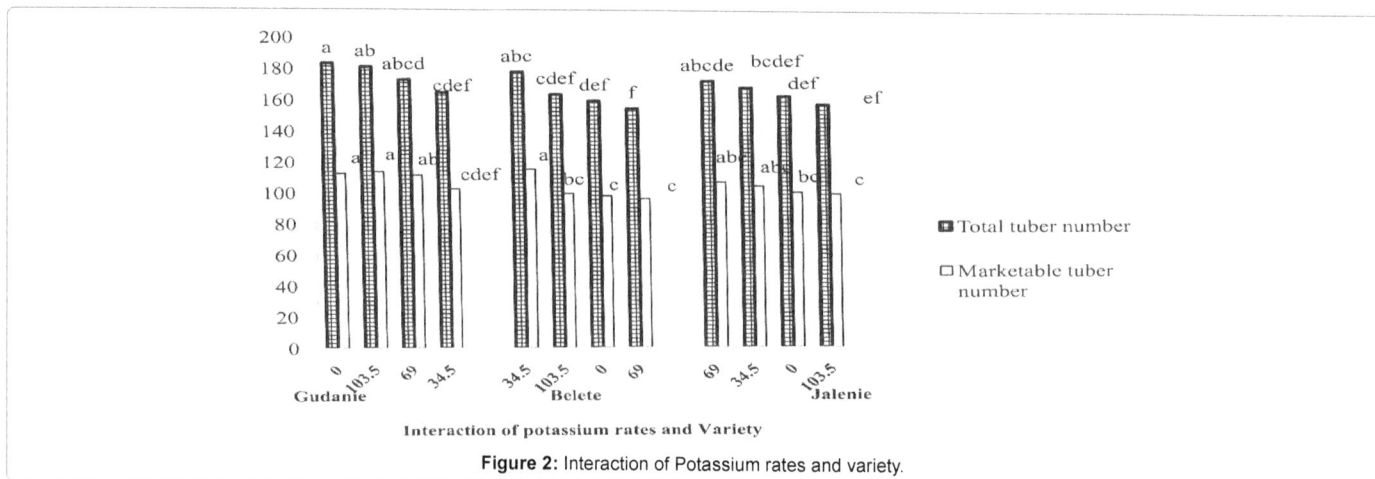

Figure 2: Interaction of Potassium rates and variety.

Nitrogen rates Kg ha⁻¹	Variety	Total Tuber Number/plot	Marketable Tuber Number/plot
87	Belete	154c**	93c**
	Gudanie	172b**	110ab**
	Jalenie	175b**	109ab**
110	Belete	168ab**	107ab**
	Gudanie	179a**	109ab**
	Jalenie	150c**	92c**
133	Belete	166ab**	105ab**
	Gudanie	174ab**	111a**
	Jalenie	163bc**	104ab**
CV% (0.05)		20.9	26.04

Means followed by same letter(s) are not significantly different from each other at p ≤ 0.05. **- indicate means which are significantly different at 1% level of probability. CV% - Coefficient of Variance.

Table 5: Interaction effect of nitrogen rates and varieties.

Nitrogen rates kg/ha	Variety	Marketable tuber number/plot	
		2014	2015
87	Belete	100bcde*	85e*
	Gudanie	111ab*	109abc*
	Jalenie	103abcd*	116a*
110	Belete	113ab*	102abcd*
	Gudanie	102abcd*	115ab*
	Jalenie	94cde*	89de*
133	Belete	104abcd*	105abc*
	Gudanie	117a*	106abc*
	Jalenie	103abcd*	104abcd*
CV% (0.05)		26.04	

Means followed by same letter(s) are not significantly different from each other at p ≤ 0.05. * - indicate means which are significantly different at 5% level of probability. CV% - Coefficient of Variance

Table 6: Interactions of year, nitrogen and variety.

obtained from Gudanie at 133 kg/ha nitrogen in 2014 and Jalenie at 87 kg/ha nitrogen in 2015. The lowest marketable tuber number was recorded from Jalenie at 110 kg/ha in 2014 and Belete at 87 kg/ha in 2015. The interaction of year, variety, potassium and nitrogen rates was significantly (P<0.05) affected marketable yield ton/ha. The maximum marketable yields for Belete, Gudanie and Jalenie in 2014 were 27.31, 27.04 and 23.97 ton/ha at 34.5 kg potassium and 110 kg/nitrogen, 69 kg/ha potassium and 133 kg/ha nitrogen, 69 kg/ha potassium and 133 kg/ha nitrogen, respectively.

Moreover, maximum marketable tubers of 2015 indicated that 29.68, 30.53 and 27.87 ton/ha at 34.5 kg/ha potassium and 133 kg/ha nitrogen, 69 kg/ha potassium and 110 kg/ha nitrogen, and 0 kg/

ha potassium and 133 kg/ha nitrogen in 2015 for Balete, Gudanie and Jalenie, respectively. The lowest marketable yield ton/ha (18.92 ton/ha) in 2014 was produced by Jalenie variety at 34.5 kg/ha potassium and 87 kg/ha nitrogen while it was 16.16 ton/ha at 0 kg/ha potassium and 110 kg/ha nitrogen in 2015.

The interaction of growing year and location was also significant (P<0.05) (Table 6). It affected all parameters measured. As indicated in Table 7, maximum total tuber number (175) and yield (37.55 ton/ha), unmarketable tuber number (68) and unmarketable tuber weight (4.31 ton/ha), and plant height (77.02 cm) were produced at Holetta location in 2015 while maximum marketable tuber yield (28.75 ton/ha) and 114 tuber number were obtained in 2014 growing year from the

Location	Total tuber Number/ plot		Total Yield ton/ha		Marketable Tuber Number		Marketable Tuber weight ton/ha		Unmarketable Tuber Number/plot		Unmarketable Tuber weight ton/ha		Plant Height cm	
Growing year	2014	2015	2014	2015	2014	2015	2014	2015	2014	2015	2014	2015	2014	2015
Holetta	165b**	185a**	31.8b**	43.3a**	114a**	99c**	28.75a**	27.04b**	49b**	87a**	2.7b**	5.9a**	74.44b**	79.61a**
Jeldu	169b**	148c**	20.06d**	27.18c**	96c**	109ab**	16.64d**	19.67c**	45bc**	39d**	0.9d**	1.3c**	37.94d**	47.07c**
CV%(0.05)	20.9		20.5		26.04		24.14		30.49		23.21		14.95	

Means followed by same letter(s) are not significantly different from each other at p ≤ 0.05. ** - indicate means which are significantly different at 1% level of probability. CV% - Coefficient of Variance

Table 7: Interaction of growing year and location.

same location. Minimum values measured were obtained from Jeldu location in 2014 year.

Discussion

Based on 2014 data result, the potassium rates affected the total tuber yield and marketable tuber yield significantly (P<0.05). The highest tuber yield of (21.72 kg/ha) and marketable yield (18.58 kg/ha) was obtained at 103 kg/ha potassium than other rates. This result is similar with the finding of Shahid and Moinuddin [12]. It also agrees with finding of which mentioned yield increment due to applied potassium through increase of number and size of tubers [11]. According to Bansal and Trehan [11] there is significant yield variability in relation to variety and growing location which make it consistent with present experiment as variety, growing year and location affected highly significantly yield and yield component of the potato product. Response of potato to NPK varies with variety, soil characteristics and geographical escarpment [16]. These results again correlated with the investigations results of Lamberti et al.; Vreugdenhil et al.; Trehan; Gumul et al. [17-20]. The interaction of potassium and nitrogen was also produced significantly different marketable tuber number and plant height while potassium and variety interaction provided significantly different total and marketable tuber number which have similar concept with the experimental results of Tally and Berug et al. [21,22]. The interaction of nitrogen and variety was also highly significant. Kathryn [23] reported that, increment of yield of potato with applied K and N. Similar concept was also noticed by Allison et al. [24]. Supporting investigations results were found in Anabausi et al.; Tawfik; Al-Moshileh et al.; Sharmila and Santhu [25-28]. On other hand, Locascio et al. [29] did not find an effect on the crop yield with increasing K rate which may be due to varieties used response to potassium fertilizer and growing location soil and climatic condition variation with present experiment. Ismail and Abu-Zinada [30] indicated that interaction of potassium and nitrogen significantly increased the tuber number and yield. There was variability to applied K by variety as mentioned in experimental results of Moinuddin et al.; Trehan [19,31]. According to Singh and Lal [32] the interaction of N and K has significantly affected the plant height and yield components. It also further mentioned ways of boosting yields such as increment of tuber size and number as well as total yield as a result of potassium and nitrogen rate applied. In addition, potato produced by potassium application has less weight loss and highest resistance to diseases. Moinuddin et al.; Ummar and Moinuddin [33,34] also observed increase in potato tuber yield due to potassium application up to 120 kg K_2O ha^{-1}. The report of Eleiwa [35] indicated increase yield with increasing NPK, the highest yield was attained at (120:80:100) rates, respectively. Moreover, significant response of Jalenie potato variety to potassium fertilizer is identified Geremew et al. [36].

Conclusion and Recommendation

According to these results, the main effect nitrogen was not affected any measured parameters of the varieties under experiment. But the interaction of potassium and nitrogen affected the marketable tuber number and plant height highly significantly. The interaction of potassium and variety showed significant influence on total and marketable tuber number. The interaction of growing year, nitrogen and variety also caused significant effect on marketable tuber number per plot. The interaction of growing year, potassium and nitrogen rates with variety was also brought significantly effect on tuber yield ton/ha. The maximum marketable tuber yield was attained in 2015 at 69 kg/ha potassium and 110 kg/ha nitrogen from Gudanie variety. Therefore, it is better to apply 69 kg/ha potassium and 110 kg/ha nitrogen to potato production for reasonable yield at sites similar to experimental locations. However, further research on time of application will be required in relation to locations and the rates of potassium and nitrogen.

References

1. FAOSTAT (2015) Data base of agricultural production. Food and Agriculture Organization, Rome, Italy.

2. Israel Z, Ali M, Solomon T (2012) Effect of Different Rates of Nitrogen and Phosphorus on Yield and Yield Components of Potato (Solanum tuberosum L.) at Masha District, Southwestern Ethiopia. International Journal of Soil Science, 7: 146-156.

3. CSA (2014) Agricultural sample survey: Report on area and production and farm management practice of belg season crops for private peasant holdings. Statistical Bulletin 532. Central Statistical Agency of Ethiopia (CSA), Addis Abeba, Ethiopia.

4. Yilma S (1991) The potential of true potato seed in potato production in Ethiopia. Actae Hurticultre 270: 389-394.

5. Woldegiorgis G (2013) Potato variety development strategies and methodologies in Ethiopia. In: Woldegiorgis G, Schulz S, Berihun B, (eds.), Seed Potato Tuber Production and Dissemination Experiences, Challenges and Prospects. EIAR and ARARI, Bahir Dar, Ethiopia, p: 45-59.

6. Recke H, Schnier HF, Nabwile S, Qureshi JN (1997) Responses of Irish potatoes (Solanum tuberosum L.) to mineral and Organic fertilizer in various agro-ecological environments in Kenya. Exp Agric 33: 91-102.

7. Kiiya WW, Mureithi JG, Kiama JM (2006) Improving production of Irish potato (Solanum tuberosum, L.) in Kenya: The use of green manure legumes for soil fertility Improvement. In: Mureithi JG, Gachene CKK, Wamuongo JW, Eilitta M (eds). Development and up scaling of Green manure legumes Technologies in Kenya. KARI.

8. Ochapa CO (1984) Introduction to Tropical soil science. Macmillan Intermediate Agriculture series.

9. Zelalem A, Tekalign T, Nigussie D (2009) Response of potato (Solanum tuberosum L.) to different rates of nitrogen and phosphorus fertilization on vertisols at Debre Berhan, in the central highlands of Ethiopia. African Journal of Plant Science 3: 16-24.

10. Honisch M, Hellmeier C, Weiss K (2002) Response of surface and sub-surface water quality to land use changes. Geoderma 105: 277-298.

11. Bansal SK, Trehan SP (2011) Effect of potassium on yield and processing quality attributes of potato. Karnataka J Agric Sci 24: 48-54.

12. Shahid U, Moinuddin M (2001) Effect of Sources and Rates of Potassium

Application on Potato Yield and Economic Returns. Better Crops International 15: 13-15.

13. Girma A, Ravishanker H (2008) The nutritive value of Potato tuber and N and P Effect on its Crude protein and Dry Matter Production. Ethiopian Journal of Crop Science 1: 28-37.

14. Tarriessa J (1997) A simple guide to potato production in Estern Ethiopia. Alemaya University of Agriculture, p: 50.

15. SAS Institute Inc. (2009) SAS 9.2. stored processes developer's guide. Carry. SAS. Institute Inc 2009 Cary, NC, USA.

16. Naz F, Ali A, Iqbal Z, Akhtar N, Asghar S, et al. (2011) Effect of different levels of NPK fertilizers on the proximate composition of potato crop at Abbottabad. Sarhad J Agric 27: 353-356.

17. Lamberti M, Geiselmann A, Conde PB, Escher F (2004) Starch transformation and structure development in production and reconstitution of potato flakes. LWT-Food Sci ence and Technology 37: 417-427.

18. Vreugdenhil D, Bradshaw J, Gebhardt C, Govers F, Mackerron DKL, et al. (2007) Potato biology and biotechnology. Advances and perspectives, p: 857.

19. Trehan SP (2007) Efficiency of potassium utilization from soil as influenced by different potato cultivars in the absence and presence of green manure (Sesbania aculeata). Advances Horti Sci 21: 156-164.

20. Gumul D, Ziobro R, Noga M, Sabat R (2011) Characterisation of five potato cultivars according to their nutritional and pro-health components. Acta Sci Pol Technol Aliment 10: 77-81.

21. Talley EA (1983) Protein nutritive values of Potatoes are improved by fertilization with nitrogen. American Potato journal 60: 35-39.

22. Berug R, Roer L, Tor T (1979) Amino acid composition of potato tuber s as influenced by nitrogen and potassium fertilization, year, location and variety. Meldinger fra Noges landbrukshogskole 58: 1-24.

23. Kathryn G (2014) Effect of Nitrogen and Potassium on Potato Yield, Quality and Acrylamide-Forming Potential. Electronic Theses and Dissertations, Paper 2170.

24. Allison MF, Flowler JH, Allen EJ (2001) Response of Potato (Solanum tuberosum. L) to Potassium Fertilizers. The Journal of Agricultural Sciences 136: 407-426.

25. Anabousi OAN, Hattar BI, Suwwan MA (1997) Effect of Rate and Source of

Nitrogen on Growth, Yield and Quality of Potato (Solanum tuberosum L) under Jordan Valley Conditions. Dirasat - Agriculture-Science 24: 242-259.

26. Tawfik AA (2001) Potassium and Calcium Nutrition Improves Potato Production in Drip-irrigation Sandy Soil. African Crop Science Journal 9: 147-155.

27. Al-Moshileh AM, Errebhi MA, Motawei MI (2005) Effect of various potassium and nitrogen rates and splitting methods on potato under sandy soil and arid environmental conditions. Emir J Food Agric 17: 1-9.

28. Sharmila BSM, Malarvizi PP, Thiyagarajan TM, Nagendrarao T (2006) Effect of Nitrogen, Potassium and Magnesium on Tuber Yield Grade and Quality of Potato Cv. Kufri Giriraj. 18th World Congress of Soil Science, Philadelphia, Pennsylvania, USA.

29. Locascio SJ, Bartz JA, Weingartner DP (1992) Calcium and potassium fertilization of Potato Grown in North Florida I. Effect on Potato Yield and Tissue Ca and K Concentration. American Potato Journal 69: 95-104.

30. Ismail AI, Abu-Zinada A (2009) Potato Response to Potassium and Nitrogen Fertilization Under Gaza Strip Conditions. Journal of Al Azhar University - Gaza (Natural Sciences) 11: 15-30.

31. Moinuddin SK, Bansal SK, Pasricha NS (2003) Influence of graded levels of potassium on growth, yield and economic parameters of potato. J Plant Nutr 35: 164-172.

32. Singh SK, Lal SS (2012) Effect of Potassium Nutrition on Potato Yield, Quality and Nutrient Use Efficiency under Varied Levels of Nitrogen Application. Potato J 39: 155-165.

33. Moinuddin SK, Bansal SK (2005) Growth yield and economics of potato in relation to progressive application of potassium fertilizer. J of Plant Nutr 28: 183-200.

34. Umar S, Moinuddin (2001) Effect of sources and rates of potassium application on potato yield and economic return. Better Crops International 15: 13-15.

35. Eleiwa ME, Ibrahim SA, Mohamed MF (2012) Combined effect of NPK levels and foliar nutritional compounds on growth and yield parameters of potato plants (Solanum tuberosum L.). African Journal of Microbiology Research 6: 5100-5109.

36. Geremew T, Ayalew A, Getachew A (2015) Response of Potato (Solanum tuberosum L.) to Potassium Fertilizer on Acid Soils of Wolmera and Gumer Weredas, in the High Lands of Ethiopia. Biology Agriculture and Healthcare 5: 1-6.

35

Production Potential of Faba Bean (*Vicia faba L.*) Genotypes in Relation to Plant Densities and Phosphorus Nutrition on Vertisols of Central Highlands of West Showa Zone, Ethiopia, East Africa

Tekle Edossa Kubure, Cherukuri V Raghavaiah* and Ibrahim Hamza

Department of Plant Sciences, College of Agriculture and Veterinary Sciences, PO Box 19, Ambo University, Ambo, West Shoa Zone, Ethiopia

Abstract

Faba bean is an important grain legume grown on vertisols of central high lands in Ethiopia, and is constrained by low yielding varieties, soil acidity, besides scanty information on optimum plant density and phosphorus nutrition. Field experiment was therefore conducted at Ambo University research farm during 2014 rainy season with the objective to determine optimum P rate and population densities for Faba bean (*Vicia faba* L.) genotypes grown on vertisols under rain fed conditions. The treatments comprised three genotypes (Hachalu, Walki and Local), three spacings (30 cm × 7.5 cm, 40 cm × 5.0 cm and 60 cm × 5.0 cm) and two phosphorus levels (0 kg P_2O_5/ha and 46 kg P_2O_5/ha), which were combined factorially and laid out in a split –split plot design with three replications. The results showed that the improved genotype, Walki (3,407 kg/ha) being comparable with Hachalu (3,037 kg/ha) gave substantially greater seed yield than the local cultivar (2,833 kg/ha). Seeding at 44 plants/m² resulted in significantly higher seed and biological yields (3,815 kg/ha and 7,894 kg/ha) than 50 plants/m² (3,074 kg/ha and 6,570 kg/ha) and 33 plants/m² (2,388 kg/ha and 4,696 kg/ha); although the harvest index was unaltered. Fertilization of faba bean with 46 kg P_2O_5/ha resulted in substantial increase in seed (3,531 kg/ha) and biological yields (7,172 kg/ha) over no fertilizer check (2,654 kg/ha seed and 5,602 kg/ha haulm yield). The harvest index tended to improve with P nutrition (49.7) over no phosphorus (47.4). Correlations worked between yield and growth and yield components showed a significant positive relation between seed yield and plant height, leaf area/plant, leaf area index, biological yield and seed yield/plant. Biomass yield is correlated with leaf area/plant, leaf area index, and plant height. Cultivation of improved varieties of faba bean Welki and Hachalu with a plant density of 44 plants/m² (30 cm × 7.5 cm spacing) was found to be better than the local cultivar in terms of yield and yield attributes. Phosphorus fertilizer application at 46 kg P_2O_5/ha improved the growth, yield and yield components of faba bean on Vertisols of high lands.

Keywords: Faba bean; Genotypes; Plant densities; Phosphorus nutrition; Root growth; Shoot growth; Nodulation; Yield; Yield components

Introduction

Faba bean *(Vicia faba L.)* is an important pulse crop grown in the highlands (1800-3000 masl) of Ethiopia, where the soil and weather are considered to be congenial for better growth and development of the crop. The crop takes the largest share of the area under pulses production in Ethiopia. The Central Statistical Agency [1] reported that faba bean is planted to 4.34% (about 5.38 lakh ha), of the grain crop area with an annual production of about 99.17 lakh quintals, 3.94% of the total grain production and yield of 18.42 q/ha in Ethiopia. It is a crop of manifold merits in the economy of the farming communities in the highlands of Ethiopia and serves as a source of food and feed and a valuable and cheap source of protein, apart from playing a significant role in soil fertility restoration in crop rotation through fixation of atmospheric nitrogen. It is a reliable source of income to the farmers, and earns foreign exchange to the country. It is mainly produced in Tigray, Gondar, Gojjam, Wollo, Wollega, Shoa and Gamo-Gofa regions of Ethiopia.

Ethiopia's faba bean export has moved north ward since the year 2000 and the major destinations are Sudan, South Africa, Djibouti, Yemen, Russia and USA, though its share in the countries pulse export is small [2,3]. The productivity of the crop is far below the potential and is constrained by several limiting factors. Faba bean is raised by farmers at varied row spacing resulting in reduced productivity. Plant density is an important production factor that ultimately determines the yield of crop per unit area. Besides, being a legume, it needs phosphorus for better root and nodule development, which is often ignored by farmers.

The low-yield potential of the indigenous cultivars is one of the most important production constraints [4,5]. Added to this, abiotic stress like water logging have all been identified as important production constraints [4]. Besides, in Vertisols of Ethiopian highlands, phosphorous is fixed and its non-availability is a limiting factor for better crop growth and development. It is known that Phosphorous nutrition plays a prime role in growth and development of roots and its role in nodulation, dry matter production, N fixation, and protein synthesis of leguminous crops is vital [6,7], though the nutrition for N is met through rhizobial fixation of atmospheric nitrogen. Hence a balanced nutrition of legumes gains significance to reap better yields, particularly under rain fed cropping conditions, where rain fall quantum and its distribution controls the crop production system. This warrants a need to generate information on the P needs of faba bean genotypes for better expression of their genetic potential in terms of growth, development and productivity [7]. Nitrogen and Phosphorous interact closely in affecting plant maturity. Phosphorous is implicated in speeding up maturity and enhancing root-shoot growth ratio, the formation of glycol-phosphate involved in photosynthesis,

*Corresponding author: Cherukuri V Raghavaiah, Department of Plant Sciences, Ambo University, Ethiopia, E-mail: cheruraghav@yahoo.in

respiratory metabolism, apart from being a part of nucleotides (RNA, DNA) and phospholipids of membranes and play a role in energy transfer metabolism (ATP, ADP, AMP, Pyro-phosphates) [8].

The high yield potential of faba bean has not been exploited in Ethiopia and the yield in the Southwestern Ethiopian highlands is generally low (1.3 ton ha^{-1}. compared to 1.8 ton ha^{-1} of world average) [9] This is largely attributed to raising low yielding local varieties, low soil pH and low P-availability in Nitisols [10,11] However, faba bean has the capacity to mobilize soil phosphorus by secretion of acids from its rhizosphere, and is therefore of important value in low-input crop rotation systems [12]. This apart, there is scanty information on the optimum plant density to reap better harvests. Therefore, the current investigation was made to evaluate the performance of faba bean genotypes as influenced by varied plant densities and phosphorus levels, in terms of growth parameters, yield and its components and root nodulation under rain-fed vertisol conditions of Ethiopia.

Materials and Methods

Description of study area

A field experiment was conducted under rain-fed conditions at Ambo University research farm during the main cropping season from July to December, 2014 on vertisol. Ambo is located in the West Shoa Zone of Oromia Regional State, Western Ethiopia, at about 115 km west of Addis Ababa, located within Coordinates: 8°59'N 37°50'E, and an altitude of 2068 m a s l. The seasonal total rain fall of the area during the crop season was 570 mm, with the average minimum and maximum temperature of 9.2°C and 27.08°C, mean relative humidity of 58.02% and a mean sun shine hours of 5.62 day^{-1}, respectively (Figure 1). The soil was characterized by Pellic vertisol [13]. The farm land preceding the current faba bean experiment was a fallow.

Genotypes tested

In the experiment, two improved high yielding genotypes of faba bean viz; Hachalu and Walki, which are adapted to the vertisols of the highland areas, were used. These varieties are recommended for highland vertisols of Ethiopia (Ambo, Adadi, Arsi, Robe, Sinja and etc.) with altitudes of 1900-2800 masl, having a rain fall of 700-1000 mm for planting in mid-June to early July, moderately resistant to chocolate spot and rust, released from HARC /EIAR in 2010 and 2008, respectively. The days to maturity of Hachalu and Walki are 122-156 and 133-146, respectively. The potential yields of Hachalu and Walki variety were 32-45 and 24-52 quintal ha^{-1} on research stations and 24-35 and 20-42 quintal ha^{-1} on farmer's field, respectively [14]. These two genotypes were compared with a local variety.

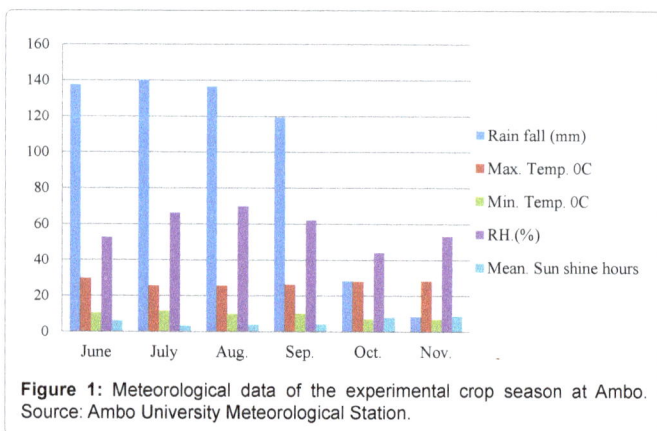

Figure 1: Meteorological data of the experimental crop season at Ambo. Source: Ambo University Meteorological Station.

Treatments and design

The treatments comprised three faba bean genotypes (Hachalu, Walki and a local cultivar) as main- plot treatments; three spacing's (30 cm × 7.5 cm, 40 cm × 5 cm and 60 cm × 5 cm) as sub- plot treatments and two phosphorous levels (0 and 46 kg P$_2$O$_5$ ha^{-1}) assigned to the sub-sub plot treatments, and were tested in split-split plot design with three replications.

Soil and plant analysis

Initially soil samples were collected from randomly selected sites of the experimental plots from a depth of 0–30 cm prior to cultivation and fertilizer application. The composite soil samples were analyzed for physical and chemical properties, using standard procedures for pH, CEC, organic carbon, total N, available P and K to evaluate the initial nutrient status. After the crop harvest, the soil of each treatment was analyzed for N, P, and K status. The soil physicochemical and plant tissue analysis was carried out at Holetta Agricultural Research Center (HARC), Soil and Plant Tissue Analysis Laboratory. The soil samples were air dried and ground to pass through 0.2 mm sieve for total N. Organic carbon was determined by wet digestion method as described [15]. Total N was estimated by Kjeldahl method as described[16]. Available P in soil was determined by Olsen method [17]. Soil texture was analyzed by Bouyoucos hydrometer method and Soil pH was measured by glass electrode pH meter. The plant samples (seed and haulm) were analyzed for N and P contents to calculate the nutrient uptake treatment wise. Protein content of seed was calculated based on N content of seed.

Field operations and crop management

After preparatory tillage, planting was carried out on 7 July 2014. Two seeds of each genotype were planted per hill at a depth of 2–3 cm using three spacing's (30 cm × 7.5 cm, 40 cm × 5 cm and 60 cm × 5 cm) to obtain 444,444, 500,000 and 333,333 plants ha^{-1}, respectively. Thinning was carried out two weeks after germination to maintain one plant/hill. The source of phosphorus was Di-ammonium phosphate which was applied pre planting as per treatments. Nitrogen was applied uniformly at 18 kg N ha^{-1} as a starter dose with urea and DAP being the sources of N. In the current experiment chocolate spot and rust diseases were observed which were managed using a fungicide Mancozeb 80 WP (Dithane M-45), at the rate of 2.5 kg a.i/ha at weekly intervals 3 times as foliar spray. The crop was harvested at physiological maturity on 19 November 2014, and subjected to sun drying to standardize the seed moisture content to 10 percent. Net plots were harvested leaving border rows to determine the per plot yields of beans and haulms.

Data collected

Data were collected on days to emergence, days to 50% flowering, days to maturity, plant height, leaf area /plant, leaf area index, leaf number/plant; root parameters like root length, nodule number, nodule biomass; bean yield and yield components like pods/plant, seeds/pod, pod length, pod weight, 1000 seed weight, and haulm yield from five randomly selected and tagged representative plants from each net plot. Leaf area was determined by multiplying leaf length and maximum leaf breadth of all fully opened leaves on five tagged plants. It was adjusted by a correction factor of LA estimation model is LA=-1.6923+(L*0.0161)+(W*0.0929)+(0.0062*L*W), where LA is leaf area (cm^2), L is leaflet length (cm), W is the leaflet width (cm). The leaf area of faba bean was calculated using the method formulated [18]. Leaf area index (LAI) was calculated as the ratio of total leaf area of five plants^{-1} (cm^2)/ area of land occupied by the plants.

Harvest Index (HI) was calculated as below.

Harvest index HI=Economic yield (kg/ha)/Biological yield (kg/ha) × 100

The data collected were subjected to the analysis of variance using SAS version 9.1.3 (2009), with model described below:

Model: $Y_{ijkl}=\mu+r_i+A_j+e_{ij}(a)+B_k+(AB)_{jk}+e_{ijk}(b)+C_l+(CA)_{lj}+(CB)_{lk}+(CAB)_{jkl}+e_{ijkl}(c)$

Where, μ=Population mean; r=replication; A=Main plot; ea=Main plot error; B=Sub plot; eb=Sub plot error; (AB, CA, CB, CAB)=Interaction; C=Sub-sub plot; ec=Sub-sub plot error. Wherever, the treatments showed a significant effect, the Duncan's multiple range test (DMRT) was used for means separation. The treatments were compared for their significance using calculated least significant difference (LSD) values at p=0.05.

Results and Discussion

Initial physico-chemical properties of the experimental soil

The pre- planting soil analysis showed that the texture of the soil is dominated by the clay fraction. On the basis of particle size distribution, the soil contains Sand 2.5%, Silt 22.5%, and Clay 75% (Table 1). The soil reaction (pH) of the experimental site is 6.79, which was near neutral. According to FAO, suitable pH range for most crops is between 6.5 and 7.5 in which total N availability is optimum (Table 1).

The organic carbon content of the soil was 1.17%. The soil has low organic carbon content, indicating moderate potential of the soil to supply nitrogen to plants through mineralization of organic carbon, low (0.070%) level of total N, indicating that the nutrient was not optimum for crop growth, low (5.94 ppm) available phosphorus, the K content was 1.63 meq/100 g; while the Cation exchange capacity was 1.17 meq/100 g soil [18].

Differential response of faba bean genotypes to plant density and phosphorus in terms of growth

Days to flowering: The number of days to 50% flowering of genotypes differed significantly, where the local genotype flowered late (47 days), in comparison with improved genotypes Hachalu (42 days) and Walki (43 days). This indicates that under rain fed conditions improved genotypes tend to flower earlier than traditional cultivars to complete life cycle early and escape any probable drought after anthesis, which is a desirable trait (Table 2). These results are in agreement with Gemechu et al. [19]. The days to 50% flowering was not altered either by different plant Densities or by phosphorus levels.

Physical properties				Chemical properties					
Particle size distribution (%)				pH	OC (%)	CEC (Meq/100 g)	Total N (%)	Av. P (ppm)	K (Meq/100 g)
Sand	Silt	Clay	Textural class						
			Clay						
2.5	22.5	75		6.79	1.17	1.17	0.07	5.94	1.63

Table 1: Selected physico-chemical properties of the experimental soil before sowing.

Treatments	Growth parameters									Flowering		Root parameters		
	Leaf number/plant at days			Plant height (cm) at days			Basal shoots/plant	Leaf area (cm)	LAI	Days to 50% flowering	Days to maturity	No. of nodules/plant	Nodule dry wt./plant (g)	Tap-root length (cm)
	60	90	110	60	90	110								
Genotype														
Hachalu	19.61a	29.33a	32.17ab	70.68a	97.87a	116.94a	1.33b	54.81a	5.83a	42.28b	123a	25.28a	0.56a	18.98b
Walki	19.55a	29.11a	31.61a	71.84a	99.33a	118.33a	1.05c	52.75a	5.71a	43.00b	123a	24.28a	0.57a	18.33b
Local	19.61a	29.39a	33.00a	67.71a	93.55a	111.40a	1.61a	44.75a	4.70a	46.67a	123a	24.94a	0.55a	21.37a
Mean	**19.59**	**29.28**	**32.26**	**70.08**	**96.92**	**115.56**	**1.33**	**50.77**	**5.41**	**43.98**	**123**	**24.83**	**0.57**	**19.56**
LSD (0.05 %)	NS	NS	NS	NS	NS	NS	0.25	NS	NS	3.07	NS	NS	NS	2.15
CV%	5.57	7.68	9.16	9.55	9.59	9.92	19.89	27.77	27.35	7.54		20	17.85	11.88
Pl.density/m²														
44	19.56a	29.44a	32.67a	69.61a	96.44a	115.29a	1.30a	52.82a	5.79b	43.94a	123a	24.72ab	0.56a	19.82a
50	19.39a	28.89a	31.28b	70.23a	96.74a	114.69a	1.33a	47.37a	6.63a	43.94a	123a	25.44a	0.57a	19.3a
33	19.83a	29.50a	32.83a	70.39a	97.56a	116.70a	1.36a	52.12a	3.81c	44.06a	123a	24.33b	0.54a	19.56a
Mean	**19.59**	**29.28**	**32.26**	**70.08**	**96.92**	**115.56**	**1.33**	**50.77**	**5.41**	**43.98**	**123**	**24.83**	**0.57**	**19.56**
LSD (0.05 %)	NS	NS	0.99	NS	NS	NS	NS	NS	0.74	NS	NS	0.96	NS	NS
CV%	5.26	5.32	4.23	2.8	2.72	3.06	13.02	17.15	18.74	0.6		5.33	17.85	9.55
P₂O₅ (kg/ha)														
0	19.3b	28.81b	31.78b	66.34b	91.64b	109.07b	1.30a	44.56b	4.78b	44.00a	123a	23.74b	0.54a	19.16a
46	19.93a	29.74a	32.74a	73.82a	102.2a	122.04a	1.36a	56.98a	6.05a	43.96a	123a	25.92a	0.58a	19.96a
Mean	**19.59**	**29.28**	**32.26**	**70.08**	**96.92**	**115.56**	**1.33**	**50.77**	**5.41**	**43.98**	**123**	**24.83**	**0.57**	**19.56**
LSD (0.05 %)	0.45	0.52	0.76	2.29	3.04	3.76	NS	6.42	0.71	NS	NS	4.28	NS	NS
CV%	4.05	3.12	4.11	5.73	5.49	5.69	16.36	22.1	22.79	0.92	0	5.62	14.13	11.35
Interaction														
G × Pl.				*	*	*								

Means in same columns followed by the same letter(s) are not significantly different, G=Genotype; Pl=plant density; *.=significant. LA calculated per individual leaf for genotype 2, 1.7 and 1.4 cm² and per plant leaves 54.81, 52.75 and 44.75 cm² for Hachalu, Walki and Local cultivar respectively and also the same procedure for plant densities and phosphorus level.

Table 2: Effect of genotypes, plant density and P levels on growth, flowering and root parameters of faba bean.

Days to maturity: Days to physiological maturity and grain filling period (from flowering to physiological maturity) for all the three test genotypes was non-significant and remained same at 123 days. Faba bean has indeterminate growth habit in that it flowers in phases depending up on the soil moisture. These results are in agreement with the findings of Gemechu et al. [19] reported that days to maturity and grain filling of faba bean genotypes ranged from 130-143 and 75-88 days, respectively. The days to maturity were not discernible in relation to plant densities and phosphorus nutrition (Table 2).

Plant height (cm): There was a significant interaction between genotypes and spacing on plant height at 60, 90 and 110 DAS (Table 2) where the genotype Welki with a spacing of 40 cm × 5.0 cm produced significantly taller plants at 60, 90 and 110 days after sowing than Hachalu and local varieties. At 60, 90 and 110 DAS Hachalu showed greater plant height at dense stands (30 cm × 7.5 cm) than at sparse plant stands. Welki possessed taller plants with 40 cm × 5.0 cm spacing than with 30 cm × 7.5 cm. Whereas, the local variety grew taller at sparse stand 60 cm × 5.0 cm than with dense stands of 30 cm × 7.5 cm and 40 cm × 5.0 cm. In this case, the local genotype grew better with wider spacing, while the improved genotypes responded to higher densities/ narrow spacing of 40 cm × 5.0 cm and 30 cm × 7.5 cm. Thus the taller stature of plants in dense crop stand is due to competition of plants for sun light (Table 3). The different plant densities could not bring about significant variations in plant height as measured at 60, 90 and 110 days after seeding. Application of 46 kg P_2O_5/ha resulted in substantial enhancement in plant stature as compared with no phosphorus application at 60, 90 and 110 days after sowing. This positive response could be attributed to better root growth due to phosphorus application that facilitated better absorption of soil moisture and nutrients resulting in taller plants in comparison with no phosphorus application (Table 2). These results are in agreement with Gemechu et al. [19].

Leaf area per plant (cm): The leaf area/plant, an indicator of assimilatory surface exposed to sunlight, varied with faba bean genotype, and the improved types Hachalu and Walki being comparable spread out more leaf area than that of local cultivar. Thus the improved genotypes could have more photosynthesis that could have larger bearing on productivity in terms of grain and biological yield. The rate of leaf production increased linearly as temperature increased where water supply is non-limiting [20]. Different spacing's exerted significant influence on the leaf area of genotypes where in 30 cm × 7.5 cm, 40 cm × 5.0 cm and 60 cm × 5.0 cm possessed leaf areas of 1.8, 1.54 and 1.63 cm² per leaf respectively, (52.82, 47.37 and 52.12 cm² per plant, respectively). There was relatively less leaf area in higher density stand obtained with 40 cm × 5.0 cm spacing compared to other densities. This could be due to plant competition for below and above ground growth factors, like sun light, soil moisture and nutrient ions in dense stands. Application of phosphorus fertilizer showed a significant influence on the leaf area

where in application of 46 kg P_2O_5/ha resulted in distinct enhancement in leaf area when compared with no Phosphorus fertilization (Table 2). This could be due to better growth and development of plants with application of phosphorus that ultimately lead to more leaf area than the unfertilized check.

Leaf area index (LAI): The genotype Hachalu (5.83) being on par with Walki (5.71) exhibited significantly greater leaf area index than the local cultivar (4.70) which in turn was at par with Walki. Thus higher LAI in improved genotypes which could lead to greater photosynthesis, is a desirable physiological trait that has the potential to enhance crop productivity; particularly under rain-fed conditions where often soil moisture is a limiting factor. The LAI of faba bean has been reported to vary from 3.5 to 7.0 until mid-June and there after they stabilized at 4-5 which supports our result [6]. Regarding the effect of plant densities on LAI, it was found that dense plant stands of 40 cm × 5.0 cm (6.63) being comparable with 30 cm × 7.5 cm (5.79) resulted in significant enhancement in LAI than sparse plant stand obtained with 60 cm × 5.0 cm spacing (3.81). This can be explained based on the fact that, though the leaf area/plant in dense stand was lower, the greater number of plants per unit land area could have contributed to more LAI; an indicator of more assimilatory surface per unit ground area. Higher densities were reported to favour early canopy development and increased light interception, and densities of 40 to 160 plants/m² produced LAI varying between 3.5 to 7.0 [6]. Application of 46 kg P_2O_5/ha resulted in substantially higher LAI (6.84) than that obtained without phosphorus application (5.73) (Table 2). Phosphorus nutrition often could have complimentary effect on better uptake of nitrogen from soil through its effect on better root proliferation that could lead to greater shoot growth in terms of leaf area and LAI that could have beneficial effects on source-sink relationships and greater dry matter partitioning efficiency. Faba beans have adopted to acquire P from low P soils and may indirectly make more P and K available for subsequent crops [21].

Number of leaves plant⁻¹: There was no significant difference between genotypes in the number of leaves/plant at 60, 90, 110 DAS. In general the leaf number/plant varied from 19-33 at 60 days through 110 days after sowing. The Faba bean plant population densities did not also bring about significant variation in the number of leaves/plant. However, application of Phosphorus showed a significant difference between phosphorus treated and untreated plots (Table 2).

Number of basal shoots plant⁻¹: The local cultivar produced significantly greater number of basal shoots (1.60) than Hachalu (1.33) and Walki (1.00). Different plant densities could not account for significant variations in the number of basal shoots/plant, though wider spacing tended to produce more shoots (1.36) than close spacings (1.30). Application of 46 kg P_2O_5/ha resulted in significant enhancement in the number of basal shoots/plant (1.40) in comparison with no phosphorus application (1.30) (Table 2).

Plant height (cm)												
Genotype	60 DAS				90 DAS				110 DAS			
	44 pl/m²	50 pl/m²	33 pl/m²	Mean	44 pl/m²	50 pl/m²	33 pl/m²	Mean	44 pl/m²	50 pl/m²	33 pl/m²	Mean
Hachalu	72.13	69.80	70.10	70.68	99.92	96.46	97.23	97.87	119.47	114.97	116.40	116.94
Walki	69.37	74.83	71.33	71.84	96.08	103.00	98.90	99.33	114.77	121.80	118.43	118.33
Local	67.33	66.07	69.73	67.71	93.32	90.78	96.56	93.55	111.63	107.30	115.27	111.40
Mean	69.61	70.23	70.39	70.08	96.44	96.74	97.56	96.92	115.29	114.69	116.70	
LSD (0.05)	1.75			2.35			3.16					

Table 3: Interaction effect of genotype and plant density on plant height of faba bean.

Root parameters of faba bean as influenced by genotypes, plant density and phosphorus

Root nodule number plant[-1]: The genotype Hachalu produced highest number of nodules/plant (25.3), closely followed by the local cultivar (24.9) and Walki (24.3) which had recorded the least number of nodules (Table 2). Thus the test genotypes were similar in nodulation and atmospheric nitrogen fixation, which is of greater significance for meeting the nitrogen requirement of the crop. Variability of faba bean genotypes for nutrient uptake and yield response has been reported by Balaban [22] increased root and shoot dry weight with fertilizer application [23]. A density of 50 plants/m² resulted in more nodulation (25.4) than lower density (24.3) on vertisols under rain fed conditions. Application of phosphorus played a significant role in enhancing the root nodulation (25.9) of faba bean in comparison with no phosphorus (23.7). Although in legumes, nodulation ability is a genetic character, it is often influenced by crop nutrition, especially of phosphorus which is implicated in better growth and development of root system. These findings corroborate with the results of Asfaw [24] reported variability of faba bean genotypes in terms of number of nodules/plant (>30) and nodule dry weight/plant (>2 g) under rain-fed situations of Ethiopia. Increased nodulation and yield due to application of 50 kgP₂O₅/ha has also been reported in soils having 3.5 and 2.0 ppm P at ICARDA [25]. These findings are in consonance with those reported by Haque et al. [26] observed an increase in dry matter, nodulation, N fixation, P-uptake and protein yield of legumes. Faba bean requirement of P is reported to be high due to strong energy expenditure utilized during nodule formation and operation [21].

Nodule dry weight/plant (g): The dry weight of nodules/plant was more in improved genotypes Hachalu (0.56 g) and Welki (0.57 g) than the local cultivar (0.55 g) (Table 2). Looker also reported that root growth differs between varieties of faba bean, and both drought and water logging leads to fewer nodules on roots and hence less N fixation [27,28]. The root nodule biomass of faba bean has not been distinctly influenced by the plant densities either, which is in tandem with the nodule density per plant. Application of phosphorus has no discernible influence on the dry biomass of root nodules as compared with no phosphorus application.

Tap-root length (cm): It was observed that the local cultivar produced significantly longer tap roots (21.4 cm) than those of improved genotypes Hachalu (18.9 cm) and Walki (18.3 cm), indicating its adaptation to drought (Table 2). With regards to plant density, the variations in tap root length were not discernible. Application of Phosphorus however, tended to produce marginally longer tap root (19.9 cm) than no Phosphorus (19.1 cm) though it was statistically non-significant.

Productivity and its components of faba bean as influenced by genotypes, plant density and phosphorus

Pods plant[-1]: The local cultivar produced more number of pods (21.6) compared to the improved genotypes that produced 19.2 and 18.6 pods/plant in Hachalu and Walki, respectively. With regard to plant densities, the number of pods/plant did not differ significantly, though 30 × 7.5 cm spacing tended to produce more pods/plant. Faba bean did not respond to phosphorus application in terms of pod number/plant. The pod number/plant is a genetic character and is less influenced by the environment in terms of plant density and P nutrition (Table 4) results are in agreement with Gemechu et al. [19] reported 3 to 15 pods/plant for faba bean genotypes in Ethiopia. There was report of an increase in plant height, fresh weight, and pod number with 80 kgP₂O₅/ha [29].

Treatments	Pods/plant	Pod length (cm)	Pod weight/ plant (g)	No. of Seeds/ pod	1000 Seed wt. (g)	Seed yield (kg/ ha)	Biological yield (kg/ha)	Harvest index (%)
Genotype								
Hachalu	19.17b	6.19a	24.27a	2.89a	650.06a	3037.0ab	6361.4a	48.17a
Welki	18.61b	6.04a	23.47a	2.83a	523.89b	3407.4a	6973.5a	48.42a
Local	21.56a	4.58b	20.90a	2.94a	344.06c	2833.3b	5825.7a	49.10a
Mean	**19.78**	**5.61**	**22.88**	**2.89**	**506**	**3092.6**	**6386.9**	**48.57**
LSD (0.05 %)	**2.16**	**0.99**	NS	NS	**34.7**	**564.4**	NS	NS
CV%	11.79	19.19	17.25	15.08	7.41	19.72	19.92	4.58
Pl. density/m²								
44	20.00a	5.61a	22.87ab	2.72b	508.89a	3814.8a	7894.2a	48.93a
50	19.50a	5.52a	21.46b	2.94a	517.33a	3074.1b	6570.2b	47.32a
33	19.83a	5.69a	24.31a	3.00a	491.78b	2388.9c	4696.3c	49.45a
Mean	**19.78**	**5.61**	**22.88**	**2.89**	**506**	**3092.6**	**6386.9**	**48.57**
LSD (0.05 %)	NS	NS	1.75	0.22	16.93	167.98	367.29	NS
CV%	10.53	8.17	10.55	10.38	4.6	7.48	7.91	9.9
P₂O₅ (kg/ha)								
0	19.33a	5.43b	21.75b	2.81a	492.16b	2654.3b	5601.7b	47.45a
46	20.22a	5.78a	24.01a	2.96a	519.82a	3530.9a	7172.1a	49.69a
Mean	**19.78**	**5.61**	**22.88**	**2.89**	**506**	**3092.6**	**6386.9**	**48.57**
LSD (0.05 %)	NS	0.23	2.4	0.19	17.5	251.14	465.01	NS
CV%	11.45	7.35	18.35	11.54	6.05	14.2	12.73	11.64
Interaction								
G × Pl						*	*	
G × P		*				*		
Pl × P								
G × Pl × P		*		*				

Means in same columns followed by the same letter(s) are not significantly different, *.=Significant.

Table 4: Effect of genotypes, plant density and P levels on yield and yield components of Faba bean.

Pod length (cm): Improved genotypes Hachalu (6.2 cm) and Walki (6.00) produced distinctly longer pods than the local cultivar (4.6 cm). Pod length is a heritable genetic character which has a bearing on ultimate seed yield of faba bean. Akin to the number of pods/plant, the length of pods was not substantially influenced by the different plant densities and varied marginally from 5.5 cm to 5.7 cm with the different spacings. Application of 46 kg P_2O_5/ha resulted in significant enhancement in the length of pods (5.8 cm) as compared with no phosphorus application (5.43 cm); thus indicating that pod length can be altered by P fertilization in faba bean. Significant interaction between genotypes and phosphorus on pod length showed that the genotype Walki produced significantly longer pods (6.49 cm) than the rest of the treatment combinations with the application of 46 kg P_2O_5/ha (Table 5).

There was also significant interaction between genotypes, plant density and phosphorus on pod length. The genotype Walki produced significantly longer pods (6.83 cm) than the rest of the treatment combinations with the application of 46 kg P_2O_5/ha and a plant density of 44 plants/m² (Table 6).

Pod weight/plant (g): Improved genotype Hachalu produced highest pod weight/plant (24.3 g) followed by Walki (23.5 g) and the local cultivar (20.9 g) (Table 4). The plant densities showed significant influence on pod weight/plant where wider spacing of 60 cm × 5.0 cm resulted in substantially greater pod weight/plant (24.3 g) than 30 cm × 7.5 cm (22.9 g) and 40 cm × 5.0 cm spacing (21.5 g). This indicates that pod weight/plant can be altered by plant spacing. The greater pod weight/plant recorded with low plant density could be attributed to less competition for growth resources like soil moisture, nutrients and sun light as compared to the dense stands. Faba bean exhibited significant response in terms of greater pod weight/plant with application of 46 kg P_2O_5/ha (24.0 g) compared to 21.7 g obtained with no phosphorus. This signifies the beneficial role of phosphorus in improving the pod weight of faba bean (Table 4).

Seeds pod[-1]: The seeds/pod did not vary significantly among the genotypes, while it tended to vary with plant density and phosphorus nutrition. Among the spacings, wider spacing tended to improve the seeds/pod (3.0) as compared with narrow spacings (2.7). On the other hand, phosphorus application tended to improve seeds/pod (3.0) when compared with no phosphorus (2.8). The number of seeds/pod varied distinctly when Hachalu fertilized with 46 kg P_2O_5/ha and sown at wider spacing of 60 cm × 5.0 cm (3.33) than when, sown at 30 cm × 7.5 cm (2.33) as evident from interaction (Table 7). Interaction effect of genotype, plant density and phosphorus levels on seed pod[-1] of faba bean (Table 7) showed that local cultivar was relatively superior (2.94) to Hachalu (2.89) and Welki (2.83). With the application of phosphorus fertilizer and sowing at a spacing of 60 cm × 5.0 cm Hachalu produced more seeds/pod (3.33) than Walki [3] and local cultivar. In general, as the seeds/pod is a genetic character, it is less influenced by either management or P nutrition. These results are in agreement with Gemechu et al. [19] reported that seeds pod[-1] of faba bean genotypes ranged from 2-3.

Test weight of seed (g): Among the genotypes Hachalu recorded substantially greater test seed weight (650 g) compared to Walki (524 g) and the local cultivar which recorded the least weight (344 g) (Table 4). This elucidates the greater source –sink relation of the improved

Genotype	Phosphorus level		Mean
	0 P_2O_5 kg/ha	46 P_2O_5 kg/ha	
Hachalu	6.14	6.24	6.19
Walki	5.59	6.49	6.04
Local	4.56	4.61	4.58
Mean	5.43	5.78	
LSD (0.05)	0.29		

Table 5: Interaction effect of Genotype with P level on Pod length (cm) of faba bean.

Genotype	Pod length (cm)						Mean
	0 kg P_2O_5/ha			46 kg P_2O_5/ha			
	44 plants/m²	50 plants/m²	33 plants/m²	44 plants/m²	50 plants/m²	33 plants/m²	
Hachalu	6.60	5.67	6.17	5.77	6.33	6.63	6.19
Walki	5.50	5.73	5.53	6.83	6.53	6.10	6.04
Local	4.47	4.40	4.80	4.47	4.43	4.93	4.58
Mean	5.52	5.27	5.50	5.69	5.77	5.89	
LSD (0.05)	0.69						

Table 6: Interaction effect of genotype × plant density × phosphorus levels on pod length of faba bean.

Genotype	0 kg P_2O_5			46 kg P_2O_5			Mean
	44 plants/m²	50 plants/m²	33 plants/m²	44 plants/m²	50 plants/m²	33 plants/m²	
Hachalu	3.00	2.67	3.00	2.33	3.00	3.33	2.89
Walki	2.33	3.00	2.67	3.00	3.00	3.00	2.83
Local	2.67	3.00	3.00	3.00	3.00	3.00	2.94
Mean	2.67	2.89	2.89	2.78	3.00	3.11	
LSD (0.05)	0.41						

Table 7: Interaction effect of genotypes, plant density and phosphorus levels on Seeds pod[-1] of faba bean.

genotypes than that of traditional cultivars. The plant densities have shown significant influence on the test weight of seed, where high density of 44 plants/m² (508 g) and 50 plants/m² (517 g) had seeds of greater weight than low density planting at 33 plants/m² (492 g). Phosphorus fertilization at 46 kg P_2O_5/ha significantly improved the test seed weight (520 g) over no phosphorus (492 g). These results are in agreement with Gemechu et al. [19] found that 1000 seed weight of faba bean genotypes ranged from 249-553 g

Biological yield (kg ha⁻¹): The biological yield followed a trend similar to that of seed yield/plot. Improved genotype Walki (6973.5 kg/ha) remaining comparable with Hachalu (6361.4 kg/ha) produced significantly higher biological yield/plot than that of local variety (5825.7 kg/ha). Higher plant densities represented by closer spacing of 30 cm × 7.5 cm (7894.2 kg/ha) and 40 cm × 5.0 cm (6570.2 kg/ha) produced superior biological yield kg/ha to that of wider spacing 60 cm × 5.0 cm (4696.3 kg/ha). Significantly greater biological yield kg/ha has been obtained with the application of 46 kgP_2O_5/ha (7172.1 kg/ha) than that obtained in no phosphorus plots (5601.7 kg/ha). Significant interaction between genotype and spacing on biological yield/ha showed that irrespective of genotype there was reduction in biological yield/ha with reduced plant density. The genotype Walki grown with a spacing of 30 cm × 7.5 cm produced significantly higher biological yield of 8,367 kg/ha than the other genotype spacing combinations. The next best is Hachalu raised with 30 cm × 7.5 cm spacing (7,915.77 kg/ha) and the local cultivar (7399.74 kg/ha) (Table 8).

Seed yield (kg ha⁻¹): Among the faba bean genotypes, Walki (3407 kg/ha) and Hachalu (3037 kg/ha) gave significantly higher productivity than the local genotype (2833 kg/ha). The percentage yield enhancement of Walki and Hachalu over local cultivar was 20 and 7.2%, respectively. The superior performance of Walki could be attributed to more length of pods, greater pod weight/plant, higher seed weight/plant, more seeds/plant, higher test seed weight, ultimately leading to substantial enhancement in seed yield/plot. Bianchi et al. also reported that the number of seeds/pod and seed weight are most stable components and seed weight varies between cultivars and range from 0.1 g to 2.4 g/seed. Among the plant densities, seeding at 30 cm × 7.5 cm (44 plants/m²) resulted in superior seed productivity (3814.8 kg/ha) than that obtained with 40 cm × 5.0 cm (50 plants/m²) (3074.1 kg/ha) and 60 cm × 5.0 cm (33 plants/m²) (2388.9 kg/ha). Significant interaction between genotype × plant density on seed yield revealed that by and large, all the genotypes yielded maximum with 30 cm × 7.5 cm spacing, closely followed by 40 cm × 5.0 cm spacing, while their yields significantly dwindled with wider spacing of 60 cm × 5.0 cm. The genotype Walki seeded at a spacing of 30 cm × 7.5 cm surpassed (4166.7 kg/ha) the rest of the genotype × spacing combinations in seed productivity (Table 9). The next best was Hachalu grown at 30 cm × 7.5 cm (3777.8 kg/ha) in terms of productivity.

Fertilizing the crop with 46 kg P_2O_5/ha resulted in significantly greater seed yield (3531 kg/ha) than that without P. fertilizer (2654 kg/ha) in vertisols. Application of 80 kgP_2O_5/ha has been reported to give 13 t/ha green pods of faba bean [15,19,22,29]. Based on results of 31 fertilizer trials (1967-1973) on faba bean concluded that response to phosphorus was high, increasing P from 36 to 72 kg/ha increased yield by 9.8% and 15.7% over control [14]. There was significant interaction between genotype × phosphorus on seed yield, where Walki fertilized with 46 kg P_2O_5/ha gave greater productivity (4074 kg/ha) than the rest of the combinations. The next best was Hachalu grown with 46 kg P_2O_5/ha (3407 kg/ha). Thus the new genotypes responded better to phosphorus application than the local cultivar (3111 kg/ha) (Table

Genotype	Biological yield (kg/ha)			
	44 plants/m²	50 plants/m²	33 plants/m²	Mean
Hachalu	7915.77	6465.6	4702.74	6361.37
Welki	8367	7630.58	4923.01	6973.53
Local	7399.74	5614.34	4463.11	5825.73
Mean	7894.17	6570.17	4696.29	
LSD (0.05)	400.85			

Table 8: Interaction effect of genotype with plant density on biological yield of faba bean.

Genotype	Population density			
	44 plants/m²	50 plants/m²	33 plants/m²	Mean
Hachalu	3777.78	2944.44	2388.89	3037.04ab
Walki	4166.67	3555.56	2500	3407.41a
Local	3500	2722.22	2277.78	2833.33b
Mean	3814.81a	3074.07b	2388.89c	
LSD (0.05)	205.76			

Table 9: Interaction effect of genotype and plant density on seed yield (kg/ha) of faba bean.

Genotype	Phosphorus level		
	0 P_2O_5 kg/ha	46 P_2O_5 kg/ha	Mean
Hachalu	2666.67	3407.41	3037
Walki	2740.74	4074.08	3407.4
Local	2555.55	3111.11	2833.33
Mean	2654.3	3530.9	
LSD (0.05)	198.4		

Table 10: Interaction effect of genotype with phosphorus on seed yield (kg/ha) of faba bean.

10). Seed productivity, the culmination of vegetative and reproductive growth and developmental metabolic processes that have been taken place since the time of seeding through the maturity phases in the crop life cycle, is the economic product in which the farmer is interested. Seed yield of faba bean is a product of number of plants/m², number of pod bearing nodes/plant, pods/node, seeds/pod and seed weight [30-35].

Harvest Index (HI): The harvest index, a measure of translocates partitioning efficiency, revealed that the genotypes did not differ in harvest index. Low density planting 60 × 5.0 cm (33 plants/m²) resulted in higher harvest index (49.45) over high density seeding at 30 cm × 7.5 cm (44 plants/m²) (48.93) and 40 × 5.0 cm (50 plants/m²) (47.32) which gave comparable harvest index (Table 4). The application of phosphorus tended to improve the harvest index (49.69) of faba bean when compared with no P application (47.45) though the variation was not discernible. Application of 50 kgP_2O_5 has been reported to enhance nodulation and yield of faba bean in soils having 3.5 and 2.0 ppm Pat ICARDA [25].

Correlation between Seed yield, growth and yield components: Correlations computed between growth, yield and yield components showed a significant positive relation between seed yield and plant height at different stages, leaf area/plant, leaf area index, biological yield and seed yield/plant. Biomass yield was correlated with leaf area/plant, LAI and plant height (Table 11).

Conclusion

From the foregoing discussion, it could be concluded that the

Variables	LFA	LFB	LFC	PHA	PHB	PHC	NTILL	LA	LAI	DF	NNOD	DwtNW	TRL	PPPL	PL	Pwt	Swt	SPP	SPPL	BMY	SYLDP	TSW	SYLDH	HI
LFA	—	0.920***	0.697***	0.343*	0.365**	0.389**	-0.027ns	0.176ns	0.017 ns	0.057 ns	0.147ns	0.087 ns	0.035 ns	0.039ns	0.219ns	0.50***	0.492**	0.292*	0.307*	0.056ns	0.087ns	0.006ns	0.087ns	0.039ns
LFB		—	0.773***	0.328*	0.349*	0.368**	0 ns	0.159ns	0.031ns	0.052ns	0.272*	0.159ns	0.048ns	0.052ns	0.184ns	0.447**	0.429**	0.160ns	0.280*	0.106ns	0.128ns	0.023ns	0.128ns	0.036ns
LFC			—	0.292*	0.328*	0.374**	0.133ns	0.089ns	-0.099ns	0.161ns	0.288*	0.270*	0.154ns	0.144ns	0.012ns	0.320*	0.332*	0.177ns	0.380**	0.008ns	0.057ns	-0.229ns	0.057ns	0.186ns
PHA				—	0.993***	0.958***	0.121 ns	0.63***	0.517***	-0.36ns	0.273*	0.341*	0.031ns	0.042ns	0.404**	0.52**	0.549***	0.120ns	0.291*	0.565***	0.603***	0.320*	0.603***	-0.15ns
PHB					—	0.985***	0.128ns	0.660***	0.520***	-0.381**	0.296*	0.332*	0.032ns	0.045ns	0.425**	0.568***	0.594***	0.130ns	0.304*	0.560***	0.597***	0.329*	0.597***	-0.144ns
PHC						—	0.135ns	0.677***	0.515***	-0.396**	0.323*	0.311*	0.030ns	0.045ns	0.448**	0.620***	0.646***	0.146ns	0.317*	0.539***	0.574***	0.335*	0.574***	-0.131ns
NTILL							—	0.159ns	0.071ns	0.268ns	0.162ns	-0.075ns	0.204ns	0.205ns	-0.27*	0.037ns	0.044ns	0.106ns	0.369**	0.034ns	0.007ns	-0.225ns	0.007ns	-0.125ns
LA								—	0.772***	-0.529***	0.389**	0.065ns	-0.116ns	-0.117ns	0.503**	0.690***	0.687***	0.088ns	0.209ns	0.496**	0.525***	0.498**	0.525***	-0.136ns
LAI									—	-0.467**	0.409**	0.158ns	-0.14ns	-0.151ns	0.378**	0.405**	0.404**	-0.024ns	0.101ns	0.742***	0.708***	0.469**	0.708***	-0.427***
DF										—	-0.247ns	-0.301*	0.310*	0.315*	-0.775***	-0.387**	0.396**	0.087ns	0.356**	-0.262ns	-0.264ns	-0.815***	-0.264ns	0.045ns
NNOD											—	0.450**	-0.026ns	-0.031ns	0.188ns	0.361**	0.351**	0.054ns	0.016ns	0.239ns	0.267ns	0.184ns	0.267ns	-0.066ns
DwtNW												—	-0.036ns	-0.052ns	0.178ns	0.080ns	0.091ns	0.032ns	0.023ns	0.194ns	0.175ns	0.074ns	0.175ns	-0.175ns
TRL													—	0.994***	-0.454**	-0.219ns	-0.194ns	-0.19ns	0.264ns	0.075ns	0.041ns	-0.402**	0.041ns	-0.204ns
PPPL														—	-0.434**	-0.204ns	-0.179ns	-0.188ns	0.270*	0.072ns	0.039ns	-0.391**	0.039ns	-0.204ns
PL															—	0.584***	0.582***	0.282*	-0.2ns	0.245ns	0.303*	0.756***	0.303*	0.116ns
Pwt																—	0.989***	0.281*	0.315*	0.258ns	0.291*	0.452**	0.291*	-0.044ns
Swt																	—	0.298*	0.318*	0.252ns	0.285*	0.446**	0.285*	-0.041ns
SPP																		—	0.264ns	-0.2ns	-0.173ns	-0.05ns	-0.173ns	0.101ns
SPPL																			—	-0.074ns	-0.048ns	-0.487**	-0.048ns	0.066ns
BMY																				—	0.959***	0.255ns	0.959***	-0.545***
SYLDP																					—	0.260ns	1***	-0.311*
TSW																						—	0.260ns	-0.057ns
SYLDH																							—	-0.311*
HI																								—

LFA, LFB, LFC=No of leaves at 60, 90, 110 DAS (Days after sowing); PA, PHB, PHC=plant height at 60, 90, 110 DAS; NTILL=Basal shoot (Tiller); LA=Leaf area; LAI=Leaf area Index; DF=50 % Flowering; DM=Days to Maturity; NNOD= No. of Nodule/plant; DwtNW=Nodule Dry wt./plant; TRL=Tap-root length; PPP= Pod/plant; PL=Pod Length; Pwt=Pod Weight/Plant; Swt=Seed wt./Plant; SPP=No. of Seed/pod; SPPL=No. of Seed/plant; BMY=Biological Yield/plot; SYLDP=Seed Yield/plot; TSW=1000 Seed wt; SYLDH=Seed Yield/ha; HI=Harvest Index

Table 11: Correlation coefficient (r) relationship between grain yields with various phenological, growth and yield components of faba bean.

new improved genotypes Welki and Hachalu out yielded the local traditional variety of faba bean, the percent yield enhancement being 20 and 7.2, respectively. Faba bean genotypes responded positively to phosphorus, where Welki fertilized with 46 kgP_2O_5/ha gave greater yield than Hachalu at the same level of phosphorus, both being superior to local variety. The genotypes exhibited differential behaviour in terms of seed yield when raised at different plant densities, where all the genotypes yielded maximum when raised with 44 pl/m^2, followed by 50 pl/m^2 while their yield dwindled with 33 pl/m^2. All the genotypes performed better when sown at 44 pl/m^2 density along with application of 46 kg P_2O_5/ha under rain fed vertisol conditions of central high lands of Ethiopia.

References

1. Central Statistical Agency (CSA) (2013) Agricultural sample survey. Report on area and production for major crops (private peasant holdings, meher season). Addis Ababa, Ethiopia.

2. Amanuel G, Tanner DG, Assefa T, Duga D (1993) Observation on wheat and barley-based cropping sequences trials conducted for eight years in South-eastern Ethiopia. Paper presented at the 8th Regional Workshop for Eastern, Central and Southern Africa, Kampala, Uganda.

3. Lupwayi NZ, Kennedy AC, Chirwa RM (2011) Grain legume impacts on soil biological processes in sub-Saharan Africa. African Journal of Plant sciences 5: 1-7.

4. Asfaw T, Tesfaye G, Beyene D (1994) Genetics and breeding of faba bean. Cool-season food legumes of Ethiopia. Proc. First National cool-season food legumes review conference, Addis Ababa, Ethiopia.

5. Yohannis D, Kiros H, Yirga W (2015) Inoculation, phosphorous and zinc fertilization effects on nodulation, yield and nutrient uptake of Faba bean (Vicia faba L.) grown on calcaric cambisol of semiarid Ethiopia. J of soil science and environmental management 6: 9-15.

6. Igwilo N (1982) Nodulation and Nitrogen accumulation in field beans (Vicia faba L). J Agric Sci Camb 98: 269-288.

7. Tekle EK, Raghavaiah CV, Chavan A, Ibrahim H (2015) Effect of faba bean (Vicia faba L.) genotypes, plant densities and phosphorus on productivity, nutrient uptake, soil fertility changes and economics in central high lands of Ethiopia. Int J of Life Sciences 3: 287-305.

8. Salisbery FB, Ross CW (1992) Plant Physiology. 4th edn. Wadsworth Cengage Learning, New Delhi.

9. FAOSTAT (2010) Statistical databases and data sets of the Food and Agriculture Organization of the United Nations.

10. Agegnehu G, Ghizaw A, Sinebo W (2006) Yield performance and land-use efficiency of barley and faba bean mixed cropping in Ethiopian highlands. European Journal of Agronomy 25: 202–207.

11. Agegnehu G, Chilot Y (2009) Integrated nutrient management in faba bean and wheat on Nitisols. Research Report No. 78. EIAR, Addis Ababa, Ethiopia.

12. Nuruzzaman M, Lambers H, Bolland MDA, Veneklaas EJ (2005) Phosphorus uptake by grain legumes and subsequently grown wheat at different levels of residual phosphorus fertiliser. Austr J Agricult Res 56: 1041-1047.

13. Tesfaye Balemi T (2012) Effect of integrated use of cattle manure and inorganic fertilizers on tuber yield of potato in Ethiopia. Journal of Soil Science and Plant Nutrition 12: 257-265.

14. Ethiopian Institute of Agricultural Research (EIAR) (2011) Faba bean producing manual. Holetta Agricultural Research Center. Addis Ababa, Ethiopia.

15. Walkley A, Black CA (1934) An examination of the Degtjareff method for determining organic carbon in soils: Effect of variations in digestion conditions and of inorganic soil constituents. Soil Sci 63: 251-263.

16. Jackson ML (1970) Soil chemical analysis. Prentice hall, England, New Jersey, p: 498.

17. Olsen SR, Dean LA (1965) Phosphorous. In Black CA (ed.). Methods of Soil analysis. Agronomy No. 9, Am Soc Agron Madisen, Wisconsin, USA, pp: 1044-1046.

18. Erdogan C (2012) A Leaf Area Estimation Model for Faba Bean (Vicia faba L.) Grown in the Mediterranean type of Climate. Kabul 7: 58-63.

19. Gemechu K, Mussa J, Tezera W (2006) Faba bean (Vicia faba L) Genetics and Breeding Research in Ethiopia: A Review. In: Ali K, Kenneni G, Ahmed S, Malhotra R, Beniwal S, Makkouk K, Halila MH (eds.) Food and Forage legumes of Ethiopia: Progress and prospects. Proceedings of the workshop on Food and Forage Legume, Addis Ababa, Ethiopia.

20. Dennetti MD, Auld BA (1980) The effect of position and temperature on the expansion of leaves of Vicia faba L. Annals of botany 46: 511-517.

21. Kopke U, Nemecek T (2010) Ecological services of faba bean. Field Crops Research 115: 217-233.

22. Balaban M, Sepetoğlu H (1991) Growth, nutrients uptake and grain yield in faba bean genotypes under various plant densities. J Ege Univ 26: 181-197.

23. Babiker EE, Elsheikh AG, Osman AH, El-Tinay (1995) Effect of nitrogen fixation, nitrogen fertilization and viral infection on yield, tannin and protein contents and In vitro protein digestibility of faba bean. Plant Foods Hum Nutr 47: 257-263.

24. Asfaw H, Angaw T (2006) Food and Forage legumes of Ethiopia progress and prospects. Biological nitrogen fixation research on food legumes in Ethiopia, pp: 172-175.

25. Murinda MV, Saxena MC (1983) Agronomy of faba beans, lentils and chickpeas. In: Saxena MC, Verma S (eds.) Proceedings of the International Workshop on Faba Beans, Kabuli Chickpeas and Lentils in the 1980s. ICARDA, Aleppo, Syria, pp: 229-244.

26. Haque A (1986) Improved management of vertisols for sustainable crop-livestock production in Ethiopia 4 Nutrient Management. Plant Science Division Working Document 13. ILCA, Addis Ababa, Ethiopia.

27. Sprent JI (1972) The effects of water stress on Nitrogen fixing root nodules IV. Effects on whole plants of Vicia faba L. and Glyciene maximum. New Phytol 71: 603-611.

28. Sprent JI (1976) Water deficits and Nitrogen fixing root nodules. In Water defects and Plant growth Vol. IV. Soil water measurements plant responses and breeding for drought resistance. J Agric Sci Cambridge, pp: 103-152.

29. Davood H (2013) Phosphorus fertilizers effect on the yield and yield components of faba bean (Vicia faba L.). Annals of biological research 4: 181-184.

30. Thompson R, Taylor H (1977) Yield components and cultivars sowing date and density in faba bean. Ann Appl Biol 86: 313-320.

31. Ishag HM (1973) Physiology of seed yield in field beans (Vicia faba L.). J Agric Sci Camb 80: 191-199.

32. Hamissa MR (1973) Fertilizer requirements for Broad beans and Lentils Improvement and Production of Field crops, First FAO/SIDA Seminar for Plant Scientists from Africa and Near East, Cairo, Egypt.

33. Kay D (1979) Crop and Product Digest No.3 - Food legumes, London: Tropical products institute UK. Kluwer Academic Publishers, Dordrecht, Netherlands, pp: 26-47.

34. Salih FA, Ali AM, Elmubarak AA (1986) Effect of phosphorus application and time of harvest on the seed yield and quality of faba bean. FABIS Newsletter 15: 32-35.

35. Tekalign T (1991) Soil, plant, water, fertilizer, animal manure and compost analysis. Working Document No. 13. International Livestock Research Center for Africa, Addis Ababa, Ethiopia.

36

Production Potential of Tef (*Eragrostis tef* (Zucc.) Trotter) Genotypes in Relation to Integrated Nutrient Management on Vertisols of Mid High lands of Oromia Region of Ethiopia, East Africa

Yonas Mebratu, Cherukuri V Raghavaiah* and Habtamu Ashagre

Department of Plant Science, College of Agriculture and Veterinary Sciences, Ambo University, Ambo, Ethiopia

Abstract

Tef is a highly valued nutritious cereal crop which plays an important role in the diet of Ethiopians. Soil fertility depletion pose a serious threat to tef production in high lands of Ethiopia which are characterised by high rainfall, soil acidity, soil erosion, leaching and the attendant non availability of plant nutrients to the crop. In view of this a field experiment was carried out during 2014/15 cropping season on a field belonging to Hommicho Ammunition Engineering Industry with the objective to evaluate the response of Tef genotypes to integrated nutrient management in terms of productivity and yield components in Guder, Toke kuttai district. The treatments consisted of six levels of integrated nutrient management practices: 1) 0-0-0 (check) 2) 40-60-0 NPK (RDF) 3) 50%RDF+50% N (FYM) 4) 75%RDF+25% N (FYM) 5) 100% RDF+5 t FYM/ha and 6)RDF through new complex fertilizer (19-38-7 NPS) tested on five genotypes (Magna, Simoda, Quncho, Dz-Cr-409, Local variety). The experiment was laid out in a randomized complete block design with factorial arrangement with three replications. The results revealed that there was significant interaction between genotypes and integrated nutrient management practices where in application of 75% RDF+25% FYM and 100% RDF+5 t/ha FYM to genotypes DZ-CR-387 and DZ-01-196 delayed days to flowering and days to maturity, but in other genotypes these were not altered due to fertilizer application. In plant height, variety DZ-CR-385 and DZ-CR-409 responded better to 100%RDF+5 t/ha FYM combination, while DZ-01-196 and local variety was significantly affected over all fertilizer treatments. Significantly higher initial tiller capacity and fertile tiller production were obtained with application of 75% RDF+25% FYM and 100% RDF+5 t/ha FYM. Local variety had significantly higher number of leaves with 100% RDF+5 t/ha FYM followed closely by DZ-CR-409 with 75% RDF+25% FYM, and DZ-CR-385 and DZ-CR-387 with 75% RDF+25%FYM. Length of panicle and panicle weight were significantly affected where integrated nutrient management in new varieties DZ-CR-409 and DZ-CR-387 gave higher seed weight with 50%RDF+50% FYM and 75% RDF+25%FYM. There was significant interaction between varieties and integrated nutrient management on grain and straw yield, where DZ-01-196 recorded maximum grain and straw yield with application of 100% RDF+5 t/ha FYM which was comparable with 75%RDF+25% FYM. Therefore application of 50% RDF+50%FYM, 75%RDF+25%FYM and 100% RDF+5 tha^{-1} to DZ-01-196, DZ-CR-409 and Local varieties of Tef, respectively exhibited best production performance on vertisols of mid high lands of Ethiopia.

Keywords: Tef; Genotypes; Nutrient management; Vertisols; Midhigh lands; Productivity; Yield components

Introduction

Agriculture is the basis of the Ethiopian economy accounting for 46% of Ethiopia GDP and 90% of its export earnings and employ 85% of the countries labour force [1,2]. Cereals including tef are the most important crops for human consumption. Tef is a superior cereal grain crop solely produced and is considered as the noble grain of Ethiopia. Most of the areas used for production of grains especially tef, wheat and barley fall under the low fertility [3]. Soils in the highlands of Ethiopia usually have low levels of essential plant nutrients and organic matter content, especially low availability of nitrogen and phosphorus has been demonstrated to be major constraint to cereal production [4]. There are now Ethiopian restaurants in USA that are flourishing due to demand for ethnic foods, enjera and watt (stew) for human consumption. It grows well in most of the agro-ecological zones of the country.

Tef [*Eragrostis tef* (Zucc.) Trotter] is one among the major cereals of Ethiopia and occupies about 2.7 million hectares (27% of the grain crop area) of land which is more than any other major cereals such as maize (22.7%), sorghum (19%) and wheat (16%) (5,36). It is an indigenous cereal crop to Ethiopia and it has been recognized that Ethiopia is the centre of origin and diversity of tef. It is a C4 self pollinated chasmogamous annual cereal which belongs to the family poacea, sub family Eragrostidae and genus *Eragrostis*. Of the 82% gross grain production (about 17 million tonnes) contributed by cereals, tef accounts for 19.9% during the main season of 2010/11 [5].

Ethiopian farmers grow tef due to a number of merits, which are mainly attributed to the socio-economic, cultural and agronomic benefits [6]. Tef has more food value than the major grains such as wheat, barley and maize. Tef grain contains 14-15% proteins, 11-33 mg iron, 100-150 mg calcium, and rich in potassium and phosphorus nutrients [6]. Tef has got many prospects outside of Ethiopia due to its gluten free nature, tolerance to biotic and abiotic stresses, animal feed value and soil erosion control quality [7]. Small-scale commercial production of tef has begun in areas of the wheat belts of the USA, Canada and Australia (34). Tef has been introduced to South Africa

*Corresponding author: Cherukuri V Raghavaiah, Department of Plant Science, College of Agriculture and Veterinary Sciences, Ambo University, Ambo, Post Box No 19, Ethiopia, E-mail: cheruraghav@yahoo.in

and cultivated as a forage crop, and in recent years cultivated as a cereal crop in Northern Kenya [7].

Tef production has been increasing from year to year and so does the demand for it as staple grain in both rural and urban areas of Ethiopia [8]. Although tef is found in almost all cereal growing areas of Ethiopia, the major areas of production are Shewa, Gojam, Gonder, Wellega and Wello with central highlands of the country [9]. In those areas where it is consumed as a staple food, tef contributes about two-thirds of the dietary protein intake [7].

Tef is adapted to diverse agro-ecological zones which are marginal to most other crops [10]. Tef suffers less from diseases and gives better grain yield and possesses higher nutrient content especially protein when grown on Vertisols rather than Andosols [8]. Since tef tolerates water logging, it is sown during the wetting part of the rainy season, that is, from late July to mid August. It is mainly cultivated as a mono-crop. It is also among the most suitable crops for multiple cropping systems such as double and relay cropping.

Tef is predominantly cultivated on sandy-loam to black clay soils. Moreover, tef withstands low moisture conditions and it is often considered as a rescue crop that survives and grows well in the season when early planted crops (e.g., maize) fail due to low moisture. In addition, its ability to tolerate drainage problems makes it a preferred cereal by farmers. It is a highly valued crop primarily grown for its grain that is used for making injera. It is typically hand-broadcast on the field and, in most cases, seeds are left uncovered. Tef can produce a crop in a relative short growing season and will produce both grain for human food and fodder for cattle [7]. Regardless of its high area coverage, adaptation to different environmental conditions and requirement as a staple food in Ethiopia, the yield of tef grain is not increasing above the national average grain yield of 1.2 t/ha [5]. Its low productivity may be due to several production constraints like growing on marginal soils, traditional method of seed sowing, low application of fertilizers, and late planting of the crop exposing it to moisture stress.

Moreover, tef yields are almost stagnant probably because of the occurrence of accelerated soil erosion and lack of appropriate cultural practices on farmer's fields [11]. Lodging being among the most important factors threatening production and productivity of tef [12] up to 22% grain yield reduction and reduces straw quality [6]. Likewise, lodging restricts the use of high doses of nitrogen fertilizer in order to enhance grain yield [13].

Increasing agricultural productivity is absolutely necessary to feed the ever growing demography by enhancing land productivity. The national average yield of tef is currently below 1 ton per hectare and the present production system of tef cannot satisfy the consumer's demand as the current farming system that farmer adopts is traditional and is at subsistence level which is not supported by modern technologies [14].

Improved tef varieties have been developed since the mid-1950s, only about 20 improved varieties have been released [11]. Its grain is mainly used for making enjera, spongy flatbread, the main national dish in Ethiopia. Tef is also valued for its fine straw, which is used for animal feed as well as mixed with mud for building purposes. Tareke found sowing tef with application of 46 kg N/ha^{-1}, 46 kg P_2O_5/ha, 32 kg K_2O /ha, 12 kg S/ha and 0.3 kg Zn /ha provided a grain yield of up to 6 t /ha [14]. However, this result has not been validated under farmer's fields.

The principal factors for low productivity of tef are: improper use of recommended fertilizer rates, lack of information on response of different varieties, non-availability of genotypes suitable for this area and paucity of information on integrated nutrient management practices in tef. Therefore, the current investigation was made to evaluate the response of tef varieties in terms of growth, yield and yield parameters to integrated nutrient management practices on vertisols of mid high lands of Ethiopia.

Materials and Methods

Description of the study area

The field experiment was conducted under rain fed condition during the main cropping season from July to December, 2014 at Guder located in the central high lands of west Shoa zone of Oromiya Regional state, Ethiopia. Guder is one of the main Tef growing areas of West Shoa zone in the Ethiopia. It is situated at 8°56'30"-8°59'30" N latitude and 37°47'30"-37°55'15" longitude the altitude of the area ranges from 1380-3030 masl, characterized by warm temperate weather which is locally called Bada-dare (mid altitude). The temperature ranges from 15°C-29°C with an average of 22°C. It receives a mean annual rain fall ranging from 800-1000 mm with an average of 900 mm. The highest rainfall occurs from June to September, and the mean monthly relative humidity varies from 64.6% in August to 35.8% in December. The soil is clay loam in texture with good moisture holding capacity.

Seed material of the varieties

Four improved varieties of tef which are adapted to the agro-ecology of high lands were evaluated. The variety Quncho (Dz-CR-387), Magna (DZ-01-196) Simada (DZ-CR-385) and DZ-CR-409 were tested for their performance and compared with a local variety.

Farmyard manure and soil analysis

The field selected for the study was analysed for initial soil nutrient status in terms of physical (Texture), Chemical parameters (PH, EC, OC, Available N, P, K). The farm yard manure was analysed for available N, P, and K content before its application as an organic source. Auger samples were taken from 10 spots of the experimental area at a depth of 0-30 cm and composite sample of approximately 1 kg soil was made separately before sowing. After crop harvesting, soil sample was taken from each plot and the same treatments from each block were composited and 1 kg soil sample was made for each treatment. The composited soil was air dried, ground and sieved through 2 mm mesh sieve before laboratory analysis. The analysis for specific soil parameters was carried out at the National Soil Laboratory. Soil colour was determined using the Munsell soil colour chart, whereas soil pH was determined in a 1: 2.5 soil water suspension using glass electrode pH meter [15]. Determination of particle size distribution (texture) was carried out using the hydrometer method [16]. Based on the oxidation of organic carbon with acid potassium dichromate, organic matter was determined using the Walkley and Black wet digestion method as described by Nelson and Sommers, and total nitrogen was analyzed using the Kjeldhal method [17]. Available and total phosphorus were determined using the Olsen (NaHCO$_3$) extraction method [18]. Cation exchange capacity (CEC) of the soil was determined from ammonium-saturated sample that was subsequently replaced by sodium (Na) from a percolating sodium chloride solution. The excess salt was removed by washing with alcohol and the ammonium that was displaced by sodium was measured by Kjeldahl method. Exchangeable K was determined with flame photometer.

Layout of the experiment

After analysing the soil for chemical and physical parameters, land preparation with two times ploughing, harrowing and levelling were done to obtain a fine tilth. The field was then marked out into 90 plots of 3.2 m × 2.0 m². After preparing the land the layout of the experiment was done as per the treatments randomly in factorial randomized block design with 3 replications. Farm yard manure was applied to the plots as per the treatments 20 days before application of inorganic fertilizer. Before seeding, inorganic fertilizer as per treatments was applied. Urea was top dressed 2 times, once before sowing as basal dose and the other 7 days after emergence.

Treatments and design

There were six nutrient management treatments and five varieties of tef. The experiment was laid out in 5 × 6 factorial randomized Complete Block Design with three replications.

Varieties of Tef: 1. Magna (DZ-01-196), 2. Simada (DZ-Cr-385), 3. Quncho (Dz-Cr-387) 4. Dz-Cr-409), 5. Local variety.

Nutrient management practices: T1-0-0-0 (check), T2-40-60-0 NPK (RDF), T3-50% RDF+50% N (FYM), T4-75% RDF+25% N (FYM), T5-100% RDF+5 t FYM/ ha, T6- RDF through new complex fertilizer (19-38- 7 NPS). The net plot size was 3.2 m × 2.0 m=6.2 m². Sowing was done on 23 July 2015, adapting a row spacing of 20 cm using a seed rate of 5 Kg/ha⁻¹. The crop was harvested at physiological maturity in November 2014.

Data collection

Plant growth parameters: The following parameters were recorded at harvest from 5 randomly selected and tagged plants in the net plot.

Plant height (cm): It was measured from the base of the main stem to the base of the fully opened top leaf until panicle emergence. Later it was measured from the base of plant to the collar of flag leaf from five randomly selected plants from demarcated area in the net plot.

Number of total tillers: The average number of total tillers with and without panicle excluding the main shoot from five randomly selected plants in demarcated area.

Number of leaves: The average number of leaves from five randomly selected plants in demarcated area.

Number of fertile tillers: The average number of tillers with panicle excluding the main shoot from five randomly selected plants in demarcated area.

Leaf area (cm²): It was recorded as length × breath × 0.73 from five randomly selected plants in demarcated area.

Yield and yield components: The panicles from five randomly selected and tagged plants from the net plot at the time of harvest were used.

Days to 50% flowering: This parameter was determined by counting the number of days from sowing to the time when 50% of the plants started flowering through visual observation.

Days to maturity: It was determined as the number of days from sowing to the time when the plants reached maturity based on visual observation. It was indicated by senescence of the leaves as well as free threshing of grain from the glumes when pressed between the forefinger and thumb.

Number of Panicles/m²: The number of panicles per meter was counted from the net plot area.

Panicle length (cm): It is the length of the panicle from the node where the first panicle branches emerge to the tip of the panicle which was determined from an average of five selected plants per plot.

Panicle weight(g): The average panicle weight of the main panicle at harvest was recorded from the average of five randomly selected pre-tagged plants from net plot.

1000 seed weight (g): The weight of 1000 seeds was determined by carefully counting the grains and weighing them using a sensitive balance.

Grain yield (Kg/ha-1): Grain yield was measured by harvesting the crop from the net plot area of 3.2 × 2 m to avoid border effects.

Straw yield (Kg/ha⁻¹): After threshing and recording the grain yield, the straw yield was measured by drying the straw to a constant weight.

Lodging index: It was scored visually at the time of harvest according to the procedure described by Caldicott and Nuttall in which lodging percentage was taken as the sum of the product of each scale of lodging (0-5 scale) and its respective percentage divided by five, where 0 stands for upright stand, 1 for slightly slant, 2 for medium slant, 3 for very slant and 4 for extremely slant and 5 stands for 100% plants lodged.

Statistical analysis: The crop data collected were subjected to analysis of variance (ANOVA) using SAS software program version 9.0 [19]. Significant differences among treatment means were separated using the least significant difference (LSD) at 5% level of probability [20].

Results and Discussion

Initial soil physic-chemical properties

The soil particle distribution has 2.5% Sand, 22.5% Silt and 75%Clay which can be classified as clay loam soil with better moisture holding capacity suitable for rising a successful crop of tef. The soil pH of the experimental field was 6.79 and neutral in reaction (Table 1). The organic matter and organic carbon content of 2.91 and 1.69%, respectively are considered to be medium. Total N of 0.12% which is low, available Phosphorous content of 12.8 ppm which is medium; available Potassium content of 1.63 mg/100 g of soil which is high and cation exchange capacity of 1.17 mg/100 g of soil which is very low.

Chemical composition of farmyard manure (FYM): The FYM used in the current study has organic matter content of (51.76 g/kg⁻¹), Organic carbon content of 33.21 g/kg⁻¹; Nitrogen (2.24 g/kg⁻¹); Phosphorous (58.29 mg/kg⁻¹) and exchangeable K content of 2.55 cmol/kg⁻¹ (Table 1).

Days to flowering: Tef flowered earlier (37 days) without fertilizer application. Integration of inorganic fertilizer with farmyard manure (FYM) resulted in delayed flowering by a week (45 - 49 days) due to slow release of nutrients and better vegetative growth of the crop this result is in agreement with the report of Seyfu [6]. The days to flowering of varieties differed significantly, with DZ-CR-385(37 days) and DZ-CR-409(34 days) were early to flowering as compared with DZ-01-196 and DZ-CR-387 which are late to flowering (51 days), whereas the local variety was intermediate (46 days) in anthesis and flowering. Seyfu also reported genetic variation in flowering in tef in Ethiopia [6]. There was significant interaction between varieties and nutrient management where DZ-CR-387 and DZ-01-196 fertilized with 75% RDF+25% FYM remained comparable with 100% RDF+5 t FYM /ha⁻¹ in days to flowering(58 days) in comparison with the rest of the treatments (Table 2).

Chemical property	pH	OM (%)	O.C (%)	Total N (%)	Av.P (ppm)	Av.K (mg/100 g)	CEC (mg/100 g)	Physical Properties			Textural classification Clay loam
Soil	6.97	2.91	1.69	0.12	12.8	163	1.17	Sand %	Silt%	Clay%	
FYM	-	51.76	33.21	2.24	58.29	2.55	-	2.5	22.5	75	

Table 1: The initial physico-chemical properties of the experimental soil and analysis of FYM.

	T_1	T_2	T_3	T_4	T_5	T_6	Mean
DZ-01-196	43.3[e]	38.3[fg]	52.0[d]	58.0[ab]	58.0[ab]	56.0[bc]	50.9
DZ-CR-385	39.0[fg]	38.0[fgh]	32.0[j]	36.0[hi]	38.0[fgh]	36.0[hi]	36.5
DZ-CR-387	32.0[j]	45.0[e]	56.0[bc]	59.0[a]	58.0[ab]	54.0[cd]	50.6
DZ-CR-409	32.0[j]	34.0[ij]	33.0[j]	34.0[ij]	38.3[fg]	32.0[j]	33.8
Local Varity	39.7[f]	45.3[e]	58.3[a]	39.3[fg]	54.0[cd]	37.3[gh]	45.6
Mean	37.18	40.12	46.2	45.2	49.2	43	43.4
LSD (5%) 2.2							
CV (%) 1.56							

Where, T_1=Control (0-0-0); T_2=RDF(40-60-0); T_3=50% RDF+50% FYM; T_4=75% RDF+25% FYM; T_5=100% RDF+5 t/ha FYM and T_6=RDF NPS 23-10-5. Value within a column followed by the same letter is not significantly different at LSD 5% probability level.

Table 2: Interaction effect of genotypes and fertilizer on days to flowering.

	T_1	T_2	T_3	T_4	T_5	T_6	Mean
DZ-01-196	109.3[d]	98.0[f]	112.0[b]	114.0[a]	110.0[cd]	112.0[b]	109.1
DZ-CR-385	89.0[ij]	89.0[ij]	91.0[h]	87.0[k]	87.0[k]	83.0[m]	87.6
DZ-CR-387	100.0[e]	110.0[cd]	112.0[b]	114.0[a]	115.0[a]	114.0[a]	110.8
DZ-CR-409	84.0[lm]	85.0[l]	90.0[hi]	84.0[lm]	88.0[jk]	94.0[l]	86
Local Varity	112.0[b]	110.0[cd]	111.7[bc]	98.0[f]	110.0[cd]	94.0[g]	106
Mean	98.8	98.4	103.2	99.4	102	97.6	99.6
LSD (5%) 1.96							
CV (%) 0.6							

Where, T_1=Control (0-0-0); T_2=RDF(40-60-0); T_3=50% RDF+50% FYM; T_4=75% RDF+25% FYM; T_5=100% RDF+5 t/ha FYM and T_6=RDF NPS 23-10-5. Value within a column followed by the same letter is not significant different at LSD 5%.

Table 3: Interaction effect of genotype and fertilizer on days to maturity.

Days to maturity: The number of days taken for the crop to attain physiological maturity too exhibited distinct variation with nutrient management. It was observed that the crop matured earlier when supplied with only inorganic fertilizer and no fertilizer (97-98.8 days). Integrated nutrient supply through inorganic fertilizer with farmyard manure delayed the crop maturity due to prolonged vegetative growth and balanced nutrition (99-102 days) in comparison with unfertilized control. Delay in crop maturity due to INM has also been reported by Brady and Weil [21]. Early flowering varieties (DZ-CR-385, DZ-CR-409) have taken 96 days to mature, while the late maturing varieties (DZ-01-196, DZ-CR-387) matured in 110 days, whereas the local variety matured in 106 days, indicating differing maturing groups which is a genetic character and not much altered by growing environment.

Significant interaction between varieties and fertilizer revealed that DZ-CR-387 and DZ-01-196 when fertilized with 75% RDF+25% FYM and 100%RDF+5 t/ha FYM (Table 3) delayed maturity (114 days) in comparison with control and only inorganic fertilizer (110 days). But in DZ-CR-385, DZ-CR-409 and local Varity the days to maturity was not altered much due to fertilizer management practices (Table 3).

Lodging index: Tef being a weak stalked plant, often is prone for lodging, especially at reproductive stage due to weight of developing spikes. It was found that the crop exhibited less lodging without fertilizer (1.1%), whereas the lodging percent increased with the application of either inorganic fertilizer alone or in integration with organic manure (1.3-1.5%) (Table 2). The role of P in providing strength of straw and thus preventing lodging has been reported [21,22]. This calls for a need for balanced/optimum fertilization of tef to obtain less lodging so as to avoid pre and postharvest loss of grain. Different varieties did not exhibit

discernible variation in lodging percent; however local variety tended to lodge more (1.87%) than improved tef varieties (1.2%). Traditional varieties are tall in stature and prone to lodging in comparison with the improved genotypes, which are medium in stature and have stiff straw and consequently are less prone to lodging.

Plant height (cm): Application of nutrient either in the form of inorganic or organic forms resulted in enhanced plant height (76.7 to 82.3 cm) in comparison with unfertilized crop (61.5 cm) elucidating better vegetative growth. Enhanced crop growth due to N fertilizer treatment has also been reported by Haftom et al. [23]. Local tef variety grew significantly taller(83 cm), while DZ-01-196 and DZ-CR-387 were comparable (80 cm) followed closely by DZ-CR-409 (72 cm), whereas DZ-CR-385 was of short stature (65 cm). The plant height is a genetic attribute and is little influenced by growing environment as observed in the current study.

The varieties and nutrient management practices showed significant interaction on plant height (Table 4). Tef variety DZ-CR-385 and DZ-CR-409 responded to 100% RDF+5 t FYM /ha[-1] in comparison with rest of the fertilizer treatments. The genotypes DZ-01-196, DZ-CR-387 and local Varity showed significantly different effect with all fertilizer management practices over unfertilized check (Table 4).

Tillering capacity: Tillering capacity at the vegetative growth stage and the productive tillers at reproductive stage represent yielding ability of the crops, as it is a major yield component. The integration of inorganic fertilizer with farmyard manure resulted in production of significantly greater number of tillers per unit area (23-25.5) than that could be produced due to application of inorganic fertilizer alone

	T_1	T_2	T_3	T_4	T_5	T_6	Mean
DZ-01-196	64.7[gh]	80.3[bcd]	88.0[ab]	85.5[ab]	83.3[abc]	81.1[abcd]	80.5
DZ-CR-385	52.5[i]	65.5[fgh]	63.2[ghi]	62.3[ghi]	82.8[abc]	64.0[gh]	65
DZ-CR-387	55.1[ij]	82.9[abc]	86.8[ab]	82.5[abc]	82.5[abc]	85.5[ab]	79.1
DZ-CR-409	62.0[hi]	76.9[cde]	76.7[cde]	69.3[efgh]	80.0[bcd]	70.6[efg]	72.5
Local Varity	73.2[def]	86.3[ab]	84.7[abc]	84.0[abc]	83.1[abc]	89.1[a]	83.3
Mean	61.5	78.4	79.8	76.7	82.3	78	76.08
LSD (5%) 8.5							
CV (%) 3.47							

Where, T_1=Control(0-0-0); T_2=RDF(40-60-0); T_3=50% RDF+50% FYM; T_4=75% RDF+25% FYM; T_5=100% RDF+5 t/ha FYM and T_6=RDF NPS 23-10-5. Value within a column followed by the same letter is not significant different at LSD 5% probability level.

Table 4: Interaction effect of genotypes and fertilizer on plant height (cm).

	T_1	T_2	T_3	T_4	T_5	T_6	Mean
DZ-01-196	11.1[h]	16.4[g]	24.3[abcd]	25.0[abcd]	25.9[ab]	24.4[abcd]	21.1
DZ-CR-385	9.4[h]	21.0[ef]	23.4[bcd]	25.2[abc]	27.6[a]	24.3[abcd]	21.8
DZ-CR-387	10.9[h]	21.4[def]	24.5[abcd]	26.1[ab]	26.7[ab]	23.2[bcde]	22.1
DZ-CR-409	10.1[h]	17.9[fg]	23.1[bcde]	25.7[ab]	23.7[bcde]	26.1[ab]	21.1
Local Varity	8.6[h]	17.9[fg]	21.5[cdef]	23.3[bcde]	24.8[abcd]	21.5[cdef]	19.6
Mean	10	18.9	23.3	25	25.8	23.8	21.1
LSD (5%) 3.8							
CV (%)5.53							

Where, T_1=Control(0-0-0); T_2=RDF(40-60-0); T_3=50% RDF+50% FYM; T_4=75%RDF+25% FYM; T_5=100%RDF+5 t/ha FYM and T_6=RDF NPS 23-10-5. Value within a column followed by the same letter is not significant different at LSD 5% probability level.

Table 5: Interaction effect of genotypes and fertilizer on initial tillering.

	T_1	T_2	T_3	T_4	T_5	T_6	Mean
DZ-01-196	9.1[i]	14.1[h]	22.0[abcdef]	22.7[abcde]	23.7[abc]	22.1[abcdef]	18.9
DZ-CR-385	7.2[i]	18.7[fg]	20.9[bcdef]	22.7[abcde]	24.7[ab]	21.3[abcdef]	19.2
DZ-CR-387	8.4[i]	18.8[fg]	22.3[abcdef]	23.7[abc]	24.9[a]	20.8[cdef]	19.8
DZ-CR-409	7.9[i]	15.3[gh]	20.8[cdef]	23.4[abc]	21.7[abcdef]	23.1[abcd]	18.6
Local Varity	6.5[i]	15.5[gh]	19.3[cdef]	21.0[bcdef]	22.1[abcdef]	19.1[efg]	17.2
Mean	7.8	16.4	21	25	25.8	21.2	18.7
LSD (5%) 3.8							
CV (%) 6.33							

Where, T_1=Control(0-0-0); T_2=RDF(40-60-0); T_3=50% RDF+50% FYM; T_4=75% RDF+25% FYM; T_5=100% RDF+5 t/ha FYM and T_6=RDF NPS 23-10-5. Value within a column followed by the same letter is not significant different at LSD 5% probability level.

Table 6: Interaction effect of genotype and fertilizer on fertile tillers.

(18.9) or no fertilizer check (10,0). This result corroborates with the findings of Haftom et al., Al Abdul Salam and Warraich et al. [23-25]. The different genotypes did not show discernible variation in the initial tiller counts, though the traditional cultivar tended to produce less tillers than the improved test varieties.

Significant interaction between varieties and nutrient management practices (Table 5) showed that all the improved varieties produced significantly greater number of tillers than local variety with the application of 75% RDF+25% FYM and 100% RDF+5 t/ha FYM when compared with RDF, 50% RDF+50% FYM and control. Application of FYM improved tillering capacity.

Effective/productive tillers: Effective tillers are those bearing panicles that contribute to the grain yield. There was a decrease in the number of tillers at reproductive stage in comparison with those observed at vegetative stage owing mainly to mortality and variable source to sink relationships. The effective tillers followed a trend akin to the vegetative tillers in relation to the nutrient management practices; in that integrated nutrient management had an edge (21-23.4 tillers)over exclusive application of inorganic fertilizer(16.4 tillers) or no fertilizer control (7.8 tillers) in manifestation of tillering capacity. Enhancement in productive tillers due to application of nitrogen has also been reported by Al-Abdul Salam and Warraich et al. [25]. Tef varieties showed distinct variation in panicle bearing tillers where variety DZ-CR-387(20 tillers) remaining comparable with DZ-CR-385(19 tillers), DZ-01-196 and DZ-CR-409 produced greater number of tillers than local cultivar (17 tillers). Variation in productive tillers has also been reported by Belay and Baker [26].

Significant interaction between varieties and nutrient management revealed that all the improved varieties exhibited significant improvement in effective tillers over local variety with the application of 75% RDF+25% FYM (Table 6). Application of 100% RDF+5 t/ha FYM for all varieties produced higher number of effective tillers. The fertile tillers were significantly lower with RDF through inorganic fertilizer, which in turn was superior to unfertilized control.

Leaf number plant[-1]: Application of fertilizer showed substantial improvement in the number of leaves/plant (42.9-56.2) over unfertilized check (25.3). Integrated application of inorganic fertilizer with farmyard manure produced significantly higher number of leaves/ plant (49 to 56) than with the application of inorganic fertilizer (42.9) or with no fertilizer check (25.3) which produced least leaf number / plant. Tef varieties differed significantly in the number of leaves/ plant where Local variety possessed grater leaf number (57) than the

	T_1	T_2	T_3	T_4	T_5	T_6	Mean
DZ-01-196	31.9^{no}	37.0^{klmn}	35.3^{lmno}	38.9^{ijkl}	36.3^{klmn}	43.7^{ghij}	37.1
DZ-CR-385	22.3^{p}	44.3^{ghi}	39.5^{ijkl}	38.6^{jklm}	52.9^{f}	36.3^{klmn}	39
DZ-CR-387	21.0^{p}	41.5^{hijk}	46.6^{gh}	43.0^{hij}	49.0^{fg}	33.3^{mno}	39
DZ-CR-409	30.5^{o}	53.6^{ef}	59.0^{de}	67.0^{bc}	61.6^{cd}	62.7^{cd}	55.6
Local Varity	21.0^{p}	38.3^{jklm}	70.3^{b}	61.0^{d}	81.0^{a}	72.3^{b}	57.3
Mean	25.3	42.9	50.2	25	56.2	49.6	45.6
LSD (5%) 5.5							
CV (%)3.75							

Where, T_1=Control(0-0-0); T_2=RDF (40-60-0); T_3=50% RDF+50% FYM; T_4=75% RDF+25% FYM; T_5=100% RDF+5 t/ha FYM and T_6=RDF NPS 23-10-5. Value within a column followed by the same letter is not significant different at LSD 5% probability level.

Table 7: Interaction effect of genotype and fertilizer on leaf number/plant.

	T_1	T_2	T_3	T_4	T_5	T_6	Mean
DZ-01-196	7.66^{cde}	6.27^{ghi}	8.57^{c}	6.42^{fghi}	7.30^{defg}	5.03^{j}	6.87
DZ-CR-385	7.88^{cd}	3.94^{kl}	3.56^{lm}	8.54^{c}	3.28^{lm}	6.42^{fghi}	5.6
DZ-CR-387	6.42^{fghi}	6.75^{efgh}	7.84cd	8.54^{c}	7.15^{defgh}	10.78^{b}	7.91
DZ-CR-409	3.93^{kl}	7.04^{defgh}	4.70^{jk}	5.53^{ij}	2.82^{m}	6.87^{defgh}	5.14
Local Varity	7.40^{def}	4.89^{jk}	6.14^{hi}	6.25^{hi}	9.82^{b}	12.00^{a}	7.74
Mean	6.64	5.78	6.16	7.05	6.07	8.22	6.65
LSD (5%) 1.05							
CV (%)4.87							

Where, T_1=Control(0-0-0); T_2=RDF(40-60-0); T_3=50% RDF+50% FYM; T_4=75% RDF+25% FYM; T_5=100% RDF+5 t/ha FYM and T_6=RDF NPS 23-10-5. Value within a column followed by the same letter is not significant different at LSD 5%.

Table 8: Interaction effect of genotype and fertilizer on leaf area (cm).

	T_1	T_2	T_3	T_4	T_5	T_6	Mean
DZ-01-196	22.6^{h}	29.5^{bcdef}	32.9^{abcde}	33.9^{abcd}	35.8^{a}	35.7^{a}	31.7
DZ-CR-385	23.5^{gh}	28.6^{defg}	32.7^{abcde}	32.5^{abcde}	33.8^{abcd}	33.2^{abcde}	30.7
DZ-CR-387	24.1^{fgh}	32.8^{abcde}	34.7^{ab}	35.1^{a}	37.0^{a}	35.7^{a}	33.1
DZ-CR-409	23.9^{gh}	28.9^{cdefg}	34.0^{abcd}	36.1^{a}	37.2^{a}	34.4^{abc}	32.4
Local Varity	24.3^{fgh}	27.9^{efgh}	36.0^{a}	37.8^{a}	36.6^{a}	36.7^{a}	33.1
Mean	23.6	29.5	34	35	36	35.1	32.2
LSD (5%) 5.5							
CV (%)5.27							

Where, T_1=Control(0-0-0); T_2=RDF(40-60-0); T_3=50% RDF+50% FYM; T_4=75% RDF+25% FYM; T_5=100% RDF+5 t/ha FYM and T_6=RDF NPS 23-10-5. Value within a column followed by the same letter is not significant different at LSD 5% probability level.

Table 9: Interaction effect of genotype and fertilizer on panicle length (cm).

improved varieties DZ-CR-409(56) which in turn had higher leaf number in comparison with the rest of the genotypes which had almost similar leaf number/plant^{-1} (37-39).

Interaction of varieties and nutrient management practices significantly affected mean number of leaf plant (Table 7) where Local variety produced large number of leaves/plant with 100% RDF+5 t/ha FYM, followed by DZ-CR-409 with 75%RDF+25% FYM.

Leaf area (cm): Application of N at stem elongation stage has been reported to greatly stimulate leaf area growth resulting in significantly greater assimilation capacity, both before and after antithesis. Application of new complex fertilizer (NPS 23-10-5) resulted in substantially higher leaf area (8.2 cm) than the rest of the fertilizer treatments. However, 75% RDF+25% FYM was found superior to the other treatments in leaf area; whereas inorganic fertilizer alone (5.78 cm) showed least leaf area.

Leaf area, which is an indicator of assimilatory surface, varied with varieties; where the variety DZ-CR-387(7.9 cm) remaining at a par with local cultivar (7.7 cm) possessed grater leaf area than DZ-01-196(6.87 cm), DZ-CR-385(5.60 cm) and DZ-CR-409(5.1 cm).

The varieties DZ-CR-385 and DZ-CR-387 showed significant enhancement in leaf area (8.54 cm) when fertilized with 75% RDF+25% FYM as revealed by varieties and fertilizer interaction.

Yield components

Panicle length (cm): Application of fertilizer either in organic or inorganic form (29.5 cm) and their integrated application (35.0 cm) brought about discernible variation in the length of panicle in comparison with unfertilized control (23.7 cm). Higher number of tillers in fertilized plots could also produce longer panicles due to less competition for sinks in comparison with unfertilized plots. This is in accordance with the reports that combined application of half dose of inorganic and half dose of organic source resulted in more panicle length apart from plant height and tiller production [23]. Panicle length is an indicator of sink capacity which differed significantly with the varieties; where the varieties DZ-CR-387(33.2 cm) remaining comparable with the local cultivar (33.2 cm) and DZ-CR-409 (32.4 cm) produced distinctly longer panicles than DZ-01-196 (31.7 cm) and DZ-CR-385(30.7 cm). Variation in panicle length among tef varieties was also reported by Belay and Baker [26].

Interaction of varieties with nutrient management practice was significant for panicle length (Tables 8 and 9) where Tef varieties showed distinct improvement in length of panicle with integrated

	T_1	T_2	T_3	T_4	T_5	T_6	Mean
DZ-01-196	1.1[ij]	1.3[hij]	1.8[abcdefg]	1.9[abcde]	1.9[abcd]	1.3[ghij]	1.53
DZ-CR-385	1.1[ij]	1.5[efghi]	1.9[abcde]	1.8[abcdefg]	1.9[abcd]	1.4[fghij]	1.59
DZ-CR-387	1.1[ij]	1.6[cdefgh]	1.9[abcde]	2.1[a]	1.9[abcde]	1.8[abcdef]	1.71
DZ-CR-409	1.0[j]	1.5[defgh]	2.0[abc]	1.9[abcd]	2.0[a]	1.5[efghi]	1.66
Local Varity	1.3[hij]	1.6[cdefgh]	1.9[abcd]	1.9[abcde]	1.8[abcdef]	1.6[bcdefgh]	1.68
Mean	1.1	1.49	1.88	1.9	1.92	1.53	1.63
LSD (5%) 0.4							
CV (%) 8.07							

Where, T_1=Control(0-0-0); T_2=RDF(40-60-0); T_3=50% RDF+50% FYM; T_4=75% RDF+25% FYM; T_5=100% RDF+5 t/ha FYM and T_6=RDF NPS 23-10-5. Value within a column followed by the same letter is not significant different at LSD 5% probability level.

Table 10: Interaction effect of genotype and fertilizer on panicle weight (g).

	T_1	T_2	T_3	T_4	T_5	T_6	Mean
DZ-01-196	0.22[k]	0.28[ghij]	0.28[ghijk]	0.35[abcdef]	0.35[abcde]	0.38[abc]	0.30
DZ-CR-385	0.28[ghijk]	0.29[fghi]	0.33[bcdef]	0.32[defgh]	0.38[abcd]	0.29[efghi]	0.31
DZ-CR-387	0.27[hijk]	0.28[ghij]	0.36[abcd]	0.38[abcd]	0.39[ab]	0.35[abcdef]	0.33
DZ-CR-409	0.25[ijk]	0.35[abcdef]	0.38[abcd]	0.40[a]	0.33[cdefgh]	0.29[ghi]	0.33
Local Varity	0.22[jk]	0.40[a]	0.33[bcdefg]	0.37[abcd]	0.39[ab]	0.28[ghijk]	0.33
Mean	0.24	0.31	0.33	0.36	0.36	0.31	0.32
LSD (5%) 0.06							
CV (%)5.7							

Where, T_1=Control(0-0-0); T_2=RDF(40-60-0); T_3=50% RDF+50% FYM; T_4=75% RDF+25% FYM; T_5=100% RDF+5 t/ha FYM and T_6=RDF NPS 23-10-5. Value within a column followed by the same letter is not significant different at LSD 5%.

Table 11: Interaction effect of genotype and fertilizer on 1000 seed weight (g).

	T_1	T_2	T_3	T_4	T_5	T_6	Mean
DZ-01-196	659.0[ijkl]	893.7[fghij]	1256.0[bcd]	1626.0[a]	1644.0[a]	953.0[efghi]	1171.8
DZ-CR-385	680.3[ijkl]	758.3[hijk]	985.3[defgh]	1113.0[defg]	1156.0[cdef]	783.3[hijk]	912.6
DZ-CR-387	447.7[l]	945.0[fghi]	1027.0[cdefgh]	1167.0[bcdef]	978.7[defgh]	846.3[ghij]	901.8
DZ-CR-409	749.0[hijk]	965.7[efgh]	1300.0[b]	1157.0[cdef]	1237.0[bcde]	749.0[hijk]	1025.8
Local Varity	560.0[kl]	999.3[defgh]	1100.0[cdefg]	1442.0[ab]	1008.0[defgh]	999.7[defgh]	1018.1
Mean	619	912.4	1133.6	1301	1204.2	866.2	1006.2
LSD (5%)281.5							
CV (%) 8.6							

Where, T_1=Control(0-0-0); T_2=RDF(40-60-0); T_3=50% RDF+50% FYM; T_4=75% RDF+25% FYM; T_5=100% RDF+5 t/ha FYM and T_6=RDF NPS 23-10-5. Value within a column followed by the same letter is not significant different at LSD 5% probability level.

Table 12: Interaction effect of genotype and fertilizer on grain yield (kg/ha).

nutrient management and application of new complex fertilizer over RDF and unfertilized control. Application of RDF through inorganic source was superior to no fertilizer control in all varieties.

Panicle weight (g): The panicle weight has been significantly higher with integrated use of inorganic fertilizer with organic manure (1.9 g) as compared with sole application of inorganic nutrient (1.5 g) or no fertilizer (1.1 g). This is in agreement with the finding of Tekalign et al. [13]. The panicle weight tended to be in accordance with the length of the panicle. Among the varieties, DZ-CR-387 possessed panicles of greater weight (1.72 g), closely followed by local cultivar (1.68 g) and DZ-CR-409(1.66 g); while the lower panicle weight was obtained from DZ-CR-385(1.59 g) and DZ-01-196(1.54 g). These findings are in agreement with the report of Blum and Belay; Baker [26,27].

The panicle weight of all the varieties improved substantially with integrated nutrient management in comparison with application of RDF through inorganic source and no fertilizer check. The least panicle length was recorded in all the varieties with no fertilizer.

Thousand seed weight (g): Application of fertilizer significantly improved thousand seed weight (0.318 g) over no fertilization check (0.248 g). Further, integration of inorganic fertilizer with farm yard manure in different proportions had a synergistic effect on thousand

seed weight (0.337 g-0.368 g) in comparison with sole inorganic fertilizer or no fertilizer application. Improvement in thousand seed weight due to fertilizer application has also been reported by AL-Abdul Salam [24]. The tef varieties differed significantly in their thousand seed weight where DZ-CR-387 had superior thousand seed weight (0.33 g) followed by DZ-CR-409(0.331 g) and local cultivar (0.331 g) which in turn were comparable. The variety DZ-CR-385(0.316 g) was found superior to DZ-01-196 (0.309 g) which gave the least thousand seed weight.

The interaction of varieties with nutrient management was significant on thousand seed weight of tef where DZ-CR-409 and DZ-CR-387 possessed higher thousand seed weight with 50% RDF+50% FYM and 75% RDF+25% FYM over RDF and unfertilized control (Tables 10-13). The thousand seed weight of DZ-01-196, DZ-CR-385 and Local cultivar improved with 100% RDF+5 t FYM/ha[-1]. All the varieties recorded least thousand seed weight with no fertilizer treatment.

Grain yield (kg/ha): Application of 50% recommended dose of fertilizer through inorganic source and 50% through farmyard manure resulted in significant improvement in grain yield (1133 kg/ha[-1]) over no fertilizer check (619 kg/ha). Application of RDF through complex fertilizer (866 kg/ha[-1]); and remained comparable with 75%

	T_1	T_2	T_3	T_4	T_5	T_6	Mean
DZ-01-196	2790[jk]	3205[efghi]	3949[c]	4771[a]	4824[a]	2810[jk]	3724.7
DZ-CR-385	2834[jk]	2980[ghij]	3333[ef]	3352[e]	3955[c]	2995[fghij]	3241.4
DZ-CR-387	1139[l]	3251[efgh]	3440[de]	4045[bc]	3296[efg]	3311[efg]	3080.3
DZ-CR-409	2909[hij]	3292[efg]	4333[b]	3841[c]	3955[c]	2904[ij]	3538.9
Local Varity	2553[k]	3332[ef]	3771[cd]	3930[c]	3470[de]	3342[e]	3399.6
Mean	2444.9	3212.1	3764.9	3987.7	3899.8	3072.5	3397.03
LSD (5%) 342.9							
CV (%)3.18							

Where, T_1=Control(0-0-0); T_2=RDF(40-60-0); T_3=50% RDF+50% FYM; T_4=75% RDF+25% FYM; T_5=100% RDF+5 t/ha FYM and T_6=RDF NPS 23-10-5. Value within a column followed by the same letter is not significant different at LSD 5% probability level.

Table 13: Interaction effect of genotype and fertilizer on straw yield (kg/ha).

	T_1	T_2	T_3	T_4	T_5	T_6	Mean
DZ-01-196	23.62[kl]	27.89[fghij]	31.80[bcde]	34.07[b]	34.09[b]	33.91[bc]	30.89
DZ-CR-385	24.00[kl]	25.44[jk]	29.57[efg]	33.19[bcd]	29.21[efgh]	26.15[ghijk]	27.92
DZ-CR-387	39.31[a]	29.06[efghi]	29.84[def]	28.85[efghij]	29.69[def]	25.55[ijk]	30.38
DZ-CR-409	25.75[hijk]	29.33[efg]	30.34[def]	30.12[def]	31.20[bcdef]	25.79[hijk]	28.74
Local Varity	21.88[l]	29.99[def]	29.17[efgh]	30.48[cdef]	29.06[efghi]	28.90[efghij]	28.24
Mean	26.9	28.33	29.17	31.33	30.64	28.05	29.23
LSD (5%) 3.54							
CV (%) 3.7							

Where, T_1=Control(0-0-0); T_2=RDF(40-60-0); T_3=50% RDF+50% FYM; T_4=75% RDF+25% FYM; T_5=100% RDF+5 t/ha FYM and T_6 =RDF NPS 23-10-5. Value within a column followed by the same letter is not significant different at LSD 5% probability level.

Table 14: Interaction effect of genotype and fertilizer on Harvest index of Tef.

RDF through inorganic fertilizer+25% N through farmyard manure (1300 kg/ha[-1]). Thus the superior performance of integrated nutrient management comprising 50%+50% through inorganic and farmyard manure of nutrients could be attributed to enhancement in various growth parameters and increased seed yield. In line with the present finding, at Holleta Research centre on Nitosols, incorporation of organic mustard meal at 31 kg/ha[-1] 20 days ahead of sowing tef resulted in yield increase of 42, 32 and 25% over control [28]. Improvement in yield due to fertilizer application has also been reported by Haftom et al., Al-Abdul Salam and Warriach et al., ZARC [23-25,29]. NFIU reported response of tef to 60 kg N/ha[-1] on high land Vertisols of Ethiopia [30]. Kenea et al. reported that the recommended fertilizer for tef is 100 kg DAP and 100 kg urea/ha[-1] [31].

Grain yield is a product of all the yield attributes and is the principal economic output of the crop. Tef varieties exhibited significant variations in the grain yields. Improved tef variety DZ-01-196 (1172 kg/ha[-1]) remaining comparable with DZ-CR-409(1026 kg/ha[-1]) and local cultivar (1018 Kg/ha[-1]) offered substantially greater grain yield than DZ-CR-385 (913 Kg/ha) and DZ-CR-387 (902 Kg/ha) which in turn were comparable. In the present study, the lower yield of DZ-CR-385 and DZ-CR-387 could probably be attributed to genetic potential of varieties as compared to others. Differential performance tef varieties with varied grain yield was also reported by Kelsa on three tef varieties at Awassa and Areka areas of Ethiopia [32]. Considerable genetic variability in grain yield and dry matter accumulation has been reported to exist between and within crop species [27]. Belay and Baker also reported variation in biological yield of tef varieties [26].

There was significant interaction between varieties and nutrient management on grain yield of tef (Table 12) where DZ-01-196 yielded higher with 100% RDF+5 tFYM/ha[-1].(1644 kg/ha[-1].) which was comparable with 75% RDF+25%FYM (1626 kg/ha[-1]) and local variety fertilized with 75% RDF+25% FYM (1442 kg/ha[-1]). The variety DZ-CR-409 yielded better with 50% RDF+50% FYM (1300 kg/ha[-1]). Tef varieties DZ-CR-385 and DZ-CR-387 performed well with 75%

RDF+25% FYM application. Abuhay also reported variation in grain yield of tef [33].

Straw yield (kg/ha): Application of 75% recommended dose of fertilizer through inorganic source and substitution of 25% RDF with farm yard manure resulted in significantly higher straw yield (3988.6 kg ha[-1]) than other treatments. This was followed by RDF+5 t/ha FYM (3900.9 kg ha[-1]) and 50% RDF+50% through FYM (3765 kg ha[-1]). The least straw production was obtained from unfertilized control (2445.1 kg ha[-1]) and the RDF through new complex fertilizer (3073.1 kg ha[-1]). Thus integration of inorganic and farmyard manure had a beneficial effect on production of biological yield of tef in comparison with use of inorganic fertilizer alone or no fertilizer application. Tef variety DZ-01-196 gave significantly higher straw yield (3725 kg ha[-1]) over other varieties. This was followed by DZ-CR-409(3539 kg ha[-1]), local variety (3399 kg ha[-1]), DZ-CR-385(3241 kg ha[-1]) and DZ-CR-387 (3080 kg ha[-1]) [34].

Significant interaction between varieties and nutrient management elucidated that the improved variety DZ-01- 196 produced higher straw yield of 4824 kg/ha[-1] with 100% RDF+5 t/ha FYM which was comparable with 75% RDF+25% FYM (4771 kg ha[-1]) (Table 13). The variety DZ-CR-409 and DZ-CR-385 gave higher straw yield with 50% RDF+50% FYM (4333 kg ha[-1]) and (3955 kg ha[-1]), whereas DZ-CR-387 with 75% RDF+25% FYM gave 4045 kg ha[-1].

Harvest index: Harvest index of tef was significantly higher with 75% RDF+25%FYM (31.3) which was comparable to 100% RDF+5 t/ha FYM (30.64) and superior to the rest of the fertilizer treatments. The second best treatments was 50% RDF+50% FYM (30.14) which in turn was superior to RDF (28.34) and RDF through complex fertilizer (28.05). The least harvest index of 26.91 was recorded with unfertilized control [35].

Among the tef varieties, DZ-01-196 gave the highest harvest index of 30.89 which was comparable with DZ-CR-387(30.38) and superior to the other varieties. Tef variety DZ-CR-409 remaining at a par with local variety (28.24) was found superior to DZ-CR-385 (27.92).

Interaction of varieties of Tef with nutrient management practice on harvest index was significant (Table 14) where Tef variety DZ-01-196 with integrated nutrient management practice produced significantly greater harvest index (34.09), which was comparable with DZ-CR-385 with 75% RDF+25% FYM (33.19) and DZ-CR-409 with 100% RDF+5 t/ha FYM (31.20). There was distinct improvement of harvest index of local variety with fertilizer use (31.33) over no fertilizer (21.88).

Conclusion

From the foregoing account it can be inferred that in rain fed Tef crop raised on Vertisols, Integrated use of FYM in conjunction with inorganic fertilizer is more efficient than use of RDF through inorganic source and unfertilized crop employing selected improved Tef genotype can considerably improve grain yields. In this study, application of 50% RDF+50% FYM, 75% RDF+25% FYM and 100% RDF+5 t/ha FYM using DZ-01-196, DZ-CR-409 and Local variety, respectively exhibited best yield performance in the mid high lands of West Shoa zone in Oromia region of Ethiopia.

References

1. Mulat D (1999) Agricultural Technology, Economic Viability and Poverty Alleviation in Ethiopia, Paper presented to the Agricultural Transformation Policy Workshop, Kenya.

2. UNDP (2002) UNDP assistance in the fifth country program to the agricultural sector.

3. Yihenew G (2002) Selected chemical and physical characteristics of soils of Adet Research Centre and its testing sites in north western Ethiopia. Ethiopian Journal of Natural Resources 4: 199-215.

4. Tekalign M (1998) Effect of source, rate and timing of nitrogen applied to wheat on soil N level in a vertisol in central highland of Ethiopia. In: Crop Management Options to sustain Food Security, pp: 85-100.

5. Central Statistic Authority (2008) Agricultural Sample Survey. Report on area and production for major crops (private peasant holdings Meher season) Statistical Bulletin 417, Addis Ababa.

6. Seyfu K (1993) Tef [Eragrostis tef (Zucc.)Trotter]. Breeding, Genetic Resources, Agronomy, Utilization and Role in Ethiopian Agriculture. Institute of Agricultural Research, Addis Ababa, Ethiopia.

7. Seyfu K (1997) Tef [Eragrostis tef (Zucc.) Trotter]. Promoting the conservation and use of underutilized and neglected crops. Biodiversity institute, Addis Ababa, Ethiopia.

8. Mitiku H, Fassil K (1996) Soil and moisture conservation in Semi arid areas of Ethiopia. In: Proceedings of the Third Conference of Ethiopian Soil Science Society (ESSS). Addis Ababa, Ethiopia.

9. Piccinin D (2010) More About Ethiopian Food: Teff. Ethno Med: Ethiopian food. Online. Internet available.

10. Hailu T, Seyfu K (1990) Variability and genetic advance in tef (Eragrostis tef) cultivars. Tropical Agriculture 67: 317-320.

11. Fufa H, Tesfa B, Hailu T, Kebebew A, Tiruneh K, et al. (2001) Agronomy Research in tef. In: Hailu Tefera, Getachew Belay and M. Sorrels (ed.), Narrowing the Rift: Tef Research and Development, Proceedings of the International Workshop on Tef Genetics and Improvement, Addis Ababa, Ethiopia, pp: 167-176.

12. Yu JK, Graznak E, Breseghello F, Hailu T, Sorrells ME (2007) QTL mapping of agronomic traits in tef [Eragrostistef (Zucc) Trotter]. BMC Plant Biol 7: 13.

13. Tekalign M, Teklu E, Balesh T (2001) Soil Fertility and Plant Nutrition Research on Tef in Ethiopia. In: Hailu T, Getachew B, Mark S (eds.), Narrowing the Rift. Tef Research and Development, Proceedings of the International Workshop on Tef Genetics and Improvement, Addis Ababa, Ethiopia, pp: 167-176.

14. Tareke B (2008) Increasing Productivity of Tef, Eragrostis tef (Zucc.) Trotter: New Approaches with Dramatic results (Unpublished Report), Addis Ababa, Ethiopia.

15. Van Reeuwijk LP (1992) Procedures for soil analysis. 6th edn. International soil reference and information center, Wageningen (ISRIC).

16. Day PR (1965) Hydrometer method of particle size analysis. In: Back CA (ed.), Method of Soil Analysis. Amer. Soc. Agron. Madison, Wisconsm Agron, p: 562.

17. Nelson DW, Sommers LE (1982) Total carbon, organic carbon and organic matter. In: Page AL, Miller RH, Keeney DR (eds.), Methods of Soil Analysis. American Society of Agronomy, Madison, USA, pp: 539-579.

18. Olsen S, Sommer L (1982) Phosphorus method for soil analysis. Chemical and microbiological properties. ASA Monograph number 9: 403-430.

19. SAS Institute (2004) JMP 5.1.1 Users Guide. SAS Institute Inc. Cary, NC, USA.

20. Steel RGD, Torrie JH (1984) Principles and Procedures of Statistics. McGraw Hill Book Co., Inc., Singapore, p: 481.

21. Brady NC, Weil RR (2002) The Nature and Properties of Soils. 13th edn. Person Education Ltd, USA, p: 740.

22. Miller RW, Donahue RL (1995) Soils in Our Environment. 7th edn. Prentice-Hall Englewood Cliffs, New Jersey, USA.

23. Haftom G, Mitiku H, Yamoah CH (2009) Tillage frequency, soil compaction and N-fertilizer rate effects on yield of tef (Eragrostis tef (Zucc.) Trotter). Ethiopia Journal of Science 1: 82-94.

24. Al-Abdul Salam MA (1997) Influence of nitrogen fertilization rates and residual effect of organic manure rates on the growth and yield of wheat. Arab Gulf J Sci Res 15: 647-660.

25. Warraich EA, Ahmad N, Basra SM, Afzal I (2002) Effect of nitrogen on source-sink relationship in wheat. Int Journal Agri Biol 4: 300-302.

26. Belay S, Baker DB (1996) Agronomy and Morphological response of tef to drought. Trop sci 36: 41-50.

27. Blum (1989) Considerable genetic variation in grain yield and dry matter accumulation exist between and within in crop species.

28. Balesh T, Bernt Aune J, Breland T (2007) Availability of organic nutrient sources and their effects on yield and nutrient recovery of tef [Eragrostis tef (Zucc.) Trotter] and on soil Properties. Journal of Plant Nutrition and Soil Science 170: 543-550.

29. DzARC (Debrezeit Agricultural Research Center) (1988) Annual research report for 1987/88. Debrezeit, Ethiopia, p: 147.

30. NFIU (National Fertilizer and Input Unit) (1993) Agronomic feasibility of the proposed recommendation and comparison with the previous recommendation. NFIU general paper No. 17, Addis Ababa.

31. Kenea Y, Getachew A, Workneh N (2001) Farming Research on Tef: Small Holders Production Practices. In: Hailu T, Getachew B, Sorrels M (eds.), Narrowing the Rift: Tef Research and Development.Proceeding of the International Work shop on tef genetics and improvement, Addis Ababa, Ethiopia.

32. Kelsa K (1998) Effect of DAP and Urea fertilizers on grain yield of three tef varieties in Awassa and Areka. In: Tadele G, Selasssie S, Sahlemedhin S (eds.). Proceedings of the Fourth Conference of the Ethiopian Society of Soil Science. Addis Ababa, Ethiopia, pp: 122-127.

33. Abuhay T (1997) Genetic variability in dry matter production, partitioning and grain yield of tef [Eragrostis tef (zucc) Trotter] under moisture deficit. SINET: Ethiopia J Sci 20: 177-188.

34. National Academy of Sciences (1996) Lost crops of Africa Volume 1, Grains. Bostid National Research Council. National Academy Press. Washington DC, USA.

35. Landon JR (1991) Booker Tropical Soil Manual: A Handbook for Soil Survey and Agricultural Land Evaluation in the Tropics and Sub-tropics. Longman Scientific and Technical, Essex, New York, USA, p: 474.

37

Productivity, Yield Attributes and Weed Control in Wheat (*Triticum aestivum* L.) as Influenced by Integrated Weed Management in Central High Lands of Ethiopia, East Africa

Tesfay Amare, Cherukuri V. Raghavaiah* and Takele Zeki

Department of Plant Science, College of Agriculture and Veterinary Sciences, Ambo University, Ambo, Post Box No: 19, Ethiopia

Abstract

Wheat (*Triticum aestivum* L.) crop is often fraught with weeds in central high lands of Ethiopia which are characterized by high rainfall, low humidity and low temperatures favorable for development of diseases and insect pests resulting in dwindled productivity. Field experiment was conducted at the Research farm of Plant Science department, Ambo University, for two consecutive years (2014 and 2015) to delineate the effect of 2, 4-D alone, hand weeding alone and their integration on weed control and wheat productivity in comparison with un weeded check in a randomized complete block design with six replications. The experiment comprised four treatments (hand weeding, hand weeding + 2, 4-D @ 2.0 kgha⁻¹, 2, 4-D application @ 2.0 kg/ha and un-weeded check). The experimental site was predominantly infested with different weed species belonging to different families such as grasses, broadleaved weeds and sedges. It was found that Integrating hand weeding + 2, 4-D at 2.0 kgha⁻¹ significantly reduced weed density and dry biomass of weeds in both 2014 and 2015 cropping years compared with the other weed control methods. Highest grain yield (4322, 3989 kg ha⁻¹) was recorded with hand weeding + 2, 4-D at 2.0 kgha⁻¹, followed by hand weeding (3500, 2851 kg ha⁻¹), whereas the lowest yield was recorded from un weeded check (1167, 1082 kg ha⁻¹) in 2014 and 2015 cropping seasons, respectively. Uncontrolled weed growth throughout the crop growth period caused a yield reduction of 72% in both cropping seasons. Application of Post- emergence herbicide 2, 4-D and /or hand weeding and hoeing at tiller stage could further reduce the deleterious effect of weeds on wheat crop raised in central high lands of Ethiopia.

Keywords: 2, 4-D; Hand weeding; Wheat yield; Weed density; Weed biomass; Weed control efficiency; Relative yield loss

Introduction

Wheat, since time immemorial, played a pivotal role in development of civilization and predominant in antiquity as a source of human staple food around the world. It is the cereal of the temperate regions of the world and at high altitudes in tropics and subtropics between 1600-3000 m. Wheat occupies about 17% of the world's cropped land and contributes 35% of the staple food is next only to rice, so its increased production is essential for food security [1,2]. Wheat is one of the major cereal crops grown on the Ethiopian highlands. Despite its importance in Ethiopia, the mean national yield is 1.3 tons ha⁻¹ which is 24% below the mean yield of Africa and 48% below the global mean yield of wheat [2]. Yield reducing factors in wheat are soil fertility decline, weeds, disease, and insects. Weeds compete with crop plants for essential growth factors like light, moisture, nutrients and space. Weeds can also increase harvesting costs, reduce quality of product [3]. Apart from increasing the production cost, weeds also intensify the disease and insect pest problem by serving as alternative hosts, and uncontrolled weed growth throughout the crop growth caused a yield reduction of 57.6 to 73.2% [2]. Though manual and physical methods of weed control are very effective in Ethiopia, however, non availability of labor during peak period under intensive farming, high labor cost; regeneration of weeds which require frequent operation and weeds cannot effectively be managed merely due to crop mimicry [4]. Therefore, the use of chemical weed control has become necessary [5] and this has created a scope for using herbicides and they are becoming more popular in developing countries like Ethiopia. Weed management systems that depend heavily on herbicides are now accepted as unsustainable and it has also created a problem of evolution of herbicide resistant weeds [6,7]. Hence, development of more comprehensive and sustainable weed management system is warranted for economic production of wheat. Moreover, control of weeds by a single method usually does not give positive results and may also not be socio-economically acceptable. An integrated weed management involves specific control measures to be directed not only against one weed species, but also for all the species affecting a crop in a particular area [8], and crop species and cultivars that compete better is an important component of IWM [9]. Currently there is scanty information on integrated weed management approach in wheat crop in Ethiopia. Therefore, the present investigation has been made with an objective to delineate the influence of hand weeding alone, post- emergence 2, 4-D application alone and the integration of hand weeding with herbicide at low dose in comparison with un weeded check on weed control, yield and yield components of rain fed wheat raised on clay loam soils in central high lands of Ethiopia.

Materials and Methods

The present experiment was conducted for two consecutive cropping years (2014 and 2015) at Ambo University research farm. The site is located at a latitude of 9°11'0" North, 38°20'0" East and an altitude of 1980 m.a.s.l. The area received an average annual rainfall of 780 mm. The mean annual minimum and maximum temperatures are 8.25 and 23.4°C, respectively. The field experiment comprised four

*Corresponding author: Cherukuri V. Raghavaiah, Department of Plant Science, College of Agriculture and Veterinary Sciences, Ambo University, Ambo, Post Box No: 19, Ethiopia, E-mail: cheruraghav@yahoo.in

treatments: one hand weeding (25 days after sowing), 2, 4-D at 2.0 kgha^{-1} (25 days after sowing), 2, 4-D at 2.0 kgha$^-$ (25 days after sowing)+ hand weeding (40 days after sowing) and compared with a weedy check arranged in a randomized complete block design with six replications. 2, 4-D was applied at 25 days after sowing as post-emergence with the help of Knapsack/Backpack sprayer. The spray volume was 600 L of water per ha. The size of each plot was 1.0 m × 2.0 m. The distance between adjacent replications and plots was 1 m and 0.5 m, respectively.

Wheat variety HAR 604 was planted at recommended seed rate of 150 kg ha^{-1} in plots. Fertilizer was used at the rate of 64 kg N ha^{-1} and 46 kg P$_2$O$_5$ ha^{-1} through di-ammonium phosphate (DAP) and urea. Half of nitrogen and full dose of phosphorus was drilled in rows at the time of sowing and the remaining N through urea was applied at shoot elongation stage of crop. The weed population count was taken with the help of 0.25 m × 0.25 m quadrant thrown randomly at three places in each plot and was identified and converted to population density per m^2 at 60 days after sowing. After recording weed population the biomass was harvested from each quadrant. The harvested weeds were placed into paper bags separately and then dried in oven at a 65°C temperature for 24 h. till constant weight and subsequently the dry weight was measured and converted into gm^{-2}. Weed Control Efficiency (WCE) was calculated from weed control treatments in controlling weeds:

$$WCE = \frac{WDC - WDT}{WDC} \times 100;$$

Where WDC: Weed Dry Matter in Weedy Check; WDT: Weed Dry Matter in a Treatment. Tillers per meter row length, plant height, grains per spike, thousand kernel weight, grain yield and relative yield loss were recorded. The final produce was measured and adjusted to 12.5% moisture content with the help of the following formula:

$$\text{Adjusted grain yield} \left(kg\, ha^{-1}\right) = \frac{\text{Actual yield} \times 100 - M}{100 - D}$$

Where, M is the measured moisture content in grain and D is the designated moisture content.

Relative yield loss due to weeds was calculated based on the maximum yield obtained from a treatment /treatment combination.

$$\text{Relative yield loss} = \frac{MY - YT}{MY} \times 100,$$

Where, MY= maximum yield from a treatment, YT = yield from a particular treatment.

Harvest index (%): It was calculated by

$$HI = \frac{\text{Grain yield}}{\text{Total aboveground dry biomass yield}} \times 100$$

Population density of weeds was subjected to square root transformation $\left(\sqrt{(X+0.5)}\right)$ to have data normal distribution. Data were subjected to analysis of variance and mean separation was conducted for significant treatment means using Least Significance Differences (LSD) at 5% probability level.

Results and Discussion

Weed floral composition of the experimental site

The experimental site was infested with different weed species belonging to different families. The predominant wed flora comprised *Avena fatua* L., *Cynodon dactylon* L., *Phalaris minor* L., *Poa annua* and *Snowdenia polystachia* L were among the grass weeds, and *Amaranthus hybridus* L., *Biden pilosa* L. *Chenopodium album* L., *Commelina benghalensis* L, *Commelina arvensis* L, *Datura stramonium* L., *Galinsoga palviflora*, *Nicandra physelodes* *Oxalis latifolia* HBK, Polygonum *nepalense* L., and *Raphanus raphanistrum* L., were among broadleaved weeds and *Cyperus esculentus* L. was the only sedge weed. This indicated that a species-rich weed community existed in the experimental field.

Weed density

Effects of different weed control methods on weed density were significant. The results showed (Table 1) that the lowest weed density (10.45, 7.49 m^{-2}) was recorded in plot treated with 2,4-D at 2.0 kgha^{-1}+ hand weeding followed by hand weeding (11.89, 14.77 m^{-2}) whereas the maximum was recorded in weedy check (14.16, 27.4 m^{-2}), respectively during 2014 and 2015. However, no significant difference was observed between hand weeding and 2,4-D at 2.0 kgha^{-1}+ hand weeding .These results are in agreement with Raize et al. [6] and Bibi et al. [3] who reported herbicides supplemented with cultural practices (hand weeding) improved weed controlling ability. When there is sufficient moisture up to grain filling, 2,4-D was found most economical under high broad leaf infestation in barley [10].

Weed biomass

The different weed control methods exhibited significant influence on the dry weight of weeds at both growing periods. The lowest dry weight of weeds (5.1, 6.93 gm^{-2}) was recorded in plot treated with 2,4-D at 2.0 kgha^{-1}+ hand weeding, followed by hand weeding (16.0 gm^{-2}, 26.7 gm^{-2}). Whereas the highest biomass was observed in weedy check (50. gm^{-2}, 207.47a gm^{-2}) during 2014 and 2015, respectively (Table 1). These results are in tandem with those reported by Raize et al. [6] and Tesfay [2] who reported post- emergence herbicides and /or hand weeding and hoeing at tillering stage reduced the dry weight of weeds as compared to herbicides alone or weedy check.

Weed control efficiency

The data presented in Table 1 revealed that during 2014, the minimum weed control efficiency was observed in weedy check

Treatments	Weed density (m^{-2})		Weed dry biomass (gm^{-2})		WCE (%)	
	2014	2015	2014	2015	2014	2015
Hand weeding	11.89(141.3)bc	14.77(218.67)c	16.0c	26.67bc	68.2c	87.1b
2,4-D at 2.0 kgha^{-1} + Hand weeding	10.45(108.8)c	7.49(58.67)b	5.1d	6.93c	89.9d	96.7a
2,4-D at 2.0 kgha^{-1}	12.57(158.7)ab	18.31(336.00)b	32.8b	44.8b	34.8b	78.4c
Weedy check	14.16(200.3)a	27.4(752.0)a	50.3a	207.47a	0.0a	0.0a
LSD(0.05)	1.73	3.87	7.44	25.86	11.66	2.86
CV (%)	7.04	10.86	14.31	17.24	11.48	12.09

Figures or numbers in the parenthesis are original values,
LSD: Least Significant Difference; CV: Coefficient Of Variation; WCE: Weed Control Efficiency.
Table 1: Effect of weed control methods on weed density (m^{-2}), weed dry biomass (gm^{-2}) and weed control efficiency (%) during 2014 and 2015 cropping seasons.

(0.00%) whereas the highest (100.0%) was recorded in a plot treated with hand weeding +2, 4-D at 2.0 kgha^{-1}. Similarly during 2015, the maximum weed control efficiency (96.7%) was recorded in 2,4-D at 2.0 kgha^{-1}+ hand weeding, followed by hand Weeding (52.5.) ,whereas the minimum efficiency was observed in weedy check (0.0%). This result further elucidates that herbicide application supplemented by hand weeding is more effective in reducing weed density and dry biomass weights of weeds next to hand weeding as compared to weedy check. This result was in accordance with Raize et al. [6] who reported that herbicides supplemented by hand weeding gave higher weed control efficiency which could be due to the complementary effect of hand weeding and/herbicides.

Wheat yield and yield attributes

Plant height (cm): The effect of weed management practices on plant height was not discernible (Table 2). This could be due to availability of abundant of growth promoting factors in weed free plot that allowed the plants to attain their maximum height, the competition between weeds and crop for sun light and space in unweeded plots resulted in tall stature of plants. Thus, there was no distinct variation in plant height with different treatments.

Tiller number (m^{1}): Effect of weed management practices on number of tillers per plant was significant. It was found that the highest number of tillers per meter row length (169.9,174.6) was observed with 2,4-D at 2.0 kgha^{-1}+ hand weeding, followed by hand weeding (146.6, 155.3), whereas it was the lowest in weedy check (82.7, 121). This was due to more effectiveness of these treatments on weed control that resulted in lower weed density and dry weight thus reduced crop-weed competition that contributed to more number of tillers per plant in comparison with un weeded check .Reduction in number of tillers, and productive tillers in barley with increased weed density was also reported earlier by Takele which supports the current findings [11].

Grains per spike: Different weed control methods influenced the grains per spike significantly where in the highest number of grains per spike was recorded with 2,4-D at 2.0 kgha^{-1}+ hand weeding, closely followed by hand weeding. The lowest grains per spike was recorded in weedy check during both the years of experimentation. The grains per

spike in 2015 were low compared with 2014 which can be attributed to greater weed density and weed biomass exerting greater competitive stress on wheat. Considerable enhancement in grains per spike due to integrated treatment was probably due to reduced weed competition and availability of adequate quantities of plant nutrients and moisture to crop plants. Similar result was also reported earlier by Raize et al. [6].

Thousand kernel weight: The data presented in Table 2 showed that thousand kernel weight was significantly affected by weed control methods during 2014 and 2015 cropping seasons. The highest (69.6, 76.7 g) thousand grain weight was recorded from plot treated with 2,4-D at 2.0 kgha^{-1}+ hand weeding, followed by hand weeding (54.2, 69.1 g). The minimum was recorded from weedy check (32, 51.5 g) during 2014 and 2015, respectively. The weeding at proper time employing herbicide and supplementing with hand weeding could provide favorable environment for the crop which ultimately lead to better grain filling. This is quite possible that weed free crop stand produced robust grains and ultimately resulted in more 1000-grain weight. Takele also reported a decline in test seed weight with increasing weed density in central high lands of Ethiopia [11]. Similar results have also been reported earlier by Narkhede et al. [12], Tomar et al. [13].

Grain yield (kg/ha): Grain yield was significantly influenced by different weed management practices during 2014 and 2015 cropping season (Table 3). Among weed management practices, the highest grain yield (4322, 3989 kg ha^{-1}) was recorded with 2,4-D at 2.0 kgha^{-1}+ hand weeding, followed closely by hand weeding (3500, 2851 kg ha^{-1}), while the lowest yield was recorded in weedy check (1167, 1028 kg ha^{-1}). The enhancement in wheat grain productivity with integrated approach could be attributed to suppression of weed density, weed growth and biomass that favored increase in yield attributes such as number of tillers per meter row length, grains per spike and test seed weight. Reduction in grain yield with increased weed density was also reported earlier by Takele which corroborates the present findings [11]. Application of 2, 4-D alone could only control broad leave weeds, leaving grassy weeds which competed with the crop at later stages resulting in considerable yield reduction. Significantly higher yield in weed control treatments compared to weedy check has also been reported by Pandey and Mishra [14] and Roslon and Fozelfors [15].

Treatments	Plant height (cm)		Tiller No(m^{-1})		Grains per spike (no)		TKW(g)	
	2014	2015	2014	2015	2014	2015	2014	2015
Hand weeding	96.3	74.67	146.6c	155.3b	23.1b	8.3ab	54.2b	69.1ab
2,4-D at 2.0 kgha^{-1} + Hand weeding	82.4	77.53	169.9a	174.6a	29.0a	10.8a	69.6a	76.7a
2,4-D at 2.0 kgha^{-1}	76.0	70.96	153.3b	149.67c	18.8c	7.4b	56c	65.5b
Weedy check	79.2	69.83	82.7d	121d	15.0d	5.8b	32d	51.5b
LSD(0.05)	NS	NS	2.2	4.11	2.0	3.03	5.9	8.41
CV (%)	11	7.66	20.7	20.42	13.8	17.89	12.3	6.10

LSD: Least Significant Difference; CV: Coefficient of Variation; TKW: Thousand Kernel Weight.
Table 2: Effect of weed control methods on plant height, tiller number, grains/spike and thousand kernel weight in wheat during 2014 and 2015 cropping seasons.

Treatments	Grain yield(kgha^{-1})		HI (%)		RYL (%)	
	2014	2015	2014	2015	2014	2015
Hand weeding	3500.0b	2851.1a	25.12b	26.3b	16.4c	28.2b
2,4-D at 2.0 kgha^{-1} + Hand weeding	4322.2a	3988.9a	31.56ab	29.7a	0.0d	0.00c
2,4-D at 2.0 kgha^{-1}	2444.4c	2526.7b	22.21b	20.6c	41.5b	36.3b
Weedy check	1166.7d	1082.2c	11.4c	9.8d	72.0a	72.7a
LSD(0.05)	802.31	746.3	5.41	1.9	14.89	16.44
CV (%)	14.17	13.61	11.39	9.7	22.93	22.84

HI: Harvest Index; RYL: Relative Yield Loss
Table 3: Effect of weed control methods on grain yield, harvest index and relative yield loss during 2014 and 2015 cropping season.

Wogayehu Worku et al. [16] reported no significant difference between herbicides Tomahawk, Herb knock, Draget –SD, U-46, and wheat yield advantage of 37, 36, 32 and 32%, respectively was observed over weedy check in West Shoa zone of Ethiopia. Experiments conducted at Sinana with eleven herbicides compared with hand weeding showed that one hand weeding was economical, and in Bale 3 herbicides gave 33%yield advantage over weedy check as reported by SARC [17].

Harvest index: The effect of different weed control methods on the harvest index of wheat, an indicator of assimilate partitioning efficiency, was statistically significant during both years of study. The data presented in Table 3 showed that the highest harvest index (31.56, 29.7%) was recorded with 2, 4-D at 2.0 kgha^{-1} supplemented by hand weeding, whereas the lowest harvest index was from weedy check (11.4, 9.8%). The significant increase in harvest index with integrated weed management may be attributed to suppression of weed growth resulting in more availability of plant nutrients, soil moisture and space to wheat crop, which favored utilization of photo synthates for better grain yield formation.

Relative yield loss: While comparing the loss in wheat yield due to the weed management practices, it was observed that the lowest yield loss (0.0, 0.0%) was recorded with 2,4-D at 2.0 kgha^{-1}+ hand weeding as compared to the rest of the treatments. This was followed by hand weeding (16.4, 28.2) and 2, 4-D at 2.0 kgha^{-1} (41.5, 36.3%), Whereas the loss was highest (72.0, 72.7) in weedy check (Table 3) during 2014 and 2015, respectively. The higher yield loss in 2, 4-D at 2.0 kgha^{-1} may be due to greater density of grassy weeds in the current field experiment. By and large, herbicides minimize crop yield loss when supplemented with had weeding or other cultural practices. These results are in corroboration with Tesfay who reported that uncontrolled weed growth throughout the crop growth caused a yield reduction to the tune of 57.6 to 73.2%in Ethiopia [18-22].

Conclusion

From the results of the two year field experiment it could be inferred that, among the weed management practices, post emergence application of 2, 4-D at 2.0 kgha^{-1}at 25 days after sowing, supplemented with hand weeding at 40days after sowing reduced weed density and dry biomass of weeds significantly, closely followed by hand weeding. These treatments also enhanced yield and yield components significantly and reduced relative yield loss of wheat. Application of 2,4-D alone could not prove effective in controlling weeds, thereby resulting in more yield loss in comparison with hand weeding and integrated approach. Uncontrolled weed growth throughout the crop growth period caused a yield reduction of 72% and 72.7% in wheat during 2014 and 2015 cropping seasons, respectively.

References

1. Kumar S, Agarwal A (2010) Effect of weed management practices on nitrogen removal by Phalaris minor and wheat (Triticum aestivum). Asian Journal of Experimental Biological Science 81-84.

2. Tesfay A, Sharma JJ, Kassahun Z (2014) Effect of Weed Control Methods on Weeds and Wheat (Triticum aestivum L.) Yield. World Journal of Agricultural Research 2: 124-128.

3. Bibi KBM, Hassan G, Noor MK (2008) Effect of herbicides and wheat population on control of weeds in wheat. Pakistan Journal of Weed Science Research 14: 111-119.

4. Kebede D (2000) Weed Control Methods Used in Ethiopia.

5. Marwat KB, Muhammad S, Zahid H, Gul B, Rashid H (2008) Study of various weed management practices for weed control in wheat under irrigated conditions. Pak J Weed Sci Res 14: 1-8.

6. Riaz M, Azim MM, Mahmood TZ, Jamil M (2006) Effect of Various Weed Control Methods on Yield and Yield Components of Wheat under Different Cropping Patterns. International Journal of Agriculture & Biology 8: 636–640.

7. Amit JJ, Shah SC, Paresh HR, Hitesh B (2008) Integrated Effect of Seed Rates and Weed Management Treatments in Wheat (Triticum aestivum L). Research Journal of Agriculture and Biological Sciences 4: 704-711.

8. Syed HA, Saleem M, Maqzood M, Yaqub MM, Hassan M et al. (2009) Weed density and grain yield of wheat as affected by spatial Arrangements and weeding techniques under rain fed Conditions of pothowar. Pak J Agri Sci 46.

9. Tanner DG (1999) Principles and practices of integrated weed management in the wheat production system of Ethiopia. Arem 27-65.

10. Kedir N, Tilahun G, Feyissa T (2005) Alternative weed control methods for food barley in the Bale high lands. Arem 6: 69-78.

11. Takele N (2001) Competition effects of major weeds on yield and yield components of barley. M.Sc thesis, Alemaya university, Ethiopia.

12. Narkhede TN, Wadile SC, Attarde DR, Surya WRJ (2000) Integrated weed management in Sesame under rain-fed condition. Indian J Agric Res 34: 247–250.

13. Tomar RK, Singh JP, Garg RN, Gupta VK, Sahoo RN et al. (2003) Effect of weed management practices on weed growth and yield of wheat in rice based cropping system under varying levels of tillage. Annals Pl Protect Sci 11: 123–128.

14. Pandey J, Mishra BN (2002) Effects of weed management practices in rice-wheat mungbean cropping system on weeds and yield of crops. Annals Agric Res 23: 646–50.

15. Roslon E, Fogelfors H (2003) Crop and weed growth in a sequence of spring barley and winter wheat crops established together from a spring sowing (relay cropping). J Agron Crop Sci 189: 185–190.

16. Wogayehu W, Taye T, Tariku H, Takele N (2009) Preliminary evaluation of Tomahawk, and 2,4-D herbicides against broad leaf weeds in wheat. Ethiopian journal of weed management 3: 77-78.

17. SARC (1992) Sinana progress report 1992.

18. Tesfay A (2014) Effect of weed management methods on weeds and wheat (Triticum aestivum L.) yield. African Journal of Agricultural Research 9: 1914-1920.

19. Rezene F (1986) A review of weed science research activities on wheat and barley in Ethiopia 121-148. In:Tsedeke Abeta.(ed) A review of crop protection research in Ethiopia. Addis Ababa, Ethiopia.IAR.

20. Hailu GM (1991) Wheat production and research in Ethiopia. In:Hailu et al.(eds) wheat research in Ethiopia : a historical perspective. Addis Ababa, IAR/CIMMYT.

21. Tanner DG, Giref S (1991) Weed control research conducted in wheat in Ethiopia.pp.235-275.In: Hailu Gebre –Mariam,Tanner DG, Mengistu Hulluka (eds).Wheat research in Ethiopia: a historical perspective. Addis Ababa, IAR/CIMMYT.

22. Tanner DG (1999) Principles and practices of integrated weed management in the wheat production system of Ethiopia. Arem 27-65.

Permissions

List of Contributors

Qi Zhang, Nan Jiang, Yahui Hong and Zhilong Wang
Hunan Provincial Key Laboratory of Crop Germplasm Innovation and Utilization, College of Biological Science and Technology, College of Agronomy, Hunan Agricultural University, Changsha, Hunan 410128, China

Guo-Liang Wang
Hunan Provincial Key Laboratory of Crop Germplasm Innovation and Utilization, College of Biological Science and Technology, College of Agronomy, Hunan Agricultural University, Changsha, Hunan 410128, China Department of Plant Pathology, Ohio State University, Columbus, Ohio 43210, USA

Al-Hmoud G and Al-Momany A
University of Jordan, Department of Plant Protection, Amman, Jordan

Alemayehu Hailu and Getaneh Woldeab
Ethiopian Institute of Agricultural Research, Plant Protection Research Center P.O.Box 37, Ambo, Ethiopia

Woubit Dawit
Ambo University, P.O.Box 19, Ambo, Ethiopia

Endale Hailu
Ambo University, P.O.Box 19, Ambo, Ethiopia Ethiopian Institute of Agricultural Research, Plant Protection Research Center P.O.Box 37, Ambo, Ethiopia

Hadush Hagos and Abuhay Takele
Ethiopian Sugar Corporation, Research and Training Division, Sugarcane Production Research Directorate, Agronomy and Protection Research Team, Wonji Research Center, P.O.Box 15, Wonji, Ethiopia

Walelign Worku
Hawassa University, College of Agriculture, Hawassa, Ethiopia

Hadush Hagos and Kidane Tesfamicheal
Ethiopian Sugar Corporation, Research and Training, Agronomy and Crop Protection Research Team, Wonji Research Center, P.O.Box 15, Wonji, Ethiopia

Leul Mengistu
Ethiopian Sugar Corporation, Research and Training, Sugarcane Production Research Directorate, Wonji Research Center, P.O.Box 15, Wonji, Ethiopia

Yusuf Kedir
Ethiopian Sugar Corporation, Research and Training, Soil and Irrigation Research Team, Wonji Research Center, P.O.Box 15, Wonji, Ethiopia

Zeleke Teshome
Sugar Corporation, Research and Training, P. O. Box 15, Wonji, Ethiopia

Kibebew Kibret
Haramaya University, P.O.Box 138, Dire Dawa, Ethiopia

Diwash Tamang, Rajib Nath and Kajal Sengupta
Department of Agronomy, Bidhan Chandra Krishi Viswavidyalaya, Mohanpur, Nadia-741252, West Bengal, India

Adeyeye AS
Department of Crop Production and Protection, Federal University Wukari, Taraba State, Nigeria

Togun AO
University of Ibadan, Ibadan, Oyo State, Nigeria

Olaniyan AB and Akanbi WB
Ladoke Akintola University of Technology, Ogbomoso, Oyo State, Nigeria

Egata Shunka Tolessa
Ethiopian Institute of Agricultural Research, Holetta Research Center, Horticulture Research Division, Ethiopia

Derbew Belew, Adugna Debela and Beshir Kedi
Jimma University College of Agriculture and Veterinary Medicine, Ethiopia

Uma Maheswari M and Karthik A
Department of Agronomy, Agricultural College and Research Institute, Coimbatore, Tamil Nadu, India

Solomon Ali, Wassu Mohammed and Beneberu Shimelis
Department of Plant Science, Debre Markos University, Ethiopia

Zeleke Teshome, Girma Abejehu and Hadush Hagos
Sugar Corporation, Research and Training, P.O. Box 15, Wonji, Ethiopia

Okunlola A Ibironke
Department of Crop, Soil and Pest Management, The Federal University of Technology, PMB 704 Akure, Ondo State, Nigeria

Hadush Hagos and Luel Mengistu
Ethiopian Sugar Corporation, Research and Training Division, Sugarcane Production Research Directorate, Agronomy and Protection Research Team, Wonji Research Center, P.O.Box 15, Wonji, Ethiopia

Yohannes Mequanint
Ethiopian Sugar Corporation, Research and Training Division, Sugarcane Production Research Directorate, Wonji Research Center, P.O.Box 15, Wonji, Ethiopia

Hayfa Jabnoun-Khiareddine, Mouna Gueddes-Chahed, Ahmed Hajlaoui, Samir Ben Salem, Wissem Ben Dhia and Mejda Daami-Remadi
UR13AGR09- Integrated Horticultural Production in the Tunisian Centre- East, Regional Center of Research on Horticulture and Organic Agriculture, University of Sousse, 4042, Chott-Mariem, Tunisia

Rania Aydi Ben Abdallah
UR13AGR09- Integrated Horticultural Production in the Tunisian Centre- East, Regional Center of Research on Horticulture and Organic Agriculture, University of Sousse, 4042, Chott-Mariem, Tunisia
National Agronomic Institute of Tunisia, 1082 Tunis Mahrajène, University of Carthage, Tunisia

Fakher Ayed
Technical Center of Organic Agriculture, 4042, Chott-Mariem, Tunisia

Endale Hailu and Getaneh Woldaeb
Ambo Plant protection Research center, 37Ambo, Ethiopia

Worku Danbali
Kulumisa Agricultural research center, 489, Asella, Ethiopia

Wubishet Alemu
Sinana Agricultural Research Center, 208, Bale-Robe, Ethiopia

Teklay Abebe
Alamata Agricultural research center, 56, Alamata, Ethiopia

Umeri C, Moseri H and Onyemekonwu RC
Department Agricultural Science Education, College of Education, Agbor, Delta State, Nigeria

Mulugeta Hailu
Department of Biotechnology, College of Dry Land Agriculture and Natural Resources, Mekelle University, Mekelle, Ethiopia

Melaku Tesfa
Ethiopian Sugar Corporation Research and Training Division, Biotechnology Research Team, Wonji Research Center, Wonji, Ethiopia

Belayneh Admassu
National Agricultural Biotechnology Research Program, Holetta Agricultural Research Center, Ethiopia

Kassahun Bantte
Jimma University College of Agriculture and Veterinary Medicine, Jimma, Ethiopia

Vinod Kumar and A.K. Chopra
Agro-ecology and Pollution Research Laboratory, Department of Zoology and Environmental Science, Gurukula Kangri University, Haridwar-249404 (Uttarakhand), India

Alemayehu Dengia
Sugar Corporation, Research and Training, Wonji Research Center, Wonji-Ethiopia, Ethiopia

Egbert Lantinga
Department of Plant Sciences, Biological Farming Systems Group, Wageningen University, Radix West, 2nd floor, Droevendaalsesteeg, 16708 PB Wageningen, The Netherlands

Rajeshwari N, Kavya Muduvala R, Bhargavi MK and Ramesh Babu HN
Sahyadri Science College, Kuvempu University, Shivamogga, Karnataka, India

Bitew B and Tigabie A
Debre Birhan Agricultural Research Center, PO Box 112, Debre Birhan, Ethiopia

Ayaz Ahmad
Department of Agriculture Extension Gujranwala, Punjab, Pakistan

Ghulam Farid
Department of Agriculture Extension Gujranwala, Punjab, Pakistan
Institute of Soil and Environmental Sciences, University of Agriculture, Faisalabad 38040, Pakistan

Nadeem Sarwar
Nuclear Institute for Agriculture and Biology (NIAB), Faisalabad 38000, Pakistan

Institute of Soil and Environmental Sciences, University of Agriculture, Faisalabad 38040, Pakistan

Saifullah and Abdul Ghafoor
Institute of Soil and Environmental Sciences, University of Agriculture, Faisalabad 38040, Pakistan

Mariam Rehman
Lahore College for Women University, Lahore, Pakistan

Getnet Yitayih Alemu and Yehizbalem Azmeraw Tadele
Department of Plant Sciences, Debre Tabor University, Ethiopia

Makled KHM and Abou El Enin MM
Agronomy, Department of Agriculture, Al-Azhar University, Cairo, Egypt

Anteneh Ademe
Sekota Dry-land Agricultural Research Center, Sekota, Ethiopia

Esmelalem Mehiretu
Adet Agricultural Research Center, Adet, Ethiopia

Abraham Mulu
Department of Dryland Crop Science, College of Dryland Agriculture, Jigjiga University, Ethiopia

Andergachew Gedebo and Hussien Mohammed
Hawassa University, Awassa, Ethiopia

Narendra Kumar
Amity Institute of Biotechnology, Amity University Haryana, Manesar, Gurgaon, Haryana, India

Belay Feyisa, Alemu Lencho and Thangavel Selvaraj
Department of Plant Sciences, College of Agriculture and Veterinary Sciences, Ambo University, Ambo, Post Box No: 19, Ethiopia, East Africa

Gezehegne Getaneh
Addis Ababa University, Salale Campus, Addis Ababa, P. B. No: 2003, Ethiopia

Okunlola AI
Department of Crop, Soil and Pest Management, Federal University of Technology, Nigeria

Ogungbite OC
Centre for Continuing Education, Federal University of Technology, Nigeria

Ezena GN
African Regional Postgraduate Programme in Insect Science, PMB L59, University of Ghana, Legon, Ghana

Akotsen-Mensah C
African Regional Postgraduate Programme in Insect Science, PMB L59, University of Ghana, Legon, Ghana Forest and Horticultural Crops Research Centre, University of Ghana, Kade, Ghana

Fening KO
African Regional Postgraduate Programme in Insect Science, PMB L59, University of Ghana, Legon, Ghana Soil and Irrigation Research Centre, University of Ghana, Kpong, Ghana

Matheri F, Runo S, Ngugi M, Fred Teya, Mawia AM and Kioko FW
Department of Biochemistry and Biotechnology, School of Pure and Applied Sciences, Kenyatta University, Nairobi, Kenya

Mwangi M and Kirubi DT
Department of Agricultural Science and Technology, School of Agriculture and Enterprise Development, Kenyatta University, Nairobi, Kenya

Kamau DN
Department of Microbiology, School of Pure and Applied Sciences, Kenyatta University, Nairobi, Kenya

Egata Shunka, Abebe Chindi, Gebremedhin W/giorgis, Ebrahim Seid and Lema Tessema
Ethiopian Institute of Agricultural Research, Holetta Agricultural Research Center, Ethiopia

Tekle Edossa Kubure, Cherukuri V Raghavaiah and Ibrahim Hamza
Department of Plant Sciences, College of Agriculture and Veterinary Sciences, PO Box 19, Ambo University, Ambo, West Shoa Zone, Ethiopia

Yonas Mebratu, Cherukuri V Raghavaiah and Habtamu Ashagre
Department of Plant Science, College of Agriculture and Veterinary Sciences, Ambo University, Ambo, Ethiopia

Tesfay Amare, Cherukuri V. Raghavaiah and Takele Zeki
Department of Plant Science, College of Agriculture and Veterinary Sciences, Ambo University, Ambo, Post Box No: 19, Ethiopia

Index

www.ingramcontent.com/pod-product-compliance
Lightning Source LLC
Chambersburg PA
CBHW070153240326

41458CB00126B/4545